INTERNATIONAL SERIES
OF
MONOGRAPHS ON PHYSICS

International Series of Monographs on Physics

140. M. Alcubierre: *Introduction to 3 + 1 numerical relativity*
139. A. L. Ivanov, S. G. Tikhodeev: *Problems of condensed matter physics—quantum coherence phenomena in electron-hole and coupled matter-light systems*
138. I. M. Vardavas, F. W. Taylor: *Radiation and climate*
137. A. F. Borghesani: *Ions and electrons in liquid helium*
136. C. Kiefer: *Quantum gravity, Second edition*
135. V. Fortov, I. Iakubov, A. Khrapak: *Physics of strongly coupled plasma*
134. G. Fredrickson: *The equilibrium theory of inhomogeneous polymers*
133. H. Suhl: *Relaxation processes in micromagnetics*
132. J. Terning: *Modern supersymmetry*
131. M. Mariño: *Chern-Simons theory, matrix models, and topological strings*
130. V. Gantmakher: *Electrons and disorder in solids*
129. W. Barford: *Electronic and optical properties of conjugated polymers*
128. R. E. Raab, O. L. de Lange: *Multipole theory in electromagnetism*
127. A. Larkin, A. Varlamov: *Theory of fluctuations in superconductors*
126. P. Goldbart, N. Goldenfeld, D. Sherrington: *Stealing the gold*
125. S. Atzeni, J. Meyer-ter-Vehn: *The physics of inertial fusion*
123. T. Fujimoto: *Plasma spectroscopy*
122. K. Fujikawa, H. Suzuki: *Path integrals and quantum anomalies*
121. T. Giamarchi: *Quantum physics in one dimension*
120. M. Warner, E. Terentjev: *Liquid crystal elastomers*
119. L. Jacak, P. Sitko, K. Wieczorek, A. Wojs: *Quantum Hall systems*
118. J. Wesson: *Tokamaks, Third edition*
117. G. Volovik: *The Universe in a helium droplet*
116. L. Pitaevskii, S. Stringari: *Bose-Einstein condensation*
115. G. Dissertori, I.G. Knowles, M. Schmelling: *Quantum chromodynamics*
114. B. DeWitt: *The global approach to quantum field theory*
113. J. Zinn-Justin: *Quantum field theory and critical phenomena, Fourth edition*
112. R. M. Mazo: *Brownian motion—fluctuations, dynamics, and applications*
111. H. Nishimori: *Statistical physics of spin glasses and information processing—an introduction*
110. N. B. Kopnin: *Theory of nonequilibrium superconductivity*
109. A. Aharoni: *Introduction to the theory of ferromagnetism, Second edition*
108. R. Dobbs: *Helium three*
107. R. Wigmans: *Calorimetry*
106. J. Kübler: *Theory of itinerant electron magnetism*
105. Y. Kuramoto, Y. Kitaoka: *Dynamics of heavy electrons*
104. D. Bardin, G. Passarino: *The Standard Model in the making*
103. G. C. Branco, L. Lavoura, J. P. Silva: *CP Violation*
102. T. C. Choy: *Effective medium theory*
101. H. Araki: *Mathematical theory of quantum fields*
100. L. M. Pismen: *Vortices in nonlinear fields*
99. L. Mestel: *Stellar magnetism*
98. K. H. Bennemann: *Nonlinear optics in metals*
96. M. Brambilla: *Kinetic theory of plasma waves*
94. S. Chikazumi: *Physics of ferromagnetism*
91. R. A. Bertlmann: *Anomalies in quantum field theory*
90. P. K. Gosh: *Ion traps*
88. S. L. Adler: *Quaternionic quantum mechanics and quantum fields*
87. P. S. Joshi: *Global aspects in gravitation and cosmology*
86. E. R. Pike, S. Sarkar: *The quantum theory of radiation*
83. P. G. de Gennes, J. Prost: *The physics of liquid crystals*
73. M. Doi, S. F. Edwards: *The theory of polymer dynamics*
69. S. Chandrasekhar: *The mathematical theory of black holes*
51. C. Møller: *The theory of relativity*
46. H. E. Stanley: *Introduction to phase transitions and critical phenomena*
32. A. Abragam: *Principles of nuclear magnetism*
27. P. A. M. Dirac: *Principles of quantum mechanics*
23. R. E. Peierls: *Quantum theory of solids*

Introduction to 3+1 Numerical Relativity

MIGUEL ALCUBIERRE

Nuclear Sciences Institute,
Universidad Nacional Autónoma de México

OXFORD
UNIVERSITY PRESS

OXFORD
UNIVERSITY PRESS

Great Clarendon Street, Oxford OX2 6DP

Oxford University Press is a department of the University of Oxford.
It furthers the University's objective of excellence in research, scholarship,
and education by publishing worldwide in

Oxford New York

Auckland Cape Town Dar es Salaam Hong Kong Karachi
Kuala Lumpur Madrid Melbourne Mexico City Nairobi
New Delhi Shanghai Taipei Toronto

With offices in

Argentina Austria Brazil Chile Czech Republic France Greece
Guatemala Hungary Italy Japan Poland Portugal Singapore
South Korea Switzerland Thailand Turkey Ukraine Vietnam

Oxford is a registered trade mark of Oxford University Press
in the UK and in certain other countries

Published in the United States
by Oxford University Press Inc., New York

First published 2008

British Library Cataloguing in Publication Data
Data available

Library of Congress Cataloging in Publication Data
Data available

Printed in Great Britain
on acid-free paper by
Biddles Ltd. www.biddles.co.uk

ISBN 978–0–19–920567–7 (Hbk)

1 3 5 7 9 10 8 6 4 2

To Miguel, Raul and Juan,
for filling my life with beautiful moments.

To María Emilia and Mili,
for being such a wonderful gift.

ACKNOWLEDGEMENTS

There have been a large number of people that, both directly and indirectly, have helped me get to a point where I could write this book. First of all, I would like to thank my mentors and advisors, who put me in the path of studying physics in general, and relativity in particular. Among them I would like to mention Luis de la Peña and Ana María Cetto who, perhaps somewhat reluctantly, accepted my change of heart when I decided to leave the study of the foundations of quantum mechanics in favor of general relativity. I also thank Eduardo Nahmad for pointing the way, Bernard F. Schutz for guiding me and being there up to the present time, and Ed Seidel for always being a friend.

I am also in debt to many friends and colleagues who have been there along the road. My old friends from Cardiff, Gabrielle Allen, Gareth Jones, Mario Antonioletti, Nils Anderson, Kostas Kokkotas, as well as my friends from Potsdam, John Baker, Werner Benger, Steve Brandt, Bernd Bruegmann, Manuela Campanelli, Peter Diener, Tony Font, Tom Goodale, Carsten Gundlach, Ian Hawke, Scott Hawley, Frank Herrmann, Daniel Holz, Sasha Husa, Michael Koppitz, Gerd Lanfermann, Carlos Lousto, Joan Masso, Philippos Papadopoulos, Denis Pollney, Luciano Rezzolla, Alicia Sintes, Nikolaus Stergioulas, Ryoji Takahashi, Jonathan Thornburg and Paul Walker. There have been other people who have helped me in many ways, either by giving direct answers to my questions, or from whom I have simply learned a lot of what has contributed to this book throughout the years, Carles Bona, Matt Choptuik, Eric Gourgoulhon, Pablo Laguna, Luis Lehner, Roy Maartens, Mark Miller, Jorge Pullin, Oscar Reula, Olivier Sarbach, Erik Schnetter, Deirdre Shoemaker, Wai-Mo Suen, Manuel Tiglio and Jeff Winicour, among others.

I also want to thank my friends and colleagues in Mexico, Alejandro Corichi, Jose Antonio Gonzalez, Francisco S. Guzman, Tonatiuh Matos, Dario Nuñez, Luis Ureña, Marcelo Salgado, Daniel Sudarsky, Roberto Sussman and Hernando Quevedo. And of course the students of our numerical relativity group, most of whom have proof-read different sections of the original manuscript, Antonio Castellanos, Juan Carlos Degollado, Cesar Fuentes, Pablo Galaviz, David Martinez, Martha Mendez, Jose Antonio Nava, Bernd Reimann, Milton Ruiz and Jose Manuel Torres.

Finally, I wish to thank my editor, Sonke Adlung, who suggested the idea of writing this book in the first place and has been pushing hard ever since. I wouldn't have made it without him.

PREFACE

General relativity is a highly successful theory. Not only has it radically modified our understanding of gravity, space and time, but it also possesses an enormous predictive power. To date, it has passed with extraordinary precision all the experimental and observational tests that it has been subjected to. Among its more important results are the predictions of exotic objects such as neutron stars and black holes, and the cosmological model of the Big Bang. Also, general relativity has predicted the existence of gravitational waves, which might be detected directly for the first time before this decade is out.

General relativity, for all its conceptual simplicity and elegance, turns out to be in practice a highly complex theory. The Einstein field equations are a system of ten coupled, non-linear, partial differential equations in four dimensions. Written in fully expanded form in a general coordinate system they have thousands of terms. Because of this complexity, exact solutions of the Einstein equations are only known in cases with high symmetry, either in space or in time: solutions with spherical or axial symmetry, static or stationary solutions, homogeneous and/or isotropic solutions, *etc.* If we are interested in studying systems with astrophysical relevance, which involve strong and dynamical gravitational fields with little or no symmetry, it is simply impossible to solve the field equations exactly. The need to study this type of system has given birth to the field of numerical relativity, which tries to solve the Einstein field equations using numerical techniques and complex computational codes.

Numerical relativity appeared as an independent field of research in the mid 1960s with the pioneering efforts of Hahn and Lindquist [158], but it wasn't until the mid 1970s when the first truly successful simulations where carried out by Smarr [271] and Eppley [123, 124] in the context of the head-on collision of two black holes. At that time, however, the power of the available computers was very modest, and the simulations that could be performed where limited to either spherical symmetry or very low resolution axial symmetry. This situation has changed, and during the decades of the 1980s and 1990s a true revolution has taken place in numerical relativity. Researchers have studied ever more complex problems in many different aspects of general relativity, from the simulation of rotating stars and black holes to the study of topological defects, gravitational collapse, singularity structure, and the collisions of compact objects. Maybe the most influential result coming from numerical relativity has been the discovery by Choptuik of critical phenomena in gravitational collapse [98] (for a more recent review see [153]). A summary of the history of numerical relativity and its more recent developments can be found in [186].

Numerical relativity has now reached a state of maturity. The appearance of powerful super-computers, together with an increased understanding of the underlying theoretical issues, and the development of robust numerical techniques, has finally allowed the simulation of fully three-dimensional systems with strong and highly dynamic gravitational fields. And all this activity is happening at precisely the right time, as a new generation of advanced interferometric gravitational wave detectors (GEO600, LIGO, VIRGO, TAMA) is finally coming on line. The expected gravitational wave signals, however, are so weak that even with the amazing sensitivity of the new detectors it will be necessary to extract them from the background noise. As it is much easier to extract a signal if we know what to look for, numerical relativity has become badly needed in order to provide the detectors with precise templates of the type of gravitational waves expected from the most common astrophysical sources. We are living a truly exciting time in the development of this field.

The time has therefore arrived for a textbook on numerical relativity that can help as an introduction to this promising field of research. The field has expanded in a number of different directions in recent years, which makes writing a fully comprehensive textbook a challenging task. In particular, there are several different approaches to separating the Einstein field equations in a way that allows us to think of the evolution of the gravitational field in time. I have decided to concentrate on one particular approach in this book, namely the 3+1 formalism, not because it is the only possibility, but rather because it is conceptually easiest to understand and the techniques associated with it have been considerably more developed over the years. To date, the 3+1 formalism continues to be used by most researchers in the field. Other approaches, such as the characteristic and conformal formulations, have important strengths and show significant promise, but here I will just mention them briefly.

This book is aimed particularly at graduate students, and assumes some basic familiarity with general relativity. Although the first Chapter gives an introduction to general relativity, this is mainly a review of some basic concepts, and is certainly not intended to replace a full course on the subject.

Miguel Alcubierre
Mexico City, September 2007.

CONTENTS

1

BRIEF REVIEW OF GENERAL RELATIVITY

1.1 Introduction

The theory of general relativity, postulated by Einstein at the end of 1915 [120, 121], is the modern theory of gravitation. According to this theory, gravity is not a force as it used to be considered in Newtonian physics, but rather a manifestation of the *curvature* of spacetime. A massive object produces a distortion in the geometry of spacetime around it, and in turn this distortion controls the movement of physical objects. In the words of John A. Wheeler, "matter tells spacetime how to curve, and spacetime tells matter how to move" [298].

When Einstein introduced special relativity in 1905 it became clear that Newton's theory of gravity would have to be modified. The main reason for this was that Newton's theory implies that the gravitational interaction was transmitted between different bodies at infinite speed, in clear contradiction with one of the fundamental results of special relativity: No physical interaction can travel faster than the speed of light. It is interesting to note that Newton himself was never happy with the existence of this *action at a distance*, but he considered that it was a necessary hypothesis to be used until a more adequate explanation of the nature of gravity was found. In the years from 1905 to 1915, Einstein focused his efforts on finding such an explanation.

The basic ideas that guided Einstein in his quest towards general relativity were the *principle of general covariance*, which says that the laws of physics must take the same form for all observers, the *principle of equivalence*, which says that all objects fall with the same acceleration in a gravitational field regardless of their mass, and *Mach's principle*, formulated by Ernst Mach at the end of the 19th century, which states that the local inertial properties of physical objects must be determined by the total distribution of matter in the universe. The principle of general covariance led Einstein to ask for the equations of physics to be written in tensor form, the principle of equivalence led him to the conclusion that the natural way to describe gravity was identifying it with the geometry of spacetime, and Mach's principle led him to the idea that such geometry should be fixed by the distribution of mass and energy.

The discussion that follows will serve to present some of the basic concepts of general relativity, but it is certainly not intended to be a detailed introduction to this theory, it is simply too short for that. Readers with no training in relativity are well advised to read any of the standard textbooks on the subject (for example: Misner, Thorne and Wheeler [206], Wald [295] and Schutz [259]).

1.2 Notation and conventions

A comment about notation and conventions is in order. Throughout the book I will use the conventions of Misner, Thorne and Wheeler (MTW) [206]. That is, spacetime indices will go from 0 to 3, with 0 representing the time coordinate. Greek indices $(\alpha, \beta, \mu, \nu, ...)$ always refer to four-dimensional spacetime and can take values from 0 to 3, while Latin indices $(i, j, k, l, ...)$ refer to three-dimensional space and take values from 1 to 3. The signature of the spacetime metric will be taken to be $(-1, +1, +1, +1)$ (see Section 1.3 below). I will also use Einstein's summation convention: Unless otherwise stated, repeated indices are summed over all their possible values.

The system of units used in this book will be the so-called *geometric units*, in which the speed of light c and Newton's gravitational constant G are taken to be equal to one. Conventional metric (SI) units can always be recovered by multiplying a given quantity with as many powers of c and G as are needed in each case in order to get back to their correct SI dimensions (for example, making the substitutions $t \to ct$ and $M \to GM/c^2$). In geometric units all physical quantities have dimensions of length to some power. In particular, time will be measured in meters, a meter of time being equal to the time it takes light to travel one meter (roughly 3×10^{-9} seconds). Mass will also be measured in meters, with a meter of mass equal to the mass of a point particle that in Newton's theory has an escape velocity equal to the speed of light at a distance of two meters (the reason for using two meters instead of one comes from the factor 2 in the expression for the kinetic energy $E_K = mv^2/2$). A meter of mass is a fairly large mass, equal to about 1.35×10^{27} kilograms, *i.e.* roughly 200 times the mass of the Earth (or about 1.4 times the mass of Jupiter). In other words, in these units the mass of the Earth is about half a centimeter, and the mass of the Sun about one and a half kilometers.

Finally, for the partial derivatives of a geometric quantity T, I will use indistinctively the symbols $\partial T/\partial x^\mu = \partial_\mu T$, and for the covariant derivatives the symbol $\nabla_\mu T$. When considering covariant derivatives in only three dimensions (as in the 3+1 formalism), I will denote them instead by $D_i T$.

1.3 Special relativity

Before going into general relativity it is important to discuss some of the basic concepts and results of special relativity. Special relativity was introduced by Einstein in 1905 [118, 119] as a way of reconciling Maxwell's electrodynamics with the Galilean principle of relativity that states that the laws of physics are the same for all inertial observers. It is based on two basic postulates, the first of which is the principle of relativity itself, and the second the empirical fact that the speed of light is the same for all inertial observers, a fact elevated by Einstein to the status of a physical law. The invariance of the speed of light was established by the null experiment of Michelson and Morley in 1887, thought it is unclear how much influence this experiment had in Einstein's development of special relativity.[1]

One can ask what is "special" about special relativity. It is common to hear, even among physicists, that special relativity is special because it can not describe accelerating objects or accelerating observers. This is, of course, quite wrong. Special relativity is essentially a new kinematic framework on which we can do dynamics, that is, we can study the effects of forces on physical objects, so that accelerations are included all the time. Accelerating observers, or more to the point accelerating coordinate systems, can also be dealt with, though the mathematics becomes more involved. What makes special relativity "special" is the fact that it assumes the existence of global inertial frames, that is, reference frames where Newton's first law holds: Objects free of external forces remain in a state of uniform rectilinear motion. Inertial frames play a crucial role in special relativity. In fact, one of the best known results from this theory are the Lorentz transformations that relate the coordinates in one inertial frame to those of another. If we assume that we have two inertial frames $\mathcal{O}$ and $\mathcal{O}'$, with $\mathcal{O}'$ moving with respect to $\mathcal{O}$ with constant speed v along the x axis, then the Lorentz transformations are:

$$t' = \gamma \left(t - vx \right) \; , \qquad x' = \gamma \left(x - vt \right) \; , \qquad y' = y \; , \qquad z' = z \; , \tag{1.3.1}$$

with $\gamma := 1/\sqrt{1-v^2}$ the *Lorentz factor*. These transformations had been derived by Lorentz before the work of Einstein as transformations that left Maxwell's equations invariant. In more compact notation, the Lorentz transformations can be written as (notice the use of the summation convention)

$$x^{\mu'} = \Lambda^{\mu'}_{\nu} x^{\nu} \; , \tag{1.3.2}$$

with $\{x^\mu\} = \{t, x, y, z\}$ the coordinates in $\mathcal{O}$, $\{x^{\mu'}\}$ the corresponding coordinates in $\mathcal{O}'$, and $\Lambda^{\mu'}_{\nu}$ the Jacobian matrix:

$$\Lambda^{\mu'}_{\nu} = \begin{pmatrix} \gamma & -\gamma v & 0 & 0 \\ -\gamma v & \gamma & 0 & 0 \\ 0 & 0 & 1 & 0 \\ 0 & 0 & 0 & 1 \end{pmatrix} . \tag{1.3.3}$$

Notice that the Lorentz transformations mix space and time components, something that was difficult to interpret before special relativity.

The Lorentz transformations have a number of important consequences. The first of these can be easily derived by asking where the events that happen at $t = 0$ according to $\mathcal{O}$ end up in the frame $\mathcal{O}'$. From the equations above we see that these events will have coordinates in $\mathcal{O}'$ such that $t' = -vx'$. That is, events that happen *at the same time* $t = 0$ in frame $\mathcal{O}$ and are thus simultaneous, happen at times that depend on their spatial positions according to $\mathcal{O}'$ and are then *not* simultaneous. Simultaneity is therefore relative, or in other words it has

[1]Einstein did not cite Michelson and Morley in his original papers on relativity [223].

no absolute physical meaning – it is just a convention. Notice that this implies, in particular, that the time order of events is not always fixed: If two events are simultaneous in a given inertial frame $\mathcal{O}$, then in a frame $\mathcal{O}'$ moving with a non-zero speed v with respect to $\mathcal{O}$, one of the two events will happen at an earlier time than the other, while in a frame $\mathcal{O}''$ moving with speed $-v$ with respect to $\mathcal{O}$ it will be the other way around.

The theory of special relativity was put on a more geometric foundation by Herman Minkowski in 1908. Although Einstein initially perceived this as an unnecessarily abstract way of rewriting the same theory, he later realized that Minkowski's contribution was in fact crucial for the transition from special to general relativity. Minkowski realized that we could rewrite Einstein's second postulate about the invariance of the speed of light in geometric terms if we first defined the so-called *interval* Δs^2 between two events as:[2]

$$\Delta s^2 := -\Delta t^2 + \Delta x^2 + \Delta y^2 + \Delta z^2 \,. \tag{1.3.4}$$

Einstein's second postulate then turns out to be equivalent to saying that the interval defined above between any two events is absolute, that is, all inertial observers will find the same value of Δs^2. This means that we can define a concept of invariant distance between events, and once we have a measure of distance we can do geometry. Notice that the ordinary three-dimensional Euclidean distance $\Delta l^2 = \Delta x^2 + \Delta y^2 + \Delta z^2$ between two events is not absolute, nor is the time interval Δt^2, as can be easily seen from the Lorentz transformations. Only Minkowski's four-dimensional spacetime interval is absolute. In Minkowski's own words "... henceforth space by itself, and time by itself, are doomed to fade away into mere shadows, and only a kind of union of the two will preserve an independent reality".

A crucial property of the spacetime interval defined in (1.3.4) is the fact that, because of the minus sign in front of the first term, it is not positive definite. Rather than this being a drawback, it has an important physical interpretation as it allows us to classify the separation of events according to the sign of Δs^2:

$$\Delta s^2 > 0 \qquad \text{spacelike separation}\,, \tag{1.3.5}$$

$$\Delta s^2 < 0 \qquad \text{timelike separation}\,, \tag{1.3.6}$$

$$\Delta s^2 = 0 \qquad \text{null separation}\,. \tag{1.3.7}$$

Spacelike separations correspond to events separated in such a way that we would have to move faster than the speed of light to reach one from the other (they are separated mainly in space), timelike separations correspond to events that can be reached one from the other moving slower than light (they are separated mainly in time), and null separations correspond to events that can be reached one from the other moving precisely at the speed of light. An important consequence of

[2]The overall sign in the definition of Δs^2 is a matter of convention, and many textbooks choose the opposite sign to the one used here.

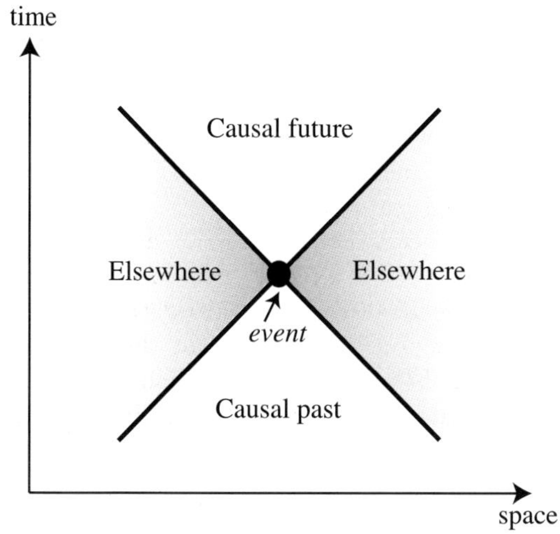

Fig. 1.1: The light-cone of a given event defines its causal relationship with other events, and divides space into three regions: the causal past (those events that can influence the event under consideration), the causal future (those events that can be influenced by the event under consideration), and *elsewhere* (those events with which there can be no causal relation).

the Lorentz transformations is that the time order of events is in fact absolute for events with timelike or null separations – it is only for events with spacelike separations that there is no fixed time order. This allows us to define a notion of causality in an invariant way: Only events separated in a timelike or null way can be causally related. Events separated in a spacelike way must be causally disconnected, as otherwise in some inertial frames the effect would be found to precede the cause. In particular this implies that no physical interaction can travel faster than the speed of light as this would violate causality – this is one of the reasons why in relativity nothing can travel faster than light. In fact, in relativity all material objects move following timelike trajectories, while light moves along null trajectories. Null trajectories also define the *light-cone* (see Figure 1.1), which indicates which events can be causally related with each other.

From the invariance of the interval, or equivalently from the Lorentz transformations, we can derive two other important results. Let us start by defining the *proper time* between two events as the time measured by an ideal clock that is moving at constant speed in such a way that it sees both events happen at the same place. From the point of view of this clock we have $\Delta l^2 = 0$, which implies $\Delta s^2 = -\Delta t^2$. If we use the Greek letter τ to denote the proper time we will then have $\Delta\tau = \sqrt{-\Delta s^2}$. We clearly see that the proper time can only be defined for timelike or null intervals – it has no meaning for spatial intervals (which makes sense since no physical clock can travel faster than light). Having defined the proper time, it is not difficult to show that the interval of time Δt measured between two events in a given inertial frame is related to the proper time between those events in the following way:

$$\Delta t = \gamma \Delta \tau \geq \Delta \tau \, . \tag{1.3.8}$$

This effect is known as *time dilation*, and implies that in a given reference frame all moving clocks are measured to go slow. The effect is of course symmetrical: If I measure the clocks of a moving frame as going slow, someone in that frame will measure my clocks as going slow.

Another consequence of the Lorentz transformations is related with the measure of spatial distances. Let us assume that we have a rod of length l as measured when it is at rest (the *proper length*). If the rod moves with speed v we will measure it as being contracted along the direction of motion. The length L measured will be related to l and the speed of the rod v according to:

$$L = l/\gamma \leq l \, . \tag{1.3.9}$$

This is known as the *Lorentz–Fitzgerald contraction.*[3]

Up until this point we have been considering finite separations in spacetime, but it is more fundamental to consider infinitesimal separations. In the case of special relativity, the fact that spacetime is homogeneous, isotropic and time independent allows us to consider finite intervals in an unambiguous way, but in general relativity this will no longer be the case. Because of this, from now on we will write the spacetime interval of special relativity as

$$ds^2 = -dt^2 + dx^2 + dy^2 + dz^2 \, . \tag{1.3.10}$$

Moreover, we will use the following compact form

$$ds^2 = \eta_{\mu\nu} \, dx^\mu dx^\nu \, , \tag{1.3.11}$$

where $\eta_{\mu\nu}$ is the *Minkowski metric tensor* defined as

$$\eta_{\mu\nu} = \begin{pmatrix} -1 & 0 & 0 & 0 \\ 0 & +1 & 0 & 0 \\ 0 & 0 & +1 & 0 \\ 0 & 0 & 0 & +1 \end{pmatrix} . \tag{1.3.12}$$

The infinitesimal interval allows us to think of the length of curves in spacetime. For spacelike curves the length is simply defined as the integral of $\sqrt{ds^2}$

[3]This contraction had been derived by Lorentz and Fitzgerald before special relativity, but it was considered a dynamical effect due to the interaction between the moving object and the *luminiferous aether*. However, the fact that the contraction was independent of the physical properties of the moving object was difficult to justify. In special relativity the contraction is not dynamical in origin, but is instead a purely kinematical consequence of the invariance of the interval.

along the curve, while for timelike curves it is defined instead as the integral of $d\tau = \sqrt{-ds^2}$ and corresponds directly to the time measured by an ideal clock following that trajectory. Null curves, of course, have zero length by definition. We can then easily see that for spacelike curves a straight line between two events is the one of minimum length, while for timelike curves the straight line has in fact maximum length. In any case straight lines are always extremal curves, also known as *geodesics*. We can use this fact to rewrite Newton's first law in geometric terms. Notice first that when thinking about spacetime, Newton's first law is simply stated as: *Objects free of external forces move along straight timelike trajectories in spacetime.* This simple statement captures both the fact that the trajectories are straight in three-dimensional space, and also that the motion is uniform. Using the notion of a geodesic we can rewrite this as: *Objects free of external forces move along timelike geodesics in spacetime.* We can even extend Newton's first law to light and say: *Light rays (photons) in vacuum move along null geodesics of spacetime.* At this point, we should also mention that it is customary in relativity to refer to the trajectory of a particle through spacetime as its *world-line*, so Newton's first law states that a free particle moves in such a way that its world-line corresponds to a geodesic of spacetime.

1.4 Manifolds and tensors

When making the transition from special to general relativity it is important to first learn a few basic concepts of differential geometry. Differential geometry is in fact also extremely useful in special relativity, as Minkowski himself showed, but it becomes essential in the general theory.

Differential geometry starts from the notion of a differentiable manifold, which is the formal mathematical way of defining what intuitively is a continuous and smooth space of n dimensions. We are familiar with many manifolds in physics. For example, Euclidean three-dimensional (3D) space is a 3-manifold, while the spacetime of special relativity is a 4-manifold. But there are many other interesting manifolds. The surface of a sphere is a curved, and closed, 2-manifold. The configuration space for a system of two point particles is a 6-manifold, as we need six coordinates to describe the system. There are also some common manifolds where there is no notion of distance – the phase space of classical mechanics is one such manifold. Another interesting manifold is the space of rotations of a rigid body in three dimensions. Since we can describe any such rotation using the three Euler angles, the space of rotations is a 3-manifold, but it has a complicated structure: It is not only closed, as rotating an angle of 2π about any axis brings us back to the original situation, but it also has a non-trivial topology.

Mathematically, a manifold M is defined as a space that can be covered by a collection of *charts*, that is, one-to-one mappings from $\Re^n$ to M. Euclidean 3D space can be clearly covered by one such mapping from $\Re^3$ in a trivial way, while for the surface of a sphere we actually need at least two mappings from $\Re^2$ (standard maps fail at two points, namely the poles, but a stereographic map fails at only one point, so we need two of those). In more physical language, a mapping is nothing more than a set of coordinates that label the different points in M. Notice that a coordinate system on a given patch of M is not unique, coordinate systems are in fact arbitrary.

Once we have a manifold, we can consider *curves* in this manifold defined as functions from a segment of the real line into the manifold. It is important to distinguish between the image of a curve (*i.e.* its trajectory), and the curve itself: The curve is the function and contains information both about which points on M we are traversing, and how fast we are moving with respect to the parameter. In terms of a set of coordinates $\{x^\alpha\}$ on M, a curve is represented as

$$x^\alpha = x^\alpha(\lambda) , \qquad \lambda \in \Re . \tag{1.4.1}$$

Notice that a change of coordinates will alter the explicit functional form above, but will not alter the curve itself. Working with coordinates helps to make many concepts explicit, but we should always be careful not to confuse the coordinate representation of a geometric object with the object itself.

Vectors are defined as derivative operators along a given curve. The precise definition is somewhat abstract, and although this is certainly convenient from a mathematical point of view, here we will limit ourselves to working with the components of a vector. The components of a vector $\vec{v}$ tangent to a curve $x^\alpha(\lambda)$ are given simply by

$$v^\alpha = dx^\alpha / d\lambda . \tag{1.4.2}$$

Vectors are defined at a given point, and on that point they form a vector space known as the *tangent space* of M (one should really think of vectors as representing only infinitesimal displacements on M).

Since vectors form a vector space, we can always represent them as linear combinations of some basis vectors $\{\vec{e}_\alpha\}$, where here α is an index that identifies the different vectors in the basis and not their components. For example, for an arbitrary vector $\vec{v}$ we have

$$\vec{v} = v^\alpha \vec{e}_\alpha . \tag{1.4.3}$$

A common basis choice (though certainly not the only possibility) is the so-called *coordinate basis* for which we take as the basis those vectors that are tangent to the coordinate lines, using as parameters the coordinates themselves. It is precisely in this basis that the components of a vector are given as defined above. From here on we will always work with the coordinate basis (but see Chapter 8 where we will discuss the use of a non-coordinate basis known as a *tetrad frame*).

Let us now consider functions of vectors on the tangent space. A linear, real-valued function of one vector is called a *one-form*. We will denote one-forms with a tilde and write the action of a one-form $\tilde{q}$ on a vector $\vec{v}$ as $\tilde{q}(\vec{v})$.[4] It is not

[4]The name one-form comes from the calculus of differential forms and the notion of exterior derivatives. Forms are very important in the theory of integration in manifolds. Here, however, we will not consider the calculus of forms, and we will take "one-form" to be just a name –

difficult to show that one-forms also form a vector space of the same dimension as that of the manifold – this is known as the *dual* tangent space, and for this reason one-forms are often called *co-vectors*.

The components of a one-form are defined as the value of the one-form acting on the basis vectors:

$$q_\alpha := \tilde{q}(\vec{e}_\alpha) \; . \tag{1.4.4}$$

Notice that while the components of a vector are represented with indices up, those of a one-form have the indices down.

We can also define a basis for the space of one-forms, known as the *dual basis* $\tilde{\omega}^\alpha$, and defined as those one-forms such that, when acting on the basis vectors $\vec{e}_\alpha$, give us the identity matrix

$$\tilde{\omega}^\alpha(\vec{e}_\beta) = \delta^\alpha_\beta \; , \tag{1.4.5}$$

where δ^α_β is the Kronecker delta. The dual basis has one very important property, namely that the components of a one-form defined above are precisely the ones that allow us to write the one-form in terms of the dual basis:

$$\tilde{q} = q_\alpha \tilde{\omega}^\alpha \; . \tag{1.4.6}$$

In terms of the basis of vectors and one-forms, the action of an arbitrary one-form $\tilde{q}$ on a vector $\vec{v}$ can be represented as:

$$\tilde{q}(\vec{v}) = q_\alpha \tilde{\omega}^\alpha (v^\beta \vec{e}_\beta) = q_\alpha v^\beta \tilde{\omega}^\alpha (\vec{e}_\beta) = q_\alpha v^\beta \delta^\alpha_\beta = q_\alpha v^\alpha \; . \tag{1.4.7}$$

where we have used the linearity of the one-forms and the definition of the dual basis. The last expression shows that the action of $\tilde{q}$ on $\vec{v}$ is just the sum of the components of $\tilde{q}$ times the components of $\vec{v}$. This operation is called a *contraction*, and it gives us a real number that is in fact independent of the coordinate system or basis we are using.

One-forms and vectors are dual in one other respect: We can invert the definition and define a vector $\vec{v}$ as a real-valued function of a one-form, namely the function that, given a one-form $\tilde{q}$, gives us back the real number $\tilde{q}(\vec{v})$. We can then generalize this idea and think of real-valued functions of m one-forms and n vectors that are linear in all their arguments. This is what defines a *tensor* of type $\binom{m}{n}$. The components of a tensor are then just the values of the tensor applied to the elements of the basis of vectors and one-forms. For example, for a tensor T of $\binom{2}{2}$ type, we have

$$T^{\alpha\beta}{}_{\mu\nu} := T\left(\tilde{\omega}^\alpha, \tilde{\omega}^\beta; \vec{e}_\mu, \vec{e}_\nu\right) \; . \tag{1.4.8}$$

We can then think of vectors and one-forms as $\binom{1}{0}$ and $\binom{0}{1}$ tensors respectively. Just like vectors, tensors are defined at each point, but we can easily construct tensor fields by a rule that gives a tensor at each point of the manifold M.

readers interested in this subject can consult any standard text on differential geometry (see *e.g.* [258]).

A particularly important one-form field is the gradient of a scalar function. Consider a function f from M into the real numbers. This is called a *scalar function* and can be thought of as a $\binom{0}{0}$ tensor field. The gradient of f is denoted by $\tilde{d}f$, and is the one-form with components $\partial f/\partial x^\alpha \equiv \partial_\alpha f$. The notation used here should already make us pay attention – the indices for the components of the gradient are down, which suggests that it is a one-form and not a vector. It becomes clear that it is a one-form since we can apply it to an arbitrary vector $\vec{v}$ to get a real number, namely $v^\alpha \partial_\alpha f$, which is nothing more than the directional derivative of f along the vector $\vec{v}$. We will see below that the gradient also transforms with the rules of a one-form and not of a vector.

An important characteristic of tensors has to do with their symmetry properties with respect to exchange of their arguments. We say that a tensor is *completely symmetric* if it remains the same with respect to exchange of any pair of arguments. Similarly, a tensor is said to be *completely antisymmetric* if it changes sign with respect to exchange of any pair of arguments.[5] The symmetric and antisymmetric parts of an arbitrary tensor can be easily constructed by adding together all possible permutations of their arguments with appropriate signs: all positive for the symmetric part, and positive or negative for the antisymmetric part depending on whether the permutation is even or odd. For example, for a $\binom{0}{2}$ tensor T we have:

$$T_{(\alpha\beta)} := \frac{1}{2!}\left(T_{\alpha\beta} + T_{\beta\alpha}\right) , \tag{1.4.9}$$

$$T_{[\alpha\beta]} := \frac{1}{2!}\left(T_{\alpha\beta} - T_{\beta\alpha}\right) , \tag{1.4.10}$$

where the round and square brackets are standard notation for the symmetric and antisymmetric parts, respectively, and the $1/2!$ is a normalization factor. Generalizations to tensors of higher rank are straightforward. The symmetries of tensors are quite important in general relativity where many of the most important tensors have very specific symmetry properties.

1.5 The metric tensor

As we have seen, the crucial contribution of Minkowski to the theory of relativity was the realization that there exists a measure of invariant distance between two events in four-dimensional spacetime. More generally, in a given manifold M a notion of distance is given by the existence of a symmetric, non-degenerate, $\binom{0}{2}$ tensor g that defines the *dot* or *scalar* product between two vectors

$$g(\vec{v}, \vec{u}) \equiv \vec{v} \cdot \vec{u} = g_{\alpha\beta}\, v^\alpha u^\beta , \tag{1.5.1}$$

with $g_{\alpha\beta}$ the components of g. By non-degenerate we mean that if we have a vector $\vec{w}$ such that $g(\vec{w}, \vec{v}) = 0$ for all $\vec{v}$, then we must necessarily have that

[5] A p-form is defined as a completely antisymmetric $\binom{0}{p}$ tensor.

$\vec{w} = 0$ (this also implies that the components $g_{\alpha\beta}$ form an invertible matrix). The tensor g giving this scalar product is called the *metric tensor*, and allows us to define the magnitude or norm of a vector as

$$|v|^2 := g(\vec{v}, \vec{v}) = \vec{v} \cdot \vec{v} = g_{\alpha\beta}\, v^\alpha v^\beta \; . \tag{1.5.2}$$

In particular, we can calculate the magnitude of the displacement vector $d\vec{x}$ between two infinitesimally close points in the manifold as

$$ds^2 = g_{\alpha\beta}\, dx^\alpha dx^\beta \; , \tag{1.5.3}$$

which allows us to define a notion of distance between these two points. Not all manifolds have such a metric tensor defined on them, the obvious physical example of a manifold with no metric being the phase space of classical mechanics. The metric tensor gives an extra degree of structure to a manifold. In the particular case of special relativity we do in fact have such a notion of distance, with the metric tensor g given by Minkowski's interval: $g_{\alpha\beta} = \eta_{\alpha\beta}$. This also means that in special relativity we can not only calculate distances, but in fact we can also construct the scalar product of two arbitrary vectors.

As already mentioned, the metric of special relativity is not positive definite. In general, the *signature* of a metric is given by the signs of the eigenvalues of its matrix of components. We call a positive definite metric, *i.e.* one with all eigenvalues positive, *Euclidean*, while a metric like the one of special relativity with signature $(-,+,+,+)$ is called *Lorentzian*. For a Lorentzian metric like the one in relativity, we classify vectors in the same way as intervals according to the sign of their magnitude and talks about spacelike, null and timelike vectors.

The metric tensor can be used to define a one-to-one mapping between vectors and one-forms. Assume, for example, that we are given a vector $\vec{v}$. If we use that vector in one of the slots of the tensor g, we will have an object $g(\vec{v}, \;)$ that takes one arbitrary vector $\vec{u}$ as an argument and gives us a real number, namely $\vec{v} \cdot \vec{u}$. That is, $g(\vec{v}, \;)$ defines a one-form. Since this one-form is associated with the vector $\vec{v}$, we will denote it simply by $\tilde{v}$. Its components are easy to obtain from the standard definition

$$v_\alpha := \tilde{v}\,(\vec{e}_\alpha) = g\,(\vec{v}, \vec{e}_\alpha) = g_{\mu\nu}\, v^\mu (\vec{e}_\alpha)^\nu = g_{\mu\nu} v^\mu \delta^\nu_\alpha \; , \tag{1.5.4}$$

and finally,

$$v_\alpha = g_{\alpha\beta} v^\beta \; . \tag{1.5.5}$$

Since by definition g is non-degenerate, we can invert this relation to find

$$v^\alpha = g^{\alpha\beta} v_\beta \; , \tag{1.5.6}$$

where $g^{\alpha\beta}$ are the components of the inverse matrix to $g_{\alpha\beta}$, that is $g^{\alpha\mu} g_{\mu\beta} = \delta^\alpha_\beta$. This mapping between vectors and one-forms means that we can think of them as the same geometric object, that is, just a "vector". The operations given by (1.5.5) and (1.5.6) are then simply referred to as *lowering* and *raising* the index of the vector. But it is important to stress the fact that this mapping between vectors and one-forms can only be defined if we have a metric tensor.

In the case of Euclidean space, the metric tensor is given by the identity matrix, which implies that the components of a vector and its associated one-form are identical. This explains why the distinction between vectors and one-forms is usually not even introduced for Euclidean space. In special relativity, however, we need to be more careful as the components of a vector and its associated one-form are no longer the same since the metric is now given by $\eta_{\alpha\beta}$:

$$v_\alpha = \eta_{\alpha\beta}\, v^\beta \quad \Rightarrow \quad (v_0, v_1, v_2, v_3) = (-v^0, v^1, v^2, v^3) \,, \tag{1.5.7}$$

We see then that lowering and raising indices of vectors in special relativity changes the sign of the time components.

Notice that once we have defined the operation of lowering indices, the dot product between two vectors $\vec{v}$ and $\vec{u}$ can be simply calculated as

$$g(\vec{v}, \vec{u}) = g_{\alpha\beta} v^\alpha u^\beta = v_\alpha u^\alpha \,, \tag{1.5.8}$$

that is, the direct contraction between one of the vectors and the one-form associated with the other one.

We can easily generalize the notion of raising and lowering indices to tensors of arbitrary rank by simply contracting a given index with the metric tensor $g_{\alpha\beta}$ or its inverse $g^{\alpha\beta}$. For example:

$$T^\mu{}_\nu = g^{\mu\alpha}\, T_{\alpha\nu} \,, \qquad T^{\mu\nu} = g^{\mu\alpha} g^{\nu\beta}\, T_{\alpha\beta} \,, \qquad S_{\sigma\mu\nu\rho} = g_{\sigma\lambda} S^\lambda{}_{\mu\nu\rho} \,.$$

In this way we will think of tensors as objects of a given rank indicated by the total number of indices, irrespective of where those indices are (but the position of the indices *is* important once we assign explicit values to the components).

The existence of a metric tensor also allows us to introduce the notion of orthogonality of vectors. We simply say that two vectors are orthogonal if their scalar product vanishes: $\vec{v} \cdot \vec{u} = 0$. Having a notion of orthogonality allows us to define the *projection operator*. Consider a spacelike unit vector $\vec{s}$, the projection operator onto a hypersurface orthogonal to $\vec{s}$ is defined as

$$P^\mu_\nu := \delta^\mu_\nu - s^\mu s_\nu \,. \tag{1.5.9}$$

Notice how in this expression we have already used the components of one-form s_μ associated with the vector $\vec{s}$. If we now apply the projection operator to an arbitrary vector $\vec{v}$ and calculate its dot product with $\vec{s}$ we find

$$(P^\mu_\nu v^\nu)\, s_\mu = (\delta^\mu_\nu - s^\mu s_\nu)\, v^\nu s_\mu = 0 \,, \tag{1.5.10}$$

which implies that the projection of the vector $\vec{v}$ is always orthogonal to $\vec{s}$. The projection operator also corresponds to the *induced* metric tensor on the hypersurface orthogonal to $\vec{s}$. To see this, consider the norm of a projected vector:

$$g_{\alpha\beta}\left(P^{\alpha}_{\mu}v^{\mu}\right)\left(P^{\beta}_{\nu}v^{\nu}\right) = \left(g_{\mu\nu} - s_{\mu}s_{\nu}\right)v^{\mu}v^{\nu} = P_{\mu\nu}v^{\mu}v^{\nu} \; , \qquad (1.5.11)$$

that is, the norm of a projected vector can be calculated directly with $P_{\mu\nu}$.

If instead of a unit spacelike vector $\vec{s}$ we consider a unit timelike vector $\vec{n}$, then the projection operator takes the slightly different form

$$P^{\mu}_{\nu} := \delta^{\mu}_{\nu} + n^{\mu}n_{\nu} \; . \qquad (1.5.12)$$

In this case, for a manifold with Lorentzian signature, the induced metric

$$\gamma_{\mu\nu} := P_{\mu\nu} = g_{\mu\nu} + n_{\mu}n_{\nu} \; , \qquad (1.5.13)$$

will necessarily be positive definite: $\gamma_{\mu\nu}v^{\mu}v^{\nu} \geq 0$ for arbitrary $\vec{v}$.

We can also use the scalar product to define the angle θ between two vectors through the usual relation from Euclidean geometry

$$\vec{v} \cdot \vec{u} = |v|\,|u| \cos\theta \; . \qquad (1.5.14)$$

Notice that the angle between two vectors defined above remains invariant if we change the metric tensor in the following way:

$$g_{ij} \quad \rightarrow \quad \bar{g}_{ij} = \Phi \, g_{ij} \; , \qquad (1.5.15)$$

with Φ some scalar function. Such a change of the metric is called a *conformal transformation* (since it preserves angles), and the function Φ is called the *conformal factor*.

The metric tensor can also be used to measure volumes and not just linear distances. Consider for a moment a two-dimensional manifold, and assume that we want to find the area element associated with the infinitesimal coordinate square defined by dx^1 and dx^2. If the coordinate lines are orthogonal at the point considered, the area element will clearly be given by $dA = |e_1|\,|e_2|\,dx^1dx^2$, with $\vec{e}_1$ and $\vec{e}_2$ the corresponding basis vectors. Of course, in the general case, the coordinate lines will not be orthogonal, but it is clear that the general expression will be given by the formula for the area of a parallelogram, $dA = |e_1|\,|e_2|\sin\theta\,dx^1dx^2$, with θ the angle between $\vec{e}_1$ and $\vec{e}_2$. Using now the definition of the angle θ given above we find

$$\begin{aligned} dA &= |e_1|\,|e_2|\sin\theta\,dx^1dx^2 = |e_1|\,|e_2|\left[1 - \cos^2\theta\right]^{1/2} dx^1dx^2 \\ &= \left[\left(|e_1|\,|e_2|\right)^2 - \left(\vec{e}_1 \cdot \vec{e}_2\right)^2\right]^{1/2} dx^1dx^2 \; , \end{aligned}$$

and from the definition of the metric components this becomes

$$dA = \left[g_{11} g_{22} - (g_{12})^2 \right]^{1/2} = [\det(g)]^{1/2} \, dx^1 dx^2 \; ,$$

where $\det(g)$ is the determinant of the matrix of metric components. It turns out that this relation can be generalized to an arbitrary number of dimensions, and the volume element of an n-dimensional manifold is always given by

$$dV = |g|^{1/2} \, dx^1 \cdots dx^n \; . \tag{1.5.16}$$

In the last expression we have introduced the standard convention of using simply g to denote the determinant of the metric. Also, the absolute value is there to allow for the possibility of having a non-positive definite metric.

A particularly useful tensor related to the determinant of the metric g is the *Levi–Civita completely antisymmetric tensor* ϵ, which in four dimensions is defined as[6]

$$\epsilon_{\alpha\beta\mu\nu} = \begin{cases} +|g|^{1/2} & \text{for even permutations of 0,1,2,3} \\ -|g|^{1/2} & \text{for odd permutations of 0,1,2,3} \\ 0 & \text{if any indices are repeated} \end{cases} \tag{1.5.17}$$

The factor $|g|^{1/2}$ guarantees that, in an orthonormal frame, the components of ϵ are always ± 1. We can define the Levi–Civita tensor with two or three indices by considering lower dimensional manifolds within our four-dimensional spacetime, but notice that the Levi–Civita tensor with five indices is identically zero.

1.6 Lie derivatives and Killing fields

Like vectors, tensors are defined at a specific point on a manifold, and although we can define tensor fields easily enough there is no natural way to compare tensors at different points in the manifold since they live on different spaces. This means, in particular, that there is no natural notion of the derivative of a tensor, since that requires us to compare the values of a tensor field at nearby points in the manifold. We will see in the next Sections that, given a metric, we can introduce a notion of derivative known as the *covariant derivative*. However, there is a more primitive notion of derivative that does not depend on the existence of a metric and that is often quite useful. This is known as the *Lie derivative*, and is based on the existence of a vector field $\vec{v}$ defined on the manifold.

Consider a smooth vector field $\vec{v}$ on a region of a manifold. The congruence of integral curves of this vector field defines a mapping ϕ of the manifold into itself – we simply identify a given point p with another point $q = \phi_\lambda(p)$ that lies on the same integral curve as p at a *distance* λ, as defined by the change in the

[6] In an n-dimensional manifold, the Levi–Civita tensor is strictly speaking an n-form known as the *natural volume element*. It is also only defined for *orientable* manifolds, and the standard definition corresponds to a right handed basis.

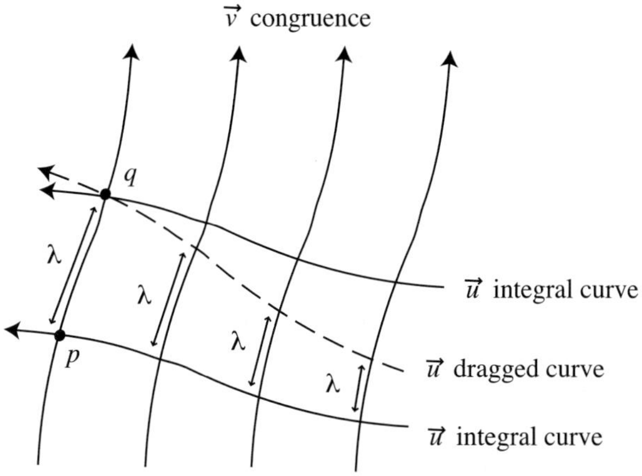

Fig. 1.2: Lie dragging of vectors. The integral curves of the vector $\vec{u}$ are dragged along the congruence associated with the vector field $\vec{v}$ an equal distance as measured by the parameter λ.

parameter that defines the curve. We can then use this mapping to drag tensors from the point p to the point q. This is known as *Lie dragging*.

For a scalar function it is easy to see how Lie dragging works: We say that the dragged function $\phi_\lambda(f)$ is such that its value on q is equal to $f(p)$, that is $\phi_\lambda(f)(q) = f(p)$. For a vector field $\vec{u}$ we define the dragging by looking at the curve associated with $\vec{u}$, and dragging the curve from p to q using the congruence associated with $\vec{v}$ (see Figure 1.2). The dragged vector $\phi(\vec{u})$ at q will be the tangent to the new curve.

The Lie derivative of a vector field $\vec{u}$ is then defined in the following way: Evaluate the vector at $q = \phi_\lambda(p)$, drag it back to p using $\phi_{-\lambda}^{-1}$, and take the difference with the original vector at p in the limit when λ goes to zero:

$$\pounds_{\vec{v}}\, \vec{u} := \lim_{\lambda \to 0} \left[\frac{\phi_{-\lambda}^{-1}\left(\vec{u}|_{\phi_\lambda(p)}\right) - \vec{u}|_p}{\lambda} \right] . \tag{1.6.1}$$

The notation indicates that the Lie derivative depends on the vector field used for the dragging. To find the components of the Lie derivative, let μ be the parameter for the integral curves of the vector field $\vec{u}$. If we drag an integral curve of $\vec{u}$ an infinitesimal distance λ along the vector field $\vec{v}$ we will find that

$$\begin{aligned} \phi_\lambda(x^\alpha) &= x^\alpha + v^\alpha \lambda \\ &\Rightarrow \phi_\lambda(u^\alpha) = u^\alpha + (dv^\alpha / d\mu)\, \lambda = u^\alpha + \left(u^\beta \partial_\beta v^\alpha\right) \lambda \, . \end{aligned} \tag{1.6.2}$$

This implies that

$$\begin{aligned} \pounds_{\vec{v}} u^\alpha &= \lim_{\lambda \to 0} \left[\frac{u^\alpha|_{\phi_\lambda(p)} - \lambda \left(u^\beta|_{\phi_\lambda(p)} \partial_\beta v^\alpha \right) - u^\alpha|_p}{\lambda} \right] \\ &= du^\alpha / d\lambda - u^\beta \partial_\beta v^\alpha \\ &= v^\beta \partial_\beta u^\alpha - u^\beta \partial_\beta v^\alpha \,. \end{aligned} \tag{1.6.3}$$

If we now define the *Lie bracket* or *commutator* of two vectors as

$$[\vec{v}, \vec{u}]^\alpha := v^\beta \partial_\beta u^\alpha - u^\beta \partial_\beta v^\alpha \,, \tag{1.6.4}$$

then we find that

$$\pounds_{\vec{v}} \vec{u} = [\vec{v}, \vec{u}] \,. \tag{1.6.5}$$

That is, the Lie derivative of $\vec{u}$ with respect to $\vec{v}$ is nothing more than the commutator of $\vec{v}$ and $\vec{u}$.

Once we know how to drag vectors, we can also drag one-forms by just asking for the value of the dragged one-form when applied to the dragged vector to be equal to the value of the original one-form applied to the original vector: $\phi_\lambda(\tilde{\omega})(\phi_\lambda(\vec{u})) = \tilde{\omega}(\vec{u})$. In the same way we can drag tensors of arbitrary rank. We can then define the Lie derivative of tensors in an analogous way to that of vectors. For a one-form $\tilde{\omega}$ we find

$$\pounds_{\vec{v}} \omega_\alpha = v^\beta \partial_\beta \omega_\alpha + \omega_\beta \partial_\alpha v^\beta \,. \tag{1.6.6}$$

We can do the same for tensors of arbitrary rank, just adding one more term for each index, with the adequate sign. For example

$$\pounds_{\vec{v}} T^\alpha{}_\beta = v^\mu \partial_\mu T^\alpha{}_\beta - T^\mu{}_\beta \partial_\mu v^\alpha + T^\alpha{}_\mu \partial_\beta v^\mu \,. \tag{1.6.7}$$

There is an important property of Lie derivatives that helps considerably with their interpretation. Assume that we adapt one of our coordinates, say x^1, to the integral curves of the vector field $\vec{v}$. In that case we have that $x^1 = \lambda$ and $\vec{e}_1 = \vec{v}$, which implies $v^\alpha = \delta^\alpha_1$. It is then easy to see that the Lie derivative of a tensor T of arbitrary rank will simplify to

$$\pounds_{\vec{v}} T^{\alpha\beta\ldots}{}_{\mu\nu\ldots} = \partial_1 T^{\alpha\beta\ldots}{}_{\mu\nu\ldots} \,. \tag{1.6.8}$$

This shows that the Lie derivative is a way to write partial derivatives along the direction of a given vector field in a way that is independent of the coordinates.

A particularly important application of the Lie derivative is related to the possible symmetries of a manifold that has a metric tensor defined on it. We say that the manifold has a specific symmetry if the metric is invariant under Lie dragging with respect to some vector field $\vec{\xi}$, that is, if we have

$$\pounds_{\vec{\xi}} g = 0 \,. \tag{1.6.9}$$

From the expression for the Lie derivative of a tensor we find that this implies

$$\xi^\mu \, \partial_\mu g_{\alpha\beta} + g_{\alpha\mu} \, \partial_\beta \xi^\mu + g_{\mu\beta} \, \partial_\alpha \xi^\mu = 0 \, . \tag{1.6.10}$$

If given a metric $g_{\mu\nu}$ there exists a vector field $\vec{\xi}$ that satisfies the above equation, then $\vec{\xi}$ is called a *Killing field.* It is easy to see that the existence of a Killing field implies a symmetry of the manifold: Assume as before that the coordinate x^1 is adapted to the integral curves of $\vec{\xi}$, the condition above then reduces to

$$\partial_1 g_{\alpha\beta} = 0 \, , \tag{1.6.11}$$

so the components of the metric tensor are in fact independent of the coordinate x^1. We will come back to the condition for the existence of a Killing field in Section 1.8, where we will rewrite it in a more standard way.

1.7 Coordinate transformations

Up until this point we have assumed that we are given a specific coordinate system $\{x^\alpha\}$ and its associated coordinate basis $\{\vec{e}_\alpha\}$. However, we must also consider what happens when we transform to a different set of coordinates. This is important as coordinates are in fact arbitrary labels for points in a manifold, and we might want to choose different coordinates under different circumstances. For example, in the particular case of special relativity, we might wish to use the coordinates associated with a given inertial frame, or those associated with a different inertial frame and related to the first ones via the Lorentz transformations.

More generally, we wish to consider arbitrary changes of coordinates of the form $x^{\bar{\alpha}} = f^{\bar{\alpha}}(x^\beta)$. It is easy to see that under this change of coordinates the components of the displacement vector transform as

$$dx^{\bar{\alpha}} = \frac{\partial x^{\bar{\alpha}}}{\partial x^\beta} \, dx^\beta = \partial_\beta x^{\bar{\alpha}} \, dx^\beta \equiv \Lambda^{\bar{\alpha}}_\beta \, dx^\beta \, , \tag{1.7.1}$$

where we have introduced the Jacobian matrix $\Lambda^{\bar{\alpha}}_\beta := \partial_\beta x^{\bar{\alpha}}$. In the particular case of special relativity, the Jacobian matrix for the Lorentz transformations is given by (1.3.3), but we can consider more general, even non-linear, changes of coordinates. An important property of a change of coordinates is that, in the region of interest, the transformation should be one-to-one, as otherwise the new coordinates would be useless. This implies that the Jacobian matrix is always invertible in this region.

From the definition of the components of a vector it is easy to see that they transform just as the displacement vector:

$$v^{\bar{\alpha}} = \Lambda^{\bar{\alpha}}_\beta \, v^\beta \, . \tag{1.7.2}$$

When we transform the coordinates we clearly have also changed our coordinate basis, as the new basis must refer to the new coordinates. From the fact that a vector as a geometric object is invariant under coordinate transformations we have that

$$\vec{v} = v^{\alpha} \vec{e}_{\alpha} = v^{\bar{\mu}} \vec{e}_{\bar{\mu}} \; , \tag{1.7.3}$$

from which we can derive the transformation law for the basis vectors themselves:

$$\vec{e}_{\bar{\alpha}} = \Lambda^{\beta}_{\bar{\alpha}} \, \vec{e}_{\beta} \; , \tag{1.7.4}$$

where now $\Lambda^{\beta}_{\bar{\alpha}} = \partial_{\bar{\alpha}} x^{\beta}$ is the Jacobian of the inverse transformation: $\Lambda^{\alpha}_{\bar{\mu}} \Lambda^{\bar{\mu}}_{\beta} = \delta^{\alpha}_{\beta}$. In a similar way we can find the transformation laws for the components of a one-form and the one-forms that form the dual basis:

$$p_{\bar{\alpha}} = \Lambda^{\beta}_{\bar{\alpha}} \, p_{\beta} \; , \qquad \tilde{\omega}^{\bar{\alpha}} = \Lambda^{\bar{\alpha}}_{\beta} \, \tilde{\omega}^{\beta} \; . \tag{1.7.5}$$

Because of the form of the transformation laws, we usually refer to lower indices as *co-variant*[7] since they transform just like the basis vectors (with the inverse Jacobian), and to upper indices as *contra-variant* since they transform in the opposite way (with the direct Jacobian). Some textbooks refer to vectors as *contra-variant vectors* and to one-forms as *co-variant vectors*, but this is somewhat misleading. For us, the terms co-variant and contra-variant will always refer to the position of the indices, and not to the geometric objects themselves.

As an example of these transformation laws, consider the gradient $\tilde{d}f$ of a scalar function f. We have already mentioned that the gradient is a one-form and not a vector. We can see this directly by considering the transformation of the components of the gradient. Using the chain rule we find that

$$\partial_{\bar{\alpha}} f = \partial_{\bar{\alpha}} x^{\beta} \, \partial_{\beta} f = \Lambda^{\beta}_{\bar{\alpha}} \, \partial_{\beta} f \; . \tag{1.7.6}$$

We clearly see the components of the gradient transform as those of a one-form and not as those of a vector. Why is it then that in vector calculus we are used to thinking of the gradient as a vector that points in the direction of steepest ascent? The key phrase here is *steepest ascent*, which means that f becomes larger in the smallest possible distance, *i.e.* it requires us to be able to measure distances. The gradient can only be associated with a vector if we have a metric tensor, with the components of the *gradient vector* being simply $\partial^{\alpha} f := g^{\alpha\beta} \partial_{\beta} f$, which for the Euclidean case makes no difference.

We can easily generalize the transformation laws to tensors of arbitrary rank, the rule is simply to use one Jacobian factor for each index, using either the direct or inverse Jacobian depending on the position of the index, for example:

$$T^{\bar{\alpha}}{}_{\bar{\beta}} = \Lambda^{\bar{\alpha}}_{\mu} \, \Lambda^{\nu}_{\bar{\beta}} \, T^{\mu}{}_{\nu} \; .$$

In special relativity, the change of coordinates is usually given by the Lorentz transformations between inertial frames, which are linear in their arguments.

[7]One must take care with the word *covariant* as it can mean different things. We also say that an expression is covariant if it involves only tensorial quantities, as in the covariant derivative that we will see in the following section, or in the principle of general covariance. Because of this, when I refer to indices I will use the form *co-variant* with a hyphen.

However, we often need to consider more complex non-linear coordinate transformations even on Euclidean space. Consider, for example, a transformation from Cartesian coordinates $\{x, y, z\}$ to spherical coordinates $\{r, \theta, \phi\}$ on a three-dimensional Euclidean space:

$$x = r \sin\theta \cos\phi \,, \qquad y = r \sin\theta \sin\phi \,, \qquad z = r \cos\theta \,. \tag{1.7.7}$$

By either transforming the differentials directly, or by finding first the Jacobian and then using the transformation laws derived above, we find that the components of the displacement vector transform as

$$dx = \sin\theta \cos\phi \, dr - r \sin\theta \sin\phi \, d\phi + r \cos\theta \cos\phi \, d\theta \,, \tag{1.7.8}$$
$$dy = \sin\theta \sin\phi \, dr + r \sin\theta \cos\phi \, d\phi + r \cos\theta \sin\phi \, d\theta \,, \tag{1.7.9}$$
$$dz = \cos\theta \, dr - r \sin\theta \, d\theta \,. \tag{1.7.10}$$

We can use this to find the form of the infinitesimal distance in spherical coordinates, starting from the Pythagoras rule

$$dl^2 = dx^2 + dy^2 + dz^2 \,. \tag{1.7.11}$$

Substituting the expressions for the differentials we find that the distance between two infinitesimally close points in spherical coordinates is

$$dl^2 = g_{\alpha\beta} dx^\alpha dx^\beta = dr^2 + r^2 d\theta^2 + r^2 \sin^2\theta \, d\phi^2 \equiv dr^2 + r^2 d\Omega^2 \,, \tag{1.7.12}$$

where the last equality defines the element of solid angle $d\Omega^2$. The components of the metric tensor are now clearly non-trivial. It is also clear that in order to find the finite distance between two points in these coordinates we need to do a non-trivial line integration. We can, of course, also decide to use spherical coordinates in special relativity, in which case the components of the metric will no longer be those of the Minkowski tensor. Moreover, we can even transform coordinates to those of an accelerated observer, which will again give us a non-trivial metric.

The distance element (1.7.12) above still represents the metric of a flat three-dimensional space. We can use it, however, to find distances on the surface of a sphere of radius R. It is clear that if we remain on this surface, we will have $dr = 0$, so the metric reduces to

$$dl^2 = R^2 \left(d\theta^2 + \sin^2\theta \, d\phi^2 \right) \,. \tag{1.7.13}$$

The last expression now represents the metric of a curved two-dimensional surface, namely the surface of the sphere. We then see that the fact that a metric is non-trivial does not in general help us to distinguish between a genuinely curved space, and a flat space in curvilinear coordinates (we will wait until Section 1.9 below to introduce properly the notion of curvature).

Notice also that in spherical coordinates not only is the metric non-trivial, but also the basis vectors themselves become non-trivial. Consider the coordinate basis associated with the spherical coordinates $\{r, \theta, \phi\}$. Written in the spherical coordinates themselves, the components of this basis are by definition

$$\vec{e}_r \rightarrow (1, 0, 0) , \qquad \vec{e}_\theta \rightarrow (0, 1, 0) , \qquad \vec{e}_\phi \rightarrow (0, 0, 1) . \tag{1.7.14}$$

When written in Cartesian coordinates the components of these vectors become

$$\vec{e}_r \rightarrow (\sin\theta\cos\phi, \sin\theta\sin\phi, \cos\theta) \ , \tag{1.7.15}$$
$$\vec{e}_\theta \rightarrow (r\cos\theta\cos\phi, r\cos\theta\sin\phi, -r\sin\theta) \ , \tag{1.7.16}$$
$$\vec{e}_\phi \rightarrow (-r\sin\theta\sin\phi, r\sin\theta\cos\phi, 0) \ , \tag{1.7.17}$$

as can be found from the transformation law given above. Note, for example, that not all of these basis vectors are unitary; their square-magnitudes are:

$$|\vec{e}_r|^2 = (\sin\theta\cos\phi)^2 + (\sin\theta\sin\phi)^2 + (\cos\theta)^2 = 1 \tag{1.7.18}$$
$$|\vec{e}_\theta|^2 = (r\cos\theta\cos\phi)^2 + (r\cos\theta\sin\phi)^2 + (r\sin\theta)^2 = r^2 \ , \tag{1.7.19}$$
$$|\vec{e}_\phi|^2 = (r\sin\theta\sin\phi)^2 + (r\sin\theta\cos\phi)^2 = r^2\sin^2\theta \ , \tag{1.7.20}$$

where in order to calculate those magnitudes we have simply summed the squares of their Cartesian components. As we have seen, the magnitude of a vector can also be calculated directly from the components in spherical coordinates using the metric tensor:

$$|\vec{v}|^2 = g_{\alpha\beta}\, v^\alpha v^\beta \, . \tag{1.7.21}$$

It is not difficult to see that if we use the above equation and the expression for the metric in spherical coordinates (equation (1.7.12)) we will find the same square-magnitudes for the vectors of the spherical coordinate basis given above. This shows that the coordinate basis vectors in general can not be expected to be unitary. In fact, they don't even have to be orthogonal to each other.

1.8 Covariant derivatives and geodesics

Since under general changes of coordinates we can have a non-trivial, and non-uniform, field of basis vectors, it is clear that when we consider the change in a tensor field as we move around the manifold we must take into account not only the change in the components of the tensor, but also the fact that the basis in which those components are calculated might change from one point to another. The basic problem here is that, as mentioned before, there is in general no natural way in which to compare vectors at different points of a manifold unless we give the manifold some more structure.

We can in fact start in a very abstract way by defining a derivative operator ∇ that takes $\binom{l}{m}$ tensors into $\binom{l}{m+1}$ tensors. We get one extra index down because we can take derivatives along any direction x^α, an operation we represent by ∇_α. This operator must have a series of properties in order to qualify as a derivative: It must be linear, it must obey the Leibnitz rule for the derivative of a product, it must reduce to the standard partial derivative for scalar functions, and it must be symmetric in the sense that for a scalar function f we get $\nabla_\alpha \nabla_\beta f = \nabla_\beta \nabla_\alpha f$.[8]

Once we have an operator ∇ we can consider, for example, the derivative of a vector with respect to a given coordinate:

$$\begin{aligned} \nabla_\alpha \vec{v} = \nabla_\alpha \left(v^\beta \vec{e}_\beta \right) &= \left(\nabla_\alpha v^\beta \right) \vec{e}_\beta + v^\beta \left(\nabla_\alpha \vec{e}_\beta \right) \\ &= \left(\partial_\alpha v^\beta \right) \vec{e}_\beta + v^\beta \left(\nabla_\alpha \vec{e}_\beta \right) , \end{aligned} \tag{1.8.1}$$

where in the last step we used the fact that the components v^β are scalar functions. This equation shows that the derivative of a vector is more than just the derivative of its components. We must also take into account the change in the basis vectors themselves.

Now, if we choose a fixed direction x^α the derivative $\nabla_\alpha \vec{e}_\beta$ must in itself be also a vector, since it represents the change in the basis vector along that direction. This means that it can be expressed as a linear combination of the basis vectors themselves. We introduce the symbols $\Gamma^\mu_{\alpha\beta}$ to denote the coefficients of such linear combination:

$$\nabla_\alpha \vec{e}_\beta = \Gamma^\mu_{\alpha\beta} \, \vec{e}_\mu \, . \tag{1.8.2}$$

The $\Gamma^\mu_{\alpha\beta}$ are called the *connection coefficients*, as they allow us to map vectors at different points in order to take their derivatives.

Using the above definition and rearranging indices we finally find that

$$\nabla_\alpha \vec{v} = \left(\partial_\alpha v^\beta + v^\mu \Gamma^\beta_{\alpha\mu} \right) \vec{e}_\beta \, , \tag{1.8.3}$$

which implies that the components of $\nabla \vec{v}$ are given by

$$\nabla_\alpha v^\beta = \partial_\alpha v^\beta + v^\mu \Gamma^\beta_{\alpha\mu} \, . \tag{1.8.4}$$

This is called the *covariant derivative*, and is also commonly denoted by a semicolon $\nabla_\alpha v^\beta \equiv v^\beta{}_{;\alpha}$ (in an analogous way the partial derivative is often denoted by a comma $\partial_\alpha v^\beta \equiv v^\beta{}_{,\alpha}$). We can use the above results to show that the covariant derivative of a one-form $\tilde{p}$ takes the form

$$\nabla_\alpha p_\beta = \partial_\alpha p_\beta - p_\mu \Gamma^\mu_{\alpha\beta} \, . \tag{1.8.5}$$

And if we now take our one-form to be the gradient of a scalar function we can find that the symmetry requirement reduces to

$$\Gamma^\mu_{\alpha\beta} = \Gamma^\mu_{\beta\alpha} \, . \tag{1.8.6}$$

It is important to mention the fact that this is only true for a coordinate basis – for a non-coordinate basis the connection coefficients are generally not symmetric even if the derivative operator itself is symmetric. This is because, for a

[8]The last requirement defines a manifold with no *torsion*. This requirement can in fact be lifted, but we will only consider the case of zero torsion here.

non-coordinate basis, the components of the derivatives of a scalar function correspond to directional derivatives along the basis vectors, and there is no reason why directional derivatives along two arbitrary vectors $(\vec{v}, \vec{u})$ should commute even in flat space:

$$\begin{aligned} \partial_v \partial_u f - \partial_u \partial_v f &= v^\mu \partial_\mu (u^\nu \partial_\nu f) - u^\mu \partial_\mu (v^\nu \partial_\nu f) \\ &= (v^\mu \partial_\mu u^\nu - u^\mu \partial_\mu v^\nu)\, \partial_\nu f \neq 0 \, . \end{aligned} \tag{1.8.7}$$

We can generalize the concept of covariant derivative to tensors of arbitrary rank. The rule is to add one term with $\Gamma^\mu_{\alpha\beta}$ for each free index, with the adequate sign depending on whether the index is up or down. For example:

$$\begin{aligned} T^{\mu\nu}{}_{;\alpha} &= \partial_\alpha T^{\mu\nu} + \Gamma^\mu_{\alpha\beta} T^{\beta\nu} + \Gamma^\nu_{\alpha\beta} T^{\mu\beta} \, , \\ T_{\mu\nu;\alpha} &= \partial_\alpha T_{\mu\nu} - \Gamma^\beta_{\alpha\mu} T_{\beta\nu} - \Gamma^\beta_{\alpha\nu} T_{\mu\beta} \, , \\ T^\mu{}_{\nu;\alpha} &= \partial_\alpha T^\mu{}_\nu + \Gamma^\mu_{\alpha\beta} T^\beta{}_\nu - \Gamma^\beta_{\alpha\nu} T^\mu{}_\beta \, . \end{aligned}$$

There is in fact a lot of freedom left in the derivative operator we have introduced: Given any smooth field of coefficients $\Gamma^\mu_{\alpha\beta}$ that are symmetric in their lower indices, there will be a specific derivative operator associated with them.

The map defined by the connection coefficients defines the notion of *parallel transport*, which allows us to drag a vector along a certain curve keeping it unchanged. We say that a vector $\vec{v}$ is parallel transported along a curve with tangent $\vec{u}$ if the following condition is satisfied:

$$u^\beta \nabla_\beta v^\alpha = 0 \, . \tag{1.8.8}$$

Parallel transport defines a unique map of vectors at a given point to vectors at infinitesimally close points, and the covariant derivative uses this map to calculate the change in the vector field.

When our manifold has a metric, there is an extra requirement that we can ask from our derivative operator, namely that the scalar product of two vectors be preserved under parallel transport. We can then ask for

$$u^\beta \nabla_\beta \, (\vec{v} \cdot \vec{w}) = 0 \, , \tag{1.8.9}$$

for any vector $\vec{u}$, and any pair of vectors $\{\vec{v}, \vec{w}\}$ such that

$$u^\beta \nabla_\beta v^\alpha = u^\beta \nabla_\beta w^\alpha = 0 \, . \tag{1.8.10}$$

From the definition of the scalar product we can easily show that this requirement reduces to

$$\nabla_\alpha g_{\mu\nu} = 0 \, . \tag{1.8.11}$$

That is, in order for parallel transport to preserve the scalar product the covariant derivative of the metric must vanish (which in particular means that the operation of raising and lowering indices commutes with the covariant derivatives). Using now the general expression for the covariant derivative of a tensor we find that this last condition implies that the connection coefficients must be given in terms of the metric components as

$$\Gamma^{\alpha}_{\beta\gamma} = \frac{g^{\alpha\mu}}{2}\left[\frac{\partial g_{\beta\mu}}{\partial x^{\gamma}} + \frac{\partial g_{\gamma\mu}}{\partial x^{\beta}} - \frac{\partial g_{\beta\gamma}}{\partial x^{\mu}}\right] . \tag{1.8.12}$$

These specific connection coefficients are known as *Christoffel symbols.* Since from now on we will always use this particular derivative operator, we will consider that the connection coefficients and the Christoffel symbols are the same thing. Clearly, for the case of Euclidean space in Cartesian coordinates the Christoffel symbols vanish. But they will generally not vanish if we use curvilinear coordinates, or if we have a curved manifold.

An important application of the definition of covariant derivative and parallel transport has to do with the generalization of the idea of a straight line to the case of curvilinear coordinates or curved manifolds. In Euclidean space, a straight line is a line such that it always remains parallel to itself. We can use the same idea here and define a *geodesic* as a curve that parallel transports its own tangent vector, that is, a curve whose tangent vector satisfies

$$v^{\beta}\nabla_{\beta}v^{\alpha} = 0 . \tag{1.8.13}$$

A geodesic is simply a curve that remains locally as straight as possible. From the definition above, we can rewrite the equation for a geodesic as

$$\frac{d^2x^{\alpha}}{d\lambda^2} + \Gamma^{\alpha}_{\beta\gamma}\frac{dx^{\beta}}{d\lambda}\frac{dx^{\gamma}}{d\lambda} = 0 , \tag{1.8.14}$$

where λ is the parameter associated with the curve.[9] Clearly, in Euclidean space with Cartesian coordinates this equation simply reduces to v^{α} = constant, but in spherical coordinates the equation is not that simple because not all the Christoffel symbols vanish. As we have seen, geodesics are important as they are the paths followed by objects free of external forces in special relativity. We will see later that they also play a crucial role in general relativity.

Geodesics have another very important property. We can in fact show that they are *extremal* curves, *i.e.* curves that extremize the distance between nearby points on the manifold (they are not always curves of minimum length since the metric need not be positive definite, as in the case of special relativity).

[9]Strictly speaking, the geodesic equation only has this form when we use a so-called *affine parameter*, for which we asks not only that the tangent vector remains parallel to itself but also that it has constant magnitude. But here we will take the point of view that a geodesic is a curve and not just its image. We will then say that if we don't have an affine parameter we in fact don't have a geodesic, but rather a different curve with the same image.

Up until this point I have avoided associating the connection coefficients to the components of a tensor, and for a good reason. From the number of indices in $\Gamma^\mu_{\alpha\beta}$ we could naively think that they are components of a $\binom{1}{2}$ tensor, but this is not the case. Looking at their definition in equation (1.8.2), we can see that the connection coefficients map a vector $\vec{v}$ onto a new vector $v^\beta \nabla_\beta \vec{e}_\alpha$, so they are in fact coefficients of $\binom{1}{1}$ tensors, one such tensor for each basis vector $\vec{e}_\alpha$. This becomes more clear if we write the coefficients as $(\Gamma_\alpha)^\mu_\beta$, where the index α identifies the tensor, and the indices $\{\mu, \beta\}$ identify the components of that tensor. So, contrary to what is often said, the connection coefficients are indeed components of tensors, only not the tensors we might at first think about. However, this fact is not very useful in practice since when we change coordinates we are also changing the basis vectors and hence the tensors associated with the connection coefficients.

Because of all this we can clearly not expect the $\Gamma^\mu_{\alpha\beta}$ to transform as components of a $\binom{1}{2}$ tensor, and indeed they don't. The transformation law for the connection coefficients is more complicated and takes the form:

$$\Gamma^{\bar{\mu}}_{\bar{\alpha}\bar{\beta}} = \Lambda^{\bar{\mu}}_\nu \Lambda^\gamma_{\bar{\alpha}} \Lambda^\delta_{\bar{\beta}} \, \Gamma^\nu_{\gamma\delta} + \partial_\gamma x^{\bar{\mu}} \, \partial_{\bar{\alpha}} \partial_{\bar{\beta}} x^\gamma \; . \tag{1.8.15}$$

Notice that the first term is precisely what we would expect from the components of a tensor, but there is a second term that involves second derivatives of the coordinate transformation. Another way in which we can understand that the Christoffel symbols can not transform as tensors is precisely the fact that in Cartesian coordinates they vanish. And if we have a tensor that has all its components equal to zero in a given coordinate system, then the tensor itself must be zero and its components in any other coordinate system must also vanish.

The transformation law (1.8.15) is in fact independent of any relation between the connection coefficients and the metric, *i.e.* it is valid for any derivative operator ∇. An important consequence of this is the following: Consider the connection coefficients associated with two different derivative operators ∇ and $\hat{\nabla}$, and define $\Delta^\mu_{\alpha\beta} := \Gamma^\mu_{\alpha\beta} - \hat{\Gamma}^\mu_{\alpha\beta}$. It turns out that $\Delta^\mu_{\alpha\beta}$ transforms as a tensor since the term involving second derivatives of the coordinate transformation in (1.8.15) cancels out (it is independent of the Γs). In other words, $\Delta^\mu_{\alpha\beta}$ are the components of a properly defined tensor. Similarly, assume that we have a manifold with a given coordinate system and consider two different metric tensors $g_{\mu\nu}$ and $\hat{g}_{\mu\nu}$ defined on it. The difference between the Christoffel symbols associated with the different metrics also transforms as a tensor. As we will see in later Chapters, this observation has some interesting applications in numerical relativity.

Before finishing this section, there is an important fact about the relation between covariant derivatives and Lie derivatives that deserves mention. Consider, for example, the commutator of two vectors but written in terms of covariant derivatives instead of partial derivatives:

$$
\begin{aligned}
v^\beta \nabla_\beta u^\alpha - u^\beta \nabla_\beta v^\alpha &= v^\beta \left(\partial_\beta u^\alpha + \Gamma^\alpha_{\beta\gamma} u^\gamma \right) - u^\beta \left(\partial_\beta v^\alpha + \Gamma^\alpha_{\beta\gamma} v^\gamma \right) \\
&= v^\beta \partial_\beta u^\alpha - u^\beta \partial_\beta v^\alpha = \pounds_{\vec{v}} u^\alpha \, . \qquad (1.8.16)
\end{aligned}
$$

We see that all contributions from the Christoffel symbols in the covariant derivatives have canceled out. In fact, it is easy to convince oneself that this happens for the Lie derivative of any tensor, *i.e.* the Lie derivatives can be written indistinctively in terms of partial or covariant derivatives.

We can use this last observation to rewrite the condition for the existence of a Killing field, equation (1.6.10), in the following way

$$
\begin{aligned}
\pounds_{\vec{\xi}}\, g_{\alpha\beta} = 0 &= \xi^\mu \, \partial_\mu g_{\alpha\beta} + g_{\alpha\mu} \, \partial_\beta \xi^\mu + g_{\mu\beta} \, \partial_\alpha \xi^\mu \\
&= \xi^\mu \, \nabla_\mu g_{\alpha\beta} + g_{\alpha\mu} \, \nabla_\beta \xi^\mu + g_{\mu\beta} \, \nabla_\alpha \xi^\mu \, . \qquad (1.8.17)
\end{aligned}
$$

Using the fact that the covariant derivative of the metric vanishes (so the metric can move inside the covariant derivatives), we find that this reduces to

$$
\pounds_{\vec{\xi}}\, g_{\alpha\beta} = \nabla_\alpha \xi_\beta + \nabla_\beta \xi_\alpha = 0 \, . \qquad (1.8.18)
$$

This is known as the *Killing equation.* It is interesting to notice that the condition for the existence of a Killing vector field is more compact when written in terms of the co-variant components of $\vec{\xi}$ instead of the contra-variant ones.

1.9 Curvature

As we have seen, the metric tensor is not in itself the most convenient way of describing a curved manifold as it can become quite non-trivial even in Euclidean space by considering curvilinear coordinates. The correct way to differentiate between flat and curved manifolds is by considering what happens to a vector as it is parallel transported around a closed circuit on the manifold. On a flat manifold, the vector does not change when this is done, while on a curved manifold it does. This can be clearly seen if we think about moving a vector on the surface of the Earth. Assume that we start on the equator with a vector pointing east. We move north following a meridian all the way up to the north pole, then we come back south along another meridian that is 90 degrees to the east of the first one until we get back to the equator. Finally, we move back to our starting point following the equator itself. If we do this we will find that our vector now points south. That is, although on a curved manifold parallel transport defines a local notion of parallelism, there is in fact no global notion of parallelism.

In order to define a tensor that is associated with the curvature of a manifold we must consider the parallel transport of a vector along an infinitesimal closed circuit. If we take this closed circuit as one defined by the coordinate lines themselves (see Figure 1.3), we can show that the change of the components of a vector $\vec{v}$ as it is parallel transported along this circuit is given by

$$
\delta v^\alpha = R^\alpha{}_{\beta\mu\nu} \, dx^\mu dx^\nu v^\beta \, , \qquad (1.9.1)
$$

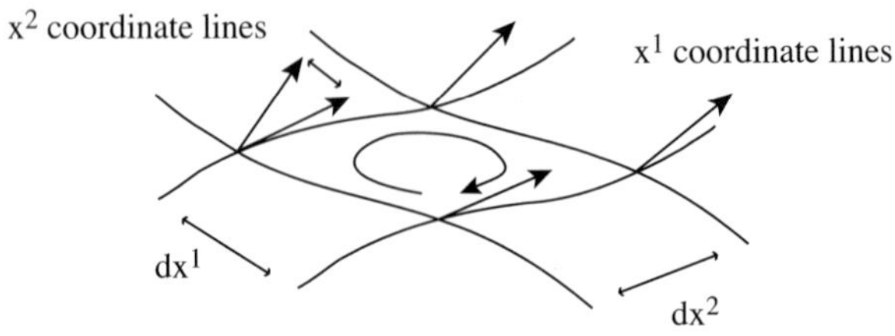

Fig. 1.3: Parallel transport of a vector around a closed infinitesimal circuit formed by coordinate lines. On a flat manifold the vector will not change when this is done, while on a curved manifold it will. The measure of the change is given by the Riemann curvature tensor.

where $R^\alpha{}_{\beta\mu\nu}$ are the components of the *Riemann curvature tensor*, which are given in terms of the Christoffel symbols as[10]

$$R^\alpha{}_{\beta\mu\nu} := \partial_\mu \Gamma^\alpha_{\beta\nu} - \partial_\nu \Gamma^\alpha_{\beta\mu} + \Gamma^\alpha_{\rho\mu}\Gamma^\rho_{\beta\nu} - \Gamma^\alpha_{\rho\nu}\Gamma^\rho_{\beta\mu} \; . \tag{1.9.2}$$

Since a non-zero Riemann tensor indicates the failure of a global notion of parallelism we define a *flat* manifold as one for which the Riemann tensor vanishes, while a *curved* manifold will be one with a non-zero Riemann tensor. Notice that for Euclidean space in Cartesian coordinates the Riemann tensor vanishes trivially. In fact, since the Riemann is a proper tensor (a fact that can easily be seen by inspecting equation (1.9.1)), it must also vanish in any curvilinear set of coordinates. The same thing is true of Minkowski spacetime.

The definition of the Riemann tensor also implies that for an arbitrary vector v^μ we must have

$$\nabla_\mu \nabla_\nu v^\alpha - \nabla_\nu \nabla_\mu v^\alpha = R^\alpha{}_{\beta\mu\nu} v^\beta \; . \tag{1.9.3}$$

This is known as the *Ricci identity* and shows that the Riemann tensor also gives the commutator of the covariant derivatives of a vector. In other words, covariant derivatives of vectors commute on a flat manifold, but fail to commute on a curved one. Of course, this is basically the same thing as before as the commutator of covariant derivatives tells us the difference when we parallel transport the vector to an infinitesimally close point following first one coordinate line and then another, or doing it in the opposite order, which is equivalent to going around a closed loop.

Note that the Riemann is a $\binom{1}{3}$ tensor, which means that in the case of four dimensional spacetime it has 256 components. However, its definition implies that it has many symmetries, in particular:

$$R_{\alpha\beta\mu\nu} = -R_{\beta\alpha\mu\nu} = -R_{\alpha\beta\nu\mu} = R_{\mu\nu\alpha\beta} \; , \tag{1.9.4}$$

[10]The overall sign in the definition of the Riemann tensor is a matter of convention and changes from one text to another. The sign used here is the same as that of MTW [206].

that is, the fully co-variant Riemann tensor is antisymmetric on the first and second pair of indices, and symmetric with respect to exchange of these two pairs. Also, it obeys a cyclic relation on the last three indices of the form

$$R^{\alpha}{}_{\beta\mu\nu} + R^{\alpha}{}_{\mu\nu\beta} + R^{\alpha}{}_{\nu\beta\mu} = 0 \; . \tag{1.9.5}$$

Using the anti-symmetry with respect to exchange of the last two indices, we find that this cyclic relation can also be expressed as

$$R^{\alpha}{}_{[\beta\mu\nu]} = 0 \; , \tag{1.9.6}$$

so the completely antisymmetric part of Riemann with respect to the last three indices vanishes.

The symmetries of the Riemann tensor imply that in the end it has only $n^2(n^2-1)/12$ independent components in n dimensions, that is 20 independent components in four dimensions. These symmetries also imply that the trace (*i.e.* the contraction) of the Riemann tensor over its first and last pairs of indices vanishes. On the other hand, the trace over the first and third indices does not vanish and is used to define the *Ricci curvature tensor*:

$$R_{\mu\nu} := R^{\lambda}{}_{\mu\lambda\nu} \; . \tag{1.9.7}$$

The Ricci tensor is clearly symmetric in its two indices, and in four dimensions has 10 independent components. Notice that in four dimensions having the Ricci tensor vanish *does not* mean that the manifold is flat, as the remaining 10 components of Riemann can still be different from zero. However, in three dimensions it does happen that the vanishing of the Ricci tensor implies that the Riemann is zero, as in that case both these tensors have only six independent components and Riemann turns out to be given directly in terms of the Ricci tensor.

Finally, the *Ricci scalar*, also known as the *scalar curvature*, is defined as the trace of the Ricci tensor itself

$$R := R^{\mu}{}_{\mu} \; . \tag{1.9.8}$$

An important result regarding the form of the metric and the Christoffel symbols is the fact that any differentiable manifold with a metric is *locally flat* in the sense that at any given point on the manifold we can always find a set of coordinates such that the metric tensor becomes diagonal with ± 1 elements, and all the Christoffel symbols vanish at that point. This means that we can always choose orthogonal coordinates at the given point p and guarantee that such coordinates will remain orthogonal in the immediate vicinity of that point so that $\partial_\mu g_{\alpha\beta}|_p = 0$. What we can not do is guarantee that all second derivatives of the metric will vanish because there are in general fewer third derivatives of the coordinate transformation that can be freely specified than second derivatives of the metric tensor itself. In four dimensions, for example, we have 80 independent third derivatives of the coordinate transformation $\partial_\mu \partial_\nu \partial_\sigma x^{\bar{\alpha}}$, while there are

100 second derivatives of the metric $\partial_\alpha \partial_\beta g_{\mu\nu}$ that we would need to cancel. This leaves 20 degrees of freedom in the second derivatives of the metric which correspond precisely with the 20 independent components of the Riemann tensor.

1.10 Bianchi identities and the Einstein tensor

Starting from the definition of the Riemann tensor it is possible to show that the covariant derivatives of this tensor have the following important property:

$$R^{\alpha}{}_{\beta\mu\nu;\lambda} + R^{\alpha}{}_{\beta\lambda\mu;\nu} + R^{\alpha}{}_{\beta\nu\lambda;\mu} = 0 \, , \tag{1.10.1}$$

or equivalently

$$R^{\alpha}{}_{\beta[\mu\nu;\lambda]} = 0 \, . \tag{1.10.2}$$

These relations are known as the *Bianchi identities*, and play a very important role in general relativity. One of their most important consequences comes from contracting them twice, which results in

$$G^{\mu\nu}{}_{;\nu} = 0 \, , \tag{1.10.3}$$

where $G^{\mu\nu}$ is the so-called *Einstein tensor* defined as

$$G^{\mu\nu} := R^{\mu\nu} - \frac{1}{2} g^{\mu\nu} R \, . \tag{1.10.4}$$

Because of its property of having zero divergence, the Einstein tensor plays a crucial role in the field equations of general relativity, as we will see below.

1.11 General relativity

As already mentioned, the equivalence principle led Einstein to the idea that gravity could be identified with the curvature of spacetime. This principle says that all objects fall with the same acceleration in a gravitational field, something that in the Newtonian theory arises from the fact that the inertial and gravitational masses of a given object are the same. Einstein's key insight was to realize that the equivalence principle implies that in a freely falling reference frame the gravitational *force* effectively vanishes and the laws of physics look just like they do in special relativity. In other words, the equivalence principle implies that there exist local inertial frames, namely those that are falling freely. In the presence of a non-uniform gravitational field, however, these local inertial frames can not be glued together to form a global inertial frame, since freely falling frames at different locations have accelerations that differ in both magnitude and direction.

But the fact that we have local inertial frames can be described geometrically by saying that locally the spacetime metric is that of Minkowski, and the fact that no global inertial frame exists will correspond to saying that there is no set of coordinates in which the spacetime metric has this simple form everywhere. In other words, spacetime must be a curved manifold, and freely falling frames must correspond to locally flat coordinates on that manifold. This remarkable insight forms the basis of the whole theory of general relativity, which takes as its basic postulate the following: *Spacetime is a four-dimensional curved manifold with a Lorentzian metric.*

General relativity therefore generalizes special relativity by considering the possibility of having a curved spacetime with no global inertial frames. Since no global inertial frame exists there is also no natural set of coordinates, so from the start we must be ready to work on a completely arbitrary coordinate system. The name "general relativity" comes precisely from the fact that we can use a general coordinate system. However, the crucial point is that curvature is allowed, the use of a general coordinate system in Minkowski spacetime is still only special relativity.

Of course, saying that spacetime is curved does not take us too far – one must say how to generalize the laws of physics to such a curved spacetime, in particular the laws of motion of test particles, and also we must say how this curvature arises in the first place. The second issue is related to the field equations, which we will consider in Section 1.13 below; here we will concentrate on the first issue: What are the laws of physics on a curved spacetime?

The first point to stress is the fact that, in general relativity, the gravitational field will not correspond to the gravitational force of Newton, as the equivalence principle tells us that such force vanishes in a freely falling frame, *i.e.* it is just an *inertial* force analogous to the centrifugal force on a rotating frame. What does not vanish, even in a freely falling frame, are the tidal forces arising from the fact that the gravitational field is not uniform. It is these tidal forces that are associated with the curvature of spacetime. The laws of physics in general relativity come from two basic postulates: the principle of general covariance that states that these laws must be written in tensor form, and the strong equivalence principle that states that in a freely falling frame they must reduce to the form that they have in special relativity. These postulates are not always completely unambiguous, but they work well in most cases and provide us with the simplest possible generalizations of the laws of physics to a curved background. In particular, these two principles suggest the rule of *minimal coupling*, which basically says that once we have a law of physics that we know holds in special relativity, we can generalize it to a curved spacetime by replacing the Minkowski metric $\eta_{\alpha\beta}$ for a general metric $g_{\alpha\beta}$, and at the same time replacing all partial derivatives ∂_μ for covariant derivatives ∇_μ, that is, there should be no explicit couplings to the curvature. Of course, in all cases, experiments should say if the generalizations obtained from these postulates are indeed the correct laws.

A first example of these principles in action corresponds to the motion of a free particle in a curved background. Since in a freely falling frame the particle must move as it does in special relativity, *i.e.* in a straight line, in the curved spacetime it must move following a geodesic. This means that in a gravitational field Newton's first law becomes: *A free particle in a curved spacetime moves following timelike geodesics, and light rays move following null geodesics.* But notice that the geodesics of a curved spacetime need not look straight when projected into three-dimensional space. Notice also that gravity is not considered a force here – a particle in the presence of a gravitational field with no other forces acting on it is considered free. This simple law is enough to describe the motion of the planets around the Sun, for example.

It is usual to parameterize a geodesic using the particle's own proper time τ, so the equation of motion for a free particle in spacetime takes the form[11]

$$\frac{d^2 x^\alpha}{d\tau^2} + \Gamma^\alpha_{\beta\gamma} \frac{dx^\beta}{d\tau} \frac{dx^\gamma}{d\tau} = 0 \; . \tag{1.11.1}$$

The geodesic equation of motion can also be written as

$$u^\mu \nabla_\mu u^\alpha = 0 \; , \tag{1.11.2}$$

where $\vec{u}$ is just the tangent vector to the geodesic curve: $u^\alpha = dx^\alpha / d\tau$. This vector is know as the 4-velocity of the particle, and is the relativistic generalization of the Newtonian concept of velocity. It is important to notice that

$$u^2 = \vec{u} \cdot \vec{u} = g_{\alpha\beta} \frac{dx^\alpha}{d\tau} \frac{dx^\beta}{d\tau} = \frac{ds^2}{d\tau^2} = -1 \; , \tag{1.11.3}$$

that is, the 4-velocity is a unit timelike vector. In an analogous way to Newtonian physics, the 4-momentum is defined simply as the vector

$$\vec{p} := m\vec{u} \; , \tag{1.11.4}$$

where m is the rest mass of the particle.

The energy of a particle with 4-momentum $\vec{p}$, as measured by an observer moving with 4-velocity $\vec{v}$ that has the same position as the particle, is defined in terms of the 4-momentum as[12]

$$E := -\vec{v} \cdot \vec{p} = -v^\alpha p_\alpha \; . \tag{1.11.5}$$

In the particular case when the observer is static in a given coordinate system its 4-velocity will have components $v^\alpha = (1/\sqrt{-g_{00}}, 0, 0, 0)$, which is easy to see from the fact that we must have $\vec{v} \cdot \vec{v} = -1$, so that for this static observer $E = p_0 / \sqrt{-g_{00}}$. In special relativity this reduces to $E = m\gamma$, with γ the Lorentz factor associated with the motion of the particle (or, in SI units, $E = m\gamma c^2$). Notice that since on a curved spacetime there is no global notion of parallelism and no preferred family of observers, in general a given observer can only define an energy for a particle at his own location but not for a particle far away. In

[11]Of course, this doesn't work for a light ray for which proper time is always zero, so in the case of a null geodesic we simply use an arbitrary affine parameter.

[12]The minus sign in this definition comes from the Lorentzian nature of the metric that implies that the time components of co-variant vectors have the opposite sign to what one would expect.

special relativity, however, given a specific inertial frame such a preferred family of observers does exist, and the energy of a particle on that inertial frame can therefore be globally defined.

There are, however, some special situations when one can define a useful notion of energy and other conserved quantities in a global sense on a curved manifold. Let us assume that we have a manifold with some symmetry. As we have seen, associated with that symmetry there will be a Killing vector satisfying equation (1.8.18). If we now take $\vec{u}$ to be the 4-velocity of a free particle, or more generally the tangent vector to a geodesic curve, we will have

$$u^\mu \nabla_\mu \left(\vec{\xi} \cdot \vec{u} \right) = u^\mu \nabla_\mu \left(\xi_\nu u^\nu \right) = \xi_\nu u^\mu \nabla_\mu u^\nu + u^\nu u^\mu \nabla_\mu \xi_\nu = u^\mu u^\nu \nabla_\mu \xi_\nu \; , \quad (1.11.6)$$

where we have used the fact that the curve is a geodesic. If we now notice that the Killing equation implies that $\nabla_\mu \xi_\nu$ is antisymmetric so that $u^\mu u^\nu \nabla_\mu \xi_\nu = 0$, we finally find

$$u^\mu \nabla_\mu \left(\vec{\xi} \cdot \vec{u} \right) = 0 \; . \quad (1.11.7)$$

This equation implies that if a Killing field $\vec{\xi}$ exists, then the scalar quantity $\vec{\xi} \cdot \vec{u}$ is constant along the geodesic, so it represents a conserved quantity associated with the motion of the particle.

In particular, assume that we have a Killing field that is everywhere timelike. In such a case the spacetime is said to be *static*, as we can always choose coordinates adapted to the Killing field for which the metric coefficients $g_{\alpha\beta}$ will be time independent. In such adapted coordinates the Killing vector field will have components $\vec{\xi} = (1,0,0,0)$, so the above result implies that on a static spacetime the quantity $\vec{\xi} \cdot \vec{u} = g_{\mu\nu} \xi^\mu u^\nu = g_{0\nu} u^\nu = u_0$ is constant along a particle's trajectory. This means that the co-variant time component of the 4-momentum $p_0 = m u_0$ is conserved in a static gravitational field and can therefore be defined as the *total energy* of the particle $\mathcal{E} := -p_0$. Notice that, in general, this quantity will not only include the *kinetic energy*, but will also include contributions from the relativistic equivalent of the *gravitational potential energy* (as clearly the kinetic energy alone will not be conserved in a gravitational field).[13] If the spacetime is not static, however, this notion of conserved energy is lost.

There are other conserved quantities associated with different symmetries. For example, on a spacetime that has axial symmetry we will find a conserved quantity that can be associated with the angular momentum of the particle.

[13]One might find confusing the fact that the conserved energy on a static spacetime is $\mathcal{E} = -p_0$, while the energy measured locally by a static observer is instead $E = -p_0/\sqrt{-g_{00}}$. The reason for the difference is that the energy measured locally only corresponds to kinetic plus rest mass energy without the potential energy contribution. This can be seen more clearly in the case of weak gravitational fields and small velocities (see Section 1.14 below), for which $g_{00} \simeq -1 - 2\phi$, with $\phi \ll 1$ the *Newtonian potential*, and where we find $\mathcal{E} \simeq m + mv^2/2 + m\phi$ and $E = \mathcal{E}/\sqrt{-g_{00}} \simeq m + mv^2/2$ (neglecting contributions quadratic in ϕ and v).

1.12 Matter and the stress-energy tensor

In the Newtonian theory, the source of the gravitational field is given by the mass density. The equivalence of mass and energy means that we should expect the energy density to be the source of the gravitational field in general relativity. The first problem we have to deal with, however, is the fact that the energy density is not a scalar quantity. This is easy to see from the fact that the Lorentz transformations imply that the volume itself is not scalar, and also from the fact that the energy of a particle is a component of the 4-momentum. In fact, it is not difficult to show that the energy density must be a component of a rank 2 tensor that includes as its other components the momentum density and the fluxes of energy and momentum. This tensor is called the *stress-energy* tensor (or sometimes the *energy-momentum* tensor), and is denoted by $T^{\alpha\beta}$. The components of this tensor have the following physical interpretations:

$$T^{00} = \text{energy density}\,, \tag{1.12.1}$$

$$T^{0i} = \text{momentum density}\,, \tag{1.12.2}$$

$$T^{ij} = \text{flux of } i \text{ momentum in direction } j. \tag{1.12.3}$$

The stress-energy tensor can be shown to be symmetric in the general case (for example, momentum density and energy flux are physically the same thing). Notice that this tensor can be defined for any continuous distribution of matter or energy, that is, it can be defined not only for matter, but also for fields other than gravity itself such as the electromagnetic field. Because of this, in general relativity it has become customary to simply call *matter* any type of field that has a stress-energy tensor associated with it.

Consider now the case of a perfect fluid, that is, a fluid with zero viscosity and zero heat conduction. Such a fluid is characterized by the energy density ρ and pressure p as measured by an observer co-moving with the fluid elements, and also by the 4-velocity field $\vec{u}$ of the fluid elements themselves. In such a case, the stress-energy tensor takes the particularly simple form

$$T^{\alpha\beta} = (\rho + p)\, u^\alpha u^\beta + p\, g^{\alpha\beta}\,. \tag{1.12.4}$$

This expression can be easily derived by noticing that in the Lorentz frame co-moving with the fluid it implies $T^{00} = \rho$, $T^{ii} = p$ and $T^{\alpha\beta} = 0$ for $\alpha \neq \beta$, which corresponds to a fluid with energy density ρ, pressure p, zero viscosity ($T^{ij} = 0$ for $i \neq j$), and zero heat conduction ($T^{0i} = 0$). Equation (1.12.4) is just the fully covariant expression for such a tensor.

The stress-energy tensor also allows us to write the differential form of the laws of conservation of energy and momentum in a simple way. The conservation law for energy in differential form, for example, states that the change in time of the energy density inside a small volume element must be balanced by the difference between the energy that enters the volume element and the energy that leaves (energy is not created or destroyed). In other words, in locally flat coordinates we should have a continuity equation of the form

$$\partial_t T^{00} + \partial_i T^{0i} = 0 \; .$$

The covariant generalization of this conservation law is clearly

$$\nabla_\beta T^{\alpha\beta} = 0 \; . \tag{1.12.5}$$

Notice that this equation includes the conservation of energy for $\alpha = 0$, and the conservation of momentum for $\alpha = i$. This conservation law for energy and momentum is a fundamental requirement of any physical theory, so we must demand the stress-energy tensor of any matter field to satisfy it. As we will see below, it also plays a crucial role in the field equations of general relativity.

There is another important comment to be made about equation (1.12.5). Strictly speaking this equation only leads to true conservation of energy, in the sense that the change of energy in a finite region equals the integrated flux of energy into that region, when we have a flat spacetime. For a curved spacetime strict conservation no longer holds as the gravitational field can do work and change the energy and momentum of the matter (the conservation law (1.12.5) actually represents local conservation as seen by freely falling observers). We might think that the obvious thing to do would be to include the stress-energy tensor for the gravitational field itself in order to recover full conservation of energy, but it turns out that in general relativity no local expression for the energy of the gravitational field exists.

The conservation laws can in fact also be derived from a variational principle. This can be done in the case when we have a matter field that can be specified by a *Lagrangian*, that is a scalar function L that depends only on the field variables, their derivatives, and the metric coefficients. Assume that our system is described by a series of field variables ϕ_a, where here a is just a label that numbers the different fields, and not a spacetime index. The Lagrangian will then be a scalar function such that $L = L(\phi_a, \partial_\alpha \phi_a, g_{\alpha\beta})$. We define the *action* as the integral of the Lagrangian over an (open) region Ω of our manifold

$$S\,(\Omega) = \int_\Omega L \, dV \; , \tag{1.12.6}$$

with $dV = \sqrt{|g|}\, d^n x$ the volume element. We then assume that the dynamical equations are obtained from a least action variational principle, that is, we demand that the action is stationary with respect to variation in the fields. Notice, in particular, that the metric $g_{\alpha\beta}$ must be considered a dynamical field in this variation. Variation with respect to the field variables ϕ_a will give us the so-called *field equations*, that is the dynamical equations for the field under consideration, while variation with respect to the metric gives us an equation that represents the equilibrium between the geometry and the fields, or in other words, the conservation laws.

Consider then a variation of the action with respect to the geometry:

$$\begin{aligned}
\delta_g S &= \delta_g \int_\Omega L \, dV \\
&= \int_\Omega \left[\sqrt{|g|} \, \frac{\partial L}{\partial g_{\alpha\beta}} \, \delta g_{\alpha\beta} + L \, \delta \left(\sqrt{|g|} \right) \right] d^n x \\
&= \int_\Omega \left[\sqrt{|g|} \, \frac{\partial L}{\partial g_{\alpha\beta}} \, \delta g_{\alpha\beta} + \frac{1}{2\sqrt{|g|}} \, L \, \delta |g| \right] d^n x \, . \qquad (1.12.7)
\end{aligned}$$

A general result from linear algebra tells us that $\delta|g| = |g| \, g^{\alpha\beta} \, \delta g_{\alpha\beta}$, so the last equation can be rewritten as

$$\delta_g S = \int_\Omega T^{\alpha\beta} \delta g_{\alpha\beta} \, dV \, , \qquad (1.12.8)$$

where we have defined the stress-energy tensor as

$$\begin{aligned}
T^{\alpha\beta} &:= \frac{\partial L}{\partial g_{\alpha\beta}} + \frac{g^{\alpha\beta}}{2} \, L \\
&= -g^{\alpha\mu} g^{\beta\nu} \, \frac{\partial L}{\partial g^{\mu\nu}} + \frac{g^{\alpha\beta}}{2} \, L \, . \qquad (1.12.9)
\end{aligned}$$

It is usually more convenient to rewrite this definition of the stress-energy tensor in terms of its co-variant components. The expression then becomes

$$T_{\alpha\beta} := -\frac{\partial L}{\partial g^{\alpha\beta}} + \frac{g_{\alpha\beta}}{2} \, L \, . \qquad (1.12.10)$$

Asking now for the action to be stationary with respect to the variation, *i.e.* asking for $\delta_g S = 0$, can be shown to directly imply the conservation laws (1.12.5).[14]

There is an important comment to be made here. Instead of working with the Lagrangian L as the basic dynamical function, we might choose to absorb the volume element and work with the *Lagrangian density* $\mathcal{L} := \sqrt{|g|} \, L$. In terms of the Lagrangian density the expression for the stress-energy tensor is more compact, and takes the simple form:

$$T_{\alpha\beta} = -\frac{1}{\sqrt{|g|}} \, \frac{\partial \mathcal{L}}{\partial g^{\alpha\beta}} \, . \qquad (1.12.11)$$

Both expressions for $T_{\alpha\beta}$ are entirely equivalent, but in practice it is faster and more transparent to use equation (1.12.10) directly to find the stress-energy tensor of a field for which we have a Lagrangian L.

[14]The proof of this involves assuming that the field equations are satisfied, and then considering an arbitrary change of coordinates inside the region Ω. By making use of the fact that this change of coordinates can not affect the action integral, and looking at the change in the form of the integrand itself, the conservation laws follow.

Consider, for example, a Klein–Gordon field ϕ with mass m, also known as a scalar field. Its Lagrangian is given by

$$L = -\left(\nabla_\alpha \phi \nabla^\alpha \phi + m^2 \phi^2\right) = -\left(g^{\alpha\beta} \nabla_\alpha \phi \, \nabla_\beta \phi + m^2 \phi^2\right) . \tag{1.12.12}$$

Using (1.12.10) we find immediately that the corresponding stress-energy tensor has the form

$$T_{\alpha\beta} = \nabla_\alpha \phi \, \nabla_\beta \phi - \frac{g_{\alpha\beta}}{2} \left(\nabla_\mu \phi \nabla^\mu \phi + m^2 \phi^2\right) . \tag{1.12.13}$$

Fundamental scalar fields are not known to exist in nature. Still, the simplicity of their stress-energy tensor and dynamical equation (see below) means that they are an ideal test system for studying properties of solutions to the field equations of general relativity, and because of this they are often used in numerical relativity. Also, fundamental theories of particle physics demand the existence of scalar fields, an example of this is the Higgs field of the standard model.

Consider now the case of an electromagnetic field described by the potential 4-vector $\vec{A} := (\varphi, A^i)$, with φ the *scalar potential* and A^i the *vector potential* (but notice that in tensorial terms neither φ is a scalar nor A^i a vector; they are in fact different components of the same 4-vector). Define now the electromagnetic field as the following antisymmetric tensor

$$F_{\alpha\beta} = 2 \, \nabla_{[\alpha} A_{\beta]} \; . \tag{1.12.14}$$

The electric field **E** and magnetic field **B** are then defined through

$$F^{\alpha\beta} = \begin{pmatrix} 0 & E_x & E_y & E_z \\ -E_x & 0 & B_z & -B_y \\ -E_y & -B_z & 0 & B_x \\ -E_z & B_y & -B_x & 0 \end{pmatrix} . \tag{1.12.15}$$

This shows that the electric and magnetic fields are in fact not two separate 3-vectors, but rather the six independent components of an antisymmetric rank 2 tensor. In particular, under Lorentz transformations the components of the electric and magnetic fields mix together. The Lagrangian for the electromagnetic field turns out to be

$$L = -\frac{1}{8\pi} \, F^{\alpha\beta} F_{\alpha\beta} = -\frac{1}{8\pi} \, g^{\alpha\mu} g^{\beta\nu} F_{\mu\nu} F_{\alpha\beta} \; , \tag{1.12.16}$$

which is not surprising since, apart from normalization factors, it is the only scalar we can form that is quadratic in the field. From this Lagrangian we find the following stress-energy tensor:

$$T_{\alpha\beta} = \frac{1}{4\pi} \left(F_{\alpha\mu} F_\beta{}^\mu - \frac{g_{\alpha\beta}}{4} \, F^{\mu\nu} F_{\mu\nu}\right) . \tag{1.12.17}$$

We can readily verify that this gives the standard expressions for the energy density and momentum density (Poynting vector) for the electromagnetic field.

A Lagrangian can in fact also be defined for a perfect fluid, but the derivation of the corresponding stress-energy tensor from such a Lagrangian is considerably less transparent than in the cases of the Klein-Gordon and electromagnetic fields, so we will not consider it here.

The conservation laws (1.12.5) are often enough to find the dynamical equations for the system under consideration. For example, in the case of a perfect fluid they reduce to the first law of thermodynamics (conservation of energy) and the relativistic Euler equations (conservation of momentum). In the case of a scalar field, the conservation laws imply directly the Klein–Gordon equation:

$$\Box\phi - m^2\phi = 0 \, , \tag{1.12.18}$$

where $\Box := g^{\mu\nu}\nabla_\mu\nabla_\nu$ is the d'Alambertian operator.

1.13 The Einstein field equations

Once we know how to generalize the laws of physics to curved spacetimes, the final building block still missing from the theory of general relativity is the one that relates the geometry of spacetime with the distribution of mass and energy. This last element is contained in the Einstein field equations, or simply Einstein's equations. These equations can be derived in several different ways, either looking for a consistent relativistic generalization of Newton's theory (the road followed by Einstein), or deriving them in a formal way from a variational principle starting with an adequate Lagrangian (the road followed by Hilbert).

The more intuitive way to derive these equations is to go back for a while to Newton's theory of gravity. Assume we have a Newtonian gravitational potential ϕ. The gravitational force will be given by $d\mathbf{v}/dt = -\nabla\phi$.[15] In general relativity, on the other hand, the motion of free particles is described by the geodesic equation, which can be written in the form

$$\frac{du^\alpha}{d\tau} = -\Gamma^\alpha_{\mu\nu}\frac{du^\mu}{d\tau}\frac{du^\nu}{d\tau} \, . \tag{1.13.1}$$

From this it is clear that the Christoffel symbols play the role of $\nabla\phi$, so the metric coefficients must play the role of the potential ϕ. However, as mentioned before, the metric and Christoffel symbols are not the best way to describe the curvature of spacetime since they can be made to become trivial by choosing locally flat coordinates, which physically means going to a freely falling frame where the gravitational force vanishes. We should then concentrate on the tidal forces which carry the real physical information about the gravitational field.

In Newton's theory, the tidal acceleration between two nearby freely falling particles with separation $\mathbf{z} = \mathbf{x}_2 - \mathbf{x}_1$ is given by $d^2\mathbf{z}/dt^2 = -(\mathbf{z}\cdot\nabla)\,\nabla\phi$. In general relativity the tidal acceleration will correspond to the acceleration between

[15] One should not be confused with the use of the ∇ symbol here. When referring to Newton's theory this symbol should be understood to mean just the standard three-dimensional gradient, and not a covariant derivative.

nearby geodesics, also known as *geodesic deviation*. Consider a congruence of geodesics with tangent vector u^α and parameter λ, and define the separation between two nearby geodesics as $z^\alpha(\lambda) = x_1^\alpha(\lambda) - x_2^\alpha(\lambda)$. The tidal acceleration will then be given by $a^\alpha := u^\mu \nabla_\mu (u^\nu \nabla_\nu z^\alpha)$. We can show that this tidal acceleration is given in terms of the Riemann curvature tensor as:

$$a^\alpha = -R^\alpha{}_{\mu\beta\nu} u^\mu u^\nu z^\beta \; . \tag{1.13.2}$$

We then see that, in flat space, geodesics do not deviate from each other, or in other words parallel straight lines remain parallel (which is Euclid's famous fifth postulate). In a curved space, however, geodesic lines deviate from each other.

Comparing the Newtonian and relativistic expressions for the tidal acceleration we see that $R^\alpha{}_{\mu\beta\nu} u^\mu u^\nu$ plays the role of $\partial_i \partial_j \phi$. On the other hand, the energy density measured by an observer following the geodesic is given by $\rho = T_{\mu\nu} u^\mu u^\nu$, which suggests that the relativistic version of Newton's field equation $\nabla^2 \phi = 4\pi\rho$ should have the form $R^\alpha{}_{\mu\alpha\nu} u^\mu u^\nu = 4\pi T_{\mu\nu} u^\mu u^\nu$, and since the vector u^α is arbitrary this reduces to $R_{\mu\nu} = 4\pi T_{\mu\nu}$. The last equation has the structure we would expect, as it relates second derivatives of the metric (the gravitational "potential") with the energy density, just as in Newton's theory. This field equation was in fact considered by Einstein, but he quickly realized that it can't be correct since the conservation laws $\nabla_\mu T^{\mu\nu} = 0$ would imply that $\nabla_\mu R^{\mu\nu} = 0$, which imposes a serious restriction in the allowed geometries of spacetime.

The solution to the problem of the compatibility of the field equations and the conservation laws comes from considering the contracted Bianchi identities. In particular, these identities imply that the Einstein tensor (1.10.4) has zero divergence, so a consistent version of the field equations would be

$$G_{\mu\nu} \equiv R_{\mu\nu} - \frac{1}{2} g_{\mu\nu} R = 8\pi T_{\mu\nu} \; , \tag{1.13.3}$$

where the factor of 8π is there in order to recover the correct Newtonian limit. These are the field equations of general relativity, or in short Einstein's equations. Note that they are in fact ten equations, as the indices μ and ν take values from 0 to 3, and both the Einstein and stress-energy tensors are symmetric. Notice also that, as written above, the field equations imply the conservation laws, as $\nabla_\mu T^{\mu\nu} = 0$ follows from pure geometric consistency.

It is sometimes useful to rewrite the field equations in the equivalent form

$$R_{\mu\nu} = 8\pi \left(T_{\mu\nu} - \frac{1}{2} g_{\mu\nu} T \right) \; , \tag{1.13.4}$$

with $T := T^\mu{}_\mu$ the trace of the stress-energy tensor. This equation can be obtained by simply taking the trace of the original field equations and noticing that they imply $R = -8\pi T$ (the trace of the metric is always equal to the total number of dimensions of the manifold, which in the case of spacetime is 4).

The Einstein equations just introduced could not appear to be more simple. This simplicity, however, is only apparent since each term is a short-hand for considerably more complex objects. Written in their most general form, in an arbitrary coordinate system and with all terms expanded out, the Einstein equations become a system of ten coupled, non-linear, second order partial differential equations with thousands of terms.

In the case of vacuum the stress-energy tensor vanishes, and Einstein's equations reduce to:

$$G_{\mu\nu} = 0 \,, \tag{1.13.5}$$

or equivalently,

$$R_{\mu\nu} = 0 \,. \tag{1.13.6}$$

Note that, as mentioned before, the fact that the Ricci tensor vanishes does not imply that spacetime is flat. This is as it should be, since we know that the gravitational field of an object extends beyond the object itself, which means that the curvature of spacetime in the empty space around a massive object can not be zero. The Einstein equations in vacuum have another important consequence, they describe the way in which the gravitational field propagates in empty space and, in an analogous way to electromagnetism, predict the existence of gravitational waves: perturbations in the gravitational field that propagate at the speed of light. The prediction of the existence of gravitational waves tells us that, in Einstein's theory, gravitational interactions do not propagate at an infinite speed, but rather at the speed of light (see Section 1.14 below).

The field equations (1.13.3) can also be derived from a variational principle, we only need to introduce a Lagrangian for the gravitational field itself. Since gravity is associated with the Riemann tensor, and the Lagrangian must be a scalar, the only possibility is to take the Ricci scalar as the Lagrangian:

$$L_G = R \,. \tag{1.13.7}$$

If we now consider the total Lagrangian to be $L = L_G + L_M$, with L_M the Lagrangian associated with the matter, and ask for the action integral to be stationary, Einstein's field equations follow. This was the approach followed by D. Hilbert, who derived the field equations independently of Einstein (in fact, Hilbert's derivation came first by a few days). Because of this the field equations are also known as the Einstein–Hilbert equations.

There are two important final comments to be made about the Einstein field equations. The first has to do with the relation between the field equations and the conservation laws. Since the conservation laws are a consequence of the field equations, and they in turn contain all the dynamical information in many cases, it turns out that often we do not need to postulate the equations of motion for the matter as they can be derived directly from the field equations. In particular, for a perfect fluid with zero pressure (known as *dust* in relativity), the field equations predict that the flow lines should be timelike geodesics. Also, it can be shown that a small test body with weak self-gravity will necessarily move on a geodesic. So there is really no need to postulate geodesic motion of test particles in general relativity – it is a consequence of the field equations.

The second comment has to do with the meaning of a solution to the Einstein equations. Notice that we have ten second-order differential equations for the ten components of the metric tensor $g_{\mu\nu}$, so naively we might expect that, given a matter distribution and appropriate boundary conditions, the full metric tensor would be completely determined. But this can not be since the $g_{\mu\nu}$ are components of a tensor in some specific coordinate system. If we change the coordinates the components of the metric tensor will change, but the physical content of the solution can not depend on the choice of coordinates. This means that a "solution" of the field equations should really be understood as the equivalence class of all four-dimensional metrics related to each other by a change of coordinates. Since there are four coordinates, there are four arbitrary degrees of freedom in the metric components, so the field equations should only fix the remaining six components. But this is precisely the content of the Bianchi identities $\nabla_\nu G^{\mu\nu} = 0$ – they provide us with four differential relations between the ten field equations, so there are in fact only six independent equations. In general there is no "natural" way of separating the six independent components of $g_{\mu\nu}$ in a clear way. However, such a separation can be achieved by choosing a specific set of coordinates such as the one used in the 3+1 formulation that we will study in Chapter 2.

1.14 Weak fields and gravitational waves

Newton's theory served well for over two centuries. The reason is that the gravitational fields on the Solar System, including that of the Sun, are actually very weak. In fact, Einstein's motivation for developing general relativity was purely theoretical, as at the time there were no strong observational reasons to abandon Newton's theory (the advance of Mercury's perihelion was known, but it was a very small effect and at the time there was still hope that some unknown Newtonian perturbation could account for it). The study of weak gravitational fields is therefore very important, not only because they correspond to a very common case, but also because they lead to the prediction of the existence of gravitational radiation.

Since in general relativity the gravitational field is associated with the curvature of spacetime, weak gravitational fields will correspond to nearly flat spacetimes. In such a case there must exist coordinates such that the metric tensor is nearly that of Minkowski, that is

$$g_{\mu\nu} = \eta_{\mu\nu} + h_{\mu\nu} \; , \qquad |h_{\mu\nu}| \ll 1 \; . \tag{1.14.1}$$

Notice that here we are assuming not only that the curvature is small, but also that our spatial coordinates are nearly Cartesian.

An interesting observation arises from considering what is known as a *background Lorentz transformation*. If we transform our nearly flat coordinates using the Lorentz transformations (1.3.2) we find that

$$\begin{aligned} g_{\bar{\mu}\bar{\nu}} &= \Lambda^{\alpha}_{\bar{\mu}} \Lambda^{\beta}_{\bar{\nu}} \, g_{\alpha\beta} = \Lambda^{\alpha}_{\bar{\mu}} \Lambda^{\beta}_{\bar{\nu}} \left(\eta_{\alpha\beta} + h_{\alpha\beta} \right) \\ &= \eta_{\bar{\mu}\bar{\nu}} + \Lambda^{\alpha}_{\bar{\mu}} \Lambda^{\beta}_{\bar{\nu}} h_{\alpha\beta} = \eta_{\bar{\mu}\bar{\nu}} + h_{\bar{\mu}\bar{\nu}} \, , \end{aligned} \tag{1.14.2}$$

where we have defined $h_{\bar{\mu}\bar{\nu}} := \Lambda^{\alpha}_{\bar{\mu}} \Lambda^{\beta}_{\bar{\nu}} h_{\alpha\beta}$. We then see that $h_{\mu\nu}$ transforms just as it would if it were a rank 2 tensor in special relativity. This implies that we can reinterpret a weak gravitational field as just a field in special relativity associated with a symmetric rank 2 tensor $h_{\mu\nu}$. We can then work as if we had a flat spacetime with a tensor field $h_{\mu\nu}$ defined on it, and use the Minkowski metric to raise and lower indices of tensors. The only exception to the rule of using $\eta_{\mu\nu}$ to move indices around will be the metric tensor $g_{\mu\nu}$ itself. Since $g^{\mu\nu}$ is by definition the inverse matrix to $g_{\mu\nu}$, to first order in $h_{\mu\nu}$ we must have

$$g^{\mu\nu} = \eta^{\mu\nu} - h^{\mu\nu} \, . \tag{1.14.3}$$

The linearized field equations are now easy to derive. We can show that to linear order in $h_{\mu\nu}$ the Riemann tensor turns out to be

$$R_{\alpha\beta\mu\nu} = \frac{1}{2} \left(\partial_\beta \partial_\mu h_{\alpha\nu} + \partial_\alpha \partial_\nu h_{\beta\mu} - \partial_\beta \partial_\nu h_{\alpha\mu} - \partial_\alpha \partial_\mu h_{\beta\nu} \right) \, , \tag{1.14.4}$$

and the Ricci tensor becomes

$$R_{\mu\nu} = \partial^\alpha \partial_{(\mu} h_{\nu)\alpha} - \frac{1}{2} \left(\partial_\mu \partial_\nu h + \partial_\alpha \partial^\alpha h_{\mu\nu} \right) \, , \tag{1.14.5}$$

with $h := h^{\alpha}_{\alpha}$ the trace of $h_{\mu\nu}$, and $\partial^\alpha \equiv \eta^{\alpha\beta} \partial_\beta$. If we define the *trace reversed* field tensor $\bar{h}_{\mu\nu} := h_{\mu\nu} - \eta_{\mu\nu} h/2$, the Einstein tensor turns out to be

$$G_{\mu\nu} = \partial^\alpha \partial_{(\mu} \bar{h}_{\nu)\alpha} - \frac{1}{2} \left(\partial_\alpha \partial^\alpha \bar{h}_{\mu\nu} + \eta_{\mu\nu} \partial^\alpha \partial^\beta \bar{h}_{\alpha\beta} \right) \, , \tag{1.14.6}$$

and the linearized field equations become

$$\partial^\alpha \partial_{(\mu} \bar{h}_{\nu)\alpha} - \frac{1}{2} \left(\partial_\alpha \partial^\alpha \bar{h}_{\mu\nu} + \eta_{\mu\nu} \partial^\alpha \partial^\beta \bar{h}_{\alpha\beta} \right) = 8\pi T_{\mu\nu} \, . \tag{1.14.7}$$

These equations, however, are still more complicated than they need to be. In order to simplify them further, consider for a moment a small, but arbitrary, change of coordinates of the form $x^{\bar{\mu}} = x^\mu + \xi^\mu$, where $\vec{\xi}$ is a small vector in the sense that $|\partial_\nu \xi^\mu| \ll 1$. The Jacobian matrix will then be given by

$$\Lambda^{\bar{\mu}}_{\nu} = \partial_\nu x^{\bar{\mu}} = \delta^\mu_\nu + \partial_\nu \xi^\mu \, , \tag{1.14.8}$$

and the inverse transformation will be (to first order)

$$\Lambda^{\nu}_{\bar{\mu}} = \delta^{\nu}_{\mu} - \partial_{\mu}\xi^{\nu} \; . \tag{1.14.9}$$

The transformation of the metric is $g_{\bar{\mu}\bar{\nu}} = \Lambda^{\alpha}_{\bar{\mu}}\Lambda^{\beta}_{\bar{\nu}}\, g_{\alpha\beta}$, which to first order becomes

$$g_{\bar{\mu}\bar{\nu}} = \eta_{\mu\nu} + h_{\mu\nu} - 2\partial_{(\mu}\xi_{\nu)} \; . \tag{1.14.10}$$

Since this is only a coordinate transformation, we find that a transformation of the metric of this type leaves the physical situation intact for arbitrary (but still small) ξ^{α}. This is called a *gauge transformation*, and we say that linearized gravity has a gauge freedom of the form

$$h_{\mu\nu} \rightarrow h_{\mu\nu} - 2\partial_{(\mu}\xi_{\nu)} \; , \tag{1.14.11}$$

or in terms of $\bar{h}_{\mu\nu}$:

$$\bar{h}_{\mu\nu} \rightarrow \bar{h}_{\mu\nu} - 2\partial_{(\mu}\xi_{\nu)} + \eta_{\mu\nu}\partial_{\alpha}\xi^{\alpha} \; . \tag{1.14.12}$$

This is entirely analogous to the electromagnetic case, where we find that the electromagnetic field $F_{\alpha\beta}$ (and hence the physics) remains unaffected by transformations in the potential of the form $A_{\mu} \rightarrow A_{\mu} + \partial_{\mu} f$, for an arbitrary scalar function f. In the case of gravity the word *gauge* is perhaps even more appropriate as we are in fact dealing with a change in the conventions for measuring distances, *i.e.* the coordinates.

We can now use the gauge freedom to simplify the equations by choosing a vector $\vec{\xi}$ that solves the equation

$$\partial_{\alpha}\partial^{\alpha}\xi^{\beta} = \partial_{\alpha}\bar{h}^{\alpha\beta} \; . \tag{1.14.13}$$

That this equation can always be solved is easy to see from the fact that it is nothing more than the wave equation for ξ^{β} with a source given by $\partial_{\alpha}\bar{h}^{\alpha\beta}$. If we do this we find that the gauge transformation implies that

$$\partial_{\nu}\bar{h}^{\nu\mu} \rightarrow \partial_{\nu}\bar{h}^{\nu\mu} - \partial_{\nu}\partial^{\nu}\xi^{\mu} = 0 \; , \tag{1.14.14}$$

That is, we can always find a gauge such that $\bar{h}_{\mu\nu}$ has zero divergence. Such a gauge is called the *Lorentz gauge*, a name taken from the analogous condition in electromagnetism $\partial_{\mu}A^{\mu} = 0$.

If we assume that we are in the Lorentz gauge we will have

$$\partial_{\nu}\bar{h}^{\nu\mu} = 0 \; , \tag{1.14.15}$$

and the field equations (1.14.7) will reduce to

$$\Box\bar{h}_{\mu\nu} = -16\pi T_{\mu\nu} \; , \tag{1.14.16}$$

where now $\Box$ stands for the d'Alambertian operator in flat space, or in other words the wave operator. These are the field equations for a weak gravitational field in their standard form.

One particularly important application of the weak field approximation is the Newtonian limit of general relativity. Notice that this limit will correspond not only to weak fields, but also to small velocities of the sources. Small velocities imply that the energy density T^{00} is much larger than both the momentum density and the stresses. In that case we can in fact consider only the component $\bar{h}^{00}$ and ignore all others ($\bar{h}^{0i} \simeq \bar{h}^{ij} \simeq 0$). The field equations then reduce to

$$\Box \bar{h}^{00} = -16\pi\rho \,. \tag{1.14.17}$$

Moreover, small velocities also imply that time derivatives in the d'Alambertian operator are much smaller than spatial derivatives, so the equation simplifies to

$$\nabla^2 \bar{h}^{00} = -16\pi\rho \,, \tag{1.14.18}$$

where here the symbol ∇^2 should be understood as the three-dimensional Laplace operator. This approximation implies that the speed of the sources is much smaller than the speed of light, so we might take the gravitational field to propagate essentially instantaneously. Comparing this with Newton's field equation $\nabla^2\phi = 4\pi\rho$, we conclude that in this limit $\bar{h}^{00} = -4\phi$ and $\bar{h}^{0i} = \bar{h}^{ij} = 0$. Going back to the definition of $\bar{h}_{\alpha\beta}$ in terms of $h_{\alpha\beta}$, we find that this implies $h^{00} = h^{ii} = -2\phi$. The spacetime metric in the Newtonian approximation then takes the final form

$$ds^2 = -\left(1 + 2\phi\right) dt^2 + \left(1 - 2\phi\right) \left(dx^2 + dy^2 + dz^2\right) \,, \tag{1.14.19}$$

with ϕ the standard Newtonian gravitational potential. Notice that in geometric units ϕ is dimensionless, and for the weak field approximation to be valid we must have $\phi \ll 1$, which in fact is always the case for gravitational fields in the Solar System.[16] One should also stress the fact that, even if in this limit the field equation is the same as that of Newton, the Newtonian limit of general relativity is a considerably richer theory as the metric above can be shown to directly imply the effects of gravitational time dilation and the bending of light rays in a gravitational field, both of which are absent in Newton's theory.

Another consequence of fundamental importance coming from the weak field equations is the existence of gravitational waves. Consider the weak field equations in vacuum, without assuming anything about the sources that might have produced that field. The Einstein equations in this case take the form

$$\Box \bar{h}_{\mu\nu} = \left(-\partial_t^2 + \nabla^2\right) \bar{h}_{\mu\nu} = 0 \,, \tag{1.14.20}$$

which is nothing more than the wave equation for waves propagating at the speed of light ($c = 1$). So we find that perturbations in the gravitational field behave

[16] The Newtonian gravitational field at the surface of the Sun in geometric units is $\phi \sim 10^{-6}$, while at the surface of the Earth it is $\phi \sim 10^{-9}$.

as waves that propagate at the speed of light, *i.e.* the field equations predict the existence of gravitational waves. The simplest solution to the above equation are plane gravitational waves of the form

$$\bar{h}_{\mu\nu} = A_{\mu\nu} \exp\left(i k_\alpha x^\alpha\right) \ , \tag{1.14.21}$$

with $A_{\mu\nu}$ the *amplitude tensor* of the waves and k^μ the *wave vector*. Substituting this into the wave equation we immediately find that

$$\eta^{\alpha\beta} k_\alpha k_\beta = k_\alpha k^\alpha = 0 \ , \tag{1.14.22}$$

that is, the wave vector must be null, which is just another way of saying that the waves propagate at the speed of light. We must remember, however, that the field equations only take the simplified form (1.14.20) in the Lorentz gauge (1.14.15), which in this case implies that

$$A^{\alpha\beta} k_\beta = 0 \ , \tag{1.14.23}$$

that is, the amplitude tensor $A_{\mu\nu}$ must be orthogonal to the wave vector.

At this point we might think that gravitational waves have six independent degrees of freedom: The ten independent components of the symmetric tensor $A_{\alpha\beta}$, minus the four constraints imposing the $A_{\alpha\beta}$ must be orthogonal to k^α. This is, however, not true since there is still considerable gauge freedom left within the Lorentz gauge. The reason is that when we impose the Lorentz gauge by choosing a specific gauge transformation vector ξ^α, we are in fact only restricting the value of $\Box\xi^\alpha$, so we can add to this ξ^α any new vector $\tilde{\xi}^\alpha$ such that $\Box\tilde{\xi}^\alpha = 0$ without changing anything. In particular, we can take $\tilde{\xi}^\alpha = B^\alpha \exp(i k_\beta x^\beta)$ for any arbitrary constant vector B^α. We then have four extra gauge degrees of freedom, which reduces the independent components of $A_{\mu\nu}$ to only two. In fact, we can always choose a B^α such that two further conditions are imposed on $A_{\mu\nu}$

$$A^\mu{}_\mu = 0 \ , \qquad A_{\mu\nu} u^\nu = 0 \ , \tag{1.14.24}$$

for an arbitrary constant 4-velocity u^α (*i.e.* an arbitrary timelike unit vector). The first condition means that $A_{\mu\nu}$ is traceless (which implies that in this gauge $h_{\mu\nu} = \bar{h}_{\mu\nu}$), and the second that it is orthogonal to u^μ. If we move to a Lorentz frame such that $u^\alpha = (1,0,0,0)$, then in that frame conditions (1.14.23) and (1.14.24) imply that

$$A_{\mu 0} = 0 \ , \qquad \sum_j A_{ij} k_j = 0 \ , \qquad \sum_j A_{jj} = 0 \ . \tag{1.14.25}$$

Any tensor that satisfies these conditions is called a *transverse-traceless* (TT) tensor, traceless since its trace clearly vanishes, and transverse because it is both purely spatial and orthogonal to its own direction of propagation. If we take the direction of propagation to be along the z axis, then all these conditions imply that $A_{\mu\nu}$ has the form

$$A_{\mu\nu} = \begin{pmatrix} 0 & 0 & 0 & 0 \\ 0 & A^+ & A^\times & 0 \\ 0 & A^\times & -A^+ & 0 \\ 0 & 0 & 0 & 0 \end{pmatrix} , \tag{1.14.26}$$

with the two remaining degrees of freedom given by A^+ and $A^\times$.

Having found the general solution for a plane gravitational wave in the TT gauge, it becomes important to ask how those waves can be detected, or in other words what their effect is on free particles. As a first step in answering this question we can try to look at the geodesic equation for the metric corresponding to a plane wave

$$\frac{du^\alpha}{d\tau} = -\Gamma^\alpha_{\mu\nu} u^\mu u^\nu . \tag{1.14.27}$$

Assuming the particle is initially at rest so that $u^\alpha = (1, 0, 0, 0)$, the equation reduces to

$$\left. \frac{du^\alpha}{d\tau} \right|_{t=t_0} = -\Gamma^\alpha_{00} . \tag{1.14.28}$$

Now, in the weak field approximation we have $\Gamma^\alpha_{00} = \eta^{\alpha\beta} \left(2\, \partial_t h_{0\beta} - \partial_\beta h_{00}\right)/2$. But in the TT gauge $h_{00} = h_{0\beta} = 0$, so we must conclude that the particle's 4-velocity remains constant, *i.e.* the gravitational wave does not change the particle's trajectory. This can lead us to the wrong conclusion that gravitational waves are pure gauge effects and are therefore not physical, but we must keep in mind the fact that all we have shown is that the particle remains at the same coordinate location, and coordinates are just labels. In other words, in the TT gauge we have attached our coordinate labels to freely falling particles, so of course we see them keeping constant coordinate position.

The true measure of the gravitational field is not in the metric itself, but in the tidal forces, that is in the curvature tensor. If we look at the Riemann tensor in the weak field limit (1.14.4), we find that, for a wave traveling along the z direction, the only non-zero independent components of the Riemann tensor in the TT gauge are

$$R_{\mu x \nu x} = -\frac{1}{2}\, \partial_\mu \partial_\nu h^{TT}_{xx} , \tag{1.14.29}$$

$$R_{\mu y \nu y} = -\frac{1}{2}\, \partial_\mu \partial_\nu h^{TT}_{yy} , \tag{1.14.30}$$

$$R_{\mu x \nu y} = -\frac{1}{2}\, \partial_\mu \partial_\nu h^{TT}_{xy} , \tag{1.14.31}$$

where here μ and ν only take the values (t, z). This shows that gravitational waves are not just gauge effects since they produce non-zero curvature.

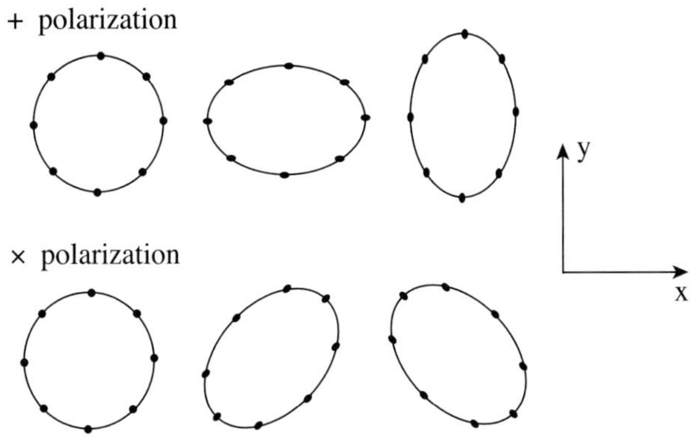

Fig. 1.4: Effect of gravitational waves on a ring of free particles. In the case of the + polarization the ring oscillates by being elongated and compressed along the x and y direction, while for a $\times$ polarization the elongations and compressions are along the diagonal directions.

Consider now the equation for geodesic deviation (1.13.2). For two nearby particles initially at rest with separation vector z^μ this becomes

$$a^\alpha = -R^\alpha{}_{0\beta 0} z^\beta \, , \tag{1.14.32}$$

or in other words

$$a^x = \frac{1}{2}\left(\ddot{h}^{TT}_{xx}\, z^x + \ddot{h}^{TT}_{xy}\, z^y\right) = -\frac{k_0^2}{2}\left(A^+\, z^x + A^\times\, z^y\right)\exp(ik_\alpha x^\alpha)\, , \tag{1.14.33}$$

$$a^y = \frac{1}{2}\left(\ddot{h}^{TT}_{xy}\, z^x + \ddot{h}^{TT}_{yy}\, z^y\right) = -\frac{k_0^2}{2}\left(A^\times\, z^x - A^+\, z^y\right)\exp(ik_\alpha x^\alpha)\, . \tag{1.14.34}$$

From this we see that gravitational waves have two independent polarizations. The first one corresponds to $A^+ \neq 0$ and $A^\times = 0$ and is called the + polarization, while the second one corresponds to $A^+ = 0$ and $A^\times \neq 0$ and is called the $\times$ polarization. We find that under the effect of a passing gravitational wave, the relative position of nearby particles changes even if the particles' coordinates do not. If we have a ring of free particles floating in space, and a gravitational wave moving along the perpendicular direction to the ring passes by, a + polarization will cause the ring to oscillate by being alternately elongated and compressed along the x and y directions, while a wave with $\times$ polarization will produce elongations and compressions along the diagonal directions (see Figure 1.4).

The effect that gravitational waves have on a ring of free particles can be used to build a gravitational wave detector. Two basic types of detectors have been considered. The first type are resonant bars, essentially large cylindrical aluminum bars that have longitudinal modes of vibration with frequencies close to those of the expected gravitational waves coming from astrophysical sources. A passing gravitational wave should excite those vibrational modes and make the bars resonate. The first of these bars were constructed by Joseph Weber in the 1960s, and several modern versions that work at cryogenic temperatures are still in use today. The second type of detectors are laser interferometers, where we track the separation of freely suspended masses using highly accurate interferometry. The first prototype interferometers where build in the 1980s, and today several advanced kilometer-scale interferometric detectors are either already working or in the last stages of development: The LIGO project in the United States, the VIRGO and GEO 600 projects in Europe, and the TAMA project in Japan.

To this day, there has been no unambiguous detection of gravitational waves, though this might change in the next few years as the new interferometers slowly reach their design sensitivities. The main reason why gravitational waves have not been observed yet has to do with the fact that gravity is by far the weakest of all known interactions. Estimates of the amplitudes of gravitational waves coming from violent astrophysical events such as the collisions of neutron stars or black holes from distances as far away as the Virgo Cluster put the dimensionless amplitude of the waves as they reach Earth at the level of $A \sim 10^{-21}$. In other words, a ring of particles one meter in diameter would be deformed a distance of roughly one millionth of the size of a proton (the effect scales with the size of the system, which explains why the large interferometric detectors have arms with lengths of several kilometers). Detecting such small distortions is clearly a huge challenge. However, with modern technology such a detection has become not only possible but even likely before this decade is out.

The prediction of the gravitational wave signal coming from highly relativistic gravitational systems is one of the most important applications of numerical relativity, and because of this we will come back to the subject of gravitational waves in later Chapters.

1.15 The Schwarzschild solution and black holes

One particularly important application of general relativity is related to the exterior gravitational field of a static and spherically symmetric object. The general metric of such a spacetime can be shown to have the simple form

$$ds^2 = -f(r)dt^2 + h(r)dr^2 + R(r)^2 d\Omega^2 \,, \tag{1.15.1}$$

where r is some radial coordinate, and f, h and R are functions of r only. We can simplify this further by choosing a radial coordinate r' such that $r' := R(r)$. In that case the metric reduces to

$$ds^2 = -f(r)dt^2 + h(r)dr^2 + r^2 d\Omega^2 \,, \tag{1.15.2}$$

where for simplicity we have dropped the prime from r'. In the general case we have therefore only two unknown functions to determine, $f(r)$ and $h(r)$. The coordinates (r, θ, ϕ) are known as *Schwarzschild coordinates*, and in particular the radial coordinate is called the *areal radius* since in this case the area of the sphere is always given by $4\pi r^2$.[17] Equation (1.15.2) for the metric allows for a huge simplification in the field equations: Instead of having to solve for 10 independent metric components we only need to find two functions of r.

The next step is to substitute the metric (1.15.2) into the Einstein field equations in order to find f and h. Since we are interested in the exterior field, we must use the vacuum field equations $R_{\mu\nu} = 0$. Doing this we find

$$0 = R_{tt} = \frac{1}{2}(fh)^{-1/2}\frac{d}{dr}\left[(fh)^{-1/2}\frac{df}{dr}\right] + \frac{1}{rfh}\frac{df}{dr} , \tag{1.15.3}$$

$$0 = R_{rr} = -\frac{1}{2}(fh)^{-1/2}\frac{d}{dr}\left[(fh)^{-1/2}\frac{df}{dr}\right] + \frac{1}{rh^2}\frac{dh}{dr} , \tag{1.15.4}$$

$$0 = R_{\theta\theta} = R_{\phi\phi} = -\frac{1}{2rfh}\frac{df}{dr} + \frac{1}{2rh^2}\frac{dh}{dr} + \frac{1}{r^2}\left(1 - \frac{1}{h}\right) , \tag{1.15.5}$$

with all other components of $R_{\mu\nu}$ equal to zero. The first two equations imply

$$\frac{d\ln f}{dr} + \frac{d\ln h}{dr} = 0 , \tag{1.15.6}$$

which can be trivially integrated to find $f = K/h$, with K some constant. Without loss of generality we can take $K = 1$, as this only requires rescaling the time coordinate. Substituting this into the third equation we find

$$-\frac{df}{dr} + \frac{1-f}{r} = 0 \quad \Rightarrow \quad \frac{d}{dr}(rf) = 1 , \tag{1.15.7}$$

whose solution is

$$f = 1 + C/r , \tag{1.15.8}$$

with C another constant. The metric then takes the form

$$ds^2 = -\left(1 + \frac{C}{r}\right) dt^2 + \left(1 + \frac{C}{r}\right)^{-1} dr^2 + r^2 d\Omega^2 . \tag{1.15.9}$$

The value of the constant C can be obtained by comparing with the Newtonian limit (1.14.19), which should correspond to $r >> 1$. Taking the Newtonian potential to be $\phi = -M/r$ we find $C = -2M$, which implies

$$ds^2 = -\left(1 - \frac{2M}{r}\right) dt^2 + \left(1 - \frac{2M}{r}\right)^{-1} dr^2 + r^2 d\Omega^2 . \tag{1.15.10}$$

[17]Notice, however, that there is no reason why the distance to the *origin* of the coordinate system $r = 0$ should be equal to r. In fact, there is no reason why the point $r = 0$ should be part of the manifold at all. For example, consider the geometry of a parabola of revolution. The areal radius in this case will measure the circumference of the surface at a given place, but there is clearly no point with $r = 0$ on the surface. In fact, the areal radius is not always well behaved. Think of a two-dimensional surface of revolution that resembles a bottle with a narrow throat (this is called a *bag of gold* geometry). It is clear that as we go from inside the bottle towards the throat and then out of the bottle, the areal radius first becomes smaller and then larger again, *i.e.* it is not monotone so it is not a good coordinate.

This is known as the Schwarzschild solution, in honor of K. Schwarzschild who discovered it in 1916 [260], a few months after Einstein first postulated the field equations of general relativity. Among the properties of the previous solution we can first mention the fact that it is *asymptotically flat*, that is, it approaches the Minkowski metric when $r \to \infty$. This is to be expected as the gravitational field becomes weaker as we move away from the source. More interesting is the fact that the metric coefficients become singular at both $r = 0$ and $r = 2M$. The radius $r_s := 2M$ is known as the Schwarzschild or *gravitational* radius, and for conventional astrophysical objects (planets, stars, *etc.*) it is usually much smaller than the actual radius of the object itself. Since the Schwarzschild solution is a vacuum solution it is no longer valid for r smaller than the radius of the object, so in those cases we do not need to worry about the metric singularities.

If, on the other hand, we wish to consider Schwarzschild's solution as that corresponding to a *point* particle, we must deal with the metric singularities at $r = 0$ and $r = 2M$. In order to study the nature of these singularities it is convenient to first calculate the Riemann tensor for this spacetime. Doing this we find that the only non-zero components of Riemann are

$$R_{trtr} = -2M/r^3 \; , \tag{1.15.11}$$

$$R_{t\theta t\theta} = R_{t\phi t\phi} = M/r^3 \; , \tag{1.15.12}$$

$$R_{\theta\phi\theta\phi} = 2M/r^3 \; , \tag{1.15.13}$$

$$R_{r\theta r\theta} = R_{r\phi r\phi} = -M/r^3 \; . \tag{1.15.14}$$

Notice that none of these components are singular at $r = 2M$, but they are all singular at $r = 0$. This means that the gravitational field is singular at $r = 0$ (as we would expect for a point particle), but perfectly regular at $r = 2M$. The only possibility is therefore that at $r = 2M$ something goes wrong with the Schwarzschild coordinates.

There is something else to notice about Schwarzschild's solution. For $r < 2M$ it turns out that the coordinates r and t change roles: r becomes a timelike coordinate ($g_{rr} < 0$), while t becomes spacelike ($g_{tt} > 0$). This implies that once an object has crossed $r = 2M$, the advance of time becomes equivalent with a decrease in r, that is, the object must continue toward smaller values of r for the same reason that time must flow to the future. As nothing can stop the flow of time, there is no force in the Universe capable of preventing the object from reaching $r = 0$, where it will find infinite tidal forces (the Riemann is singular) that will destroy it. The Schwarzschild radius represents a surface of no return: Further out it is always possible to escape the gravitational field, but move closer than $r = 2M$ and the fall all the way down to $r = 0$ becomes inevitable.

Since the regularity of Riemann implies that the problem at $r = 2M$ must be caused by a poor choice of coordinates, we can ask the question of whether or not a better set of coordinates exists to study the Schwarzschild spacetime. In fact, we can construct several coordinate systems that are regular at $r = 2M$. One of the first are the so-called Eddington–Finkelstein coordinates, discovered by Eddington in 1924 [117], and rediscovered by Finkelstein in 1958 [128]. These coordinates can be derived by considering radial null geodesics (corresponding to the motion of photons). These geodesics are very easy to find, we just take $d\theta = d\phi = 0$ in the Schwarzschild metric (1.15.10), and asks for the interval to be null, $ds^2 = 0$. We find,

$$ds^2 = 0 = -(1 - 2M/r)dt^2 + (1 - 2M/r)^{-1} dr^2 \; , \tag{1.15.15}$$

which implies

$$dt = \pm (1 - 2M/r)^{-1} dr \; . \tag{1.15.16}$$

This equation can be easily integrated to find

$$t = \pm r^* + \text{constant} \; , \tag{1.15.17}$$

where we have defined the *Regge–Wheeler tortoise coordinate* r^* as

$$r^* := r + 2M \ln \left(r/2M - 1 \right) \; . \tag{1.15.18}$$

Notice that in terms of r^* the Schwarzschild radius corresponds to $r^* = -\infty$. Let us now define a new coordinate as $\tilde{V} := t + r^*$, and make a transformation of coordinates from $\{t, r\}$ to $\{\tilde{V}, r\}$. Notice that this transformation will in fact be singular at $r = 2M$, but this is precisely what we need if we want to eliminate the singularity in the original Schwarzschild coordinates there (the tortoise coordinate is not defined for $r < 2M$, but in that case we can work with $r^* = r + 2M \ln(1 - r/2M)$ and the same results follow). In terms of the new coordinates the metric becomes

$$ds^2 = - \left(1 - 2M/r\right) \, d\tilde{V}^2 + 2 \, d\tilde{V} dr + r^2 d\Omega^2 \; . \tag{1.15.19}$$

This is the metric of Schwarzschild's spacetime in Eddington–Finkelstein coordinates. The coordinate $\tilde{V}$ is null, which can be seen from the fact that, for the metric above, the ingoing radial null lines correspond to $\tilde{V}$ = constant. A new transformation of coordinates from $\{\tilde{V}, r\}$ to $\{\tilde{t}, r\}$, where $\tilde{t} := \tilde{V} - r$, brings the metric into the *Kerr–Schild* form

$$ds^2 = - \left(1 - \frac{2M}{r}\right) d\tilde{t}^2 + \frac{4M}{r} \, d\tilde{t} \, dr + \left(1 + \frac{2M}{r}\right) dr^2 + r^2 d\Omega^2 \; . \tag{1.15.20}$$

In these coordinates, radially ingoing null lines turn out to have constant coordinate speed $dr/d\tilde{t} = -1$, just as in Minkowski's spacetime. The *outgoing* null lines, on the other hand, have a speed given by $dr/d\tilde{t} = (1 - 2M/r)/(1 + 2M/r)$, which implies that their speed is less than 1 for $r > 2M$, it is zero for $r = 2M$ ($r = 2M$ is in fact the trajectory of an "outgoing" null geodesic), and becomes negative for $r < 2M$ (*i.e.* the "outgoing" null lines move in instead of out). Since all null geodesics move in for $r < 2M$, we must conclude that this region can have no causal influence on the outside, as no physical interaction can travel faster

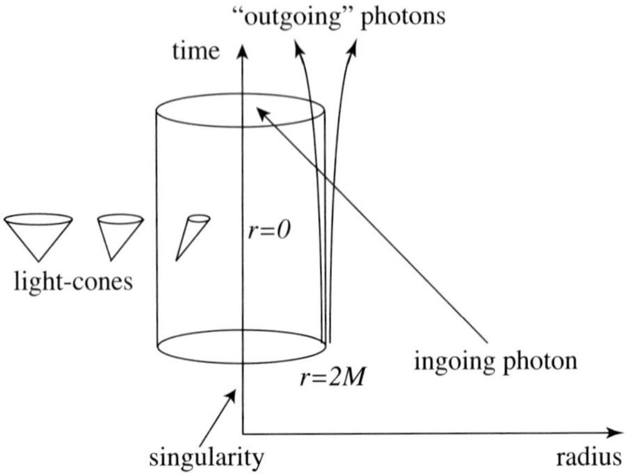

Fig. 1.5: Schwarzschild's spacetime in Kerr–Schild coordinates.

than light and light itself can not escape. These properties of the Schwarzschild spacetime in Kerr–Schild coordinates are shown in Figure 1.5.

Eddington–Finkelstein coordinates, or the closely related Kerr–Schild version, behave much better than Schwarzschild coordinates, but they have a major flaw: The time symmetry of the original solution is now gone, as ingoing and outgoing geodesics do not behave in the same way. It is in fact possible to construct coordinates of Eddington–Finkelstein type using the null coordinate $\tilde{U} := t - r^*$ instead of $\tilde{V} := t + r^*$, and in that case we finds that $r = 2M$ becomes a barrier for ingoing null lines instead of outgoing ones. We then distinguish between ingoing and outgoing Eddington–Finkelstein coordinates.

We can construct a regular coordinate system that preserves the time symmetry by going from the $\{t, r\}$ Schwarzschild coordinates to the purely null coordinates $\{\tilde{U}, \tilde{V}\}$. The metric then becomes

$$ds^2 = -(1 - 2M/r)\, d\tilde{U}\, d\tilde{V} + r^2 d\Omega^2 \; . \tag{1.15.21}$$

There is still a problem with this metric at $r = 2M$ since the first metric coefficient vanishes, but this is easy to solve. Define $\tilde{u} := -e^{-\tilde{U}/4M}$ and $\tilde{v} := +e^{+\tilde{V}/4M}$. A change of coordinates from $\{\tilde{U}, \tilde{V}\}$ to $\{\tilde{u}, \tilde{v}\}$ takes the metric into the form

$$ds^2 = -\left(32M^3/r\right)\, e^{-r/2M} d\tilde{u}\, d\tilde{v} + r^2 d\Omega^2 \; . \tag{1.15.22}$$

This expression is manifestly regular for all $r > 0$. Notice that here r still measures areas, but it is now a function of $\tilde{u}$ and $\tilde{v}$. The coordinates $\{\tilde{u}, \tilde{v}\}$ are well behaved, but being null they are not easy to visualize. However, we can use them to construct new spacelike and timelike coordinates by defining

$$\xi := (\tilde{v} - \tilde{u})/2 , \qquad \eta := (\tilde{v} + \tilde{u})/2 , \tag{1.15.23}$$

which takes the metric into the final form

$$ds^2 = \left(32M^3/r\right) e^{-r/2M} \left(-d\eta^2 + d\xi^2\right) + r^2 d\Omega^2 . \tag{1.15.24}$$

Clearly ξ is always a spacelike coordinate, while η is always timelike. The coordinates $\{\xi, \eta\}$ that we have just defined are known as *Kruskal–Szekeres* coordinates, after Kruskal and Szekeres who discovered them independently in 1960 [182, 277]. Notice that the singularity at $r = 2M$ is now completely gone, but the singularity at $r = 0$ remains. The new coordinates $\{\eta, \xi\}$ are related to the original Schwarzschild coordinates by

$$(r/2M - 1) \, e^{r/2M} = \xi^2 - \eta^2 , \qquad \tanh(t/4M) = \eta/\xi . \tag{1.15.25}$$

The allowed range for $\{\eta, \xi\}$ is given by $r > 0$, which implies $\eta^2 < \xi^2 + 1$.

In Kruskal–Szekeres coordinates the radial null lines turn out to be given by $d\xi/d\eta = 1$, *i.e.* they are lines at 45 degrees on the spacetime diagram. This means that light-cones (and the causal relationships tied to them) behave just as they do in flat space. Figure 1.6 shows a diagram of the metric (1.15.24). When looking at this diagram one must remember that only $\{\eta, \xi\}$ are represented, the angular coordinates have been suppressed so that every point in the diagram in fact corresponds to a sphere.

There are several important things to learn from the Kruskal diagram and the relations (1.15.25). First, notice that lines of constant r correspond to hyperbolae in the diagram, vertical for $r > 2M$ and horizontal for $r < 2M$. This means that a line of constant r is timelike for $r > 2M$, and hence an allowed trajectory for a particle, but becomes spacelike for $r < 2M$, so objects can not stay at constant r there. The degenerate hyperbola $r = 2M$ is in fact a null line. The two branches of the hyperbola at $r = 0$ mark the boundary of spacetime since there is a physical singularity there. On the other hand, lines of constant Schwarzschild time t correspond to straight lines through the origin in the diagram. Infinite time $t = \pm\infty$ corresponds to the lines at 45 degrees and coincides with $r = 2M$. We clearly see here the problem with Schwarzschild coordinates: At $r = 2M$ they collapse the full line into a single coordinate point.

Even more interesting is the fact that the lines $r = 2M$ separate the spacetime into four regions. In region I we have $r > 2M$ so it is an exterior region, while region II has $r < 2M$ and is clearly the interior. An object that moves from region I to region II can never get out again and must reach the singularity at $r = 0$ sometime in its future. Since neither light nor any physical influence can leave region II this region in called a *black hole*. The line at $r = 2M$ that separates the black hole from the exterior is called the *black hole horizon*.

There is a very important relationship between the area of the horizon and the mass of the Schwarzschild spacetime. Notice that, by the very definition of

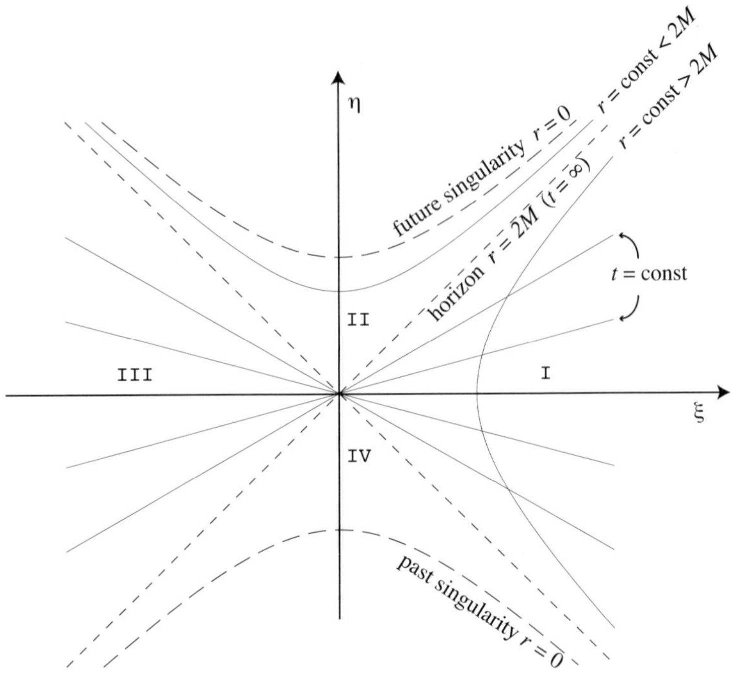

Fig. 1.6: Schwarzschild spacetime in Kruskal–Szekeres coordinates.

the areal radius, the area of a sphere of radius r is given by $A = 4\pi r^2$. At the horizon we have $r = 2M$, so that

$$M^2 = A_H/16\pi \, . \tag{1.15.26}$$

Let us consider now regions III and IV. Notice first that region IV is equivalent to II but inverted in time: The singularity is in the past, and nothing can enter region IV from the outside. This region is called a *white hole*. Finally, region III is clearly also an exterior region, but it is completely disconnected from region I: It is another exterior region, or in other words, another universe.

To understand more clearly the relationship between the two exterior regions consider the surface $t = 0$ (the horizontal line in the diagram). If we approach the origin from region I, we see that r becomes smaller and smaller until it reaches a minimum value of $r = 2M$, then as we penetrate into region III it starts growing again. The resulting geometry is known as an *Einstein–Rosen bridge* or a *wormhole*: two asymptotically flat regions joined through a narrow tunnel. There are two important things to mention about the Einstein–Rosen bridge. First, it is impossible to traverse it since objects moving from region I must always remain inside their light-cones, so if they attempt to reach region III they will find themselves first in region II and will end up in the future singularity. Second, if a black hole forms through the gravitational collapse of a star then the bridge never forms as regions III and IV would in fact be inside the star where the Schwarzschild solution is no longer valid (it is a vacuum solution). Nevertheless, as we will discuss in later Chapters, the wormhole picture has important applications in numerical relativity where it is used to construct multiple black hole initial data.

At the risk of boring the reader with yet another coordinate system for the Schwarzschild spacetime, I will introduce one last set of coordinates that is frequently used in numerical relativity. It turns out that it is possible to rewrite the metric in such a way that the spatial part is *conformally flat*, that is, the spatial metric is just the Minkowski metric times a scalar function. In order to do this we must define a new radial coordinate $\tilde{r}$ such that

$$r = \tilde{r}\,(1 + M/2\tilde{r})^2 \; . \tag{1.15.27}$$

A transformation from Schwarzschild coordinates $\{t, r\}$ to the coordinates $\{t, \tilde{r}\}$ results in the metric

$$ds^2 = -\left(\frac{1 - M/2\tilde{r}}{1 + M/2\tilde{r}}\right)^2 dt^2 + \psi^4 \left(d\tilde{r}^2 + \tilde{r}^2 d\Omega^2\right) \; , \tag{1.15.28}$$

with the conformal factor given by $\psi = 1 + M/2\tilde{r}$. In these coordinates the spatial metric is regular at the horizon, which now corresponds to $\tilde{r} = M/2$. Notice also that far away r and $\tilde{r}$ approach each other. The coordinate $\tilde{r}$ is usually called the *isotropic radius* since the spatial metric is just the flat metric, which is clearly isotropic, times a conformal factor. The metric (1.15.28) is clearly singular at $\tilde{r} = 0$, but the transformation of coordinates shows that $\tilde{r} = 0$ corresponds to $r = \infty$, so this is in fact not the physical singularity at $r = 0$. It turns out that the region $\tilde{r} \in [0, M/2]$ represents the other side of the Einstein–Rosen bridge, or in other words the whole other universe has been *compactified* to this finite region. The singularity at $\tilde{r} = 0$ is then just a coordinate singularity associated with this compactification. Notice also that this metric has an isometry (*i.e.* it is invariant) with respect to the transformation $\tilde{r} \rightarrow M^2/4\tilde{r}$. This isometry corresponds to changing a point in our universe with the corresponding point on the other side of the wormhole.

1.16 Black holes with charge and angular momentum

In the previous section we introduced the Schwarzschild metric and showed that, if taken as the solution for the full spacetime, it describes a spherically symmetric black hole. However, the Schwarzschild black hole is not the most general black hole solution known. A solution for a spherical black hole with an electric charge was found independently by Reissner and Nordstrom soon after Schwarzschild's work [240, 221]. But a non-spherically symmetric black hole solution had to wait until 1963 when Kerr found the spacetime metric for a rotating black hole [170].

Finally, in 1965 the charged rotating black hole solution, that contains all other previous solutions as special cases, was found by Newman *et al.* [217]. The metric for this so-called *Kerr–Newman* black hole is given by

$$\begin{aligned} ds^2 = &- \left(\frac{\Delta - a^2 \sin^2\theta}{\rho^2} \right) dt^2 - \frac{2a \sin^2\theta \left(r^2 + a^2 - \Delta \right)}{\rho^2} \, dt d\phi \\ &+ \left(\frac{\left(r^2 + a^2 \right)^2 - \Delta \, a^2 \sin^2\theta}{\rho^2} \right) \sin^2\theta \, d\phi^2 + \frac{\rho^2}{\Delta} \, dr^2 + \rho^2 \, d\theta^2 \,, \end{aligned} \tag{1.16.1}$$

where

$$\Delta = r^2 + a^2 + Q^2 - 2Mr \,, \qquad \rho^2 = r^2 + a^2 \cos^2\theta \,, \tag{1.16.2}$$

and with a, M and Q free parameters. For $a = Q = 0$ the metric reduces to that of Schwarzschild's spacetime, for $a = 0$ and $Q \neq 0$ it reduces to the Reissner–Nordstrom solution, while for $Q = 0$ and $a \neq 0$ it reduces to the Kerr solution. The Kerr–Newman spacetime is clearly stationary (*i.e.* the metric coefficients are time independent), and axisymmetric (the metric is independent of the angle ϕ), but the spacetime is not time-symmetric for $a \neq 0$.

Though here we will not derive equation (1.16.1), the reader can verify that for $Q = 0$ this is indeed a vacuum solution of Einstein's equations. For $Q \neq 0$, however, the metric (1.16.1) is not a vacuum solution and corresponds instead to a solution of the Einstein–Maxwell equations, *i.e.* the Einstein field equations in the presence of an electromagnetic field. The potential one-form for this solution is given by

$$A_\mu = - \frac{rQ}{\rho^2} \left(1, 0, 0, a \sin^2\theta \right) \,. \tag{1.16.3}$$

The parameters $\{a, M, Q\}$ have clear physical interpretations. The parameter Q is interpreted as the total electric charge of the black hole. To see this, consider for a moment the case $a = 0$, then the electromagnetic potential reduces to

$$A_\mu = -Q/r \; (1, 0, 0, 0) \,. \tag{1.16.4}$$

Now, since the metric is asymptotically flat, far away we can make the identification $A^\mu = (\varphi, A^i)$, with φ the electrostatic potential and A^i the magnetic vector potential. This implies $A_0 = -\varphi$, and comparing with the above equation we find $\varphi = Q/r$. This clearly shows that far away observers will measure Q to be the electric charge of the black hole. The calculation also follows for $a \neq 0$ but it becomes more involved since in that case we have a non-zero magnetic field.

For the other two parameters, we can use the results about global measures of mass and momentum discussed in Appendix A to see that M corresponds to the total mass of the spacetime, while the total angular momentum is given by $J = aM$ (so that a is the angular momentum per unit mass).

The Kerr–Newman metric has some other interesting properties. For example, it becomes singular when either $\Delta = 0$ or $\rho^2 = 0$. The singularity at $\rho^2 = 0$ can be shown to be a true curvature singularity and corresponds to

$$r^2 + a^2 \cos^2\theta = 0 , \tag{1.16.5}$$

which implies $\cos\theta = 0$, so that the singularity lies on the equatorial plane, and $r = 0$ which naively would seem to imply that it is just a point at the origin (but it would still be rather strange to have a singularity only on the equatorial plane). In fact, the singularity has the structure of a ring, a fact that can be seen more clearly if we take the case $M = Q = 0$, $a \neq 0$, for which we find that the circumference of a ring in the equatorial plane is given by $C = 2\pi\sqrt{r^2 + a^2}$, so that $r = 0$ corresponds to a ring of radius a (in fact, in this case the spacetime is Minkowski in spheroidal coordinates).

The singularity at $\Delta = 0$, on the other hand, can be shown to be only a coordinate singularity (the curvature tensor is regular there) and corresponds to

$$r^2 + a^2 + Q^2 - 2Mr = 0 \quad \Rightarrow \quad r = M \pm \sqrt{M^2 - Q^2 - a^2} . \tag{1.16.6}$$

In the general case we then have two different coordinate singularities corresponding to two distinct horizons. For the case $a = Q = 0$ the exterior horizon coincides with the Schwarzschild radius $r = 2M$, while the interior horizon collapses to $r = 0$. Notice that for $M^2 < a^2 + Q^2$ there are no horizons, and the geometry describes a *naked* singularity (*i.e.* a singularity unprotected by a horizon), a situation which is considered to be unphysical (see the following Section). The limiting case $M^2 = a^2 + Q^2$ is called an *extremal* black hole.

Finally, there is another interesting surface in the Kerr–Newman geometry. Notice that for a static observer the metric component g_{00} changes sign when

$$\Delta - a^2 \sin^2\theta = r^2 + Q^2 - 2Mr + a^2 \cos^2\theta = 0 , \tag{1.16.7}$$

corresponding to

$$r = M + \sqrt{M^2 - Q^2 - a^2 \cos^2\theta} . \tag{1.16.8}$$

This means that for smaller values of r an observer can not remain static since that would be a spacelike trajectory. Instead, inside this surface an observer with fixed r and θ must in fact rotate around the black hole in the same direction as the hole spins. This effect is known as the *dragging of inertial frames*, and it shows that to some extent general relativity does incorporate Mach's principle (local inertial properties are severely affected close to a rotating black hole). This static limiting surface lies everywhere outside the horizon, except at the poles where it coincides with it. The region between the static limiting surface and the horizon is called the *ergosphere*, as we can show that inside this region it is possible to extract rotational energy from the black hole via the so-called *Penrose process*.[18]

The relationship between the area and the radius of the exterior horizon for a Kerr–Newman black hole is not as simple as in Schwarzschild. In this case we find that the area of the exterior horizon is

$$A_H = 4\pi \left(r_+^2 + a^2\right) = 4\pi \left(2M^2 - Q^2 + 2M\sqrt{M^2 - Q^2 - a^2}\right) . \qquad (1.16.9)$$

In the particular case when $Q = 0$ this relationship implies that

$$M^2 = \frac{A_H}{16\pi} + \frac{4\pi J^2}{A_H} . \qquad (1.16.10)$$

Since the rotational energy of the black hole can be extracted, it is usual to define the *irreducible mass* as $M_I := \sqrt{A_H/16\pi}$, so that the above expression becomes $M^2 = M_I^2 + J^2/4M_I^2$.

The Kerr–Newman metric (1.16.1) is given in so-called *Boyer–Lindquist* coordinates. Another important set of coordinates are the *Kerr–Schild* coordinates, which are related to the Boyer–Lindquist coordinates by the transformations

$$\tilde{t} = \tilde{V} - r , \qquad (1.16.11)$$

$$x = \sin\theta \left(r\cos\phi - a\sin\tilde{\phi}\right) , \qquad (1.16.12)$$

$$y = \sin\theta \left(r\sin\phi + a\cos\tilde{\phi}\right) , \qquad (1.16.13)$$

$$z = r\cos\theta , \qquad (1.16.14)$$

with

$$d\tilde{V} = dt + \frac{r^2 + a^2}{\Delta}\, dr , \qquad d\tilde{\phi} = d\phi + \frac{a}{\Delta}\, dr . \qquad (1.16.15)$$

In terms of Kerr–Schild coordinates $\{\tilde{t}, x, y, z\}$, the metric of the Kerr–Newman spacetime takes the particularly simple form

$$ds^2 = g_{\mu\nu} dx^\mu dx^\nu = \left(\eta_{\mu\nu} + 2H\, l_\mu l_\nu\right) dx^\mu dx^\nu , \qquad (1.16.16)$$

with $\eta_{\mu\nu}$ the Minkowski metric, and

$$H = \frac{rM - Q^2/2}{r^2 + a^2 z^2/r^2} , \qquad l_\mu = \left(1, \frac{rx + ay}{r^2 + a^2}, \frac{ry - ax}{r^2 + a^2}, \frac{z}{r}\right) . \qquad (1.16.17)$$

In the previous expressions r is now considered a function of the new coordinates defined implicitly through

[18]The Penrose process involves separating a particle in two inside the ergosphere, then sending one part into the black hole and allowing the other to escape to infinity carrying part of the rotational energy of the hole.

$$\frac{x^2 + y^2}{r^2 + a^2} + \frac{z^2}{r^2} = 1 \,. \tag{1.16.18}$$

There are several interesting things to notice about the metric (1.16.16). First, this form of the metric is now clearly not singular at the horizon. Second, for $a = Q = 0$ it reduces precisely to the Kerr–Schild metric for the Schwarzschild spacetime (1.16.16). Finally, the one-form l_μ defined above turns out to be null with respect to the Minkowski metric, that is, $\eta^{\mu\nu} l_\mu l_\nu = 0$. In particular, this means that the inverse metric $g^{\mu\nu}$ can be written as

$$g^{\mu\nu} = \eta^{\mu\nu} - 2H \, l_*^\mu l_*^\nu \,, \tag{1.16.19}$$

with $l_*^\mu := \eta^{\mu\nu} l_\nu$.

1.17 Causal structure, singularities and black holes

As mentioned in the previous section, black holes are characterized by the fact that there exists a region of spacetime that is causally disconnected from the outside. Black holes are of fundamental importance in the study of strong gravitational fields as they are the end result of gravitational collapse, and because of this they are a recurrent theme in numerical relativity. It therefore becomes very important to develop a notion of the causal structure of spacetime and the concept of black hole in the general case. The study of these issues in general relativity is usually done in the formal mathematical context of topological spaces. The discussion here, however, will be short and considerably less formal, focusing only on the main physical ideas and results (for a more detailed discussion see *e.g.* [161] and [295]).

In special relativity, the causal structure of Minkowski's spacetime is very simple. The light-cone of a given event indicates which other events can be reached from it by physical signals. In general relativity, on the other hand, things can be considerably more complex since we are dealing with general Lorentzian manifolds. Because all spacetimes are locally flat it is clear that local causal relations will work in the same way as in special relativity, but globally things can be very different both because of the possibility of a topologically non-trivial structure, and because of the possible existence of singularities like that of Schwarzschild's spacetime. A particular problem we might face is that for some spacetimes it might not even be possible to define in a consistent way what we mean by future and past. For example the light-cones can rotate continuously as we go around a closed loop in such a way that when we get back to the starting point what we thought was the future of a given event is now the past. Such manifolds are said not to be *time orientable* and we will not consider them further, but they are a good example of the fact that in general relativity we have to take care when thinking about global causal relationships.

On a time orientable manifold M, we define the *causal future* $J^+(p)$ of an event p as the set of all events on M that can be reached from p by a future directed causal curve, where by causal curve we mean a curve that is either timelike or null. Similarly, the *chronological future* $I^+(p)$ of p is defined by replacing *causal curve* with *timelike curve* in the previous definition. The causal and chronological pasts are defined in an analogous way. The concept of causal past and future is the generalization of the light-cone of special relativity.

The notion of chronological future allows us to define what is known as an *achronal set*, which is a subset S of our manifold M such that no two events in S are inside each other's chronological futures. In more loose language, an achronal set is either null or spacelike everywhere. Given an achronal set S we define its *future domain of dependence* $D^+(S)$ as the set of all events p in the manifold such that every past directed causal curve through p intersects S. In physical terms the future domain of dependence of S are those events that are completely causally determined by events in S itself, *i.e.* knowledge of events in S is enough to predict future events in $D^+(S)$ with absolute certainty, as no physical signals from outside of S can reach them in time. Similarly, we can define the past domain of dependence $D^-(S)$ of S. The union of the past and future domains of dependence is called the *full domain of dependence* $D(S)$, and the boundary of $D(S)$ is called the *Cauchy horizon* $H(S)$.

A very important application of these concepts is the notion of a *Cauchy surface*, which is an achronal (*i.e.* spacelike or null) set Σ such that its full domain of dependence is the whole manifold, or in other words its Cauchy horizon is empty $H(\Sigma) = \emptyset$. In physical terms, given a Cauchy surface we can predict the whole history of the spacetime both to the future and to the past.[19] A Cauchy surface can be considered an *instant in time*, but notice that if a Cauchy surface exists it will necessarily be highly non-unique, as we could clearly deform it slightly everywhere to obtain a new Cauchy surface (which corresponds to the fact that in relativity there is no absolute notion of simultaneity). Not all conceivable spacetimes have a Cauchy surface. When a Cauchy surface does exist we say that the spacetime is *globally hyperbolic*. It is believed that all physically relevant spacetimes are globally hyperbolic, and indeed this is one of the basic assumptions in the formulations of general relativity used in numerical work.

An important property of general relativity is the fact that solutions to Einstein's equations can have physical singularities. Two examples of such singularities are the singularity at $r = 0$ on Schwarzschild's spacetime, and the singularity at the beginning of time in the *Big Bang* cosmological models. For some time it was hoped that such singularities were a consequence of the high degree of symmetry in such solutions, but in the 1960s the work of Penrose and Hawking showed that this was not the case and that singularities are a generic feature of general relativity under certain circumstances. We will not discuss these issues here in detail, but some results are of relevance in numerical relativity.

The first point that needs to be considered is the fact that nothing can be said

[19]This corresponds to the idea of causal determinism in classical mechanics: If we know the physical state of the whole Universe now, we can predict completely the future and the past.

about generic solutions to the Einstein field equations unless something is said first about the properties of the matter that is the source of the gravitational field. Notice that any spacetime can be considered a "solution" of Einstein's equations by simply defining the corresponding stress-energy tensor as $T_{\mu\nu} := G_{\mu\nu}/8\pi$, but doing this we would generally find a stress-energy tensor that corresponds to a completely nonsensical form of matter. There are a series of conditions that can be imposed on the stress-energy tensor that are satisfied by all known forms of matter. These so-called *energy conditions* are not physical laws as such, but they are rather assumptions about how any reasonable form of matter should behave.[20] The three most common energy conditions are the following:

1. *Weak energy condition*: The energy density seen by all observers should be non-negative. That is, $T_{\mu\nu}u^\mu u^\nu \geq 0$ for any unit timelike vector u^μ.
2. *Strong energy condition*: The energy density plus the sum of the principal pressures must be non-negative (for a perfect fluid $\rho + 3p \geq 0$). In covariant terms this condition is stated as $T_{\mu\nu}u^\mu u^\nu + T/2 \geq 0$, with u^μ an arbitrary unit timelike vector and $T \equiv T^\mu{}_\mu$. The field equations imply that this condition is equivalent to $R_{\mu\nu}u^\mu u^\nu \geq 0$.
3. *Null energy condition*: The energy density plus any of the principal pressures must be non-negative (for a perfect fluid $\rho + p \geq 0$). In covariant terms this takes the form $T_{\mu\nu}k^\mu k^\nu \geq 0$ for any null vector k^μ, which through the field equations is equivalent to $R_{\mu\nu}k^\mu k^\nu \geq 0$.

The strong and weak energy conditions are independent of each other, but they both imply the null energy condition.

In order to see the relevance of the energy conditions in the context of gravitational collapse; we will start by considering the two-dimensional boundary S of a closed region in a three-dimensional spatial hypersurface Σ. Let s^μ be the unit spacelike vector orthogonal to S in Σ, and n^μ the timelike unit vector orthogonal to Σ. Define now

$$l^\mu := (n^\mu + s^\mu)/\sqrt{2}\,, \qquad k^\mu := (n^\mu - s^\mu)/\sqrt{2}\,. \tag{1.17.1}$$

The vectors l^μ and k^μ are clearly null, and correspond to the tangent vectors to the congruence of outgoing and ingoing null geodesics through S. The projection operator onto the surface S is given by

$$P^\mu_\nu = \delta^\mu_\nu + n^\mu n_\nu - s^\mu s_\nu\,, \tag{1.17.2}$$

[20]Energy conditions, however, are notoriously problematic when dealing with quantum fields, as such fields often violate them. Worse, even quite reasonable looking classical scalar fields can violate all energy conditions (see *e.g.* [47]). The issue of the range of applicability and general relevance of the energy conditions remains a largely open problem.

where the second term projects an arbitrary vector onto Σ, and the third projects it onto S once it is in Σ. Define now the tensor $\kappa_{\mu\nu}$ as the projection of the covariant derivatives of the null vector l^μ:[21]

$$\kappa_{\mu\nu} := P^\alpha_\mu P^\beta_\nu \nabla_\alpha l_\beta \, . \tag{1.17.3}$$

The expansion of the congruence of outgoing null geodesics is then defined as the trace of $\kappa_{\mu\nu}$

$$\theta := \kappa^\mu{}_\mu \, . \tag{1.17.4}$$

The expansion θ measures the increase in the separation of the null geodesics as they leave the surface. If the expansion is negative everywhere on S, then we say that the surface is *trapped*, and if it is zero everywhere we say that the surface is *marginally trapped*. Physically this means that if light rays are sent outward from a trapped surface, then a moment later the volume enclosed by them will be smaller than the initial volume, *i.e* they are all getting closer. Notice that in flat space there can be no trapped surfaces (the outgoing light rays must be separating somewhere for any closed two-dimensional surface), but in the case of Schwarzschild's spacetime we find that all spheres with $r < 2M$ are in fact trapped ("outgoing" light rays actually fall in).

A crucial result of Penrose states that in a globally hyperbolic spacetime that satisfies the null energy condition, the existence of a trapped surface implies that a singularity will form in the future (for a formal proof see [161] or [295]). In other words, if a trapped surface ever develops in the dynamical evolution of a strong gravitational field, then gravitational collapse to a singularity is inevitable.

The previous result establishes the fact that singularities are inevitable in gravitational collapse, however, it does not guarantee that a black hole will form as the singularity can be *naked*, *i.e.* not protected by the presence of a horizon. If a horizon is absent then the singularity can be causally connected to the rest of the Universe which will mark the breakdown of predictability, as in principle anything could come out of a singularity (the field equations make no sense there). In other words the presence of a naked singularity would imply that the spacetime is not be globally hyperbolic. It turns out that in most cases studied so far naked singularities do not develop, which has given rise to the *cosmic censorship conjecture* which basically says that all physically reasonable spacetimes are globally hyperbolic, and that apart from a possible initial Big Bang type singularity all other singularities are hidden inside black hole horizons.

The cosmic censorship conjecture has not been proven except in some special cases, and there are in fact a number of counterexamples to it, but all of them are non-generic in the sense that naked singularities only seem to occur for finely tuned sets of initial data. Still, the issue has not been settled and there are

[21] The tensor $\kappa_{\mu\nu}$ corresponds to the *extrinsic curvature* of the null hypersurface generated by the outgoing null geodesics from S. See Chapter 2 for a more detailed discussion of the extrinsic curvature.

still attempts to find generic counterexamples to cosmic censorship. Numerical relativity can play a major role in guiding further developments in this area.

There is still one final issue to discuss related to the formal definition of a black hole. The presence of a black hole must imply that there is a *horizon*, that is, a boundary that separates a region where light can escape to infinity from a region where it can not. The precise definition of a black hole involves the notion of *conformal infinity*. This notion is very important in the modern theoretical framework of general relativity, but its precise definition would require a full Chapter on its own so here we will just mention some of the basic ideas. We start by considering a conformal transformation of our physical spacetime M to an unphysical, or conformal, spacetime $\bar{M}$ such that

$$\bar{g}_{\mu\nu} = \Omega^2 g_{\mu\nu} \, . \tag{1.17.5}$$

The conformal factor Ω is a smooth function that vanishes at the boundary of some finite region in $\bar{M}$, which through the conformal transformation above can be seen to correspond to events that are infinitely far away in the physical spacetime M. This is then considered to be the boundary of spacetime and is called conformal infinity. The transformation is constructed in such a way that conformal infinity has a series of properties. In the first place, all spacelike curves that are infinitely extended in M intersect the boundary at the same point i^0 in $\bar{M}$, known as *spacelike infinity.* [22] In the neighborhood of i^0 we have a region of the boundary corresponding to null curves that are infinitely extended either to the future or to the past, called *future null infinity* $\mathscr{J}^+$ and *past null infinity* $\mathscr{J}-$, respectively (the symbol $\mathscr{J}$ is referred to as *scri*). Null infinity in $\bar{M}$ is formed by a series of spheres with decreasing radius as we move away from i^0. In simple spacetimes like Minkowski, as we move to the future the spheres corresponding to $\mathscr{J}^+$ touch at a point i^+ called *future timelike infinity* where all timelike curves infinitely extended to the future intersect, and similarly as we move to the past the spheres corresponding to $\mathscr{J}^-$ touch at the point i^- called *past timelike infinity* where all timelike curves infinitely extended to the past intersect. Figure 1.7 shows the conformal diagram for Minkowski spacetime.

We can use the concept of conformal infinity to define a black hole in the following way: A globally hyperbolic spacetime is said to contain a black hole if the past domain of dependence of future null infinity $D^-(\mathscr{J}^+)$ is not the whole spacetime. In other words, there are future directed null lines that do not reach infinity, they end instead at the singularity. Figure 1.8 shows the conformal diagram of Schwarzschild spacetime. In this diagram, null geodesics are lines at 45 degrees, and we can clearly see that some of those lines end up at the singularity. The boundary of the region of spacetime not covered by the past domain of dependence of $\mathscr{J}^+$ is called the *event horizon.*

[22] Spacelike infinity is truly only one point, independently of the direction of the spatial curve considered, so that the topology of $\bar{M}$ closes at i^0 in the outgoing radial direction.

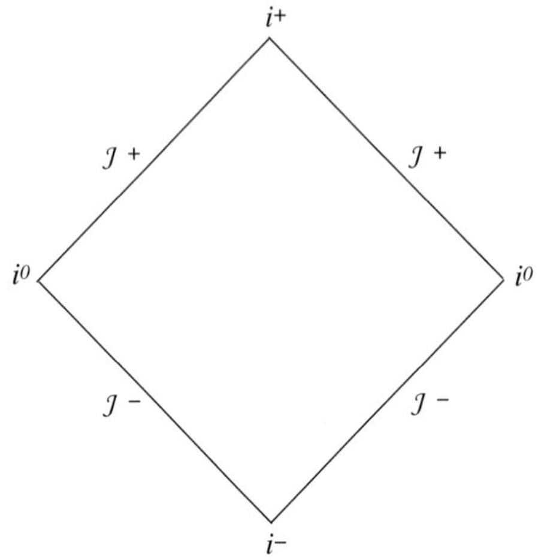

Fig. 1.7: Conformal diagram for Minkowski spacetime. The angular coordinates are suppressed, so that each point on the diagram represents a sphere in spacetime. The exception is the point i^0 which is really only one point even though it appears twice in the diagram (the correct picture is obtained by wrapping the diagram around a cylinder in such a way that the two points i^0 touch).

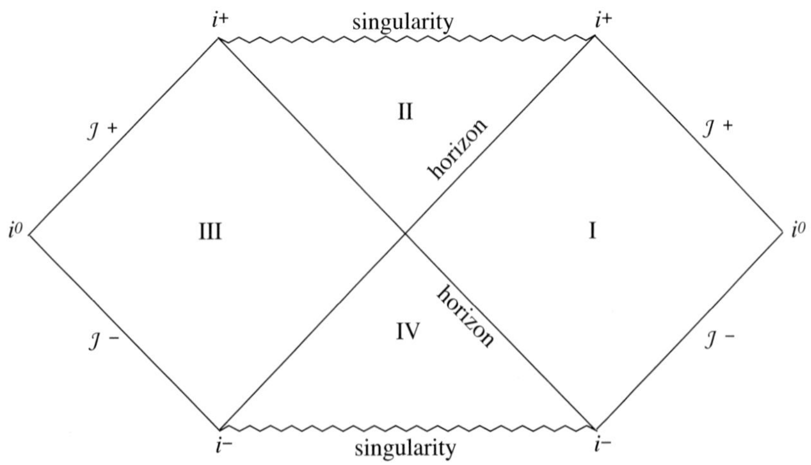

Fig. 1.8: Conformal diagram for Schwarzschild spacetime.

The event horizon marks the true boundary of a black hole, but it is clear that, if one wishes to locate an event horizon (assuming there is one), we must know the entire history of spacetime in order to be able to decide which outgoing null lines escape to infinity and which do not. The event horizon is therefore a non-local notion.

When we are considering the dynamical evolution of a spacetime from some initial data as is done in numerical relativity, it is convenient to have a more local criteria that can be used to determine the presence of a black hole at any given point in time. This is achieved by the notion of an *apparent horizon*, which is defined as the outermost marginally trapped surface on the spacetime. A crucial property of apparent horizons is the fact that if the cosmic censorship conjecture holds, and the null energy condition is satisfied, then the presence of an apparent horizon implies the existence of an event horizon that lies outside, or coincides with, the apparent horizon. In the static Schwarzschild spacetime the apparent and event horizons in fact coincide, and we can talk simply of the "horizon".

2

THE 3+1 FORMALISM

2.1 Introduction

The Einstein field equations for the gravitational field described in the previous Chapter are written in a fully covariant way, where there is no clear distinction between space and time. This form of the equations is quite natural from the point of view of differential geometry, and has important implications in our understanding of the relationship between space and time. However, there are situations when we would like to recover a more intuitive picture where we can think of the dynamical evolution of the gravitational field in "time". For example, we could be interested in finding the future evolution of the gravitational field associated with an astrophysical system given some appropriate initial data. On the other hand, we might also be interested in studying gravity as a field theory similar to electrodynamics, and define a Hamiltonian formulation that can be used, for example, as the starting point for a study of the quantization of the gravitational field.

There exist several different approaches to the problem of separating the Einstein field equations in a way that allows us to give certain initial data, and from there obtain the subsequent evolution of the gravitational field. Specific formalisms differ in the way in which this separation is carried out. Here we will concentrate on the *3+1 formalism*, where we split spacetime into three-dimensional space on the one hand, and time on the other. The 3+1 formalism is the most commonly used in numerical relativity, but it is certainly not the only one. The two main alternatives to the 3+1 approach are known as the *characteristic formalism* where spacetime is separated into light-cones emanating from a central timelike world-tube, and the *conformal formalism* where we use hyperboloidal slices that are everywhere spacelike but intersect asymptotic null infinity, *plus* a conformal transformation that brings the boundary of spacetime to a finite distance in coordinate space. Both these alternatives will be discussed briefly in Section 2.9.

We should also mention yet another approach that is based on evolving the full four-dimensional spacetime metric directly by simply expanding out the Einstein equations is some adequate coordinate system. Indeed, this was the original approach taken by Hahn and Lindquist in their pioneering work on numerical relativity [158], and has also been recently used by Pretorius with considerable success in the context of the collision of two orbiting black holes [231]. The different formalisms have advantages and disadvantages depending on the specific physical system under consideration.

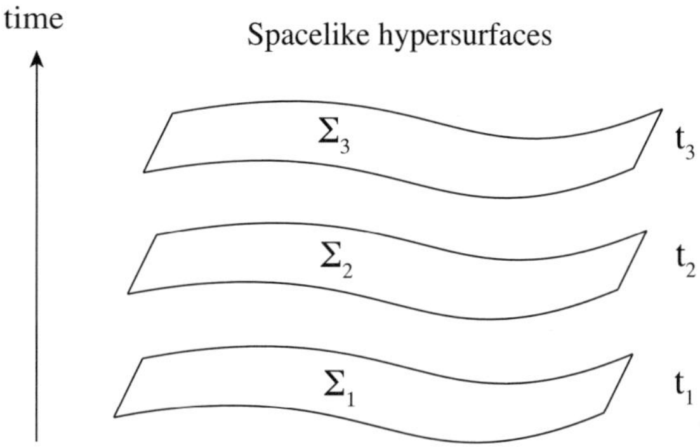

Fig. 2.1: Foliation of spacetime into three-dimensional spacelike hypersurfaces.

In the following sections I will introduce the 3+1 formalism of general relativity. The discussion found here can be seen in more detail in [206] and [305].

2.2 3+1 split of spacetime

In order to study the evolution in time of any physical system the first thing that needs to be done is to formulate such an evolution as an *initial value* or *Cauchy* problem: Given adequate initial (and boundary) conditions, the fundamental equations must predict the future (or past) evolution of the system.

When trying to write Einstein's equations as a Cauchy problem we immediately encounter a stumbling block: The field equations are written in such a way that space and time are treated on an equal footing. This covariance is very important (and quite elegant) from a theoretical point of view, but it does not allow us to think clearly about the evolution of the gravitational field in time. Therefore, the first thing we need to do in order to rewrite Einstein's equations as a Cauchy problem is to split the roles of space and time in a clear way. The formulation of general relativity that results from this splitting is known as the *3+1 formalism*.

Let us start by considering a spacetime with metric $g_{\alpha\beta}$. As already mentioned in Chapter 1, we will always assume that the spacetimes of interest are globally hyperbolic, that is, they have a Cauchy surface. Any globally hyperbolic spacetime can be completely foliated (*i.e.* sliced into three-dimensional cuts) in such a way that each three-dimensional slice is spacelike (see Figure 2.1). We can identify the foliation with the level sets of a parameter t which can then be considered a *universal time function* (but we should keep in mind that t will not necessarily coincide with the proper time of any particular observer). Because of this fact, such a foliation of spacetime into spatial hypersurfaces is often also called a *synchronization*.

Consider now a specific foliation, and take two adjacent hypersurfaces Σ_t and Σ_{t+dt}. The geometry of the region of spacetime contained between these

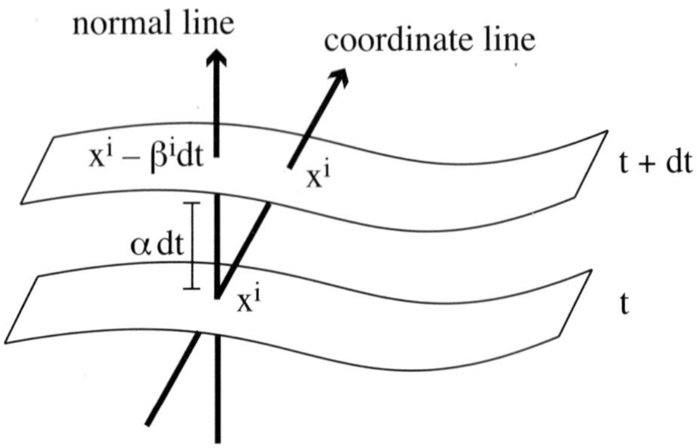

Fig. 2.2: Two adjacent spacelike hypersurfaces. The figure shows the definitions of the lapse function α and the shift vector β^i.

two hypersurfaces can be determined from the following three basic ingredients (see Figure 2.2):

- The three-dimensional metric γ_{ij} $(i, j = 1, 2, 3)$ that measures proper distances within the hypersurface itself:

$$dl^2 = \gamma_{ij}\, dx^i dx^j \,. \tag{2.2.1}$$

- The lapse of proper time $d\tau$ between both hypersurfaces measured by those observers moving along the direction normal to the hypersurfaces (the so-called *normal* or *Eulerian* observers):

$$d\tau = \alpha(t, x^i)\, dt \,. \tag{2.2.2}$$

  Here α is known as the *lapse function.*

- The relative velocity β^i between the Eulerian observers and the lines of constant spatial coordinates:

$$x^i_{t+dt} = x^i_t - \beta^i(t, x^j)\, dt \,, \qquad \text{(for Eulerian observers)} \tag{2.2.3}$$

  The 3-vector β^i is known as the *shift vector.*

Notice that both the way in which spacetime is foliated, and also the way in which the spatial coordinate system propagates from one hypersurface to the next, are not unique. The lapse function α and the shift vector β^i are therefore freely specifiable functions that carry information about our choice of coordinate system, and are known as the *gauge functions.*[23]

[23]The notation for lapse and shift used here is common, but certainly not universal. A frequently used alternative is to denote the lapse function by N, and the shift vector by N^i.

In terms of the functions $\{\alpha, \beta^i, \gamma_{ij}\}$, the metric of spacetime can be easily seen to take the following form:

$$ds^2 = \left(-\alpha^2 + \beta_i\beta^i\right) dt^2 + 2\beta_i \, dt dx^i + \gamma_{ij} \, dx^i dx^j \,, \tag{2.2.4}$$

where we have defined $\beta_i := \gamma_{ij} \beta^j$ (from here on we will assume that indices of purely spatial tensors are raised and lowered with the spatial metric γ_{ij}). The last equation is known as the 3+1 split of the metric.

More explicitly we have:

$$g_{\mu\nu} = \begin{pmatrix} -\alpha^2 + \beta_k\beta^k & \beta_i \\ \beta_j & \gamma_{ij} \end{pmatrix} , \tag{2.2.5}$$

$$g^{\mu\nu} = \begin{pmatrix} -1/\alpha^2 & \beta^i/\alpha^2 \\ \beta^j/\alpha^2 & \gamma^{ij} - \beta^i\beta^j/\alpha^2 \end{pmatrix} . \tag{2.2.6}$$

From the above expressions we can also show that the four-dimensional volume element in 3+1 language turns out to be given by

$$\sqrt{-g} = \alpha\sqrt{\gamma} \,, \tag{2.2.7}$$

with g and γ the determinants of $g_{\mu\nu}$ and γ_{ij} respectively.

Consider now the unit normal vector n^μ to the spatial hypersurfaces. It is not difficult to show that, in the coordinate system just introduced, this vector has components given by

$$n^\mu = \left(1/\alpha, -\beta^i/\alpha\right) \,, \qquad n_\mu = (-\alpha, 0) \,. \tag{2.2.8}$$

Note that this unit normal vector corresponds by definition to the 4-velocity of the Eulerian observers.

We can use the normal vector n^μ to introduce the 3+1 quantities in a more formal way that is not tied up with the choice of a coordinate system adapted to the foliation. The spatial metric γ_{ij} is simply defined as the metric induced on each hypersurface Σ by the full spacetime metric $g_{\mu\nu}$:

$$\gamma_{\mu\nu} = g_{\mu\nu} + n_\mu n_\nu \,. \tag{2.2.9}$$

Notice that written in this way the spatial metric is a full four-dimensional tensor, but when written in the adapted coordinates its time components become trivial. Also, the last expression shows that the spatial metric is nothing more than the projection operator onto the spatial hypersurfaces.

Consider now our global time function t associated with the foliation. The lapse function is defined as

$$\alpha = \left(-\nabla t \cdot \nabla t\right)^{-1/2} \,. \tag{2.2.10}$$

(The vector ∇t is clearly timelike because the level sets of t are spacelike). The unit normal vector to the hypersurfaces can then be expressed in terms of α and t as

$$n^\mu = -\alpha \nabla^\mu t \,, \tag{2.2.11}$$

where the minus sign is there to guarantee that $\vec{n}$ is future pointing.

For the definition of the shift vector we start by introducing three scalar functions β^i such that when we move from a given hypersurface to the next following the normal direction, the change in the spatial coordinates is given as before by

$$x^i_{t+dt} = x^i_t - \beta^i dt \,, \tag{2.2.12}$$

from which we can easily find

$$\beta^i = -\alpha \left(\vec{n} \cdot \nabla x^i \right) \,, \tag{2.2.13}$$

Thus defined, the β^i are scalars, but we can use them to define a 4-vector $\vec{\beta}$ by asking for its components in the adapted coordinate system to be given by $\beta^\mu = (0, \beta^i)$. The vector constructed in this way is clearly orthogonal to $\vec{n}$. We can then use the vectors $\vec{n}$ and $\vec{\beta}$ to construct a *time vector* $\vec{t}$ defined as

$$t^\mu := \alpha n^\mu + \beta^\mu \,. \tag{2.2.14}$$

The vector $\vec{t}$ is nothing more than the tangent vector to the *time lines*, *i.e.* the lines of constant spatial coordinates. Notice that, in general, we have $t^\mu \neq \nabla^\mu t$. From the above definition we find that $\vec{t}$ is such that $t^\mu n_\mu = -\alpha$, which implies

$$t^\mu \nabla_\mu t = 1 \,. \tag{2.2.15}$$

We then find that the shift is nothing more than the projection of $\vec{t}$ onto the spatial hypersurface

$$\beta_\mu := \gamma_{\mu\nu} t^\nu \,. \tag{2.2.16}$$

From this we see that we can introduce the shift vector in a completely coordinate-independent way by first choosing a vector field $\vec{t}$ satisfying (2.2.15), and then defining the shift through (2.2.16). It is important to stress the fact that the vector field $\vec{t}$ does not need to be timelike – it can easily be null or even spacelike (we will later see that this situation frequently arises in the case of black hole spacetimes). All we need to ask is that $\vec{t}$ is not tangent to the spatial hypersurfaces, and that it points to the future, which is precisely the content of equation (2.2.15). It might seem strange to allow $\vec{t}$ to be spacelike, since that would correspond to a faster than light motion of the coordinate lines, that is, a *superluminal* or *tachionic* shift. But we must remember that it is not a physical effect but only the coordinate lines that are moving "faster than light", and the coordinates can be chosen freely.

2.3 Extrinsic curvature

When considering the spatial hypersurfaces that constitute the foliation of spacetime, we need to distinguish between the intrinsic curvature of those hypersurfaces coming from their internal geometry, and the *extrinsic* curvature associated with the way in which those hypersurfaces are immersed in four-dimensional spacetime. The intrinsic curvature is given by the three-dimensional Riemann

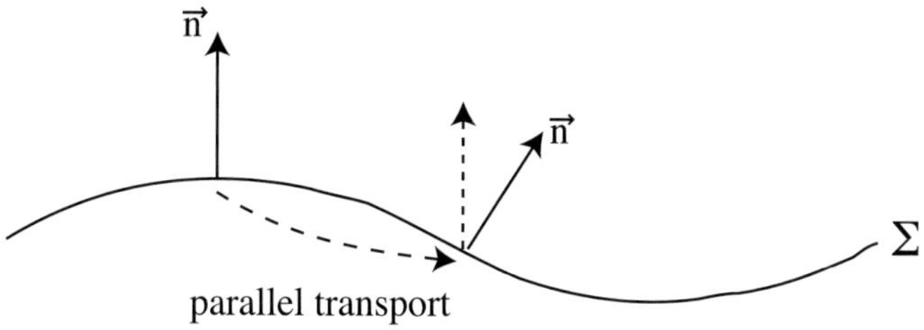

Fig. 2.3: The extrinsic curvature tensor is defined as a measure of the change of the normal vector under parallel transport.

tensor defined in terms of the 3-metric γ_{ij}. The extrinsic curvature, on the other hand, is defined in terms of what happens to the normal vector $\vec{n}$ as it is parallel-transported from one point in the hypersurface to another. In general, we will find that as we parallel transport this vector to a nearby point, the new vector will not be normal the hypersurface anymore. The *extrinsic curvature tensor* $K_{\alpha\beta}$ is a measure of the change of the normal vector under such parallel transport (see Figure 2.3).

In order to define the extrinsic curvature, we need to introduce the projection operator P^α_β onto the spatial hypersurfaces:

$$P^\alpha_\beta := \delta^\alpha_\beta + n^\alpha n_\beta \, , \tag{2.3.1}$$

which as we have seen is in fact nothing more than the induced spatial metric, $P_{\alpha\beta} = \gamma_{\alpha\beta}$. Using this projection operator, the extrinsic curvature tensor is defined as:

$$K_{\mu\nu} := -P^\alpha_\mu \, \nabla_\alpha n_\nu = -\left(\nabla_\mu n_\nu + n_\mu n^\alpha \nabla_\alpha n_\nu\right) \, , \tag{2.3.2}$$

As defined above, the tensor $K_{\mu\nu}$ is clearly a purely spatial tensor, that is, $n^\mu K_{\mu\nu} = n^\nu K_{\mu\nu} = 0$. This means, in particular, that in a coordinate system adapted to the foliation we will have $K^{00} = K^{0i} = 0$ (though in general we find that K_{00} and K_{0i} are not zero). Because of this, we will usually only consider the spatial components of K_{ij}. Moreover, the tensor $K_{\mu\nu}$ also turns out to be symmetric:

$$K_{\mu\nu} = K_{\nu\mu} \, . \tag{2.3.3}$$

A couple of remarks are important regarding the definition of $K_{\mu\nu}$. First, notice that the projection of $\nabla_\mu n_\nu$ is crucial in order to make $K_{\mu\nu}$ purely spatial. We could argue that because n^μ is unitary, its gradient is necessarily orthogonal to it. This is of course true in the sense that $n^\nu \nabla_\mu n_\nu = 0$, but $\nabla_\mu n_\nu$ is in general not symmetric and $n^\nu \nabla_\nu n_\mu \neq 0$ unless the normal lines are geodesic (which is not always the case). Let us now consider the symmetry of $K_{\mu\nu}$. As just mentioned $\nabla_\mu n_\nu$ is not in general symmetric even though n^μ is hypersurface orthogonal. The reason for this is that n^α is a unitary vector and thus in general is not equal to the gradient of the time function t, except when the lapse is unity. However, once we project onto the hypersurface, it turns out that $P^\alpha_\mu \nabla_\alpha n_\nu$ is indeed symmetric (the non-symmetry of $\nabla_\mu n_\nu$ has to do with the lapse, which is not intrinsic to the hypersurface). In order to see this, consider the congruence of timelike *geodesics* orthogonal to Σ, with unit tangent vector $\vec{\xi}$. In the neighborhood of Σ consider a new foliation of spacetime given by a time function $\tilde{t}$ such that $\xi_\mu = \nabla_\mu \tilde{t}$. Since $\vec{\xi}$ is the gradient of a scalar function we clearly will have $\nabla_\mu \xi_\nu = \nabla_\nu \xi_\mu$. Moreover, $\nabla_\mu \xi_\nu$ will be purely spatial without the need to project it since $\vec{\xi}$ is tangent to the timelike geodesics. Now, the vector field $\vec{n}$ will generally not coincide with $\vec{\xi}$ outside of Σ, but it will coincide within Σ, so its derivatives along directions tangential to Σ must be equal to those of $\vec{\xi}$, that is

$$P^\alpha_\mu \nabla_\alpha n_\nu = P^\alpha_\mu \nabla_\alpha \xi_\nu = \nabla_\mu \xi_\nu = \nabla_\nu \xi_\mu = P^\alpha_\nu \nabla_\alpha \xi_\mu = P^\alpha_\nu \nabla_\alpha n_\mu \, , \tag{2.3.4}$$

which proves the symmetry of $K_{\mu\nu}$.

Notice, in particular, that the symmetry of the projected part of $\nabla_\mu n_\nu$ also implies that when we contract it with a purely spatial antisymmetric tensor we will obtain zero, that is

$$T^{\mu\nu} \nabla_\mu n_\nu = 0 \, , \tag{2.3.5}$$

for all $T_{\mu\nu}$ such that $T_{\mu\nu} = -T_{\nu\mu}$, $n^\mu T_{\mu\nu} = 0$.

From its definition above, it is in fact not difficult to show that the extrinsic curvature $K_{\mu\nu}$ can be written in an entirely equivalent way as the Lie derivative of the spatial metric along the normal direction

$$K_{\mu\nu} = -\frac{1}{2} \pounds_{\vec{n}} \, \gamma_{\mu\nu} \, . \tag{2.3.6}$$

That this is so can be easily seen from the definition of the Lie derivative:

$$\begin{aligned}
\pounds_{\vec{n}} \, \gamma_{\mu\nu} &= n^\alpha \nabla_\alpha \gamma_{\mu\nu} + \gamma_{\mu\alpha} \nabla_\nu n^\alpha + \gamma_{\nu\alpha} \nabla_\mu n^\alpha \\
&= n^\alpha \nabla_\alpha \left(n_\mu n_\nu \right) + g_{\mu\alpha} \nabla_\nu n^\alpha + g_{\nu\alpha} \nabla_\mu n^\alpha \\
&= n^\alpha n_\mu \nabla_\alpha n_\nu + n^\alpha n_\nu \nabla_\alpha n_\mu + \nabla_\nu n_\mu + \nabla_\mu n_\nu \\
&= \left(\gamma^\alpha_\mu - g^\alpha_\mu \right) \nabla_\alpha n_\nu + \left(\gamma^\alpha_\nu - g^\alpha_\nu \right) \nabla_\alpha n_\mu + \nabla_\nu n_\mu + \nabla_\mu n_\nu \\
&= \gamma^\alpha_\mu \nabla_\alpha n_\nu + \gamma^\alpha_\nu \nabla_\alpha n_\mu = -2 K_{\mu\nu} \, ,
\end{aligned} \tag{2.3.7}$$

where we have used the fact that the covariant derivative of $g_{\mu\nu}$ is zero and also that $n^\alpha \nabla_\mu n_\alpha = 0$. We then find that the extrinsic curvature is essentially the "velocity" of the spatial metric as seen by the Eulerian observers. Notice that the extrinsic curvature depends only on the behavior of $\vec{n}$ within the slice Σ – it is therefore a geometric property of the slice itself.

Now, since $\vec{n}$ is normal to the hypersurface, it turns out that for any scalar function ϕ we have

$$\pounds_{\vec{n}} \, \gamma_{\mu\nu} = \frac{1}{\phi} \, \pounds_{\phi \vec{n}} \, \gamma_{\mu\nu} \, . \tag{2.3.8}$$

If, in particular, we take as our scalar function the lapse, we find that

$$K_{\mu\nu} = -\frac{1}{2\alpha} \pounds_{\alpha\vec{n}} \gamma_{\mu\nu} = -\frac{1}{2\alpha} \left(\pounds_{\vec{t}} - \pounds_{\vec{\beta}} \right) \gamma_{\mu\nu} , \tag{2.3.9}$$

which implies

$$\left(\pounds_{\vec{t}} - \pounds_{\vec{\beta}} \right) \gamma_{\mu\nu} = -2\alpha K_{\mu\nu} . \tag{2.3.10}$$

Concentrating now only on the spatial components, and remembering that in the adapted coordinate system we have $\pounds_{\vec{t}} = \partial_t$, we finally find

$$\partial_t \gamma_{ij} - \pounds_{\vec{\beta}} \gamma_{ij} = -2\alpha K_{ij} . \tag{2.3.11}$$

The last expression can also be rewritten as

$$\partial_t \gamma_{ij} = -2\alpha K_{ij} + D_i \beta_j + D_j \beta_i , \tag{2.3.12}$$

where here D_i represents the three-dimensional covariant derivative, that is, the one associated with the 3-metric γ_{ij}, which is in fact nothing more than the projection of the full four-dimensional covariant derivative: $D_\mu := P^\alpha_\mu \nabla_\alpha$.

This brings us half-way to our goal of writing Einstein's equations as a Cauchy problem: We already have an evolution equation for the spatial metric γ_{ij}. In order to close the system we still need an evolution equation for K_{ij}. It is important to notice that until now we have only worked with purely geometric concepts, and we have not used the Einstein field equations at all. It is precisely from the field equations that we will obtain the evolution equations for K_{ij}. In other words, the evolution equation (2.3.11) for the 3-metric is purely kinematic, while the dynamics of the system will be contained in the evolution equations for K_{ij}.

2.4 The Einstein constraints

The true dynamics of the gravitational field are contained in the Einstein field equations. In order to proceed further, we need to rewrite these equations in 3+1 language. This can be done by considering contractions of these equations with the normal vector $\vec{n}$ and with the projector operator onto the hypersurface P^μ_ν. We will proceed with this in two stages. In this Section we will consider those contractions involving the normal vector, and will leave the rest of the equations until the next Section.

The starting point is to express the four-dimensional Riemann curvature tensor $R^\alpha_{\beta\mu\nu}$ in terms of the intrinsic three-dimensional Riemann tensor of the hypersurface itself ${}^{(3)}R^\alpha_{\beta\mu\nu}$, and the extrinsic curvature tensor $K_{\mu\nu}$. The derivation of these relations is straightforward but rather long, so we will just state the final results here (the interested reader can see *e.g.* [295]). The full projection of the Riemann tensor onto the spatial hypersurfaces turns out to be given by the so-called *Gauss–Codazzi* equations

$$P^\delta_\alpha P^\kappa_\beta P^\lambda_\mu P^\sigma_\nu \, R_{\delta\kappa\lambda\sigma} = {}^{(3)}R_{\alpha\beta\mu\nu} + K_{\alpha\mu} K_{\beta\nu} - K_{\alpha\nu} K_{\beta\mu} , \tag{2.4.1}$$

Similarly, the projection onto the hypersurfaces of the Riemann tensor contracted once with the normal vector results in the *Codazzi–Mainardi* equations

$$P^\delta_\alpha P^\kappa_\beta P^\lambda_\mu n^\nu \; R_{\delta\kappa\lambda\nu} = D_\beta K_{\alpha\mu} - D_\alpha K_{\beta\mu} \,, \tag{2.4.2}$$

where as before $D_\mu = P^\alpha_\mu \nabla_\alpha$.

In order to start rewriting the Einstein field equations in 3+1 language, notice first that

$$\begin{aligned} P^{\alpha\mu} P^{\beta\nu} R_{\alpha\beta\mu\nu} &= \left(g^{\alpha\mu} + n^\alpha n^\mu\right)\left(g^{\beta\nu} + n^\beta n^\nu\right) R_{\alpha\beta\mu\nu} \\ &= R + 2n^\mu n^\nu R_{\mu\nu} \\ &= 2n^\mu n^\nu G_{\mu\nu} \,, \end{aligned} \tag{2.4.3}$$

with $G_{\mu\nu}$ the Einstein tensor. On the other hand, the Gauss–Codazzi relations imply that

$$P^{\alpha\mu} P^{\beta\nu} R_{\alpha\beta\mu\nu} = {}^{(3)}R + K^2 - K_{\mu\nu} K^{\mu\nu} \tag{2.4.4}$$

where $K := K^\mu_\mu$ is the trace of the extrinsic curvature. We then find

$$2\, n^\mu n^\nu G_{\mu\nu} = {}^{(3)}R + K^2 - K_{\mu\nu} K^{\mu\nu} \,, \tag{2.4.5}$$

which through the Einstein equations becomes

$${}^{(3)}R + K^2 - K_{\mu\nu} K^{\mu\nu} = 16\pi\rho \,, \tag{2.4.6}$$

where we have defined the quantity $\rho := n^\mu n^\nu T_{\mu\nu}$ that corresponds to the local energy density as measured by the Eulerian observers. Notice that this equation involves no explicit time derivatives, so it is not an evolution equation but rather a constraint that must be satisfied at all times. This equation is known as the *Hamiltonian* or *energy* constraint.

Consider now the mixed contraction of the Einstein tensor. We find,

$$P^{\alpha\mu} n^\nu G_{\mu\nu} = P^{\alpha\mu} n^\nu R_{\mu\nu} \,. \tag{2.4.7}$$

The Codazzi–Mainardi equations then imply

$$\gamma^{\alpha\mu} n^\nu G_{\mu\nu} = D^\alpha K - D_\mu K^{\alpha\mu} \,, \tag{2.4.8}$$

which again through the field equations becomes

$$D_\mu \left(K^{\alpha\mu} - \gamma^{\alpha\mu} K\right) = 8\pi j^\alpha \,, \tag{2.4.9}$$

where now $j^\alpha := -P^{\alpha\mu} n^\nu T_{\mu\nu}$ corresponds to the momentum density as measured by the Eulerian observers. Notice that there are in fact three equations in the last expression, as the index α is free, but the case $\alpha = 0$ is trivial. As before, there are no time derivatives in these equations so they are also constraints. They are known as the *momentum* constraints.

In a coordinate system adapted to the foliation, the Hamiltonian and momentum constraints take the final form

$$^{(3)}R + K^2 - K_{ij}K^{ij} = 16\pi\rho \,, \tag{2.4.10}$$

$$D_j\left(K^{ij} - \gamma^{ij}K\right) = 8\pi j^i \,, \tag{2.4.11}$$

with

$$\rho := n^\mu n^\nu T_{\mu\nu} \,, \qquad j^i := -P^{i\mu} n^\nu T_{\mu\nu} \,. \tag{2.4.12}$$

It is important to notice that the constraints not only do not involve time derivatives, but they are also completely independent of the gauge functions α and β^i. This indicates that the constraints are relations that refer purely to a given hypersurface.

Notice that having a set of constraint equations is not a feature of general relativity alone. In electrodynamics we have the Maxwell equations which in three-dimensional vector calculus notation, and in Gaussian units, take the form

$$\nabla \cdot \mathbf{E} = 4\pi\rho \,, \qquad \nabla \cdot \mathbf{B} = 0 \,, \tag{2.4.13}$$

$$\partial_t \mathbf{E} = \nabla \times \mathbf{B} - 4\pi\mathbf{j} \,, \qquad \partial_t \mathbf{B} = -\nabla \times \mathbf{E} \,, \tag{2.4.14}$$

where $\mathbf{E}$ and $\mathbf{B}$ are the electric and magnetic fields respectively, ρ is the charge density and $\mathbf{j}$ the current density (here ∇ stands for the ordinary flat space gradient operator and should not be confused with a four-dimensional covariant derivative). The first two equations involving the divergence of the electric and magnetic fields do not involve time derivatives, so they are in fact constraints, just as in general relativity. The main difference is that Maxwell's theory has only two constraint equations, while general relativity has four (one Hamiltonian constraint and three momentum constraints). The remaining two Maxwell equations (or rather six since they are vector-valued equations) are the true evolution equations for electrodynamics. The corresponding equations for gravity will be derived in the next Section.

The existence of the constraints implies, in particular, that in the 3+1 formulation it is not possible to specify arbitrarily all 12 dynamical quantities $\{\gamma_{ij}, K_{ij}\}$ as initial conditions. The initial data must already satisfy the constraints, otherwise we will not be solving Einstein's equations. We will come back to this issue in Chapter 3 when we discuss how to find initial data. The constraints also play other important roles. For example, they are crucial in the Hamiltonian formulation of general relativity (see Section 2.7). They are also very important in the study of the well-posedness of the system of evolution equations, something we will have a chance to discuss briefly in Section 2.8 and also in Chapter 5.

2.5 The ADM evolution equations

The Hamiltonian and momentum constraints give us four of the ten independent Einstein field equations, and as we have seen they do not correspond to the evolution equations of the gravitational field, but rather to relations between the dynamical variables that must be satisfied at all times. The evolution of the gravitational field is contained in the remaining six field equations.

In order to find these equations we still need the projection onto the hypersurfaces of the Riemann tensor contracted twice with the normal vector. This will give us the last six independent components of Riemann (the Gauss–Codazzi and Gauss–Mainardi equations give us 14 components of Riemann). These projections turn out to be given by

$$P^\delta_\mu P^\kappa_\nu n^\lambda n^\sigma R_{\delta\lambda\kappa\sigma} = \pounds_{\vec{n}} K_{\mu\nu} + K_{\mu\lambda} K^\lambda_\nu + \frac{1}{\alpha} D_\mu D_\nu \alpha \,. \tag{2.5.1}$$

The first thing to notice is that these relations do involve the lapse function α. Also, they make reference to the Lie derivative of the extrinsic curvature along the normal direction, which clearly corresponds to evolution in time.

Now, from the Gauss–Codazzi equations (2.4.1) we also find

$$P^\delta_\mu P^\kappa_\nu \left(n^\lambda n^\sigma R_{\delta\lambda\kappa\sigma} + R_{\delta\kappa} \right) = {}^{(3)}R_{\mu\nu} + K K_{\mu\nu} - K_{\mu\lambda} K^\lambda_\nu \,, \tag{2.5.2}$$

which, together with (2.5.1), implies

$$\begin{aligned} \pounds_{\vec{t}} K_{\mu\nu} - \pounds_{\vec{\beta}} K_{\mu\nu} &= -D_\mu D_\nu \alpha \\ &\quad + \alpha \left(-P^\delta_\mu P^\kappa_\nu R_{\delta\kappa} + {}^{(3)}R_{\mu\nu} + K K_{\mu\nu} - 2 K_{\mu\lambda} K^\lambda_\nu \right) \,, \end{aligned} \tag{2.5.3}$$

Using now the Einstein equations written in terms of $R_{\mu\nu}$, equations (1.13.4), we find

$$\begin{aligned} \pounds_{\vec{t}} K_{\mu\nu} - \pounds_{\vec{\beta}} K_{\mu\nu} &= -D_\mu D_\nu \alpha + \alpha \left[{}^{(3)}R_{\mu\nu} + K K_{\mu\nu} - 2 K_{\mu\lambda} K^\lambda_\nu \right] \\ &\quad + 4\pi\alpha \left[\gamma_{\mu\nu} \left(S - \rho \right) - 2 S_{\mu\nu} \right] \,, \end{aligned} \tag{2.5.4}$$

where ρ is the same as before, and where we have defined $S_{\mu\nu} := P^\alpha_\mu P^\beta_\nu T_{\alpha\beta}$ as the spatial stress tensor measured by the Eulerian observers (with $S := S^\mu_\mu$). Concentrating on the spatial components and again writing $\pounds_{\vec{t}} = \partial_t$, the last expression becomes

$$\begin{aligned} \partial_t K_{ij} - \pounds_{\vec{\beta}} K_{ij} &= -D_i D_j \alpha + \alpha \left[{}^{(3)}R_{ij} + K K_{ij} - 2 K_{ik} K^k_j \right] \\ &\quad + 4\pi\alpha \left[\gamma_{ij} \left(S - \rho \right) - 2 S_{ij} \right] \,, \end{aligned} \tag{2.5.5}$$

or expanding the Lie derivative along the shift vector

$$\begin{aligned} \partial_t K_{ij} &= \beta^k \partial_k K_{ij} + K_{ki} \partial_j \beta^k + K_{kj} \partial_i \beta^k - D_i D_j \alpha \\ &\quad + \alpha \left[{}^{(3)}R_{ij} + K K_{ij} - 2 K_{ik} K^k_j \right] + 4\pi\alpha \left[\gamma_{ij} \left(S - \rho \right) - 2 S_{ij} \right] \,. \end{aligned} \tag{2.5.6}$$

These equations give us the dynamical evolution of the six independent components of the extrinsic curvature K_{ij}. Together with equations (2.3.11) for the evolution of the spatial metric they finally allow us to write down the field equations for general relativity as a Cauchy problem. It is important to notice that we do not have evolution equations for the gauge quantities α and β^i. As we have mentioned before, these quantities represent our coordinate freedom and can therefore be chosen freely.

The evolution equations (2.5.5) are known in the numerical relativity community as the Arnowitt–Deser–Misner (ADM) equations. However, as written above, these equations are in fact not in the form originally derived by ADM [31], but they are instead a non-trivial rewriting due to York [305]. It is important to mention exactly what the difference is between the original ADM equations and the ADM equations *à la* York, which we will call from now on *standard* ADM. The two groups of equations differ in two main aspects. In the first place, the original ADM variables are the spatial metric γ_{ij} and its canonical conjugate momentum π_{ij} coming from the Hamiltonian formulation of general relativity (see the following Section), and which is related to the extrinsic curvature as [24]

$$K_{ij} = -\frac{1}{\sqrt{\gamma}} \left(\pi_{ij} - \frac{1}{2} \gamma_{ij} \, \pi \right) , \tag{2.5.7}$$

with $\pi = \pi^i_i$ and γ the determinant of γ_{ij}. This change of variables is, of course, a rather minor detail. However, even if we rewrite the original ADM evolution equations in terms of K_{ij}, they still differ from (2.5.5) and have the form

$$\begin{aligned} \partial_t K_{ij} - \pounds_{\vec{\beta}} K_{ij} &= -D_i D_j \alpha + \alpha \left[{}^{(3)}R_{ij} + K K_{ij} - 2 K_{ik} K^k_j \right] \\ &\quad + 4\pi\alpha \left[\gamma_{ij} \left(S - \rho \right) - 2 S_{ij} \right] - \frac{\alpha \gamma_{ij}}{2} \, \mathcal{H} , \end{aligned} \tag{2.5.8}$$

with $\mathcal{H}$ the Hamiltonian constraint (2.4.10) written as:

$$\mathcal{H} := \frac{1}{2} \left({}^{(3)}R + K^2 - K_{ij} K^{ij} \right) - 8\pi\rho = 0 , \tag{2.5.9}$$

and where the factor 1/2 in the definition of $\mathcal{H}$ is there for later convenience.

The difference between the ADM and York evolution equations can be traced back to the fact that the version of ADM comes from the field equations written in terms of the Einstein tensor $G_{\mu\nu}$, whereas the version of York was derived instead from the field equations written in terms of the Ricci tensor $R_{\mu\nu}$. It is clear that both sets of evolution equations for K_{ij} are physically equivalent since they only differ by the addition of a term proportional to the Hamiltonian constraint, which must vanish for any physical solution. However, the different evolution equations for K_{ij} are not *mathematically* equivalent. There are basically two reasons why this is so:

[24] One must remember that the original goal of ADM was to write a Hamiltonian formulation for general relativity that could be used as a basis for quantum gravity, and not a system of evolution equations for dynamical simulations.

1. In the first place, the space of solutions to the evolution equations is different in both cases, and only coincides for physical solutions, that is, those that satisfy the constraints. In other words, both systems are only equivalent in a subset of the full space of solutions. This subset is called the *constraint hypersurface* (but notice that this is not a hypersurface in spacetime, but instead a hypersurface in the space of solutions to the evolution equations). Of course, we could always argue that since in the end we are only interested in physical solutions, this distinction is irrelevant. This is strictly true only if we can solve the equations exactly. But in the case of numerical solutions there will always be some error that will take us out of the constraint hypersurface, and the issue then becomes not only relevant but crucial: If we move slightly off the constraint hypersurface, does the subsequent evolution remain close to it, or does it diverge rapidly away from it?
2. The second reason why both systems of evolution equations differ mathematically is related to the last point and is of greater importance. Since the Hamiltonian constraint has second derivatives of the spatial metric (hidden inside the Ricci scalar), then by adding a multiple of it to the evolution equations we are in fact altering the very structure of the differential equations.

These types of considerations take us to a fundamental observation that has today become one of the most active areas of research associated with numerical relativity: The 3+1 evolution equations are highly non-unique since we can always add to them arbitrary multiples of the constraints. The different systems of evolution equations will still coincide in the physical solutions, but might differ dramatically in their mathematical properties, and particularly in the way in which they react to small violations of the constraints (inevitable numerically). This observation is crucial, and we will come back to it both in Section 2.8, and in Chapter 5.

A final consideration about the 3+1 evolution equations has to do with the propagation of the constraints: If the constraints are satisfied initially, do they remain satisfied during the evolution? The answer to this question is, not surprisingly, yes, but it is interesting to see how this comes about. In fact it is through the Bianchi identities that the propagation of the constraints during the evolution is guaranteed. To see this, we will follow an analysis due to Frittelli [136], and define the following projections of the Einstein field equations

$$\mathcal{H} := n^\alpha n^\beta G_{\alpha\beta} - 8\pi\rho \,, \tag{2.5.10}$$

$$\mathcal{M}_\mu := -n^\alpha P^\beta_\mu G_{\alpha\beta} - 8\pi j_\mu \,, \tag{2.5.11}$$

$$\mathcal{E}_{\mu\nu} := P^\alpha_\mu P^\beta_\nu G_{\alpha\beta} - 8\pi S_{\mu\nu} \,, \tag{2.5.12}$$

with ρ, j^μ and $S_{\mu\nu}$ defined as before in terms of the stress-energy tensor $T_{\mu\nu}$. Notice that here $\mathcal{H} = 0$ corresponds precisely to the Hamiltonian constraint (2.5.9) (with the correct 1/2 factor), while $\mathcal{M}^\mu = 0$ corresponds to the momentum constraints. On the other hand, $\mathcal{E}_{\mu\nu} = 0$ reduces to the original ADM evolution equations (multiplied by 2). An important observation is that in this context York's version of the evolution equations in fact corresponds to $\mathcal{E}_{\mu\nu} - \gamma_{\mu\nu}\mathcal{H} = 0$. The Einstein field equations in terms of these quantities can then be written as

$$G_{\mu\nu} - 8\pi T_{\mu\nu} = \mathcal{E}_{\mu\nu} + n_\mu \mathcal{M}_\nu + n_\nu \mathcal{M}_\mu + n_\mu n_\nu \mathcal{H} = 0 \; , \tag{2.5.13}$$

and the (twice contracted) Bianchi identities imply

$$\nabla^\mu \left(\mathcal{E}_{\mu\nu} + n_\mu \mathcal{M}_\nu + n_\nu \mathcal{M}_\mu + n_\mu n_\nu \mathcal{H} \right) = 0 \; . \tag{2.5.14}$$

By taking the normal projection and the projection onto the hypersurface of these identities and rearranging terms, we obtain the following system of evolution equations for the constraints

$$n^\nu \nabla_\nu \mathcal{H} = -D^\nu \mathcal{M}_\nu - \mathcal{E}_{\mu\nu} D^\mu n^\nu + \mathcal{L}_{\mathcal{H}}(\mathcal{H}, \mathcal{M}_\sigma) \; , \tag{2.5.15}$$

$$n^\nu \nabla_\nu \mathcal{M}_\mu = -D^\nu \mathcal{E}_{\mu\nu} - \mathcal{E}_{\mu\nu} n^\lambda \nabla_\lambda n^\nu + \mathcal{L}_{\mathcal{M}_\mu}(\mathcal{H}, \mathcal{M}_\sigma) \; , \tag{2.5.16}$$

where the $\mathcal{L}$'s are shorthand for terms proportional to $\mathcal{H}$ and $\mathcal{M}_\mu$ that have no derivatives of these quantities. That these are evolution equations for the constraints can be seen from the fact that the terms on the left hand side are derivatives along the normal direction, *i.e.* out of the hypersurface. Notice that, if we have initial data such that $\mathcal{H} = \mathcal{M}_\mu = 0$, and we use the ADM evolution equations $\mathcal{E}_{\mu\nu} = 0$, the above equations guarantee that on the next hypersurface the constraints will still vanish. This proves that the constraints will remain satisfied. If, on the other hand, we use York's evolution equations $\mathcal{E}_{\mu\nu} = \gamma_{\mu\nu}\mathcal{H}$, the same result clearly follows. In fact, it is clear that we could take $\mathcal{E}_{\mu\nu}$ to be equal to any combination of constraints.

There is, however, an important difference in the structure of the constraint evolution equations when taking either the ADM or York's version of the 3+1 evolution equations, and we will come back to it later, in Chapter 5. For the moment, it is sufficient to mention that the constraint evolution equations are mathematically well-posed for York's system, but they are not well posed for the original ADM system, so York's system should be preferred.

2.6 Free versus constrained evolution

In the previous Sections we have separated the Einstein field equations into evolution equations and constraints equations. As already mentioned, the Bianchi identities guarantee that if the constraints are initially satisfied they will remain satisfied during the subsequent evolution. This is, however, an exact statement, and since here we are ultimately interested in numerical simulations it is important to consider the problem of how to evolve the geometric quantities γ_{ij} and K_{ij} while keeping the constraints satisfied. Ideally, we would like to have some discretized version of the 3+1 evolution equations and constraints that would guarantee that the discrete constraints remain satisfied during evolution. Unfortunately, such a discretized form of the 3+1 equations is not known to exist at this time. We must then live with the fact that not all 10 Einstein equations will remain satisfied at the discrete level during a numerical simulation. Of course, we expect that a good numerical implementation would be such that as we approach the continuum limit we would recover a solution of the full set of equations.

In practice, we can take two different approaches to the problem of choosing which set of equations to solve numerically. The first approach is known as *free evolution*, and corresponds to the case when we start with a solution of the constraint equations as initial data (see Chapter 3), and later advances in time by solving all 12 evolution equations for γ_{ij} and K_{ij}. The constraints are then only monitored to see how much they are violated during evolution, which gives a rough idea of the accuracy of the simulation. Alternatively, we can choose to solve some or all of the constraint equations at each time step for a specific subset of the components of the metric and extrinsic curvature, and evolve the remaining components using the evolution equations. This second approach is known as *constrained evolution*.

Constrained evolution is in fact ideal in situations with high degree of symmetry, like spherical symmetry for example, but is much harder to use in the case of fully three-dimensional systems. Also, the mathematical properties of a constrained scheme are much more difficult to analyze in the sense of studying the well-posedness of the system of equations (see Chapter 5). And finally, since the constraints involve elliptic equations, a constrained scheme is also far slower to solve numerically in three dimensions than a free evolution scheme. Because of all these reasons, in the remainder of the book we will always assume that we are using a free evolution scheme.

For completeness, we should mention a third alternative known as *constrained metric evolution*. In this approach, we choose some extra condition on the metric tensor, such as for example conformal flatness, and impose this condition during the whole evolution, thus simplifying considerably the equations to be solved. Such an approach has been used with some success by Wilson and Mathews in hydrodynamical simulations [300]. However, imposing extra conditions on the metric is not in general compatible with the Einstein field equations, and the results from such simulations should therefore be regarded just as approximations even in the continuum limit. For example, the condition of conformal flatness essentially eliminates the gravitational wave degrees of freedom. Whether such an approximation is good or not will depend on the specific physical system under study, and the physical information we wish to extract from the simulation.

2.7 Hamiltonian formulation

A field theoretical formulation of general relativity starts from the Hilbert Lagrangian which was already introduced in the previous Chapter:

$$L = R \,, \tag{2.7.1}$$

with R the Ricci scalar of the spacetime. As already mentioned, from a variational principle we can obtain the field equations taking this Lagrangian as the starting point.

The Lagrangian formulation of a field theory takes a covariant approach. In the first place, the Lagrangian itself must be a scalar function, and also the field equations derived from the variational principle come out in fully covariant form. A different approach is to take instead a Hamiltonian formulation of the theory. This approach has important advantages, and in particular is the starting point of quantum field theory. However, a Hamiltonian formulation requires a clear distinction to be made between space and time, so it is therefore not covariant. In field theories other than general relativity, and particularly when working on a flat spacetime background, there is already a natural way in which space and time can be split. In general relativity, on the other hand, no such natural splitting exists. However, we can take the 3+1 perspective and use this splitting as a basis to construct a Hamiltonian formulation of the theory. Of course, we can not interpret the time function t directly as a measure of the proper time of any given observer, since the spacetime metric needed in order to do this is the unknown dynamical variable under study.

The first step in a Hamiltonian formulation is to identify the configuration variables that describe the state of the field at any given time. For this purpose we will choose the spatial metric variables γ_{ij}, together with the lapse α and the co-variant shift vector β_i. We now need to rewrite the Hilbert Lagrangian in terms of these quantities and their derivatives. Notice that, from the definition of the Einstein tensor we have

$$n^\mu n^\nu G_{\mu\nu} = n^\mu n^\nu R_{\mu\nu} + \frac{1}{2} R \quad \Rightarrow \quad R = 2 \left(n^\mu n^\nu G_{\mu\nu} - n^\mu n^\nu R_{\mu\nu} \right) . \tag{2.7.2}$$

The first term on the right hand side of the last equation was already obtained from the Gauss–Codazzi relations and is given by equation (2.4.5). For the second term we use the Ricci identity that relates the commutator of covariant derivatives to the Riemann tensor (equation (1.9.3)):

$$\begin{aligned} n^\mu n^\nu R_{\mu\nu} &= n^\mu n^\nu R^\lambda{}_{\mu\lambda\nu} \\ &= n^\nu \left(\nabla_\lambda \nabla_\nu n^\lambda - \nabla_\nu \nabla_\lambda n^\lambda \right) \\ &= \nabla_\lambda \left(n^\nu \nabla_\nu n^\lambda \right) - \nabla_\nu \left(n^\nu \nabla_\lambda n^\lambda \right) - \nabla_\lambda n^\nu \nabla_\nu n^\lambda + \nabla_\nu n^\nu \nabla_\lambda n^\lambda \\ &= \nabla_\lambda \left(n^\nu \nabla_\nu n^\lambda \right) - \nabla_\nu \left(n^\nu \nabla_\lambda n^\lambda \right) - K_{\lambda\nu} K^{\lambda\nu} + K^2 . \end{aligned} \tag{2.7.3}$$

In the previous expression, we have directly identified some terms with the extrinsic curvature even though no projection operator is present. We can readily verify that the contractions in those expressions guarantee that the result follows. The Ricci scalar then takes the form

$$R = {}^{(3)}R + K_{\mu\nu} K^{\mu\nu} - K^2 - 2 \nabla_\lambda \left(n^\nu \nabla_\nu n^\lambda - n^\lambda \nabla_\nu n^\nu \right) . \tag{2.7.4}$$

The last term in this equation is a total divergence, and since in the end we are only interested in the action S which is an integral of the Lagrangian over a given volume Ω, this term can be transformed into an integral over the boundary of Ω and can therefore be ignored. The Lagrangian of general relativity in 3+1 language can then be written as

$$L = {}^{(3)}R + K_{ij}K^{ij} - K^2 \; . \tag{2.7.5}$$

Notice that L has a similar structure to that of the Hamiltonian constraint, but with the sign of the quadratic terms in the extrinsic curvature reversed.

To obtain the Lagrangian density $\mathcal{L}$ we first need to remember that the four-dimensional volume element in 3+1 adapted coordinates is given by $\sqrt{-g} = \alpha\sqrt{\gamma}$, so the Lagrangian density takes the form

$$\mathcal{L} = \alpha\sqrt{\gamma}\left({}^{(3)}R + K_{ij}K^{ij} - K^2\right) \; . \tag{2.7.6}$$

The canonical momenta conjugate to the dynamical field variables are defined now as derivatives of the Lagrangian density with respect to the velocities of those fields. For the spatial metric we have the following conjugate momenta

$$\pi^{ij} := \frac{\partial \mathcal{L}}{\partial \dot{\gamma}_{ij}} \; , \tag{2.7.7}$$

which, using the fact that $\dot{\gamma}_{ij} = -2\alpha K_{ij} + \pounds_{\vec{\beta}}\gamma_{ij}$, can be reduced to

$$\pi^{ij} = -\sqrt{\gamma}\left(K^{ij} - \gamma^{ij}K\right) \; . \tag{2.7.8}$$

By taking now the trace, we can invert this relation between π_{ij} and K_{ij} to recover equation (2.5.7):

$$K_{ij} = -\frac{1}{\sqrt{\gamma}}\left(\pi_{ij} - \frac{1}{2}\,\gamma_{ij}\,\pi\right) \; , \tag{2.7.9}$$

Since the Lagrangian density is independent of any derivatives of the lapse and shift, it is clear that the momenta conjugate to these variables are zero.

The Hamiltonian density is now the defined as

$$\mathscr{H} = \pi^{ij}\dot{\gamma}_{ij} - \mathcal{L} \; , \tag{2.7.10}$$

which in this case implies

$$\mathscr{H} = \left[-\alpha\sqrt{\gamma}\left({}^{(3)}R + K^2 - K_{ij}K^{ij}\right) + 2\pi^{ij}D_i\beta_j\right] \; . \tag{2.7.11}$$

After ignoring a total divergence, this can be rewritten as

$$\begin{aligned} \mathscr{H} &= -\sqrt{\gamma}\left[\alpha\left({}^{(3)}R + K^2 - K_{ij}K^{ij}\right) + 2\beta_i D_j\left(K^{ij} - \gamma^{ij}K\right)\right] \\ &= -2\sqrt{\gamma}\left[\alpha\mathcal{H} + \beta_i\mathcal{M}^i\right] \; , \end{aligned} \tag{2.7.12}$$

with $\mathcal{H}$ and $\mathcal{M}^i$ defined as before, but without the matter contributions (which would arise when we add the Hamiltonian density for the matter). The total Hamiltonian is now defined as

$$\mathbf{H} := \int \mathscr{H} d^3x \,. \tag{2.7.13}$$

Variation of this Hamiltonian with respect to α and β^i immediately yields the Hamiltonian and momentum constraints for vacuum, $\mathcal{H} = 0$ and $\mathcal{M}^i = 0$. In other words, the lapse and shift behave as Lagrange multipliers for a constrained system.

The evolution equations for the gravitational field now simply follow from Hamilton's equations:

$$\dot{\gamma}_{ij} = \frac{\delta \mathbf{H}}{\delta \pi^{ij}} \,, \qquad \dot{\pi}^{ij} = -\frac{\delta \mathbf{H}}{\delta \gamma_{ij}} \,. \tag{2.7.14}$$

From the first of these equations we find

$$\dot{\gamma}_{ij} = \frac{2\alpha}{\sqrt{\gamma}} \left(\pi_{ij} - \frac{1}{2} \gamma_{ij} \pi \right) + \pounds_{\vec{\beta}} \gamma_{ij} \,, \tag{2.7.15}$$

which is nothing more than the standard evolution equation for γ_{ij} written in terms of π_{ij} instead of K_{ij}, and from the second equation we recover the ADM evolution equations for π_{ij}, which are equivalent to (2.5.8).

There is an interesting observation we can make at this point, due to Anderson and York [22]. When using the Hamilton equations to derive the 3+1 evolution equations, we usually take the lapse function α as an independent quantity to be kept constant during the variation with respect to γ_{ij}. However, we might take a different point of view and assume that the independent gauge function is not the lapse as such, but rather the *densitized lapse* defined as

$$\tilde{\alpha} := \alpha / \sqrt{\gamma} \,. \tag{2.7.16}$$

This change is far from trivial, as keeping $\tilde{\alpha}$ constant alters the dependency of the Hamiltonian density on the metric during the variation (we have effectively multiplied it by a factor of $\sqrt{\gamma}$). The resulting evolution equations for π_{ij} are now different, and correspond precisely to those of York, equations (2.5.5). This observation points to the fact that perhaps the densitized lapse $\tilde{\alpha}$ is a more fundamental free gauge function than the lapse α itself. We will encounter the densitized lapse several times throughout the text.

2.8 The BSSNOK formulation

We have already mentioned that the 3+1 evolution equations are in fact non-unique since we can always add to them arbitrary multiples of the constraints to obtain new evolution equations that are just as valid. In fact, a large number of alternative formulations to ADM have been proposed in the literature. In Chapter 5, when we discuss the concept of well-posedness of a Cauchy problem and the hyperbolicity of a system of partial differential equations, we will have the opportunity to discuss some of the different reformulations and why some should be expected to be better than others in practical applications.

Here, however, we will introduce a specific reformulation that has proven to be particularly robust in the numerical evolution of a large variety of spacetimes, both with and without matter. Over the past few decades, and particularly since researchers started to work with full three-dimensional evolution codes for numerical relativity in the early 1990s, it was realized that the ADM equations lacked the necessary stability properties for long-term numerical simulations, something which is now known to be related to the fact that these equations are only weakly hyperbolic (see Chapter 5). In 1987, Nakamura, Oohara and Kojima presented a reformulation of the ADM evolution equations based on a conformal transformation that showed improved stability properties when compared to ADM [215]. This formulation evolved over the following years, though it remained largely unnoticed by the majority of researchers in numerical relativity until in 1998 Baumgarte and Shapiro systematically compared it to ADM in a series of spacetimes showing that the new formulation had far superior stability properties in all cases considered [50]. It was at this point that this reformulation became more widely noticed, and today it is used in one form or another by most large three-dimensional codes in numerical relativity. The more common version of this formulation is based on the work of Shibata and Nakamura [268], and Baumgarte and Shapiro [50], and is commonly known as the BSSN (Baumgarte, Shapiro, Shibata and Nakamura) formulation. This formulation has also been called "conformal ADM", though this name is not particularly good as it fails to make reference to the most important difference between the new formulation and ADM. A better name would probably be "conformal Γ formulation", since as we will see below the crucial element is the introduction of the auxiliary variables Γ^i. However, here we will use the name "BSSNOK formulation" (BS + Shibata, Nakamura, Oohara and Kojima) in order to make reference to the more commonly used name, and at the same time give due credit to the original authors.

In order to introduce the BSSNOK formulation, consider first a conformal rescaling of the spatial metric of the form

$$\tilde{\gamma}_{ij} := \psi^{-4} \gamma_{ij} \; . \tag{2.8.1}$$

Here ψ is a conformal factor that can in principle be chosen in a number of different ways. For example, when evolving black hole spacetimes with conformally flat initial data, we can simply take ψ to be the initial singular conformal factor and then ask for this conformal factor to remain fixed in time, something that allows us to evolve only the non-singular part of the metric and is known as the *puncture* method for evolutions (see Chapter 6). Alternatively, we can take the conformal factor to be initially given by some arbitrarily chosen scalar function and then propose some convenient evolution equation for this scalar function.

In the BSSNOK formulation, we choose the conformal factor in such a way that the conformal metric $\tilde{\gamma}_{ij}$ has unit determinant, that is

$$\psi^4 = \gamma^{1/3} \qquad \Rightarrow \qquad \psi = \gamma^{1/12} \,, \tag{2.8.2}$$

with γ the determinant of γ_{ij}. Furthermore, we ask for this relation to remain satisfied during the evolution. Now, from (2.3.11) we find that the evolution equation for the determinant of the metric is

$$\partial_t \gamma = \gamma \left(-2\alpha K + 2 D_i \beta^i \right) = -2\gamma \left(\alpha K - \partial_i \beta^i \right) + \beta^i \partial_i \gamma \,, \tag{2.8.3}$$

which implies

$$\partial_t \psi = -\frac{1}{6}\, \psi \, \left(\alpha K - \partial_i \beta^i \right) + \beta^i \partial_i \psi \,. \tag{2.8.4}$$

In practice we usually works with $\phi = \ln \psi = \frac{1}{12} \ln \gamma$, so that $\tilde{\gamma}_{ij} = e^{-4\phi} \gamma_{ij}$ and

$$\partial_t \phi = -\frac{1}{6} \, \left(\alpha K - \partial_i \beta^i \right) + \beta^i \partial_i \phi \,. \tag{2.8.5}$$

Recently, however, it has been suggested by Campanelli *et al.* [93] that evolving instead $\chi = 1/\psi^4 = \exp(-4\phi)$ is a better alternative when considering black hole spacetimes for which ψ typically has a $1/r$ singularity (so that ϕ has a logarithmic singularity), while χ is a C^4 function at $r = 0$. For regular spacetimes, of course, it should make no difference if we evolve ϕ, ψ or χ.

The BSSNOK formulation also separates the extrinsic curvature into its trace K and its tracefree part

$$A_{ij} = K_{ij} - \frac{1}{3} \gamma_{ij} K \,. \tag{2.8.6}$$

We further make a conformal rescaling of the traceless extrinsic curvature of the form[25]

$$\tilde{A}_{ij} = \psi^{-4} A_{ij} = e^{-4\phi} A_{ij} \,. \tag{2.8.7}$$

A crucial point is that BSSNOK also introduces three auxiliary variables known as the *conformal connection functions* and defined as

[25] As we will see in Chapter 3 when we discuss initial data, the "natural" conformal rescaling of the traceless extrinsic curvature is in fact $\tilde{A}^{ij} = \psi^{10} A^{ij}$, which implies $\tilde{A}_{ij} = \psi^2 A_{ij}$ (assuming we raise and lower indices of conformal quantities with the conformal metric). Since I wish to present the standard form of the BSSNOK equations, here I will continue to use the rescaling $\tilde{A}_{ij} = \psi^{-4} A_{ij}$. However, in order to avoid possible confusion later, the reader is advised to keep in mind that this rescaling is different from the one we will use in the next Chapter. It is also important to mention that if we choose to use $\tilde{A}_{ij} = \psi^2 A_{ij}$ instead, some of the equations in the BSSNOK formulation in fact simplify (most notably the momentum constraints and the evolution equations for $\tilde{\Gamma}^i$ and $\tilde{A}_{ij}$ itself), and it also becomes clear that the densitized lapse $\tilde{\alpha} = \alpha \gamma^{-1/2} = \alpha \psi^{-6}$ plays an important role (the BSSNOK equations with the natural conformal rescaling can be found in Appendix C).

$$\tilde{\Gamma}^i := \tilde{\gamma}^{jk}\tilde{\Gamma}^i_{jk} = -\partial_j \tilde{\gamma}^{ij} \;, \tag{2.8.8}$$

where $\tilde{\Gamma}^i{}_{jk}$ are the Christoffel symbols of the conformal metric, and where the second equality comes from the definition of the Christoffel symbols in the case when the determinant $\tilde{\gamma}$ is equal to 1 (which must be true by construction). So, instead of the 12 ADM variables γ_{ij} and K_{ij}, BSSNOK uses the 17 variables ϕ, K, $\tilde{\gamma}_{ij}$, $\tilde{A}_{ij}$ and $\tilde{\Gamma}^i$.[26] We can take the point of view that there are only 15 dynamical variables since $\tilde{A}_{ij}$ is traceless and $\tilde{\gamma}_{ij}$ has unit determinant, but here we will take the point of view that we are freely evolving all components of $\tilde{A}_{ij}$ and $\tilde{\gamma}_{ij}$ (however, enforcing the constraint $\tilde{A} = 0$ during a numerical calculation does seem to improve the stability of the simulations significantly, so it has become standard practice in most numerical codes).

Up to this point all we have done is redefine variables and introduce three additional auxiliary variables. The evolution equation for ϕ was already found above, while those for $\tilde{\gamma}_{ij}$, K and $\tilde{A}_{ij}$ can be obtained directly from the standard ADM equations. The system of evolution equations then takes the form[27]

$$\frac{d}{dt}\tilde{\gamma}_{ij} = -2\alpha\tilde{A}_{ij} \;, \tag{2.8.9}$$

$$\frac{d}{dt}\phi = -\frac{1}{6}\alpha K \;, \tag{2.8.10}$$

$$\frac{d}{dt}\tilde{A}_{ij} = e^{-4\phi}\left\{-D_iD_j\alpha + \alpha R_{ij} + 4\pi\alpha\left[\gamma_{ij}\left(S-\rho\right) - 2S_{ij}\right]\right\}^{\mathrm{TF}} + \alpha\left(K\tilde{A}_{ij} - 2\tilde{A}_{ik}\tilde{A}^k{}_j\right) \;, \tag{2.8.11}$$

$$\frac{d}{dt}K = -D_iD^i\alpha + \alpha\left(\tilde{A}_{ij}\tilde{A}^{ij} + \frac{1}{3}K^2\right) + 4\pi\alpha\left(\rho + S\right) \;, \tag{2.8.12}$$

with $d/dt := \partial_t - \pounds_{\vec{\beta}}$, and where TF denotes the tracefree part of the expression inside the brackets. In the previous expressions we have adopted the convention that indices of conformal quantities are raised and lowered with the conformal metric so that, for example, $\tilde{A}^{ij} = e^{4\phi}A^{ij}$. It is also important to notice that, in the evolution equation for K, the Hamiltonian constraint has been used in order to eliminate the Ricci scalar:

$$R = K_{ij}K^{ij} - K^2 + 16\pi\rho = \tilde{A}_{ij}\tilde{A}^{ij} - \frac{2}{3}K^2 + 16\pi\rho \,. \tag{2.8.13}$$

We then see how we have already started to add multiples of constraints to evolution equations.

Notice that in the evolution equations for $\tilde{A}_{ij}$ and K there appear covariant derivatives of the lapse function with respect to the physical metric γ_{ij}. These

[26] It should be noted that the formulation of [268] uses instead of the $\tilde{\Gamma}^i$ the auxiliary variables $F_i := -\sum_j \partial_j \tilde{\gamma}_{ij}$.

[27] From now on, and where there is no possibility of confusion, we will simply drop the index (3) from the three-dimensional Ricci tensor.

can be easily calculated by using the fact that the Christoffel symbols are related through:

$$\begin{aligned}\tilde{\Gamma}^k_{ij} &= \Gamma^k_{ij} - \frac{1}{3}\left(\delta^k_i \Gamma^m_{jm} + \delta^k_j \Gamma^m_{im} - \gamma_{ij}\gamma^{kl}\Gamma^m_{lm}\right) \\ &= \Gamma^k_{ij} - 2\left(\delta^k_i \partial_j \phi + \delta^k_j \partial_i \phi - \gamma_{ij}\gamma^{kl}\partial_l \phi\right) ,\end{aligned} \tag{2.8.14}$$

where $\tilde{\Gamma}^k_{ij}$ are the Christoffel symbols of the conformal metric, and where we have used the fact that $\partial_i \phi = \frac{1}{12}\partial_i \ln \gamma = \frac{1}{6}\Gamma^m_{im}$. This implies, in particular, that

$$\tilde{\Gamma}^i = e^{4\phi}\Gamma^i + 2\tilde{\gamma}^{ij}\partial_j \phi . \tag{2.8.15}$$

In the evolution equation for $\tilde{A}_{ij}$ we also need to calculate the Ricci tensor associated with the physical metric, which can be separated into two contributions in the following way:

$$R_{ij} = \tilde{R}_{ij} + R^\phi_{ij} , \tag{2.8.16}$$

where $\tilde{R}_{ij}$ is the Ricci tensor associated with the conformal metric $\tilde{\gamma}_{ij}$:

$$\begin{aligned}\tilde{R}_{ij} = &-\frac{1}{2}\tilde{\gamma}^{lm}\partial_l \partial_m \tilde{\gamma}_{ij} + \tilde{\gamma}_{k(i}\partial_{j)}\tilde{\Gamma}^k + \tilde{\Gamma}^k \tilde{\Gamma}_{(ij)k} \\ &+ \tilde{\gamma}^{lm}\left(2\tilde{\Gamma}^k_{l(i}\tilde{\Gamma}_{j)km} + \tilde{\Gamma}^k_{im}\tilde{\Gamma}_{klj}\right) .\end{aligned} \tag{2.8.17}$$

and where R^ϕ_{ij} denotes additional terms that depend on ϕ:

$$R^\phi_{ij} = -2\tilde{D}_i\tilde{D}_j\phi - 2\tilde{\gamma}_{ij}\tilde{D}^k\tilde{D}_k\phi + 4\tilde{D}_i\phi\,\tilde{D}_j\phi - 4\tilde{\gamma}_{ij}\tilde{D}^k\phi\,\tilde{D}_k\phi , \tag{2.8.18}$$

with $\tilde{D}_i$ the covariant derivative associated with the conformal metric.

We must also be careful with the fact that in the evolution equations above we are computing Lie derivatives with respect to $\vec{\beta}$ of *tensor densities*, that is tensors multiplied by powers of the determinant of the metric γ. If a given object is a tensor times $\gamma^{w/2}$, then we say that it is a tensor density of weight w. The Lie derivative of a tensor density of weight w is simply given by

$$\pounds_{\vec{\beta}} T = \left[\pounds_{\vec{\beta}} T\right]_{w=0} + w\, T\, \partial_i \beta^i , \tag{2.8.19}$$

where the first term denotes the Lie derivative assuming $w = 0$, and the second is the additional contribution due to the density factor. The density weight of $\psi = e^\phi = \gamma^{1/12}$ is clearly $1/6$, so the weight of $\tilde{\gamma}_{ij}$ and $\tilde{A}_{ij}$ is $-2/3$, and the weight of $\tilde{\gamma}^{ij}$ is $2/3$. In particular we have

$$\pounds_{\vec{\beta}}\phi = \beta^k \partial_k \phi + \frac{1}{6}\,\partial_k \beta^k , \tag{2.8.20}$$

$$\pounds_{\vec{\beta}}\tilde{\gamma}_{ij} = \beta^k \partial_k \tilde{\gamma}_{ij} + \tilde{\gamma}_{ik}\partial_j \beta^k + \tilde{\gamma}_{jk}\partial_i \beta^k - \frac{2}{3}\tilde{\gamma}_{ij}\partial_k \beta^k . \tag{2.8.21}$$

There are several motivations for the change of variables introduced above. First, the conformal transformation and the separating out of the trace of the extrinsic curvature are done in order to have better control over the slicing conditions that, as we will see in Chapter 4, are generally related with the trace of K_{ij}. On the other hand, the introduction of the conformal connection variables $\tilde{\Gamma}^i$ has the important consequence that when these functions are considered as independent variables, then the second derivatives of the conformal metric that appear on the right hand side of equation (2.8.11) (contained in the Ricci tensor (2.8.17)) reduce to the simple scalar Laplace operator $\tilde{\gamma}^{lm}\partial_l\partial_m\tilde{\gamma}_{ij}$. All other terms with second derivatives of $\tilde{\gamma}_{ij}$ have been rewritten in terms of first derivatives of the $\tilde{\Gamma}^i$.

If the $\tilde{\Gamma}^i$ are to be considered as independent variables, we are of course still missing an evolution equation for them. This equation can be obtained directly from (2.8.8) and (2.3.11):

$$\partial_t\tilde{\Gamma}^i = -\partial_j\left(\pounds_{\vec{\beta}}\tilde{\gamma}^{ij}\right) - 2\left(\alpha\partial_j\tilde{A}^{ij} + \tilde{A}^{ij}\partial_j\alpha\right) , \tag{2.8.22}$$

which after expanding out the Lie derivative term becomes

$$\begin{aligned}\partial_t\tilde{\Gamma}^i &= \tilde{\gamma}^{jk}\partial_j\partial_k\beta^i + \frac{1}{3}\,\tilde{\gamma}^{ij}\partial_j\partial_k\beta^k + \beta^j\partial_j\tilde{\Gamma}^i - \tilde{\Gamma}^j\partial_j\beta^i + \frac{2}{3}\,\tilde{\Gamma}^i\partial_j\beta^j \\ &- 2\left(\alpha\partial_j\tilde{A}^{ij} + \tilde{A}^{ij}\partial_j\alpha\right) .\end{aligned}$$

The last three terms of the first line clearly form the Lie derivative for a vector density of weight 2/3, while the extra terms involving second derivatives of the shift arise from the fact that the $\tilde{\Gamma}^i$ are not really components of a vector density, but are rather contracted Christoffel symbols. Bearing this in mind we can rewrite the last equation in more compact form as

$$\frac{d}{dt}\,\tilde{\Gamma}^i = \tilde{\gamma}^{jk}\partial_j\partial_k\beta^i + \frac{1}{3}\,\tilde{\gamma}^{ij}\partial_j\partial_k\beta^k - 2\left(\alpha\partial_j\tilde{A}^{ij} + \tilde{A}^{ij}\partial_j\alpha\right) . \tag{2.8.23}$$

We are still missing one key element of the BSSNOK formulation. In practice it turns out to be that, in spite of the motivations mentioned above, if we use equations (2.8.9), (2.8.10), (2.8.11), (2.8.12), and (2.8.23) in a numerical simulation the system turns out be violently unstable. In order to fix this problem we need to consider the momentum constraints, which in terms of the new variables take the form

$$\partial_j\tilde{A}^{ij} = -\tilde{\Gamma}^i_{jk}\tilde{A}^{jk} - 6\tilde{A}^{ij}\partial_j\phi + \frac{2}{3}\tilde{\gamma}^{ij}\partial_j K + 8\pi\tilde{j}^i , \tag{2.8.24}$$

with $\tilde{j}^i := e^{4\phi}j^i$. We can now use this equation to substitute the divergence of $\tilde{A}^{ij}$ that appears in the evolution equation for the $\tilde{\Gamma}^i$. We find:

$$\begin{aligned}\frac{d}{dt}\tilde{\Gamma}^i &= \tilde{\gamma}^{jk}\partial_j\partial_k\beta^i + \frac{1}{3}\,\tilde{\gamma}^{ij}\partial_j\partial_k\beta^k - 2\tilde{A}^{ij}\partial_j\alpha \\ &+ 2\alpha\left(\tilde{\Gamma}^i_{jk}\tilde{A}^{jk} + 6\tilde{A}^{ij}\partial_j\phi - \frac{2}{3}\tilde{\gamma}^{ij}\partial_j K - 8\pi\tilde{j}^i\right) .\end{aligned} \tag{2.8.25}$$

The final system of evolution equations is then (2.8.9), (2.8.10), (2.8.11), (2.8.12), and (2.8.25). This new system not only does not present the violent instability mentioned before, but it also turns out to be far more stable than ADM in all cases studied until now. That this is so was first shown empirically (that is, through direct comparison of numerical simulations) by Baumgarte and Shapiro [50], and was later put on somewhat firmer ground by Alcubierre *et al.* [6] by considering linear perturbations of flat space. However, a full understanding of why BSSNOK is much more robust that ADM will have to wait until Chapter 5 when we discuss the concept of hyperbolicity.

The use of the momentum constraints to modify the evolution equation for the $\tilde{\Gamma}^i$ is the key ingredient of the BSSNOK formulation, and it is used in one way or another in all current implementations of this system of equations. There are some additional tricks that have been developed by different groups to make the numerical simulations even better behaved, like actively forcing the trace of the conformal-traceless extrinsic curvature $\tilde{A}_{ij}$ to remain zero during the evolution and using the independently evolved $\tilde{\Gamma}^i$ only in terms where derivatives of these functions appear, but we will not go into such details here.

2.9 Alternative formalisms

In this Chapter we have discussed the 3+1, or Cauchy, formalism for studying the evolution of spacetime. However, it is also important to mention other approaches to the problem of constructing a solution of Einstein's equations given adequate initial data. I will consider here two alternative approaches that have important strengths, namely the characteristic formulation and the conformal approach. The discussion here will be very brief, focusing mainly on the basic ideas behind each approach.

2.9.1 *The characteristic approach*

The characteristic formulation originates from the idea of using a foliation of spacetime based not on spacelike but rather on null hypersurfaces. Here we will discuss the main ideas behind this approach only briefly; a more detailed discussion can be found in the review paper by Winicour [301].

The idea of using null hypersurfaces is very attractive if we are interested in extracting gravitational wave information from an astrophysical system. When working with spatial hypersurfaces, this extraction typically requires the use of perturbative expansions around some Schwarzschild background (see Chapter 8) that in theory only work well at infinity. On the other hand, there is a completely unambiguous and rigorous description of gravitational waves on null hypersurfaces even in the non-linear context. This is the main motivation for the use of the characteristic approach, but it has yet another important advantage. In practice, in numerical simulations we can only evolve a finite region of spacetime, so when using spatial hypersurfaces we are forced to introduce some artificial boundary condition at a finite distance. Rigorous *outgoing wave* boundary conditions can

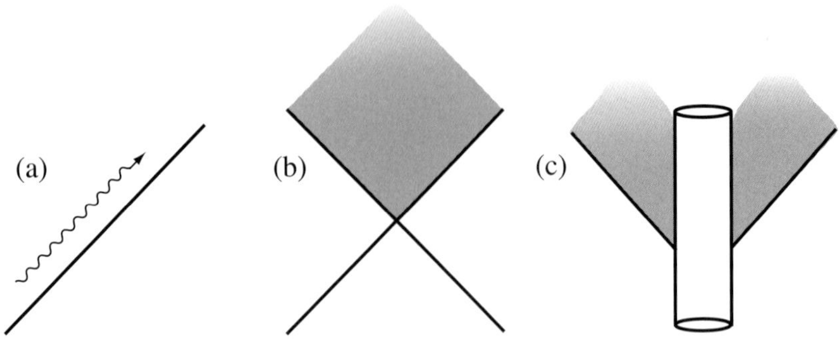

Fig. 2.4: Domain of dependence for the characteristic formulation. (a) A single null hypersurface has an empty domain of dependence as there are light rays coming from infinity that pass arbitrarily close to it but never intersect it. (b) The double null approach uses two null hypersurfaces. (c) A different approach is to use a central timelike world-tube (or world-line) to have a non-trivial domain of dependence.

in fact only be imposed at asymptotic null infinity, so the use of hypersurfaces that are null far away is ideal for this purpose.

There are, however, two important points to consider when using null hypersurfaces. The first is that even relatively small perturbations can cause the light rays generating such a hypersurface to cross, leading to the formation of caustics which are very difficult to handle numerically. The other point to consider is that the domain of dependence of a single regular null hypersurface is always empty. This is easy to understand as there can always be light rays coming from infinity that will come arbitrarily close to any point in the hypersurface without ever touching it, no matter how far the hypersurface extends. It turns out then that we must always add a second boundary surface in order to have a non-trivial future domain of dependence. Typical choices are to use either a second null hypersurface (the *double null* approach), or a central timelike world-tube that in some cases is collapsed to a single world-line (see Figure 2.4).

In the characteristic approach, we consider a foliation of null hypersurfaces corresponding to the level sets of a coordinate function u. On each hypersurface, different generating light rays are identified with angular coordinates x^A $(A = 2, 3)$, and an extra radial coordinate λ is used to parameterize these rays. In terms of these coordinates, the Einstein field equations take the following schematic form:

$$\partial_\lambda F = H_F\,(F, G) \;, \tag{2.9.1}$$

$$\partial_u \partial_\lambda G = H_G\,(F, G, \partial_u G) \;, \tag{2.9.2}$$

where F is a set of geometric variables related to the null hypersurface and G a set of evolution variables. The functions H_F and H_G are non-linear operators acting on the values of the variables on the hypersurface. The specific form of the equations depends on the choice of the form of the metric and other details.

A very common choice is to use the *Bondi–Sachs* null coordinate system, which in the general three-dimensional case corresponds to a spacetime metric of the form [71, 247]

$$\begin{aligned} ds^2 = & - \left(e^{2\beta} \frac{V}{r} - r^2 h_{AB} U^A U^B \right) du^2 \\ & - 2 e^{2\beta} du dr - 2 r^2 h_{AB} U^B du dx^A + r^2 h_{AB} dx^A dx^B \,, \end{aligned} \tag{2.9.3}$$

where here the radial coordinate r is used instead of λ. We then find hypersurface equations (*i.e.* involving only derivatives inside the hypersurface) for the metric functions $\{\beta, V, U^A\}$, and evolution equations (involving derivatives with respect to the null coordinate u) for the metric functions h_{AB}.

One important advantage of this approach is the fact that there are no elliptic constraints on the data, so the initial data is free. Additionally, there are no second derivatives in time (*i.e.* along the direction u), so there are fewer variables than in a 3+1 approach. Plus, null infinity can be compactified and brought to a finite distance in coordinate space, so that no artificial boundary conditions are required.[28]

A lot of work has been devoted to developing characteristic codes in spherical and axial symmetry, and today there are also well-developed three-dimensional codes that have been used to study, for example, scattering of waves by a black hole and even simulations of stars orbiting a black hole. A crucial development was the evolution of a black hole spacetime in a stable way for an essentially unlimited time by turning the problem around and considering a foliation of ingoing null hypersurfaces interior to an *outer* timelike world-tube [147].

The characteristic formalism has a series of advantages over traditional 3+1 approaches, but is has one serious drawback. As already mentioned, caustics can easily develop in null hypersurfaces, particularly in regions with strong gravitational fields. In those regions, a 3+1 approach should be much better behaved. This has led to an idea known as *Cauchy-characteristic matching* (see *e.g.* [57]), which uses a standard 3+1 approach based on a timelike hypersurface in the interior strong field region, matched to a null hypersurface in the exterior to carry the gravitational radiation to infinity (see Figure 2.5). Cauchy-characteristic matching has been shown to work well in simple test cases, but the technique has not yet been fully developed for the three-dimensional case for reasons related mainly to finding a stable and consistent way of injecting boundary data coming from the null exterior to the 3+1 interior.

[28] Compactifying spatial infinity is generally not a good idea since it implies a reduction in resolution at large distances: Wave packets get compressed as seen in coordinate space as they move outward. This gradual reduction in resolution acts on numerical schemes essentially as a change in the refraction index and causes waves to be "back-scattered" by the numerical grid. Compactifying null infinity, however, does not have this problem. Nevertheless, some numerical implementations do compactify spatial infinity but require the use of strong artificial damping of the waves as they travel outward to avoid this numerical back-scattering (see *e.g.* [231]).

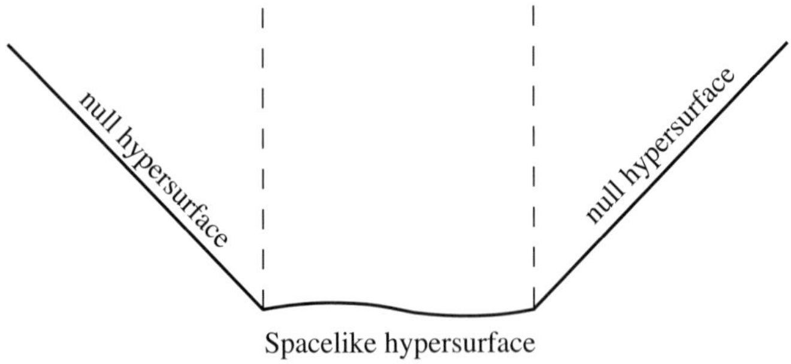

Fig. 2.5: Cauchy-characteristic matching. An interior region uses the standard 3+1 decomposition, while the exterior region uses a characteristic approach.

2.9.2 *The conformal approach*

As mentioned in the previous section, the idea of Cauchy-characteristic matching would seem to be a very promising way of having a well-behaved 3+1 interior, and at the same time a null exterior that allows us to compactify null infinity in order to both read off gravitational radiation directly and also to avoid the need to impose unphysical boundary conditions. There is, however, another approach based on a similar idea that uses a smooth hypersurface that is spacelike everywhere but nevertheless intersects null infinity. This approach evolves a conformal metric $\tilde{\gamma}_{\mu\nu}$ related to the physical metric of spacetime $g_{\mu\nu}$ through $\tilde{\gamma}_{\mu\nu} = \Omega^{-2} g_{\mu\nu}$. The conformal transformation is chosen is such a way that, in the conformal space, infinity corresponds to the boundary of a finite region where the conformal factor vanishes $\Omega = 0$.

In this approach, spacetime is foliated into spacelike hypersurfaces that reach null infinity. These hypersurfaces are called *hyperboloidal*, a name that can be understood if we consider for a moment the geometry of the hypersurface given by the hyperbola $t^2 - x^2 = a^2$ in Minkowski spacetime (we consider only the upper branch):

$$ds^2 = -dt^2 + dx^2 = \left(-\frac{x^2}{x^2 + a^2} + 1 \right) dx^2 = \frac{a^2}{x^2 + a^2}\, dx^2 \,. \qquad (2.9.4)$$

The metric of this hypersurface is clearly positive definite for all x, *i.e.* it is spatial everywhere, but it can be shown to reach null infinity. Figure 2.6 shows how such hypersurfaces look in a conformal diagram.

If we attempt to write down the Einstein field equations for the conformal metric in a straightforward way, it turns out that they are singular at places where $\Omega = 0$, that is, at null infinity. However, a regular form of the conformal field equations has been derived by Friedrich (see *e.g.* [132, 165] and references therein). The variables involved are the connection coefficients, the trace and tracefree parts of the Riemann curvature tensor (the so-called Weyl tensor, see Chapter 8), plus the conformal factor Ω and its derivatives. Friedrich's system of

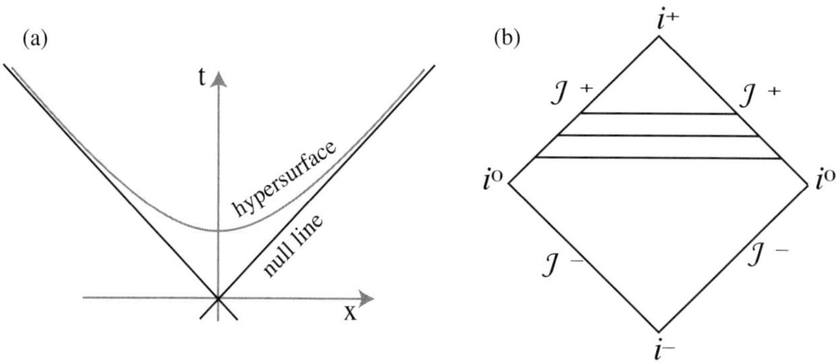

Fig. 2.6: Hyperboloidal hypersurfaces. (a) A hyperboloidal hypersurface in Minkowski spacetime. (b) A foliation of such hypersurfaces as seen in a conformal diagram.

equations can also be shown to be *symmetric hyperbolic* (see Chapter 5), which guarantees that it is mathematically well posed.

Since in the conformal formulation we are evolving the conformal factor Ω as an independent function, the position of the boundary of spacetime at null infinity $\mathscr{J}$ is not known *a priori* (except at $t = 0$). We then need to extend the physical initial data in some suitably smooth way and evolve the dynamical variables "beyond infinity". This has one important advantage, namely that it is possible to put an arbitrary (but well behaved) boundary condition at the outer boundary of the computational region without affecting the physical spacetime, as anything beyond $\mathscr{J}$ is causally disconnected from the interior.

The conformal formulation would seem to be an ideal solution to the weaknesses of both the standard 3+1 approach and the characteristic formulation. Being based on spatial hypersurfaces, it does not have to deal with the problem of caustics associated with the characteristic formulation. At the same time, by reaching null infinity, it allows clean extraction of gravitational radiation and other physical quantities such as total mass and momentum. The main problem faced by the conformal formulation today is related to the problem of constructing hyperboloidal initial data. Also, being based on spatial hypersurfaces, it will have to solve many of the same problems that standard 3+1 formulations are currently faced with, namely the choice of a good gauge and the stability of the evolutions against constraint violation. Though important progress has been made in recent years and numerical simulations of weak data in the full three-dimensional case have been carried out successfully, the conformal formulation is still considerably less developed than the standard 3+1 formulation. However, its conceptual elegance and fundamental strengths mean that this approach represents a very important promise for the future development of numerical relativity.

3

INITIAL DATA

3.1 Introduction

As we saw in the previous Chapter, out of the ten Einstein field equations only six contain time derivatives and therefore represent the true evolution equations of the spacetime geometry. The remaining four equations are constraints that must be satisfied at all times. These are the Hamiltonian and momentum constraints, which for concreteness we will rewrite here in a 3+1 adapted coordinate system

$$ {}^{(3)}R + K^2 - K_{ij}K^{ij} = 16\pi\rho \, , \tag{3.1.1} $$

$$ D_j \left(K^{ij} - \gamma^{ij} K \right) = 8\pi j^i \, , \tag{3.1.2} $$

with ρ and j^i the energy and momentum densities seen by the Eulerian (normal) observers and defined as

$$ \rho := n^\mu n^\nu T_{\mu\nu} \, , \qquad j^i := -P^{i\mu} n^\nu T_{\mu\nu} \, . \tag{3.1.3} $$

The existence of the constraint equations implies that it is not possible in general to choose arbitrarily all 12 dynamical quantities $\{\gamma_{ij}, K_{ij}\}$ as initial data. The initial data has to be chosen in such a way that the constraints are satisfied from the beginning, otherwise we will not be solving Einstein's equations. This means that before starting an evolution, it is necessary to first solve the initial data problem to obtain adequate values of $\{\gamma_{ij}, K_{ij}\}$ that represent the physical situation that we are interested in.

The constraints form a system of four coupled partial differential equations of elliptic type, and in general they are difficult to solve. Still, there are several well-known procedures to solve these equations in specific circumstances. Until a few years ago, the most common procedure was the *conformal decomposition* of York and Lichnerowicz [189, 303, 304]. More recently, the so-called *conformal thin-sandwich approach* [307] has become more and more popular when solving the constraints, as it allows for a clearer interpretation of the freely specifiable data. In this Chapter we will consider both these approaches to the problem of finding initial data. As a particular application of these techniques, we will also consider the special case of initial data for black hole spacetimes. A recent review by Cook of the initial data problem in numerical relativity can be found in [102].

3.2 York–Lichnerowicz conformal decomposition

When trying to solve the constraint equations one is immediately faced with the problem of having four differential equations for the 12 degrees of freedom associated with the spatial metric and extrinsic curvature $\{\gamma_{ij}, K_{ij}\}$. The first question that must be answered is which of those 12 quantities will be taken as free data, and which will be solved for using the constraints. Except in very simple cases like that of the linearized theory, there is no natural way of identifying which are the "true" dynamical components and which are the constrained components. One must therefore develop some procedure that chooses eight components as free data and allows one to solve for the remaining four in a clear way. The most common procedure for doing this is known as the York–Lichnerowicz conformal decomposition.

The York–Lichnerowicz conformal decomposition starts from a conformal transformation of the 3-metric of the form[29]

$$\gamma_{ij} = \psi^4 \bar{\gamma}_{ij} \,, \tag{3.2.1}$$

where the conformal metric $\bar{\gamma}_{ij}$ is considered as given. It is not difficult to show that, in terms of the conformal metric, the Hamiltonian constraint takes the form

$$8\bar{D}^2\psi - \bar{R}\,\psi + \psi^5 \left(K_{ij}K^{ij} - K^2\right) + 16\pi\psi^5\rho = 0 \,, \tag{3.2.2}$$

where $\bar{D}^2$ and $\bar{R}$ are the Laplace operator and Ricci scalar associated with $\bar{\gamma}_{ij}$.

The extrinsic curvature is also separated into its trace K and its tracefree part given by

$$A^{ij} = K^{ij} - \frac{1}{3}\gamma^{ij}\, K \,. \tag{3.2.3}$$

The Hamiltonian constraint then becomes

$$8\bar{D}^2\psi - \bar{R}\,\psi + \psi^5 \left(A_{ij}A^{ij} - \frac{2}{3}\, K^2\right) + 16\pi\psi^5\rho = 0 \,. \tag{3.2.4}$$

Notice that we have transformed the Hamiltonian constraint into an elliptic equation for the conformal factor ψ, and solving it will clearly allow us to reconstruct the full physical metric γ_{ij} from a given conformal metric $\bar{\gamma}_{ij}$.

Let us now consider the momentum constraints, which in terms of A^{ij} take the form

$$D_j A^{ij} - \frac{2}{3}\, D^i K - 8\pi j^i = 0 \,. \tag{3.2.5}$$

In order to transform the momentum constraints into three equations for three unknowns, we can now use a general algebraic result that states that any symmetric-tracefree tensor S^{ij} can be split in the following way

$$S^{ij} = S_*^{ij} + (\mathbf{L}W)^{ij} \,, \tag{3.2.6}$$

[29] In order to avoid confusion we will denote conformal quantities here with an over-bar instead of a tilde. The tilde will be used only for the specific conformal transformation that factors out the volume element for which $\psi = \gamma^{1/12}$ (as in the BSSNOK formulation discussed in Chapter 2).

where S_*^{ij} is a symmetric, traceless and *transverse* tensor (*i.e.* with zero divergence $D_j S_*^{ij} = 0$), W^i is a vector, and $\mathbf{L}$ is an operator defined as

$$(\mathbf{L}W)^{ij} := D^i W^j + D^j W^i - \frac{2}{3}\,\gamma^{ij} D_k W^k \,, \tag{3.2.7}$$

The quantity $(\mathbf{L}W)^{ij}$ is known as the *conformal Killing form* associated with the vector W^i, and its contribution is called the *longitudinal part* of S^{ij}. If the conformal Killing form vanishes, then the vector W^i is called a *conformal Killing vector*, since in that case one has

$$\pounds_{\vec{W}} \left(\gamma^{-1/3}\gamma_{ij}\right) = D_i W_j + D_j W_i - \frac{2}{3}\,\gamma_{ij} D_k W^k = (\mathbf{L}W)_{ij} = 0 \,, \tag{3.2.8}$$

that is, the conformal metric $\tilde{\gamma}_{ij} = \gamma^{-1/3}\gamma_{ij}$ with the volume element factored out is invariant along the vector field (one should not confuse $\tilde{\gamma}_{ij}$, which has unit volume element, with $\bar{\gamma}_{ij}$ which has a volume element given by $\bar{\gamma} = \gamma/\psi^{12}$).

Notice that the operator $\mathbf{L}$ can be defined using any metric tensor. Two natural choices present themselves at this point for the decomposition of A^{ij}: One can use the operator $\bar{\mathbf{L}}$ associated with the conformal metric, or the operator $\mathbf{L}$ associated with the physical metric. We will consider both these cases in turn, and later introduce a different type of tensor splitting that resolves the incompatibility between the first two approaches.

3.2.1 *Conformal transverse decomposition*

We start by performing a conformal transformation on A^{ij} in the following way

$$\bar{A}^{ij} = \psi^{10} A^{ij} \,. \tag{3.2.9}$$

The factor of ψ^{10} is chosen since one can easily show that for any symmetric-tracefree tensor S^{ij} the following identity holds

$$D_j S^{ij} = \psi^{-n} \bar{D}_j \left(\psi^n S^{ij}\right) + (10 - n)\, S^{ik} \partial_k \ln \psi \,, \tag{3.2.10}$$

where as before $\bar{D}_i$ is the covariant derivative associated with the conformal metric. The choice $n = 10$ is therefore clearly natural.[30] Notice that we will raise and lower indices of conformal tensors with the conformal metric, so that in particular we find $\bar{A}_{ij} = \psi^2 A_{ij}$. In terms of $\bar{A}^{ij}$ the momentum constraints become

$$\bar{D}_j \bar{A}^{ij} - \frac{2}{3}\,\psi^6 \bar{D}^i K - 8\pi\psi^{10} j^i = 0 \,, \tag{3.2.11}$$

We now apply the transverse decomposition to $\bar{A}^{ij}$ using the operator $\bar{\mathbf{L}}$ associated with the conformal metric $\bar{\gamma}_{ij}$:

[30] To avoid confusion, the reader is reminded that when discussing the BSSNOK formulation in the last Chapter the alternative, less "natural", rescaling $\bar{A}^{ij} = \psi^4 A^{ij}$ was used in order to recover the standard form of this formulation.

$$\bar{A}^{ij} = \bar{A}^{ij}_* + \left(\bar{\mathbf{L}}\bar{W}\right)^{ij} . \tag{3.2.12}$$

From this one can easily show that the momentum constraints reduce to

$$\bar{\Delta}_{\bar{\mathbf{L}}}\,\bar{W}^i - \frac{2}{3}\psi^6 \bar{D}^i K - 8\pi\psi^{10} j^i = 0 , \tag{3.2.13}$$

where we have defined

$$\begin{aligned}
\bar{\Delta}_{\bar{\mathbf{L}}}\,\bar{W}^i &:= \bar{D}_j(\bar{\mathbf{L}}\bar{W})^{ij} \\
&= \bar{D}^2 W^i + \bar{D}_j \bar{D}^i W^j - \frac{2}{3}\,\bar{D}^i \bar{D}_j W^j \\
&= \bar{D}^2 \bar{W}^i + \frac{1}{3}\,\bar{D}^i \bar{D}_j \bar{W}^j + \bar{R}^i_j \bar{W}^j .
\end{aligned} \tag{3.2.14}$$

The conformal Ricci tensor $\bar{R}^i_j$ appears in the last expression when we commute covariant derivatives. Equations (3.2.13) clearly form a set of three coupled elliptic equations for $\bar{W}^i$.

Let us now assume that we are given the conformal metric $\bar{\gamma}_{ij}$, the trace of the extrinsic curvature K, and the transverse-traceless part of the conformal extrinsic curvature $\bar{A}^{ij}_*$. We can then use the Hamiltonian constraint (3.2.4) and momentum constraints (3.2.13) to find the conformal factor ψ and the vector $\bar{W}^i$, and thus reconstruct the physical metric γ_{ij} and extrinsic curvature K^{ij}.

There is still, however, an important point to consider here. Even though it is a simple task to find a symmetric-tracefree tensor, it is quite a different matter to construct a transverse tensor. In order to construct such a tensor one needs to start from an arbitrary symmetric-tracefree tensor $\bar{M}^{ij}$ that is not necessarily transverse. Its transverse part can clearly be expressed as

$$\bar{M}^{ij}_* = \bar{M}^{ij} - \left(\bar{\mathbf{L}}\bar{Y}\right)^{ij} , \tag{3.2.15}$$

for some vector $\bar{Y}^i$ still to be determined. Now, since $\bar{M}^{ij}_*$ is transverse by definition, the following relation between $\bar{Y}^i$ and $\bar{M}^{ij}$ must hold

$$\bar{\Delta}_{\bar{\mathbf{L}}}\,\bar{Y}^i = \bar{D}_j \bar{M}^{ij} . \tag{3.2.16}$$

Given $\bar{M}^{ij}$, this equation must be solved to find the vector $\bar{Y}^i$, which will in turn allows us to construct the transverse tensor $\bar{M}^{ij}_*$.

The above procedure can in fact be incorporated into the solution of the constraints. Taking $\bar{A}^{ij}_* = \bar{M}^{ij}_*$ we will have

$$\bar{A}^{ij} = \bar{M}^{ij}_* + \left(\bar{\mathbf{L}}\bar{W}\right)^{ij} = \bar{M}^{ij} + \left(\bar{\mathbf{L}}\bar{V}\right)^{ij} , \tag{3.2.17}$$

with $\bar{V}^i := \bar{W}^i - \bar{Y}^i$, and where we have used the fact that $\bar{\mathbf{L}}$ is a linear operator.

One can now rewrite the Hamiltonian and momentum constraints in terms of $\bar{A}^{ij}$, $\bar{V}^i$ and $\bar{M}^{ij}$ to find

$$8\bar{D}^2\psi - \bar{R}\,\psi + \psi^{-7}\bar{A}_{ij}\bar{A}^{ij} - \frac{2}{3}\psi^5 K^2 + 16\pi\psi^5\rho = 0\,, \tag{3.2.18}$$

$$\bar{\Delta}_{\bar{\mathbf{L}}}\,\bar{V}^i + \bar{D}_j\bar{M}^{ij} - \frac{2}{3}\psi^6\bar{D}^i K - 8\pi\psi^{10}j^i = 0\,. \tag{3.2.19}$$

It is common to define $\bar{\rho} := \psi^8\rho$ and $\bar{j}^i := \psi^{10}j^i$ as the conformally rescaled energy and momentum densities. The weight of the conformal factor in the definition of $\bar{j}^i$ is chosen in order to eliminate factors of ψ from the matter terms in the momentum constraints and thus decouple them more easily from the Hamiltonian constraint. The weight of ψ in the definition of $\bar{\rho}$ is then fixed for consistency reasons (for example, ρ^2 and $j_k j^k$ must have the same power of ψ when written in terms of conformal quantities for the energy conditions to be independent of ψ).

Equations (3.2.18) and (3.2.19) are to be solved for ψ and $\bar{V}^i$, with free data given in the form of the conformal metric $\bar{\gamma}_{ij}$, a symmetric-tracefree tensor $\bar{M}^{ij}$, the trace of the extrinsic curvature K, and the energy and momentum densities $\bar{\rho}$ and $\bar{j}^i$. The physical quantities are then reconstructed as

$$\gamma_{ij} = \psi^4\bar{\gamma}_{ij}\,, \tag{3.2.20}$$

$$K^{ij} = \psi^{-10}\bar{A}^{ij} + \frac{1}{3}\gamma^{ij}K\,, \tag{3.2.21}$$

with

$$\bar{A}^{ij} = \left(\bar{\mathbf{L}}\bar{V}\right)^{ij} + \bar{M}^{ij}\,. \tag{3.2.22}$$

The equations just found are the most common way of writing the constraints in the York–Lichnerowicz approach.[31]

Notice how all four equations are coupled with each other. A way to simplify the problem considerably is to simply choose K constant, corresponding to a *constant mean curvature* spatial hypersurface, in which case the momentum constraints decouple completely from the Hamiltonian constraint. One would then start by solving the momentum constraints (3.2.19) for V^i, use this to reconstruct $\bar{A}^{ij}$, and only later solve the Hamiltonian constraint (3.2.18) for ψ.

The problem simplifies even more if, apart from taking K constant, one also takes the conformal metric to be the one corresponding to flat space (*i.e.* the physical metric is conformally flat). The Hamiltonian constraint then reduces to

$$8D^2_{\text{flat}}\psi + \psi^{-7}\bar{A}_{ij}\bar{A}^{ij} - \frac{2}{3}\psi^5 K^2 + 16\pi\psi^5\rho = 0\,, \tag{3.2.23}$$

where now D^2_{flat} is just the flat space Laplacian. In the particular case of time-symmetric initial data, corresponding to $K_{ij} = 0$ at $t = 0$, the momentum

[31] In fact, Lichnerowicz found the expression for the conformal decomposition of the Hamiltonian constraint [189], but the full decomposition of the momentum constraints is due to York [303, 304].

constraints are trivially satisfied. If, moreover, we assume that we are in vacuum the Hamiltonian constraint reduces even further to

$$D^2_{\text{flat}}\psi = 0 \,, \tag{3.2.24}$$

which is nothing more than the standard Laplace equation. We will come back to this equation later when we study black hole initial data.

3.2.2 *Physical transverse decomposition*

Let us now consider the transverse decomposition of A^{ij} (without a conformal rescaling), using the operator $\mathbf{L}$ associated with the physical metric γ_{ij}:

$$A^{ij} = A^{ij}_* + (\mathbf{L}W)^{ij} \;. \tag{3.2.25}$$

The momentum constraints then take the form

$$\Delta_{\mathbf{L}}\, W^i - \frac{2}{3} D^i K - 8\pi j^i = 0 \,. \tag{3.2.26}$$

We now need to rewrite the covariant derivative operators in this expression in terms of the conformal metric. Notice first that if we take

$$W_i = \psi^n \bar{W}_i \quad \Rightarrow \quad W^i = \psi^{(n-4)} \bar{W}^i \,, \tag{3.2.27}$$

then in general one finds that

$$\begin{aligned} (\mathbf{L}W)_{ij} &= \psi^n \left(\bar{\mathbf{L}}\bar{W}\right)_{ij} \\ &\quad + \psi^n \,(n-4) \left[\bar{W}_i \partial_j \ln\psi + \bar{W}_j \partial_i \ln\psi - \frac{2}{3}\,\bar{\gamma}_{ij} \bar{W}^k \partial_k \ln\psi\right] \,. \end{aligned} \tag{3.2.28}$$

Taking $n = 4$ therefore results in

$$(\mathbf{L}W)_{ij} = \psi^4 \left(\bar{\mathbf{L}}\bar{W}\right)_{ij} \quad \Rightarrow \quad (\mathbf{L}W)^{ij} = \psi^{-4} \left(\bar{\mathbf{L}}\bar{W}\right)^{ij} \,, \tag{3.2.29}$$

with $\bar{W}^i = W^i$ (so that the contra-variant vector W^i remains equal to itself under a conformal rescaling). Using now the relation (3.2.10) with $n = 4$ we find

$$\begin{aligned} \Delta_{\mathbf{L}}\, W^i &= D_j \,(\mathbf{L}W)^{ij} = D_j \left[\psi^{-4} \left(\bar{\mathbf{L}}\bar{W}\right)^{ij}\right] \\ &= \psi^{-4} \left[\tilde{D}_j \left(\bar{\mathbf{L}}\bar{W}\right)^{ij} + 6 \left(\bar{\mathbf{L}}\bar{W}\right)^{ij} \partial_j \ln\psi\right] \\ &= \psi^{-4} \left[\bar{\Delta}_{\bar{\mathbf{L}}}\, \bar{W}^i + 6 \left(\bar{\mathbf{L}}\bar{W}\right)^{ij} \partial_j \ln\psi\right] \,. \end{aligned} \tag{3.2.30}$$

The momentum constraints then become

$$\bar{\Delta}_{\bar{\mathbf{L}}}\, \bar{W}^i + 6 \left(\bar{\mathbf{L}}\bar{W}\right)^{ij} \partial_j \ln\psi - \frac{2}{3} \bar{D}^i K - 8\pi\psi^4 j^i = 0 \,, \tag{3.2.31}$$

where we used $\psi^4 D^i K = \psi^4 \gamma^{ij} \partial_j K = \tilde{\gamma}^{ij} \partial_j K = \tilde{D}^i K$.

Just as before, we will incorporate the procedure for obtaining the transverse part of a general symmetric-tracefree tensor into the solution of the constraints. What we want is to obtain the transverse part of A^{ij}, that is A_*^{ij}, which has not been conformally rescaled. Assuming then that we are given a symmetric-tracefree tensor M^{ij}, its transverse part can clearly be written as

$$M_*^{ij} = M^{ij} - (\mathbf{L}Y)^{ij} \; , \tag{3.2.32}$$

for some vector Y^i. As before, the fact that M_*^{ij} implies that the following relation must hold:

$$\Delta_{\mathbf{L}} \, Y^i = D_j M^{ij} \; , \tag{3.2.33}$$

but notice that all quantities here are in the physical space. Taking now $A_*^{ij} = M_*^{ij}$, we then have

$$A^{ij} = M^{ij} + (\mathbf{L}V)^{ij} \; , \tag{3.2.34}$$

with $V^i := W^i - Y^i$.

Combining all previous results, the momentum constraints become,

$$\bar{\Delta}_{\bar{\mathbf{L}}} \, \bar{V}^i + 6 \left(\bar{\mathbf{L}}\bar{V}\right)^{ij} \partial_j \ln \psi + \psi^4 D_j M^{ij} - \frac{2}{3} \bar{D}^i K - 8\pi \psi^4 j^i = 0 \; , \tag{3.2.35}$$

where as before we have used $\bar{V}^i = V^i$.

The last expression, however, still has the divergence $D_j M^{ij}$ written in terms of the physical metric. In order to express the momentum constraints completely in terms of the conformal metric we take $\bar{M}^{ij} = \psi^{10} M^{ij}$. Using again (3.2.10), we can rewrite the momentum constraints in the final form

$$\bar{\Delta}_{\bar{\mathbf{L}}} \, \bar{V}^i + 6 \left(\bar{\mathbf{L}}\bar{V}\right)^{ij} \partial_j \ln \psi + \psi^{-6} \bar{D}_j \bar{M}^{ij} - \frac{2}{3} \bar{D}^i K - 8\pi \psi^4 j^i = 0 \; . \tag{3.2.36}$$

The full system of equations to be solved is then

$$8\bar{D}^2 \psi - \bar{R}\, \psi + \psi^5 \left(A_{ij} A^{ij} - \frac{2}{3} K^2 \right) + 16\pi \psi^5 \rho = 0 \; , \tag{3.2.37}$$

$$\bar{\Delta}_{\bar{\mathbf{L}}} \, \bar{V}^i + 6 \left(\bar{\mathbf{L}}\bar{V}\right)^{ij} \partial_j \ln \psi + \psi^{-6} \bar{D}_j \bar{M}^{ij} - \frac{2}{3} \bar{D}^i K - 8\pi \psi^4 j^i = 0 \; , \tag{3.2.38}$$

with the physical quantities reconstructed as

$$\gamma_{ij} = \psi^4 \bar{\gamma}_{ij} \; , \tag{3.2.39}$$

$$K^{ij} = A^{ij} + \frac{1}{3} \gamma^{ij} K \; , \tag{3.2.40}$$

with

$$A^{ij} = \psi^{-4} \left(\bar{\mathbf{L}}\bar{V}\right)^{ij} + \psi^{-10} \bar{M}^{ij} \; . \tag{3.2.41}$$

Notice that we have the same set of initial data as before, namely $\bar{\gamma}_{ij}$, $\bar{M}^{ij}$, K and the energy and momentum densities ρ and j^i. The Hamiltonian constraint is in fact identical to equation (3.2.18) when we remember that $\bar{A}^{ij} = \psi^{10} A^{ij}$, but the momentum constraint differs from (3.2.19) in both the powers of ψ in several terms and the fact that there is an extra term involving derivatives of ψ.

The difference between the two versions of the momentum constraints, (3.2.19) and (3.2.38), can be seen more easily if we notice that

$$\psi^{-n} \bar{D}_j \left[\psi^n \left(\bar{\mathbf{L}} \bar{V} \right)^{ij} \right] = \bar{\Delta}_{\bar{\mathbf{L}}} \bar{V}^i + n \left(\bar{\mathbf{L}} \bar{V} \right)^{ij} \partial_j \ln \psi \,, \tag{3.2.42}$$

so that (3.2.38) can in fact be rewritten as

$$\bar{D}_j \left[\psi^6 \left(\bar{\mathbf{L}} \bar{V} \right)^{ij} \right] + \bar{D}_j \bar{M}^{ij} - \frac{2}{3} \psi^6 \bar{D}^i K - 8\pi \psi^{10} j^i = 0 \,. \tag{3.2.43}$$

Comparing with (3.2.19) we see that the two equations are identical except for the factor ψ^6 inside the divergence in the first term. So there is a weight difference in the conformal rescaling and as a consequence the momentum constraints are now more tightly coupled with the Hamiltonian constraint. In particular, the momentum constraints now do not decouple from the Hamiltonian constraint for K constant.

It is clear that starting from identical free data we will then obtain different physical data if we use the conformal or physical transverse decompositions. Of course, we will obtain consistent solutions of the constraints in both cases that will represent roughly the same physical situation, but there will be clear differences.

3.2.3 *Weighted transverse decomposition*

In the previous Section we considered the conformal and physical transverse decomposition of the extrinsic curvature and arrived at different forms of the momentum constraint. The conformal transverse decomposition results in somewhat simpler equations, and because of this it has been used more in practice. Still, the existence of two different methods of splitting the traceless extrinsic curvature suggests that perhaps neither is optimal. The root of the problem lies in the fact that the conformal rescaling and tensor splitting do not commute.

More explicitly, from equation (3.2.10) we learn that the natural conformal transformation for a symmetric-tracefree tensor is

$$A^{ij} = \psi^{-10} \bar{A}^{ij} \,. \tag{3.2.44}$$

However, equation (3.2.28) shows that the natural conformal transformation of the longitudinal part is instead

$$\left(\mathbf{L} W \right)^{ij} = \psi^{-4} \left(\bar{\mathbf{L}} \bar{W} \right) \,. \tag{3.2.45}$$

The mismatched powers of ψ in these transformation rules are at the root of the non-commutativity of conformal transformation and tensor splitting.

Recently, however, Pfeiffer and York have introduced a new splitting of tensors that completely resolves the problem [226]. The main idea is to split the traceless extrinsic curvature A^{ij} as

$$A^{ij} = A_*^{ij} + \frac{1}{\sigma} \left(\mathbf{L}W\right)^{ij} , \qquad (3.2.46)$$

with σ a positive definite scalar. One then introduces the following conformal transformations

$$\bar{A}_*^{ij} = \psi^{10} A^{ij} , \quad \bar{W}^i = W^i , \quad \bar{\sigma} = \psi^{-6} \sigma . \qquad (3.2.47)$$

The new splitting then has the important property that

$$\begin{aligned} A^{ij} &= A_*^{ij} + \frac{1}{\sigma} \left(\mathbf{L}W\right)^{ij} \\ &= \psi^{-10} \left[\bar{A}_*^{ij} + \frac{1}{\bar{\sigma}} \left(\bar{\mathbf{L}}\bar{W}\right)^{ij} \right] = \psi^{-10} \bar{A}^{ij} , \end{aligned} \qquad (3.2.48)$$

so conformal transformation and tensor splittings are now fully consistent.[32] With this splitting the momentum constraints become

$$\bar{D}_j \left[\frac{1}{\bar{\sigma}} \left(\bar{\mathbf{L}}\bar{W}\right)^{ij} \right] - \frac{2}{3} \psi^6 \bar{D}^i K - 8\pi \psi^{10} j^i = 0 . \qquad (3.2.50)$$

Just as in the previous two cases, we can start from an arbitrary symmetric tracefree tensor $\bar{M}^{ij}$ which we split as

$$\bar{M}^{ij} = \bar{M}_*^{ij} + \frac{1}{\bar{\sigma}} \left(\bar{\mathbf{L}}\bar{Y}\right)^{ij} . \qquad (3.2.51)$$

If we define again $\bar{V}^i := \bar{W}^i - \bar{Y}^i$, the momentum constraints take the final form

$$\bar{D}_j \left[\frac{1}{\bar{\sigma}} \left(\bar{\mathbf{L}}\bar{V}\right)^{ij} \right] + \bar{D}_j \bar{M}^{ij} - \frac{2}{3} \psi^6 \bar{D}^i K - 8\pi \psi^{10} j^i = 0 . \qquad (3.2.52)$$

The Hamiltonian constraint has the same form as before, so the final equations to solve are

$$8\bar{D}^2 \psi - \bar{R}\, \psi + \psi^{-7} \bar{A}_{ij} \bar{A}^{ij} - \frac{2}{3} \psi^5 K^2 + 16\pi \psi^5 \rho = 0 , \qquad (3.2.53)$$

$$\bar{D}_j \left[\frac{1}{\bar{\sigma}} \left(\bar{\mathbf{L}}\bar{V}\right)^{ij} \right] + \bar{D}_j \bar{M}^{ij} - \frac{2}{3} \psi^6 \bar{D}^i K - 8\pi \psi^{10} j^i = 0 . \qquad (3.2.54)$$

The free data is now given by the conformal metric $\bar{\gamma}_{ij}$, the symmetric-tracefree tensor $\bar{M}^{ij}$, the trace of the extrinsic curvature K, the energy and momentum

[32] It is important to notice that, with this new splitting, the two parts of A^{ij} are orthogonal both before and after the conformal transformation in the sense that

$$\int A_*^{ij} \left[\frac{1}{\sigma} \left(\mathbf{L}W\right)^{kl} \right] \gamma_{ik} \gamma_{jl} \, dV = \int \bar{A}_*^{ij} \left[\frac{1}{\bar{\sigma}} \left(\bar{\mathbf{L}}\bar{W}\right)^{kl} \right] \bar{\gamma}_{ik} \bar{\gamma}_{jl} \, d\bar{V} = 0 , \qquad (3.2.49)$$

with the volume elements given by $dV = \sigma \sqrt{\gamma}\, d^3x$ and $d\bar{V} = \bar{\sigma} \sqrt{\bar{\gamma}}\, d^3x$.

densities $\bar{\rho}$ and $\bar{j}^i$, *plus* the weight factor $\bar{\sigma}$. Given this data, the constraints are solved for $\bar{V}^i$ and ψ, and the physical quantities are reconstructed as

$$\gamma_{ij} = \psi^4 \bar{\gamma}_{ij} \,, \tag{3.2.55}$$

$$K^{ij} = \psi^{-10} \bar{A}^{ij} + \frac{1}{3} \gamma^{ij} K \,, \tag{3.2.56}$$

with

$$\bar{A}^{ij} = \bar{M}^{ij} + \frac{1}{\bar{\sigma}} \left(\bar{\mathbf{L}} \bar{V} \right)^{ij} \,. \tag{3.2.57}$$

The new splitting introduced here, though certainly more consistent and elegant than the previous two cases, produces yet another form of the momentum constraints that now depends on a new scalar weight $\bar{\sigma}$ that has to be chosen as free data. One is then faced with the question of what physical meaning one can give to this weight function. In the next section we will see that there is a very natural choice indeed for this function.

3.3 Conformal thin-sandwich approach

The York–Lichnerowicz conformal decomposition described in the previous section gives us a clear way to find solutions to the constraint equations, *i.e.* initial data, starting from some free data given in the form of a conformal metric $\bar{\gamma}_{ij}$, a symmetric-tracefree tensor $\bar{M}^{ij}$, the trace of the extrinsic curvature K, and the energy and momentum densities $\bar{\rho}$ and $\bar{j}^i$. However, there is in general no clear insight as to how to choose the symmetric-tracefree tensor $\bar{M}^{ij}$ to represent a particular dynamical situation, and worse, the different possible decompositions will give us different initial data starting from identical free data. Arguably, the weighted decomposition of Pfeiffer and York is the best approach, but in that case one needs to specify yet another free function in the form of the scalar weight $\bar{\sigma}$, for which there is no clear physical interpretation.

A different approach to construct initial data has also been proposed by York – the so-called *thin-sandwich* decomposition [307]. The basic idea here is to prescribe the conformal metric on each of two nearby spatial hypersurfaces (the "thin-sandwich"), or equivalently the conformal metric and its time derivative on a given hypersurface. We then start with the same conformal decomposition of the metric $\bar{\gamma}_{ij} = \psi^{-4} \gamma_{ij}$, and define

$$\bar{u}_{ij} := \partial_t \bar{\gamma}_{ij} \,. \tag{3.3.1}$$

We will further demand that the volume element of the conformal metric remains momentarily fixed (though not equal to unity necessarily), which implies

$$\bar{\gamma}^{ij} \bar{u}_{ij} = 0 \,. \tag{3.3.2}$$

Consider now the tracefree part of the evolution equation for the physical metric γ_{ij}, and define

$$
\begin{aligned}
u_{ij} &:= \partial_t \gamma_{ij} - \frac{1}{3}\,\gamma_{ij}\left(\gamma^{mn}\partial_t\gamma_{mn}\right) \\
&= -2\alpha A_{ij} + (\mathbf{L}\beta)_{ij} \;, \qquad (3.3.3)
\end{aligned}
$$

with α and β^i the lapse function and shift vector, respectively. Notice also that equation (3.3.2) implies in particular that

$$
\partial_t \ln \psi = \partial_t \ln \gamma^{1/12} \;, \qquad (3.3.4)
$$

so that one finds

$$
\begin{aligned}
\bar{u}_{ij} &= \partial_t \left(\psi^{-4}\gamma_{ij}\right) = \psi^{-4}\left(\partial_t\gamma_{ij} - 4\gamma_{ij}\partial_t \ln\psi\right) \\
&= \psi^{-4}\left(\partial_t\gamma_{ij} - \frac{1}{3}\gamma_{ij}\partial_t \ln\gamma\right) = \psi^{-4} u_{ij} \;. \qquad (3.3.5)
\end{aligned}
$$

Let us now go back to the expression for u_{ij}. Solving for A^{ij} and using the fact that $(\mathbf{L}\beta)^{ij} = \psi^{-4}\left(\bar{\mathbf{L}}\beta\right)^{ij}$ (where we remember that the natural transformation for a vector is $\beta^i = \bar{\beta}^i$) we find

$$
\begin{aligned}
A^{ij} &= \frac{1}{2\alpha}\left[(\mathbf{L}\beta)^{ij} - u^{ij}\right] \\
&= \frac{\psi^{-4}}{2\alpha}\left[\left(\bar{\mathbf{L}}\beta\right)^{ij} - \bar{u}^{ij}\right] \;, \qquad (3.3.6)
\end{aligned}
$$

and taking as before $\bar{A}^{ij} = \psi^{10} A^{ij}$ this becomes

$$
\bar{A}^{ij} = \frac{1}{2\bar{\alpha}}\left[\left(\bar{\mathbf{L}}\beta\right)^{ij} - \bar{u}^{ij}\right] \;, \qquad (3.3.7)
$$

where we have defined the conformal lapse as $\bar{\alpha} := \psi^{-6}\alpha$, which in the case when $\psi = \gamma^{1/12}$ corresponds precisely to the densitized lapse $\tilde{\alpha} = \gamma^{1/2}\alpha$. Notice that this rescaling for the lapse comes naturally out of the standard rescalings for the other quantities.

We are now almost finished with this approach to construct initial data. The Hamiltonian constraint takes the same form as in the other approaches, namely

$$
8\bar{D}^2\psi - \bar{R}\,\psi + \psi^{-7}\bar{A}_{ij}\bar{A}^{ij} - \frac{2}{3}\psi^5 K^2 + 16\pi\psi^5\rho = 0 \;. \qquad (3.3.8)
$$

The momentum constraint, on the other hand, now becomes

$$
\bar{D}_j\left[\frac{1}{2\bar{\alpha}}\left(\bar{\mathbf{L}}\beta\right)^{ij}\right] - \bar{D}_j\left[\frac{1}{2\bar{\alpha}}\,\bar{u}^{ij}\right] - \frac{2}{3}\,\psi^6\bar{D}^i K - 8\pi\psi^{10} j^i = 0 \;. \qquad (3.3.9)
$$

In this case one needs to solve (3.3.8) and (3.3.9) for the conformal factor ψ and the shift vector β^i, given free data in the form of the conformal metric $\bar{\gamma}_{ij}$, its time derivative $\bar{u}^{ij}$, the trace of the extrinsic curvature K, the conformal (densitized) lapse $\bar{\alpha}$, and the matter densities $\bar{\rho}$ and $\bar{j}^i$. One then reconstructs the physical quantities as

$$\gamma_{ij} = \psi^4 \bar{\gamma}_{ij} \; , \tag{3.3.10}$$

$$K^{ij} = \psi^{-10} \bar{A}^{ij} + \frac{1}{3} \gamma^{ij} K \; , \tag{3.3.11}$$

with

$$\bar{A}^{ij} = \frac{1}{2\bar{\alpha}} \left[\left(\bar{\mathbf{L}} \beta \right)^{ij} - \bar{u}^{ij} \right] \; . \tag{3.3.12}$$

There are several important points to mention about the conformal thin-sandwich approach. First, with this approach we end up with a physical metric γ_{ij} and extrinsic curvature K^{ij} that satisfy the constraints, *plus* a shift vector β^i that we obtain from the solution of the momentum constraint, and a physical lapse function $\alpha = \psi^6 \bar{\alpha}$. That is, we obtain initial data not only for the dynamical quantities, but also for the gauge functions. Of course, we are perfectly free to ignore these values for the gauge functions and start the evolution with arbitrary lapse and shift – the metric and extrinsic curvature obtained will still satisfy the constraints. But if we choose to keep the values for lapse and shift coming from the thin-sandwich approach, then we know that the time derivative of the physical metric will be directly given by

$$\begin{aligned} \partial_t \gamma_{ij} &= \partial_t \left(\psi^4 \bar{\gamma}_{ij} \right) = u_{ij} + \frac{2}{3} \gamma_{ij} \left(D_k \beta^k - \alpha K \right) \\ &= \psi^4 \left[\bar{u}_{ij} + \frac{2}{3} \bar{\gamma}_{ij} \left(\bar{D}_k \beta^k + 6 \, \beta^k \partial_k \ln \psi - \psi^6 \bar{\alpha} K \right) \right] \; . \end{aligned} \tag{3.3.13}$$

We then have a clear interpretation of the free data in terms of their effects on the initial dynamics of the system.

Another important point to mention is the fact that although we have obtained a value for the shift vector through the solution of the momentum constraints, the conformal lapse $\bar{\alpha}$ is still free and we might worry about what would be a good choice for this quantity. Of course, any choice would be equally valid, for example we could simply take $\bar{\alpha} = 1$, but there is a more natural way to determine $\bar{\alpha}$. We can take the point of view that the natural free data are really $\bar{\gamma}_{ij}$ and its velocity $\bar{u}_{ij} = \partial_t \bar{\gamma}_{ij}$, plus K *and* its velocity $\dot{K} = \partial_t K$. We can then use this information to reconstruct $\bar{\alpha}$ in the following way: Consider the ADM evolution equation for K, which can be easily shown to be given by

$$\begin{aligned} \partial_t K &= \beta^j \partial_j K - D^2 \alpha + \alpha \left(R + K^2 \right) + 4\pi\alpha \left(S - 3\rho \right) \\ &= \beta^j \partial_j K - D^2 \alpha + \alpha K_{ij} K^{ij} + 4\pi\alpha \left(S + \rho \right) \\ &= \beta^j \partial_j K - D^2 \alpha + \alpha \left(A_{ij} A^{ij} + \frac{1}{3} K^2 \right) + 4\pi\alpha \left(S + \rho \right) \; , \end{aligned} \tag{3.3.14}$$

where in the second line we have used the Hamiltonian constraint (3.1.1) to eliminate the Ricci scalar R. The Laplacian of the lapse can be rewritten as

$$D^2\alpha = \psi^{-4}\left[\bar{D}^2\alpha + 2\,\bar{\gamma}^{mn}\partial_m\alpha\,\partial_n\ln\psi\right] , \tag{3.3.15}$$

so that we find

$$\begin{aligned}\partial_t K &= \beta^j\partial_j K - \psi^{-4}\bar{D}^2\alpha - 2\,\psi^{-5}\bar{D}_m\alpha\,\bar{D}^m\psi \\ &+ \alpha\left(\psi^{-12}\bar{A}_{ij}\bar{A}^{ij} + \frac{1}{3}\,K^2\right) + 4\pi\alpha\,(S+\rho) \; .\end{aligned} \tag{3.3.16}$$

We now write $\alpha = \psi^6\bar{\alpha}$ and notice that

$$\begin{aligned}\bar{D}^2\left(\psi^6\bar{\alpha}\right) &= \psi^6\bar{D}^2\bar{\alpha} + 6\,\bar{\alpha}\psi^5\bar{D}^2\psi \\ &+ 30\,\bar{\alpha}\psi^4\bar{D}_m\psi\,\bar{D}^m\psi + 12\,\psi^5\bar{D}_m\bar{\alpha}\,\bar{D}^m\psi \; ,\end{aligned}$$

and

$$\bar{D}_m\left(\psi^6\bar{\alpha}\right)\bar{D}^m\psi = \psi^6\bar{D}_m\bar{\alpha}\,\bar{D}^m\psi + 6\,\bar{\alpha}\psi^5\bar{D}_m\psi\,\bar{D}^m\psi \; . \tag{3.3.17}$$

Substituting this into the time derivative of K, and using the Hamiltonian constraint in conformal form (3.3.8) to eliminate $\bar{D}^2\psi$, we finally find

$$\begin{aligned}&\bar{D}^2\bar{\alpha} + \bar{\alpha}\left[\frac{3}{4}\,\bar{R} - \frac{7}{4}\psi^{-8}\bar{A}_{ij}\bar{A}^{ij} + \frac{1}{6}\psi^4K^2 + 42\left(\bar{D}_m\ln\psi\right)^2\right] + 14\,\bar{D}_m\bar{\alpha}\,\bar{D}^m\ln\psi \\ &+ \psi^{-2}\left(\partial_t K - \beta^m\partial_m K\right) - 4\pi\bar{\alpha}\,\psi^4\,(S+4\rho) = 0 \; .\end{aligned} \tag{3.3.18}$$

This gives us an elliptic equation to solve for $\bar{\alpha}$. The equation is coupled with the other four equations coming from the constraints, so we end up with a system of five coupled equations from which we can find ψ, $\bar{\alpha}$, and β^i given $\tilde{\gamma}_{ij}$, K, and their time derivatives.

There is one final important comment to make concerning the relationship between the conformal thin-sandwich approach and the York–Lichnerowicz conformal decomposition. Notice that if, in the weighted decomposition of Section 3.2.3, we make the choices

$$\bar{V}^i = \bar{\beta}^i = \beta^i \; , \qquad \bar{\sigma} = 2\bar{\alpha} \; , \qquad \bar{M}^{ij} = -\bar{u}^{ij}/2\bar{\alpha} \; , \tag{3.3.19}$$

then the equations to solve become *identical* with those of the thin-sandwich approach. This not only reinforces the fact that the weighted decomposition is the natural decomposition of $\bar{A}^{ij}$ into transverse and longitudinal parts, but it also provides a natural interpretation for the free data in that approach. In particular, it tells us that the natural choice for the weight function $\bar{\sigma}$ is twice the conformal lapse. This observation closes the circle and we find that the York–Lichnerowicz conformal decomposition and the conformal thin-sandwich approach are completely consistent with each other.

3.4 Multiple black hole initial data

As an example of how to apply the methods described in the previous sections to the problem of finding initial data that represents a specific physical situation, we will consider the case of multiple black hole spacetimes. This type of initial data has the advantage of being a pure vacuum solution, so we will not have to worry about how to describe the matter content. Moreover, binary black hole data are astrophysically important because the merger of two black holes is considered to be a very strong source of gravitational waves, likely to be among the first gravitational wave signals detected in the next few years.

3.4.1 *Time-symmetric data*

We will start by considering the case of time-symmetric initial data, corresponding to a time when the black holes are momentarily static. Clearly this situation is not very physical, because in astrophysical systems the black holes will be orbiting each other and will therefore not be static at any time. Still, it represents an important simplification that allows us to discuss some important issues.

For time-symmetric initial data we clearly have $K_{ij} = 0$. The momentum constraints are therefore trivially satisfied, and all methods discussed in the previous sections become equivalent. We only need to solve the Hamiltonian constraint, which in this case reduces to

$$8\bar{D}^2\psi - \bar{R}\,\psi = 0 \,, \tag{3.4.1}$$

where we have also used the fact that we are in vacuum, so that $\rho = 0$.

We will simplify the problem further by choosing a flat conformal metric, so that $\bar{R} = 0$. We are then left with the simple equation

$$D^2_{\rm flat}\psi = 0 \,, \tag{3.4.2}$$

where again $D^2_{\rm flat}$ is the standard flat-space Laplace operator. The boundary conditions correspond to an asymptotically flat spacetime (far away, the gravitational field goes to zero), so that at infinity we must have $\psi = 1$. The simplest solution to this equation that satisfies the boundary conditions is clearly

$$\psi = 1 \,, \tag{3.4.3}$$

which implies that the physical metric is simply:

$$dl^2 = dx^2 + dy^2 + dz^2 \,. \tag{3.4.4}$$

That is, we have recovered initial data for Minkowski spacetime (though through a rather elaborate route).

The next interesting solution is clearly

$$\psi = 1 + k/r \,, \tag{3.4.5}$$

with k an arbitrary constant, so the physical metric (in spherical coordinates) takes the form:

$$dl^2 = (1 + k/r)^4 \left[dr^2 + r^2 d\Omega^2 \right] . \tag{3.4.6}$$

We have in fact encountered this metric before in Chapter 1 – it corresponds precisely to the spatial metric for a Schwarzschild black hole written in isotropic coordinates, equation (1.15.28), with the mass of the black hole given by $M = 2k$. We have therefore found the solution to the initial data problem corresponding to a single Schwarzschild black hole.

Notice that we have only found the spatial part of the metric – the time part is of course still free, as it corresponds to a choice of lapse and shift. In the spirit of the thin-sandwich approach we could, for example, ask for the time derivative of the trace of the extrinsic curvature to vanish. In this case equation (3.3.18) reduces to

$$D^2_{\rm flat}\bar{\alpha} + 42\,\bar{\alpha}\,\partial_m \ln\psi\,\partial^m \ln\psi + 14\,\partial_m\bar{\alpha}\,\partial^m \ln\psi = 0 . \tag{3.4.7}$$

Now, for an arbitrary scalar function f we have that in general

$$D^2 \left(f\psi^n \right) = f\psi^{n-1} D^2\psi + \psi^n \left[D^2 f + n(n-1)\, f\,\partial_m \ln\psi\,\partial^m \ln\psi + 2n\,\partial_m f \partial^m \ln\psi \right] . \tag{3.4.8}$$

Taking $n = 7$ and using the fact that $D^2\psi = 0$, the equation for the lapse becomes

$$D^2_{\rm flat} \left(\bar{\alpha}\psi^7 \right) = D^2_{\rm flat} \left(\alpha\psi \right) = 0 . \tag{3.4.9}$$

Again, we have to choose boundary conditions. If we ask for the lapse to become unity far away and vanish at the horizon, which in isotropic coordinates corresponds to $r = M/2$, we find

$$\alpha\psi = 1 - M/2r , \tag{3.4.10}$$

which implies

$$\alpha = \frac{1 - M/2r}{1 + M/2r} . \tag{3.4.11}$$

We have then also recovered the isotropic lapse of equation (1.15.28).

As Laplace's equation is linear, we can simply superpose solutions to obtain new solutions. For example, we can take the conformal factor to be given by

$$\psi = 1 + \sum_{i=1}^{N} \frac{m_i}{2\,|\vec{r} - \vec{r}_i|} . \tag{3.4.12}$$

This solution will represent N black holes that are momentarily at rest, located at the points $\vec{r}_i$. The parameters m_i are known as the *bare masses* of each black hole, and in terms of them the total ADM mass of the spacetime (see Appendix A) turns out to be simply $M_{ADM} = \sum_i m_i$. The bare mass corresponds with the individual mass only in the case of a single black hole, for more than one black hole the definition of the individual masses is somewhat trickier. One possible definition is obtained by going to the asymptotically flat end associated with each hole and calculating the ADM mass there. This can be easily done by considering spherical coordinates around the ith center and using a new radial coordinate $\tilde{r}_i = M_i^2/4r_i$ that goes to infinity at $\vec{r} = \vec{r}_i$. We then find that the ADM mass of each individual black hole is

$$M_i = m_i \left(1 + \sum_{i \neq j} \frac{m_j}{2r_{ij}} \right) , \tag{3.4.13}$$

with $r_{ij} := |\vec{r}_i - \vec{r}_j|$ the coordinate distance between the black hole centers. Another way in which we can define the mass of each hole is by locating the individual apparent horizons and using the relationship between mass and horizon area for Schwarzschild (equation (1.15.26)). Numerical experiments have in fact found that this agrees extremely well (to within numerical accuracy) with the individual ADM masses given above [289].

Since the initial data is time-symmetric, the location of the individual apparent horizons will coincide with the surfaces of minimal area, that is, with the throats of the wormholes. Strictly speaking, however, the solution will only represent N black holes if the points r_i are sufficiently far apart, as otherwise we might find that the black hole horizons have merged and there are really fewer black holes (maybe even only one) with complicated interior topologies. For example, if we have two equal-mass black holes with $m_1 = m_2 = 1$, we find that there is indeed a common apparent horizon if the centers are separated in coordinate space by less that $|r_1 - r_2| \sim 1.5$, so in that case we have in fact a single distorted black hole [79, 56, 8].

The solution (3.4.12) is known as the *Brill–Lindquist initial data* [79, 192]. As in the case of Schwarzschild, for Brill–Lindquist data each singular point represents infinity in a different asymptotically flat region, so that our Universe is connected with N different universes through Einstein–Rosen bridges (wormholes). The points $r = r_i$ are strictly speaking not part of the manifold. We can then think of this solution as given in $\Re^3$ with N points removed. These removed points are commonly know as *punctures*. Since Brill–Lindquist data represents in fact $N+1$ joined asymptotically flat universes, its topological structure is much more complex than that of Schwarzschild. Still, this non-trivial topology will be "hidden" inside the black hole horizons, so it should have no effect on the exterior Universe.

It turns out that we can in fact construct a solution that represents N black holes but that contains only two isometric (*i.e.* identical) universes (see Figure 3.1). In this case all N wormholes will connect the same two universes. This solution was found by Misner [205] and is known as *Misner initial data*. The solution involves an infinite series expansion and is constructed using a method

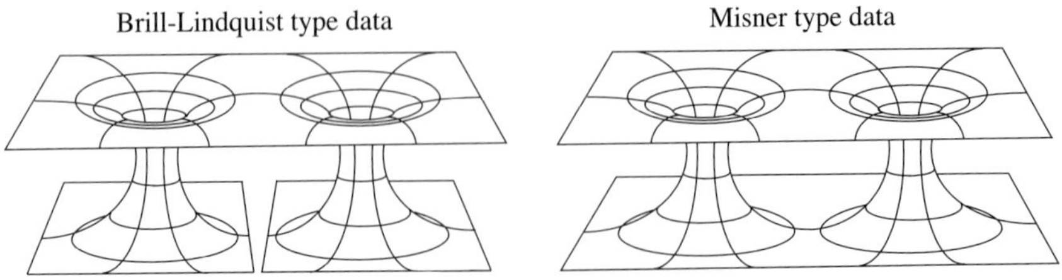

Fig. 3.1: Topology of a spacetime containing two black holes. The left panel shows the case of Brill-Lindquist data, for which our universe is joined to two distinct universes via wormholes. The right panel shows Misner type data for which there are only two isometric universes joined by two wormholes.

of images (as each hole can "feel" the others an infinite number of times through the wormholes). The construction of Misner data is rather involved, and here I will just give the result for the case of two equal-mass black holes. In this case the black hole throats are coordinate spheres whose centers are located on the z-axis by construction. The solution is given in terms of a parameter μ that is related to the position of the centers of the spheres $z = \pm z_0$ and their coordinate radius a through

$$z_0 = \coth \mu \, , \qquad a = 1/\sinh \mu \, . \tag{3.4.14}$$

In terms of this parameter, the conformal factor turns out to be

$$\psi = 1 + \sum_{n=1}^{\infty} \frac{1}{\sinh(n\mu)} \left[\frac{1}{r_n^+} + \frac{1}{r_n^-} \right] , \tag{3.4.15}$$

with

$$r_n^{\pm} = \sqrt{x^2 + y^2 + (z \pm \coth(n\mu))^2} \tag{3.4.16}$$

From this expression for the conformal factor we can find that the total ADM mass of the system is given by

$$M_{ADM} = 4 \sum_{n=1}^{\infty} \frac{1}{\sinh(n\mu)} \, . \tag{3.4.17}$$

We can also show that the proper distance L along a straight coordinate line between the throats is

$$L = 2 \left[1 + 2\mu \sum_{n=1}^{\infty} \frac{n}{\sinh(n\mu)} \right] . \tag{3.4.18}$$

Notice how, as μ increases, the centers of the holes approach each other and the coordinate radius of their throats decreases (for infinite μ we find $z_0 = 1$ and $a = 0$). At the same time, the total ADM mass becomes smaller and the two holes move closer also in terms of proper distance, but the ratio L/M_{ADM} in fact increases. For values of μ less than $\mu \sim 1.36$ we finds a single apparent horizon around the two throats [86], while numerical evolutions indicate that for values less than $\mu \sim 1.8$ there is also a common event horizon around the throats on the initial time slice [23].

Owing to its topological simplicity, for a long time Misner data was considered more "natural". It was frequently used in the numerical simulation of the head-on collision of two black holes, and as such it has become a reference case. Notice that precisely because of the property of having an isometry at each of the wormhole throats, the natural way to evolve Misner type data is to take advantage of the isometry and evolve in $\Re^3$ minus N balls on which symmetry boundary conditions are applied. This approach works well in the case of the head-on collision of two black holes, where there is rotational symmetry around the axis joining the black holes and the coordinates can be adapted to the horizons, but it becomes much more difficult in the case of orbiting black holes where there are no symmetries, and Cartesian coordinates are used. Because of this, plus the fact that it involves an infinite number of singular points inside each throat, Misner type initial data is not often used anymore and approaches based on Brill–Lindquist type data are preferred.

3.4.2 *Bowen–York extrinsic curvature*

Time-symmetric initial data are a good example of how we can solve the constraints in simple cases, and are good test cases for numerical codes. Still, they clearly have no physical relevance because astrophysical black holes would be expected to have both linear momentum and spin. In this Section we will therefore consider initial data for moving and spinning black holes, in which case the momentum constraints will no longer be trivial. Remarkably, it turns out that there is an analytical solution to the momentum constraints that has the properties we are looking for. Let us start by assuming that we are in vacuum, and also that the conformal metric is flat $\bar{\gamma}_{ij} = \delta_{ij}$ and the trace of the extrinsic curvature vanishes $K = 0$. This last condition is known as that for a *maximal* slice, as we can show that the volume of a spatial hypersurface for which $K = 0$ is a maximum with respect to small variations.

Consider the momentum constraints in the York–Lichnerowicz conformal decomposition as given by (3.2.19). If we now choose $\bar{M}^{ij} = 0$, the constraints reduce to

$$\bar{\Delta}_{\mathbf{L}} \, \bar{V}^i = \bar{D}^2 \bar{V}^i + \frac{1}{3} \, \bar{D}^i \bar{D}_j \bar{V}^j = 0 \,. \tag{3.4.19}$$

It is not difficult to show that this equation has the following solution

$$\bar{V}^i = -\frac{1}{4r} \left[7P^i + n^i n_j P^j \right] + \frac{1}{r^2} \, \epsilon^{ijk} n_j S_k \,, \tag{3.4.20}$$

with P^i and S^i constant vectors, n^i the outward-pointing unit radial vector, and ϵ^{ijk} the completely antisymmetric Levi–Civita tensor in three dimensions. To make it more transparent, the last expression can be rewritten in conventional three-dimensional vector notation as

$$\mathbf{V} = -\frac{1}{4r}\left[7\mathbf{P} + \mathbf{n}\left(\mathbf{n}\cdot\mathbf{P}\right)\right] + \frac{1}{r^2}\,\mathbf{n}\times\mathbf{S}\,. \tag{3.4.21}$$

Having found the vector $\bar{V}^i$, the conformal tracefree extrinsic curvature will simply be given by

$$\begin{aligned}\bar{A}_{ij} = \left(\bar{\mathbf{L}}\bar{V}\right)_{ij} &= \frac{3}{2r^2}\left[n_i P_j + n_j P_i + n_k P^k\left(n_i n_j - \delta_{ij}\right)\right] \\ &- \frac{3}{r^3}\left(\epsilon_{ilk} n_j + \epsilon_{jlk} n_i\right) n^l S^k \,,\end{aligned} \tag{3.4.22}$$

with $K_{ij} = \psi^{-2}\bar{A}_{ij}$ (remember that we are taking $K = 0$). This solution to the momentum constraints is known as the Bowen–York extrinsic curvature [73, 74].

The constant vectors P^i and S^i have clear physical interpretations. Using the expressions for the ADM integrals found in Appendix A, we see that the linear and angular momenta at spatial infinity can be calculated as

$$P^i = \frac{1}{8\pi}\lim_{r\to\infty}\oint\left(K^i_l - \delta^i_l K\right) n^l dS\,, \tag{3.4.23}$$

$$J^i = \frac{1}{16\pi}\lim_{r\to\infty}\oint\epsilon^{ijk} x_j K_{kl}\, n^l dS\,, \tag{3.4.24}$$

where the integrals are done over spheres of constant r, with n^i the unit outward-pointing normal vector to the sphere, and where the $\{x^i\}$ are taken to be asymptotically Cartesian coordinates (notice that in our case $K = 0$, so the first integral simplifies somewhat). Assuming now that for large r the conformal factor becomes unity, and substituting (3.4.22) in the above expressions, we find after some algebra that the vectors P^i and S^i are precisely the linear and angular momenta for this spacetime. And since the momentum constraints are linear, we can add solutions of this form at different centers $r = r_i$ to represent a set of "particles" with the given momenta and spins.

The Bowen–York extrinsic curvature can be used directly as a solution of the momentum constraints, but as it stands it is not isometric when we consider Misner-type initial data. we can, however, find a somewhat more general solution that is symmetric with respect to inversions through a coordinate sphere (see [74] for details about the inversion through a sphere). The more general expression is

$$\begin{aligned}\bar{A}_{ij} = \left(\bar{\mathbf{L}}\bar{V}\right)_{ij} &= \frac{3}{2r^2}\left[n_i P_j + n_j P_i + n_k P^k\left(n_i n_j - \delta_{ij}\right)\right] \\ &\pm \frac{3a^2}{2r^4}\left[n_i P_j + n_j P_i - n_k P^k\left(5 n_i n_j - \delta_{ij}\right)\right] \\ &- \frac{3}{r^3}\left(\epsilon_{ilk} n_j + \epsilon_{jlk} n_i\right) n^l S^k \,,\end{aligned} \tag{3.4.25}$$

The extra term proportional to a^2 does not contribute to either the linear or angular momenta, but guarantees that the solution is symmetric under inversion through a sphere of radius a, the throat of the Einstein–Rosen bridge, which can later be identified with the black hole horizon. The two different signs in the extra term correspond to the cases where the linear momentum on the other side of the wormhole is either $+P^i$ or $-P^i$. When we have more than one black hole, we can still find solutions that have inversion symmetry with respect to each throat, but just as before this requires an infinite series expression.

3.4.3 *Conformal factor: inversions and punctures*

In the last section we found the Bowen–York exact solution to the momentum constraints that represents a "particle" with a given linear momentum and spin. Of course, in order to complete the initial data specification we still need to solve the Hamiltonian constraint for the conformal factor ψ. However, the presence of a non-trivial extrinsic curvature implies that we can no longer solve this constraint exactly, so numerical solutions are required.

At this point we have to decide on the type of boundary conditions we will impose on the solution for the conformal factor. The boundary condition for large r comes from the requirement of asymptotic flatness. From our discussion of the Newtonian approximation in Section 1.14, we see that this implies

$$\psi \simeq 1 + \frac{M}{2r} \,, \tag{3.4.26}$$

or in differential terms

$$\partial_r \psi = \frac{1}{r}\left(1 - \psi\right) \,. \tag{3.4.27}$$

The inner boundary condition is more interesting. Recall that even in the case of Schwarzschild we found that $\psi = 1 + M/2r$, which is singular at $r = 0$. The solution can therefore also be expected to be singular in the more general case. There are two different approaches we can take in order to deal with this singularity.

The first approach is to insist on the inversion symmetry through the throats of the wormholes using the Bowen–York extrinsic curvature in the form (3.4.25). We then solve the Hamiltonian constraint only on the exterior of the throats, which by construction are fixed at coordinate spheres, using the following boundary condition on each of the throats (see Section 6.2)

$$\partial_r \psi|_{r=a} = - \left.\frac{\psi}{2r}\right|_{r=a} \,, \tag{3.4.28}$$

where r is the coordinate distance to the center of the throat and a is its radius.

The inversion method results in data that can be thought of as generalizations of the Misner type data for the case of non-trivial extrinsic curvature. In fact, this was the first approach used in practice, and resulted in the first initial data sets for orbiting black holes obtained by Cook in the early 1990s [101].

The generalization of Brill–Lindquist type initial data was achieved in 1997 by Brandt and Bruegmann using what is now known as the *puncture method* [76]. In this case we drop the requirement of inversion symmetry and start from the simpler form of the Bowen–York extrinsic curvature (3.4.22), adding one term of that type for each black hole. Dropping the inversion symmetry, however, implies that now there are no natural inner boundaries where we can stop the integration, so we must integrate through the complete $\Re^3$ interior and deal with the singularities in the solution for the conformal factor directly.

The basic idea behind the puncture method is to write the conformal factor with the singular piece explicitly separated as

$$\psi = \psi_{\rm BL} + u \,, \qquad \psi_{\rm BL} = \sum_{i=1}^{N} \frac{m_i}{2\,|\vec{r} - \vec{r}_i|} \,. \tag{3.4.29}$$

The singular piece is therefore assumed to have the same behavior as in Brill–Lindquist data (but notice that the additive 1 from the Brill-Lindquist conformal factor (3.4.12) has now been absorbed in the u). It is clear that the term $\psi_{\rm BL}$ has zero Laplacian on $\Re^3$ with the points $\vec{r} = \vec{r}_i$ excised, *i.e.* on a "punctured" $\Re^3$. The Hamiltonian constraint then reduces to

$$D^2_{\rm flat} u + \eta \left(1 + \frac{u}{\psi_{\rm BL}} \right)^{-7} = 0 \,, \tag{3.4.30}$$

with

$$\eta = \frac{1}{8\psi_{\rm BL}^7} \, \bar{A}_{ij} \bar{A}^{ij} \,, \tag{3.4.31}$$

and we have used the fact that $K = 0$, and also that the spatial metric is conformally flat so that $\bar{R} = 0$. The last equation must now be solved for u.

As before, we must now consider what boundary conditions must be imposed on u, both at infinity and at the punctures. At infinity, asymptotic flatness again implies that we must have $u = 1 + k/r$ for some constant k, or in differential form $\partial_r u = (1 - u)/r$.

The key observation of the puncture method is that we can in fact solve for u with no special boundary conditions at the punctures. To see this, notice that, near a given puncture, we have $\psi_{\rm BL} \sim 1/|\vec{r} - \vec{r}_i|$, while for Bowen–York extrinsic curvature of the form (3.4.22) we find that $\bar{A}_{ij}\bar{A}^{ij}$ diverges for non-zero spin as $|\vec{r} - \vec{r}_i|^{-6}$ and for zero spin as $|\vec{r} - \vec{r}_i|^{-4}$, so that η goes to zero as $|\vec{r} - \vec{r}_i|$ for non-zero spin and as $|\vec{r} - \vec{r}_i|^3$ for zero spin. The Hamiltonian constraint then reduces near the punctures to $D^2_{\rm flat} u = 0$. Brandt and Bruegmann show that under these conditions there exists a unique C^2 solution u to the Hamiltonian constraint in all of $\Re^3$, so that we can ignore the punctures when solving for u.

Another interesting property of this method is that we can show that each puncture corresponds to a separate asymptotically flat region, and that as seen from those regions the corresponding black holes have zero linear momentum. This is because the linear momentum on the other side of a given throat arises through the terms proportional to a^2 in (3.4.25), while the puncture method uses (3.4.22) instead. In fact, we can not use the inversion-symmetric extrinsic curvature (3.4.25) with this method since in that case we would find that $\bar{A}_{ij}\bar{A}^{ij} \sim |\vec{r} - \vec{r}_i|^{-8}$ near the punctures and η would diverge.

We can also show that for puncture type data the ADM masses of the individual black holes, as calculated on the other asymptotically flat regions, are given in terms of the mass parameters m_i as

$$M_i = m_i \left(1 + u_i + \sum_{i \neq j} \frac{m_j}{2r_{ij}} \right) , \tag{3.4.32}$$

with $u_i = u(\vec{r} = \vec{r}_i)$.

The puncture approach is considerably easier to implement numerically than the inversion-symmetric approach, and because of this in recent years it has become more common in practice.

3.4.4 *Kerr–Schild type data*

The methods described in the previous section for finding multiple black hole data are very general and can be used to find configurations for multiple black holes with arbitrary masses, momenta and spins. Still, the basic assumptions made, namely $K = 0$, $\bar{\gamma}_{ij} = \delta_{ij}$, and an extrinsic curvature of Bowen–York type imply that we can not generate all possible black hole initial data. In particular, it turns out that while a single Bowen–York black hole with zero linear momentum and zero spin reduces to a Schwarzschild black hole in isotropic coordinates, the case with zero linear momentum and non-zero spin does not reduce to a Kerr black hole in *any* set of coordinates. That this is so can be seen from a result of Garat and Price that shows that there is no slicing of a Kerr black hole that is axisymmetric and conformally flat [139].

As it is known that the Kerr solution is the only axisymmetric stationary black hole solution, we might wonder what a Bowen–York black hole actually describes, since by construction it is a black hole and it is axisymmetric. The obvious answer is that a Bowen–York black hole is not stationary, *i.e.* it is initial data for a dynamical spacetime that corresponds to Kerr plus some spurious gravitational waves. In fact, for the particular case of an inversion-symmetric Bowen–York black hole, we can explicitly show that it corresponds to a Kerr black hole plus a *Brill wave*, that is, a specific type of gravitational wave distortion [77, 78]. Something similar is also found for a Bowen–York black hole with linear momentum and zero spin: It does not correspond to a simple boosted Schwarzschild black hole and contains some spurious gravitational wave contribution.

The existence of this spurious gravitational wave content with no clear physical interpretation has led to the search for alternative ways of finding black hole initial data.[33] It is important to stress the fact that the problem does not come from the use of the techniques described in Sections 3.2 and 3.3 which are completely general, but rather from the specific choices made for the free data: a maximal initial slice ($K = 0$) with a conformally flat metric and a purely longitudinal Bowen–York extrinsic curvature.

Although there has been some work on trying to find alternatives to the Bowen–York extrinsic curvature while still retaining a conformally flat metric (see for example [106]), most work has centered around the idea of using Kerr black holes, originally proposed by Matzner, Huq and Shoemaker [201]. The proposal calls for the use of superposed boosted Kerr black holes as initial data, but written in Kerr–Schild form to make sure that there are no singularities at the black hole horizons. Thus, when the black holes are far apart no spurious gravitational wave content will be present.

Of course, since the Einstein equations are non-linear the principle of superposition is not satisfied, so that we must "correct" this initial data by solving the constraints. The method starts by first identifying the 3+1 quantities from the Kerr–Schild metric (1.16.16), which turn out to be

$$\alpha^2 = \frac{1}{1+2H} \,, \tag{3.4.33}$$

$$\beta^i = \frac{2Hl_*^i}{1+2H} \,, \qquad \beta_i = 2Hl_i \,, \tag{3.4.34}$$

$$\gamma_{ij} = \delta_{ij} + 2Hl_il_j \,, \quad \gamma^{ij} = \delta^{ij} - 2Hl_*^il_*^j \left(1 - \frac{2H}{1+2H}\right) \,, \tag{3.4.35}$$

with $l_*^i := \delta^{ij}l_j$. Now, for a time-symmetric situation we would want the time derivative of the metric to vanish at $t = 0$. Using this we can find that the extrinsic curvature is given by

$$\begin{aligned} K_{ij} &= \frac{1}{2\alpha}\left(D_i\beta_j + D_j\beta_i\right) \\ &= \frac{1}{\sqrt{1+2H}}\left[\partial_i\left(Hl_j\right) + \partial_j\left(Hl_i\right) + 2H\, l_*^a\partial_a\left(Hl_il_j\right)\right] \,. \end{aligned} \tag{3.4.36}$$

In the particular case of Schwarzschild we have $H = M/r$ and $l_i = x_i/r$, so that the extrinsic curvature becomes

$$K_{ij} = \frac{2M}{r^4\sqrt{1+2M/r}}\left[r^2\delta_{ij} - x_ix_j\left(2 + M/r\right)\right] \,. \tag{3.4.37}$$

[33]The total gravitational wave energy content of a single spinning Bowen–York black hole is in fact quite small. Both perturbative and fully non-linear numerical simulations indicate that for an angular momentum of $J/M^2 \sim 0.5$ the energy radiated by such a black hole is of order $10^{-4}M$ [145, 106, 91], which should be compared with $\sim 10^{-3}M$ for the energy radiated by a head-on collision of two black holes [28], and $\sim 10^{-2}M$ for the energy radiated by the inspiral collision of two orbiting black holes [42, 43].

If, on the other hand, we want to have a black hole with some initial linear momenta, we can use instead a boosted Schwarzschild or Kerr black hole as starting point and compute the corresponding extrinsic curvature.

In order to find binary black hole initial data we now choose the conformal metric $\bar{\gamma}_{ij}$ as a direct superposition of two Kerr–Schild metrics

$$\bar{\gamma}_{ij} dx^i dx^j = \left[\delta_{ij} + 2H^{(1)}(r_1)\, l_i^{(1)} l_j^{(1)} + 2H^{(2)}(r_2)\, l_i^{(2)} l_j^{(2)} \right] dx^i dx^j \,, \qquad (3.4.38)$$

where $H^{(a)}$ and $l_i^{(a)}$ are the functions corresponding to a single black hole. As before, the physical metric will be related to this metric by $\gamma_{ij} = \psi^4 \bar{\gamma}_{ij}$, and we would need to solve the Hamiltonian constraint to find ψ.

For the momentum constraints, we take as a trial extrinsic curvature simply the direct sum of the extrinsic curvatures for the two Kerr–Schild black holes. We then take the trace of the resulting extrinsic curvature as K, and the tracefree part as $\bar{M}^{ij}$, and solve the momentum constraints using the techniques described in Section 3.2 to find the physical extrinsic curvature.

Since the work of Matzner *et al.*, other approaches based on the idea of using Kerr–Schild data as a starting point have also been proposed [58, 210]. We might expect that these approaches would result in smaller amounts of spurious gravitational waves than the Bowen–York approach. Kerr–Schild initial data for binary black hole collisions have in fact been used in practice, though their total content of gravitational waves has not yet been estimated.

3.5 Binary black holes in quasi-circular orbits

Perhaps the most important motivation for constructing multiple black hole initial data is the study of the inspiral coalescence of binary black holes. This is the simplest form of the two body problem in general relativity, which is well known to have no solution in closed form. The reason for this is easy to understand: As the black holes orbit each other they continually radiate gravitational waves. As a result of this the system loses energy and the black holes spiral in, orbiting faster and faster until they eventually collide and merge.

To study this problem numerically, we would first need to find initial data for a binary system of black holes in quasi-circular orbits. This is because the emission of gravitational waves is expected to circularize an initially elliptical orbit in time-scales much shorter than the merger time-scale. To date, there are basically two approaches for constructing quasi-circular initial data for binary black holes. The first approach was introduced in the early 1990s and is based on the idea of minimizing a measure of the total energy of the binary while keeping the total angular momentum fixed. This approach is known to result in circular orbits in the case of Newtonian gravity, and is therefore expected to produce quasi-circular orbits in the general relativistic case. The second approach was proposed by Grandclement *et al.* in 2001 and is based on the idea of looking for solutions in quasi-equilibrium, for which an approximate Killing field exists.

Below we will briefly discuss each of these two approaches to finding initial data for black holes in quasi-circular orbits.

3.5.1 *Effective potential method*

As already mentioned, the effective potential method is based on the fact that minimizing the total energy of a binary system in Newtonian gravity, while keeping the total angular momentum fixed, results in circular orbits. The problem with applying this idea to the case of general relativity is twofold. First, it is clear that we can not find perfectly circular orbits since the black holes will radiate gravitational waves. However, we would expect that for a system that is not too close to the final plunge phase the time-scale for energy loss due to gravitational wave emission will be considerably larger than the orbital time-scale. The second problem has to do with the definition of the energy of the system in general relativity. Notice that although a measure of the total energy content of the spacetime can in fact be found in terms of the ADM mass as described in Appendix A, this total energy includes both the gravitational "binding energy" of the system and the mass energy of the individual black holes. We must clearly only minimize the *effective potential energy* defined as

$$E_b := M_{ADM} - M_1 - M_2 \; , \tag{3.5.1}$$

where M_{ADM} is the ADM mass of the full spacetime, and M_i are the individual masses of each black hole. The basic problem here is how to define the masses of the individual black holes, as these are not well-defined concepts for black holes that are close to each other. The usual approach is to use the relationship between the area of the horizon A and the mass M of a single Kerr black hole

$$M^2 = M_I^2 + S^2/4M_I^2 = A/16\pi + 4\pi S^2/A \; , \tag{3.5.2}$$

where $M_I = \sqrt{A/16\pi}$ is the irreducible mass and S is the spin of the black hole.

In order to find quasi-circular orbits we then have to fix the spin S_i and horizon areas A_i of the individual black holes so that their total masses remain constant, fix also the total angular momentum of the spacetime J, and then minimize the binding energy as a function of the separation L between the black hole horizons. In practice we must move the black holes to a certain separation, locate the individual horizons (see Chapter 6) and adjust the parameters of the free data until the physical parameters S_i, A_i and J reach the desired values. After that, we compute the ADM mass M_{ADM} for that separation and start again at a different separation until we find a minimum.

Quasi-circular data using the effective potential method, together with an extrinsic curvature of the Bowen–York type, were computed for isometric data by Cook [101], and later for puncture data by Baumgarte [49], with very similar results found in both cases. As we decrease the total angular momentum of the system, the minimum in the binding energy becomes shallower and is found at progressively closer separations. At a certain critical separation the minimum disappears altogether so that no quasi-circular orbit is found. This critical separation is known as the *innermost stable circular orbit* (ISCO), and if the holes are closer than this they will stop orbiting and will just plunge toward each other.[34] For non-spinning equal-mass black holes, Cook finds the ISCO at a proper separation between the black hole horizons of $L/m \sim 4.88$, with $m = 2M$ and M the mass parameter of the individual black holes, corresponding to a coordinate separation between the black hole centers of $D/m \sim 2.2$. The two black holes have equal and opposite linear momenta given by $P/m \sim 0.33$, and the total angular momentum of the system is $J/m^2 \sim 0.74$. The angular velocity can also be estimated using

$$\Omega = \frac{\partial E_b}{\partial J} \, , \tag{3.5.3}$$

which for the ISCO results in $m\Omega \sim 0.17$. A table of parameters for quasi-circular orbits scaled in terms of the ADM mass of the system can be found in [43].

The use of the effective potential method together with Bowen–York extrinsic curvature can only be expected to be a good approximation for large separations, so the position of the ISCO found with this method should not be trusted. Indeed, numerical simulations of this ISCO show that the black holes do not orbit and instead rapidly plunge toward each other [42, 11]. However, the quasi-circular orbit estimations should rapidly improve as the separation is increased.

3.5.2 *The quasi-equilibrium method*

As mentioned above, the effective potential method coupled to the use of extrinsic curvature of the Bowen–York type provides us with initial data that are a good approximation to black holes in quasi-circular orbit at large separations, but the data can not be trusted for separations close to the ISCO.

Recently, an alternative method for finding initial data for black holes in quasi-circular orbits has been proposed by Grandclement *et al.* [150], based on an approach that has been used very successfully in the past to construct initial data for orbiting neutron stars. The basic idea is that for sufficiently separated black holes the spacetime can be supposed to be quasi-stationary, which means that there should exist an approximate helical Killing field $\vec{\xi}$. We can then use this idea to construct initial data for which the conformal 3-geometry remains momentarily constant. Notice that, in order to do this consistently, we need to be on a coordinate frame corotating with the black holes. Since the idea is based on trying to reduce as much as possible the rate of change of the conformal metric, the natural framework for this approach is the conformal thin-sandwich formalism described in Section 3.3.

[34] The ISCO also exists for a test particle orbiting a Schwarzschild black hole, and in that case it is found at an areal radius of $r = 6M$, corresponding to a proper distance between the black hole horizon and the test mass of ~ 4.58. This shows a very important difference between general relativity and Newton's theory for which circular orbits exist arbitrarily close to a point mass.

Let us then start from the standard conformal decomposition of the metric $\bar{\gamma}_{ij} = \psi^{-4}\gamma_{ij}$. We will now assume that the initial conformal metric is flat $\bar{\gamma}_{ij} = \delta_{ij}$, and also that we start from a maximal slice $K = 0$. Since we want quasi-stationary initial data, it is natural to ask for

$$\bar{u}_{ij} := \partial_t \bar{\gamma}_{ij} = 0 \, , \qquad \partial_t K = 0 \, . \tag{3.5.4}$$

Notice that with these choices we have now completely exhausted the free data for the conformal thin-sandwich method. In particular, the previous choices will fix the lapse and shift through equations (3.3.9) and (3.3.18).

Even though by construction we will have momentarily stationary values for the conformal metric and extrinsic curvature, in general we will not have $\partial_t \psi = 0$ or $\partial_t \bar{A}_{ij} = 0$. This is where the existence of an approximate Killing field comes into play. It implies that it should be possible to choose coordinates such that $\partial_t \psi \sim 0$ and $\partial_t \bar{A}_{ij} \sim 0$. Of course, as already mentioned the lapse and shift have already been fixed by our previous choices, but only up to the boundary conditions used for the elliptic equations.

Consider the boundary conditions at infinity. Asymptotic flatness implies that the lapse must be such that $\lim_{r\rightarrow\infty} \bar{\alpha} = 1$. The shift, however, is more interesting. As already mentioned it is clear that if want to minimize changes in the geometry we must go to a corotating frame, so that far away the shift vector must approach a rigid rotation:

$$\lim_{r\rightarrow\infty} \vec{\beta} = \Omega \, \vec{e}_\phi \, , \tag{3.5.5}$$

where $\vec{e}_\phi$ is the basis vector associated with the azimuthal angle ϕ.

At this point, however, we still don't have information that will allow us to fix the angular velocity Ω. Grandclement *et al.* suggest that this can be done by comparing two different definitions of the mass of the spacetime, namely the ADM mass and the Komar mass. As discussed in Appendix A, the total mass and angular momentum at spatial infinity can be expressed via the ADM integrals:

$$M_{ADM} = \frac{1}{16\pi} \lim_{r\rightarrow\infty} \oint \gamma^{ij} \left[\gamma_{ik,j} - \gamma_{ij,k} \right] dS^k \, , \tag{3.5.6}$$

$$J_{ADM} = \frac{1}{8\pi} \lim_{r\rightarrow\infty} \oint K_{ij} {e_\phi}^i dS^j \, , \tag{3.5.7}$$

where the integrals are taken over coordinate spheres, and where $dS^i = s^i dA$ with s^i the unit outward-pointing normal vector to the sphere (in the first integral above, the spatial metric must be expressed in asymptotically Cartesian coordinates). In the particular case of a conformally flat spatial metric like the one we are using here, the first integral can be shown to reduce to

$$M_{ADM} = -\frac{1}{2\pi} \lim_{r\rightarrow\infty} \oint \partial_i \psi \, dS^i \, . \tag{3.5.8}$$

Now, in the case when the spacetime admits a Killing field $\vec{\xi}$, we can also define conserved quantities through the Komar integral (remember that for a Killing field we have $\nabla_{(\mu}\xi_{\nu)} = 0$):

$$I_K\left(\vec{\xi}\right) = -\frac{1}{4\pi}\oint s^\mu n^\nu \nabla_\mu \xi_\nu \, dA \,, \tag{3.5.9}$$

where here n^μ is the timelike unit normal to the spacelike hypersurfaces. If ξ^μ is timelike and such that $\xi_\mu \xi^\mu = 1$ at infinity, then the above integral corresponds to the total mass of the spacetime M, while if it is an axial vector associated with an angular coordinate ϕ it will correspond to $-2J$, with J the total angular momentum (the factor -2 is needed to obtain the correct normalization). The Komar mass and angular momentum should coincide with the ADM integrals.

In our case we have neither a timelike nor an axial Killing field, but we do have an (approximate) helical Killing field. If we choose the boundary conditions on the shift appropriately, this Killing field should correspond to the time vector $\xi^\mu = t^\mu = \alpha n^\mu + \beta^\mu$, and we would expect the following relation to hold [290]

$$I_k\left(\alpha\vec{n} + \vec{\beta}\right) = M_{ADM} - 2\Omega J_{ADM} \,, \tag{3.5.10}$$

where the factor Ω appears because far away $\vec{\beta} = \Omega \vec{e}_\phi$. Notice now that

$$s^\mu n^\nu \nabla_\mu \xi_\nu = s^\mu n^\nu \left(\alpha \nabla_\mu n_\nu + n_\nu \nabla_\mu \alpha + \nabla_\mu \beta_\nu\right) \,. \tag{3.5.11}$$

Using the fact that $n_\mu n^\mu = -1$ and $n_\mu \beta^\mu = 0$ we can rewrite this as

$$s^\mu n^\nu \nabla_\mu \xi_\nu = s^\mu \left(-\nabla_\mu \alpha - \beta^\nu \nabla_\mu n_\nu\right) = -s^\mu \left(\nabla_\mu \alpha - \beta^\nu K_{\mu\nu}\right) \,. \tag{3.5.12}$$

The Komar integral then becomes

$$\begin{aligned} I_K\left(\vec{\xi}\right) &= \frac{1}{4\pi}\oint \left(\nabla_k \alpha - \beta^j K_{jk}\right) dS^k \\ &= \frac{1}{4\pi}\oint \nabla_k \alpha \, dS^k - \frac{\Omega}{4\pi}\oint K_{jk} {e_\phi}^j \, dS^k \,. \end{aligned} \tag{3.5.13}$$

Comparing the last result with the ADM expressions given above, we see that the angular momentum part is identical while the mass part differs, being given in the Komar case in terms of the lapse α and in the ADM case in terms of the conformal factor ψ. The relation (3.5.10) then reduces to

$$\oint \nabla_k \alpha \, dS^k = -2\oint \nabla_k \psi \, dS^k \,. \tag{3.5.14}$$

The last condition must hold if $t^\mu = \alpha n^\mu + \beta^\mu$ is a Killing vector, but for this to be the case we must know the correct angular velocity in order to have the shift correspond to a true corotation frame. The key observation here is that we can turn the condition around and use it to determine the angular velocity. The idea is to solve the conformal thin-sandwich equations with a given value of Ω fixing the exterior boundary condition for the shift, and then change the value of Ω until (3.5.14) is satisfied. Of course, in our case we only have an approximate Killing field, but we would expect that this way of fixing the angular velocity will still bring us to the correct corotating frame. Notice that for Schwarzschild the asymptotic behavior of the lapse and the conformal factor is

$$\psi \simeq 1 + M/2r \; , \qquad \alpha \simeq 1 - M/r \; , \tag{3.5.15}$$

which is precisely of the form (3.5.14). In the more general case we will still have $\psi \simeq 1 + M_\psi/2r$ and $\alpha \simeq 1 - M_\alpha/r$, but M_ψ and M_α can not be expected to coincide unless we use the correct angular velocity for the shift boundary condition. This is the fundamental element of the quasi-equilibrium approach.

Of course, we still need to worry about the inner boundary conditions either at the black hole horizons or at the punctures. This issue is very technical and we will not discuss it in detail here. However, it should be mentioned that in the original approach of Grandclement *et al.* the initial data is assumed to be inversion-symmetric at the black hole throats, which also correspond to apparent horizons, a condition that requires the black holes to be corotating (*i.e.* their spin is locked with the orbital motion). However, as pointed out by Cook [103], this set of boundary conditions does not guarantee regularity, and when an artificial regularization procedure is used it leads to solutions that no longer satisfy the constraints near the throats. In [103] Cook has generalized the original proposal to allow for regular boundary conditions by giving up the inversion symmetry, and has thus been able to obtain solutions for black holes that can have arbitrary spins. Tichy *et al.* have also applied the quasi-stationary idea to puncture type data with Bowen–York extrinsic curvature [289, 290].

As a final note it is important to mention that very recently Caudill *et al.* have shown that, when using the conformal thin-sandwich approach, the effective potential method and the Komar mass method give results that are remarkably close to each other [95]. The main differences between the quasi-equilibrium and the puncture approaches should therefore be attributed to the use of the Bowen–York extrinsic curvature and not to the effective potential method itself.

4

GAUGE CONDITIONS

4.1 Introduction

As already mentioned in Chapter 2, in the 3+1 formalism the choice of the coordinate system is given in terms of the *gauge variables*: the lapse function α and the shift vector β^i. These functions appear in the evolution equations (2.3.11) and (2.5.5) for the metric and extrinsic curvature. However, Einstein's equations say nothing about how the gauge variables should be chosen. This is what should be expected, since it is clear that the coordinates can be chosen freely.

The freedom in choosing the gauge variables is a mixed blessing. On the one hand, it allows us to choose the coordinates in a way that either simplifies the evolution equations or makes the solution better behaved. On the other hand, we are immediately faced with the following question: What is a "good" choice for the functions α and β^i?

There are, of course, some guidelines that one would like to follow when choosing good gauge conditions, which can be summarized in the following "wish list":

- Whenever possible, gauge conditions should be adapted to the underlying symmetries of the problem. Ideally the gauge conditions should automatically "seek" exact or approximate symmetries of the spacetime and make them evident in the evolution.
- The gauge conditions should avoid the formation of coordinate singularities. In the case of black hole spacetimes, they should also avoid reaching the physical singularity.
- Gauge conditions should be well behaved mathematically, and whenever possible easy to implement numerically.
- Ideally, gauge conditions should be expressed in 3-covariant form, this will guarantee that we will have the same gauge when using different spatial coordinate systems (*e.g.* Cartesian vs. spherical).

In order to find gauge conditions with these properties, the natural approach is to relate the gauge choice to the evolution of certain combinations of geometric quantities in such a way that the gauge will either freeze completely the evolution of those quantities (typically by solving some elliptic equations), or will attempt to do so with some time delay (by solving parabolic or hyperbolic equations instead). Below I will describe the most common gauge conditions used in practice in numerical relativity.

4.2 Slicing conditions

In order to specify a foliation of spacetime into spacelike hypersurfaces, also known as a synchronization, we need to prescribe a way to calculate the lapse function α which measures the proper time interval between neighboring hypersurfaces along the normal direction.

Typically, the different possible lapse choices can be classified in the following way: 1) Prescribed slicing conditions where the lapse is specified as an *a priori* known function of space and time, 2) algebraic slicing conditions where the lapse is specified as some algebraic function of the geometric variables at each hyper-surface, 3) evolution-type slicing conditions where the time derivative of the lapse is specified as an algebraic function of the geometric variables and the lapse is evolved as just another dynamical quantity, and 4) elliptic slicing conditions where the lapse is obtained by solving an elliptic differential equation at every time step that typically enforces some geometric condition on the spatial hypersurfaces.

Before considering different slicing conditions we must mention a couple of results that are very important when discussing this issue. First, consider the motion of the Eulerian (normal) observers. There is no reason to assume that these observers will be in free fall, so in general they should have a proper acceleration that measures the force that would be required to keep them in a trajectory different from free fall. This acceleration is given in terms of the directional derivative of the 4-velocity of the Eulerian observers, *i.e.* the normal vector n^μ, along itself:

$$a^\mu = n^\nu \nabla_\nu n^\mu \; . \tag{4.2.1}$$

Notice first that $n^\mu a_\mu = n^\mu n^\nu \nabla_\nu n_\mu = 0$, *i.e* the vector a^μ is orthogonal to n^μ (as expected since n^μ is unitary), so the proper acceleration is a purely spatial vector. Using now the expressions for n^μ in terms of α and β^i given in equation (2.2.8), and the expressions for the 4-Christoffel symbols found in Appendix B, we can easily show that

$$a_0 = \beta^m \partial_m \ln \alpha \; , \tag{4.2.2}$$

$$a_i = \partial_i \ln \alpha \; . \tag{4.2.3}$$

We then see that the spatial components of the proper acceleration are given by the gradient of the lapse.

Another important relationship comes from the evolution of the volume elements associated with the Eulerian observers. The change in time of these volume elements is simply given by the divergence of their 4-velocities $\nabla_\mu n^\mu$. Using now the definition of the extrinsic curvature we find that:

$$\nabla_\mu n^\mu = -K \; , \tag{4.2.4}$$

that is, the rate of change of the volume elements in time is just (minus) the trace of the extrinsic curvature.

4.2.1 *Geodesic slicing and focusing*

The most obvious way to choose the lapse function would be to ask for the coordinate time t to coincide everywhere with the proper time of the Eulerian observers, that is $\alpha = 1$. After all, what could be more natural? This is an example of a prescribed slicing condition and is known as *geodesic slicing*. The name comes from the fact that according to equation (4.2.3) the proper acceleration of the Eulerian observers vanishes for constant α, which implies that in this case the Eulerian observers follow timelike geodesics (*i.e.* they are in free fall).

Geodesic slicing was in fact used in the pioneering work of Hahn and Lindquist in the mid 1960s [158]. However, it is now well known that geodesic slicing is a very poor coordinate choice. The reason for this is easy to see if we think for a moment about what will happen to observers in free fall in a non-uniform gravitational field. As the field is non-uniform, different observers will fall in different directions, and nothing can prevent them from eventually colliding with one another. When this happens, our coordinate system (which is tied to these observers) stops being one-to-one: One point has now more than one set of coordinates associated with it. This means that the coordinate system has become singular.

In order to see this, consider the evolution equation for the trace of the extrinsic curvature. From the ADM evolution equations (2.3.12) and (2.5.6) we find that

$$\partial_t K = \beta^i \partial_i K - D^2 \alpha + \alpha \left[K_{ij} K^{ij} + 4\pi \left(\rho + S \right) \right] , \tag{4.2.5}$$

where we have used the Hamiltonian constraint (2.4.10) to eliminate the Ricci scalar. For geodesic slicing this equation reduces to

$$\partial_t K - \beta^i \partial_i K = \alpha \left[K_{ij} K^{ij} + 4\pi \left(\rho + S \right) \right] . \tag{4.2.6}$$

The first term on the right hand side is clearly always positive, and so is the second term if the strong energy condition holds. This means that along the normal direction the trace of the extrinsic curvature K will increase without bound, which through (4.2.4) implies that the volume elements associated with the Eulerian observers will collapse to zero. Because of this, geodesic slicing is never used in practice except to test numerical codes. For example, it is well known that the free fall time to the singularity for an observer initially at rest at the Schwarzschild horizon is $t = \pi M$. We can then set up initial data for Schwarzschild in isotropic coordinates, evolve using geodesic slicing, and expect the code to crash at that time.

4.2.2 *Maximal slicing*

In order to look for a better slicing condition, we must think about what precisely was wrong with geodesic slicing. The main problem was that observers in free fall are "focused" by the gravitational field. This means that they will get closer to each other, so the volume elements associated with them will become smaller and smaller until they become zero. We can then try to built a better slicing condition by demanding that the volume elements associated with the Eulerian observers remain constant. From equation (4.2.4) we can see that this is equivalent to asking for

$$K = \partial_t K = 0 \; . \tag{4.2.7}$$

It is clear that we must require not only that K vanishes initially, but also that it remains zero during the evolution. Asking now for the above condition to hold, we find through (4.2.5) that the lapse function must satisfy the following elliptic equation

$$D^2 \alpha = \alpha \left[K_{ij} K^{ij} + 4\pi \left(\rho + S \right) \right] \; . \tag{4.2.8}$$

This condition is known as *maximal slicing*.[35] The name comes from the fact that we can prove that when $K = 0$ the volume of the spatial hypersurface is maximal with respect to small variations in the hypersurface itself.

Maximal slicing was suggested originally by Lichnerowicz [189], and was already discussed in the classic papers of Smarr and York [272, 305]. It has been used over the years for many numerical simulations of a number of different systems, including black holes, and is still in use today. It has the advantage of being given by a simple equation whose solution is smooth (because it comes from an elliptic equation), and also of guaranteeing that the Eulerian observers will not focus. It is important to mention also that this slicing condition can only be used for asymptotically flat spacetimes and can not be used in the case of cosmological spacetimes where volume elements always expand or contract with time (on a closed Universe only one maximal slice exists at the moment of time symmetry). However, in that case we can relax the condition and use $K =$ constant, with a different constant for each slice (in fact, in such cases K can be used as a time coordinate). For asymptotically flat spacetimes we can in fact also use the condition $K = K_0 =$ constant, $\partial_t K = 0$. This corresponds to hyperboloidal slices that reach null infinity $\mathscr{J}^{\pm}$ (depending on the sign of K_0). Such hyperboloidal slices can be useful to study gravitational radiation at infinity, and are natural choices in Friedrich's conformal approach (see Section 2.9.2). They have also been used in numerical simulations based on the standard 3+1 approach to improve the stability of the evolution [142].

A very important property of maximal slicing is that of *singularity avoidance*, by which we mean that it does not allow the spatial hypersurfaces to come arbitrarily close to a physical singularity. In the particular case of the Schwarzschild spacetime the maximal slices can be constructed analytically (see the following Section), and it is well known that inside the horizon the maximal slices approach a limiting surface given by $r = 3M/2$, with r the Schwarzschild areal radius (remember that inside the horizon the hypersurfaces $r =$ constant are spacelike), so

[35] Maximal slicing is given by an elliptic equation similar to the Helmholtz equation, which in standard 3D vector notation is $\nabla^2 \phi + k^2 \phi = 0$. However, since as long as the strong energy condition holds we have $K_{ij} K^{ij} + 4\pi(\rho + S) \geq 0$, the sign of the source term in the maximal slicing equation is opposite to that of the standard Helmholtz equation.

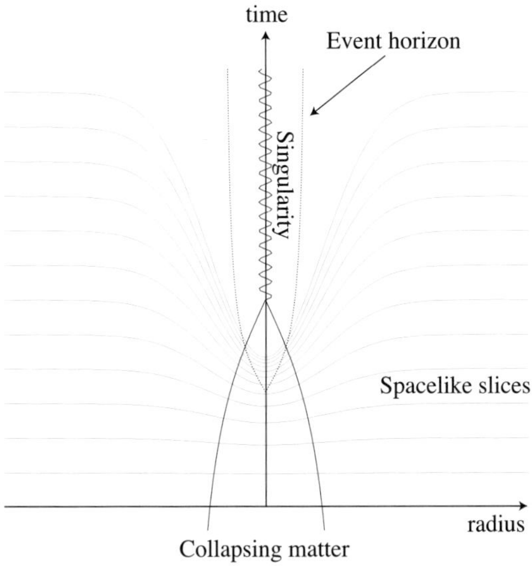

Fig. 4.1: Schematic representation of the collapse of the lapse when approaching a singularity. Time slows down in the region close to the singularity, but continues advancing away from the singularity.

that the singularity is avoided. On way to understand the behavior of maximal slices is to go back to equation (4.2.8) and use again the Hamiltonian constraint to eliminate the extrinsic curvature term. Concentrating on the vacuum case we find

$$D^2\alpha = \alpha R \, , \tag{4.2.9}$$

with R the Ricci scalar of the spatial hypersurface, which through the Hamiltonian constraint must be positive if $K = 0$. In their seminal paper [272], Smarr and York have constructed an analytical model considering a case where $R = R_0$ inside a spherical volume and $R = 0$ outside, and have shown that the lapse collapses exponentially to zero at the center as $\alpha \sim e^{-R_0}$. This exponential collapse of the lapse when approaching a singularity has also been observed numerically in many simulations of black hole spacetimes, and in fact is used in practice as an indicator of the formation of a black hole in gravitational collapse.[36]

Figure 4.1 shows a schematic representation of the singularity avoidance property of maximal slicing in the case of gravitational collapse. The *collapse of the lapse* has the consequence that time keeps advancing in the regions exterior to the black hole horizon, but effectively freezes inside the horizon. This allows us to cover a large portion of the exterior spacetime without reaching the singularity, which is ideal for numerical evolutions. There is however, a price to be

[36]However, it is known that maximal slices do not always avoid singularities. In particular, they fail to do so in the case of the spherical collapse of self-similar dust [115].

paid. As time advances outside and freezes inside, the spatial slices become more and more distorted, leading to a phenomenon known as *slice stretching* which results in a rapid growth of the radial metric component and the development of large gradients, which eventually cause numerical codes to fail.[37] As pointed out by Reimann in [237, 236], slice stretching is in fact a combination of two separate effects, referred to by Reimann as *slice sucking* and *slice wrapping*. Slice sucking refers to the fact that the mere presence of the black hole results in the differential infall of coordinate observers, as those closer to the black hole fall faster, resulting in their radial separations increasing and the radial metric component growing. This means that slice stretching will occur even for geodesic slicing. The second effect of slice wrapping is the one that one usually has in mind when thinking about singularity avoidance and is precisely a consequence of the collapse of the lapse and the slices "wrapping" around the singularity. Slice stretching results in a power law growth of the radial metric. For the maximal slicing of Schwarzschild, Reimann finds that at late times the peak in the radial metric grows as $\sim \tau^{4/3}$, with τ proper time at infinity (the loose statement often found in the numerical literature indicating that slice stretching results in "exponential" growth of the radial metric is therefore incorrect).

If there is a major disadvantage to maximal slicing it is the fact that solving elliptic equations numerically in three dimensions is a very slow process. Even with fast elliptic solvers, we can find that over 90% of the processor time is used to solve just the maximal slicing equation (with zero shift; if we also have elliptic shift conditions things get much worse). Also, setting up boundary conditions for complicated inner boundaries such as those that can be found when excising black hole interiors (see Chapter 6) can be a very hard problem. Because of this in the past few years maximal slicing has been giving place to hyperbolic-type slicing conditions like those that we will discuss in Section 4.2.4. Still, when computer time restrictions are not an issue and good boundary conditions are known, maximal slicing is probably the best slicing condition available.

Since maximal slicing requires the solution of an elliptic problem, the issue of boundary conditions is very important (here I will consider only the exterior boundaries and ignore the possible presence of complicated interior boundaries). For asymptotically flat spacetimes we can assume that far from the sources we should approach the Schwarzschild solution, in which case the lapse function behaves asymptotically as

$$\alpha = 1 - c/r \, , \tag{4.2.10}$$

with c some constant (in static coordinates we find $c = M$) . We can eliminate the unknown constant by taking a derivative with respect to r to find

$$\partial_r \alpha = (1 - \alpha) / r \, . \tag{4.2.11}$$

[37] In older literature one often finds that this is called *grid stretching*, but the name is misleading as this is a geometric effect that happens at the continuum level, and is quite independent of the existence of a numerical grid.

This is known as a *Robin* boundary condition and is the standard condition used when solving maximal slicing.

4.2.3 *Maximal slices of Schwarzschild*

In the case of a Schwarzschild black hole, the maximal slicing equation can in fact be solved analytically [51, 52, 126, 224, 238, 239]. This allows us to understand many of the properties of this slicing in detail, and even make quantitative statements about the late time behavior of the lapse. In the discussion presented here I will follow closely the work of Beig and O'Murchadha [51, 52] (there are other ways to derive the same results – in particular Estabrook *et al.* take a different route that starts from the constraint equations [126]). We start from the Schwarzschild metric in standard Schwarzschild coordinates:

$$ds^2 = -\left(1 - \frac{2M}{r}\right) dt^2 + \left(1 - \frac{2M}{r}\right)^{-1} dr^2 + r^2 d\Omega^2 \; . \tag{4.2.12}$$

Consider for a moment a hypersurface $r =$ constant. We know that for $r < 2M$ these hypersurfaces will be spacelike. The trace of the extrinsic curvature K for such a hypersurface can be easily obtained. By taking t to be a space coordinate and r to be the time coordinate (but moving backwards as r becomes smaller with proper time), the spatial volume elements are found to be given by $\gamma = r^4 \sin^2\theta \, (2M/r - 1)$ and the lapse function becomes $\alpha = 1/\sqrt{2M/r - 1}$. In that case K can be found through the ADM equations by

$$\frac{d}{dr} \ln \gamma = 2\alpha K \quad \Rightarrow \quad K = \frac{3M - 2r}{r^2 \sqrt{2M/r - 1}} \; . \tag{4.2.13}$$

Notice that here r is taken to be constant, so K is the same along the slice. In particular, we see that the hypersurface $r = 3M/2$ is a maximal slice. We will see below that this will be the limiting slice for the maximal foliation of Schwarzschild.

Consider now some arbitrary spherically symmetric spatial hypersurface in the Schwarzschild spacetime given as

$$\sigma(t, r) = t - F(r) = \text{constant} \quad \Rightarrow \quad t = F(r) + \sigma \; . \tag{4.2.14}$$

Let us now construct the normal vector n^μ to this hypersurface

$$n_\mu = -N \, \nabla_\mu \sigma = -N \, (1, -F') \; , \tag{4.2.15}$$

with $F' := dF/dr$, and where N is a normalization coefficient fixed by asking for n^μ to be a unit vector

$$n_\mu n^\mu = N^2 \left[-\left(1 - \frac{2M}{r}\right)^{-1} + \left(1 - \frac{2M}{r}\right) F'^2 \right] = -1 \; . \tag{4.2.16}$$

Solving the last equation for F' we find

$$F'^2 = \left(1 - \frac{2M}{r}\right)^{-2} - \frac{1}{N^2}\left(1 - \frac{2M}{r}\right)^{-1} . \tag{4.2.17}$$

Let us now ask for our slice to be maximal, that is

$$K = -\nabla_\mu n^\mu = -\frac{1}{|g|^{1/2}}\,\partial_\mu \left(|g|^{1/2} n^\mu\right) = 0 . \tag{4.2.18}$$

Using the Schwarzschild metric above and the expression for n^μ we find

$$0 = \frac{1}{r^2}\,\partial_r \left[r^2 N F' \left(1 - \frac{2M}{r}\right)\right] \quad \Rightarrow \quad r^2 N F' \left(1 - \frac{2M}{r}\right) = C , \tag{4.2.19}$$

with C some constant (C is often called the *Estabrook–Wahlquist time* as it can be used to parameterize the different slices [126]). Inserting now the expression for F' in the equation above and solving for N we find

$$N = \frac{1}{r^2}\,\left(r^4 - 2Mr^3 + C^2\right)^{1/2} . \tag{4.2.20}$$

And substituting this back into the expression for F' we finally find

$$F' = \pm \frac{C}{(1 - 2M/r)(r^4 - 2Mr^3 + C^2)^{1/2}} . \tag{4.2.21}$$

There are a couple of comments to be made here before proceeding any further. First, consider the sign in the previous expression. As we have seen, the timelike normal vector is given by $n_\mu = -N\ (1, -F')$, with $N > 0$. For $r > 2M$ the coordinate t is timelike, and the vector n_μ will clearly be future pointing regardless of the sign of F'. For $r < 2M$, however, the timelike coordinate is now r, and r becomes smaller to the future, so that in order for n_μ to be future pointing we must ask for $n_r > 0$, that is $F' > 0$. We therefore take the minus sign in the expression for F' above.

The second point has to do with the fact that we are looking for slices that do not reach the singularity at $r = 0$. That is, as seen in the Kruskal diagram, we want slices that come in from infinity, reach a minimum value of r at the *throat* of the Einstein–Rosen bridge, and then go back to infinity on the other asymptotically flat region. At the throat the slices should in fact be tangential to a surface $r =$ constant, so that $dt/dr = F'$ becomes infinite. Looking at (4.2.21), we see that F' becomes infinite both at the horizon $r = 2M$ and at any root of the quartic polynomial

$$P(r) = r^4 - 2Mr^3 + C^2 . \tag{4.2.22}$$

As we will see below, at $r = 2M$ the infinite value of F' is related to the singularity of the Schwarzschild coordinates, and the slices in fact go smoothly through the horizon. We then conclude that we must have a non-trivial real root of $P(r)$ if we want to obtain slices that do not reach the singularity at $r = 0$. Looking at the form of the polynomial $P(r)$ it is not difficult to show that there will be two distinct real roots for $C^2 < 27M^4/16$ (both roots are such that $r > 0$), and no roots if C^2 exceeds this value.[38] We will therefore ask for $C^2 < 27M^4/16$. Define now r_C as the position of the largest real root of $P(r)$. The particular case $C^2 = 0$ corresponds to $r_C = 2M$ and in fact reduces to the Schwarzschild $t = 0$ slice (or $t =$ constant for a non-zero value of σ). On the other hand, as C^2 approaches the critical value $27M^4/16$, r_C becomes smaller and approaches the value $3M/2$ corresponding to the limiting maximal slice of Schwarzschild.

The function F will now be given by integrating (4.2.21):

$$F(r, C) = -\int_{r_C}^{r} \frac{C}{(1 - 2M/x)\,[P(x)]^{1/2}}\, dx \; , \tag{4.2.23}$$

where the integral over the pole at $x = 2M$ is taken in the sense of the principal value. In order to see that the surfaces are indeed smooth at $r = 2M$, define first

$$H(r, C) := F(r, C) + 2M \ln\left(r/2M - 1\right) \; . \tag{4.2.24}$$

This implies that

$$\frac{dH}{dr} = -\frac{Cr - 2M\,[P(r)]^{1/2}}{(r - 2M)\,[P(r)]^{1/2}} \; . \tag{4.2.25}$$

Taking now the limit of the last expression when r approaches $2M$ we find

$$\lim_{r \to 2M} \frac{dH}{dr} = \frac{8M^4}{C^2} - 1 \; . \tag{4.2.26}$$

That is, $H(r, C)$ is a smooth function of r at the horizon. But notice now that

$$H(r, C) = t - \sigma + 2M \ln\left(1 - r/2M\right) = \tilde{t} - \sigma \; , \tag{4.2.27}$$

with $\tilde{t} = t + 2M \ln\left(1 - r/2M\right)$, the Eddington–Finkelstein time used to transform the Schwarzschild metric to the regular Kerr–Schild form (1.15.20). We then find that in terms of the regular coordinate $\tilde{t}$, the slices defined above cross the horizon smoothly.

Let us now go back to the expression for $F(r, C)$, equation (4.2.23). There are two ways we can use this to derive a maximal foliation of the Schwarzschild spacetime. The first is to simply take $C = 0$ in which case the foliation reduces to $t = \sigma =$ constant, so that we recover the standard Schwarzschild slices (which are of course maximal since the metric is time independent and the shift vector vanishes, which implies $K_{ij} = 0$). We call these the *odd maximal slices* of Schwarzschild since they correspond to the lapse going to 1 at one of the asymptotically flat regions and to -1 at the other asymptotically flat region.

[38] Notice that since there are two roots with $r > 0$, we in fact have two families of solutions, one from $r = 0$ to the left root, and another from the right root to $r = \infty$. The first family will clearly reach the singularity at $r = 0$ so we are not interested in it.

A more interesting situation is obtained by varying the value of C, starting from 0 and reaching the limiting value $C_{\rm lim} = 3\sqrt{3}M^2/4$. In this case we take $\sigma = 0$ since changing the value of σ will give us precisely the same foliation only starting at a different value of the Schwarzschild time t. Instead of parameterizing these slices with the value of the constant C, we will use as parameter the proper time measured at infinity, which is clearly given by

$$\tau(C) = F(\infty, C) = -\int_{r_C}^{\infty} \frac{C}{(1-2M/x)\,[P(x)]^{1/2}}\,dx \; . \tag{4.2.28}$$

The lapse function α associated with this foliation is then defined through

$$n_\mu = -\alpha \nabla_\mu \tau \; . \tag{4.2.29}$$

Notice now that $\vec{e}_t \cdot \vec{n} = n_t = -N$, with $\vec{e}_t$ the coordinate basis vector associated with t, and also that $\vec{e}_t \cdot \nabla\tau = \partial\tau/\partial t$. This implies

$$N = \alpha \, \frac{\partial \tau}{\partial t} \; . \tag{4.2.30}$$

Solving for α we find

$$\alpha = N \, \frac{\partial t}{\partial \tau} = N \, \frac{\partial F}{\partial \tau} = N \, \frac{\partial F}{\partial C} \frac{dC}{d\tau} \; . \tag{4.2.31}$$

The problem now is to find $\partial F/\partial C$, which is a non-trivial task. To see why this is so, notice that the lower limit of integration in (4.2.23) is r_C which clearly depends on C. We therefore pick up a boundary term when differentiating with respect to C, given essentially by the integrand itself evaluated at r_C, which clearly diverges (remember that r_C is a root of $P(r)$). There is, however, a trick that can be used to calculate this derivative (the derivation is rather long so we will not include it here; the interested reader can see Appendix B of [51]). We find that

$$\frac{\partial F(r,C)}{\partial C} = \frac{r^2}{2\,(r-3M/2)\,P(r)^{1/2}} - \frac{1}{2} \int_{r_C}^{r} \frac{x\,(x-3M)}{(x-3M/2)^2\,P(x)^{1/2}}\,dx \; . \tag{4.2.32}$$

Using now the expression for N, the lapse finally becomes

$$\alpha = \frac{1}{2}\frac{dC}{d\tau}\left[\frac{1}{r-3M/2} - \frac{P(r)^{1/2}}{r^2} \int_{r_C}^{r} \frac{x\,(x-3M)}{(x-3M/2)^2\,P(x)^{1/2}}\,dx\right] . \tag{4.2.33}$$

Notice that this expression is regular at $r = r_C$. We can in fact also show that both α and N are linearly independent spherically symmetric solutions of the maximal slicing equation $(D^2 - K_{ij}K^{ij})f = 0$, but with different boundary conditions: Both N and α go to 1 at the right infinity, but at the left infinity N goes to -1 while α goes to 1.

We can now use expression (4.2.33) to study the late time behavior of the lapse at different interesting locations. Notice first that, as C goes to $C_{\rm lim} = 3\sqrt{3}M^2/4$, $\tau(C)$ goes to infinity and the areal radius r goes to $3M/2$. Define now

$$\delta := r_C - 3M/2 \,, \tag{4.2.34}$$

that is, δ is the difference between the radius at the throat and its limiting value. In terms of δ, the parameter C can be rewritten as

$$C = (\delta + 3M/2)^{3/2} \, (M/2 - \delta)^{1/2} \,. \tag{4.2.35}$$

A long calculation (see [51]) shows that, for small δ, the following relation holds

$$\frac{\tau}{M} = -\Omega \ln \frac{\delta}{M} + A + \mathcal{O}(\delta) \,, \tag{4.2.36}$$

with Ω and A constants given by

$$\Omega = \frac{3\sqrt{6}}{4} \simeq 1.8371 \,, \tag{4.2.37}$$

$$A = \frac{3\sqrt{6}}{4} \ln \left[18 \left(3\sqrt{2} - 4 \right) \right] - 2 \ln \left[\frac{3\sqrt{3} - 5}{9\sqrt{6} - 22} \right] \simeq -0.2181 \,. \tag{4.2.38}$$

We then have

$$\frac{d\tau}{dC} = \frac{d\tau}{d\delta} \frac{d\delta}{dC} \simeq \left(-\frac{3\sqrt{6}}{4} \frac{M}{\delta} \right) \left(-\frac{1}{\sqrt{2}\sqrt{6}\,\delta} \right) = \frac{3}{4\sqrt{2}} \frac{M}{\delta^2} \,. \tag{4.2.39}$$

Notice also that from (4.2.33) the lapse at the throat can be seen to be given by

$$\alpha(r_C) = \frac{dC}{d\tau} \left(\frac{1}{2\delta} \right) \,, \tag{4.2.40}$$

so that we finally find

$$\alpha(r_C) \simeq \frac{2\sqrt{2}}{3} \frac{\delta}{M} \simeq \frac{2\sqrt{2}}{3} \exp\left(A/\Omega \right) \exp\left(-\tau/\Omega M \right) \,. \tag{4.2.41}$$

This shows that for the Schwarzschild spacetime the value of the lapse at the throat collapses exponentially with time as the slices approach the limiting surface $r = 3M/2$. The time-scale of this exponential collapse, normalized with respect to the mass M, is given by $\Omega = 3\sqrt{6}/4 \simeq 1.8371$. This decay rate was in fact discovered very early on. For example, by using a combination of numerical results and a model problem, Smarr and York estimated this time-scale to be ~ 1.82 [272].

It is also interesting to study the late time behavior of the lapse at the black hole horizon. For a long time it has been known empirically that for maximal slicing the lapse at the horizon approaches the value $\alpha \sim 0.3$, and in fact in numerical simulations there is a "rule of thumb" that says that once the collapse of the lapse is well under way the level surface $\alpha \sim 0.3$ is a rough indicator of the position of the apparent horizon. In [238], Reimann and Bruegmann have studied the late time behavior of the lapse at the horizon using the techniques described above. Notice that from equation (4.2.28) we find

$$\frac{d\tau}{dC} = \frac{\partial F(\infty, C)}{\partial C} \; . \tag{4.2.42}$$

The lapse (4.2.33) can therefore be rewritten as

$$\alpha = -\frac{1}{K_C(\infty)} \left[\frac{1}{r - 3M/2} - \frac{P(r)^{1/2}}{r^2} \, K_C(r) \right] , \tag{4.2.43}$$

with

$$K_C(r) := \int_{r_C}^{r} \frac{x\,(x - 3M)}{(x - 3M/2)^2 \, P(x)^{1/2}} \, dx \; . \tag{4.2.44}$$

The lapse at the horizon $r = 2M$ will then be

$$\alpha_{r=2M} = -\frac{1}{K_C(\infty)} \left[\frac{2}{M} - \frac{C}{4M^2} \, K_C(2M) \right] . \tag{4.2.45}$$

Reimann and Bruegmann show that at late times $K_C(\infty)$ blows up as $1/\delta^2$, and $K_C(2M) \simeq K_C(\infty) + \eta + \mathcal{O}(\delta^2)$, with η a numerical constant whose exact value is known but is of no consequence here. This implies that at late times

$$\alpha_{r=2M} = \frac{C_{\rm lim}}{4M^2} = \frac{3\sqrt{3}}{16} \sim 0.3248 \; , \tag{4.2.46}$$

which is very close to the value ~ 0.3 found empirically.

There is one last important issue to discuss regarding the maximal slicing of Schwarzschild. Notice that for the symmetric (even) slices we have been discussing here, the lapse collapses at the throat but remains equal to one at the two asymptotic infinities. If we evolve puncture data then this means that the lapse will remain equal to one at the puncture, which is in fact a point in the middle of the computational domain corresponding to $\tilde{r} = 0$ with $\tilde{r}$ the isotropic radial coordinate. However, if we solve the maximal slicing condition numerically over the whole computational domain using Robin boundary conditions on the outer boundary, we find that the lapse in fact collapses at the puncture. That is, the numerical algorithm does not settle on the "even" solution, but rather on a solution that corresponds to having $\nabla_i \alpha = 0$ at the puncture, the so-called *zero gradient at puncture*, or simply the *puncture* lapse. This solution has also been

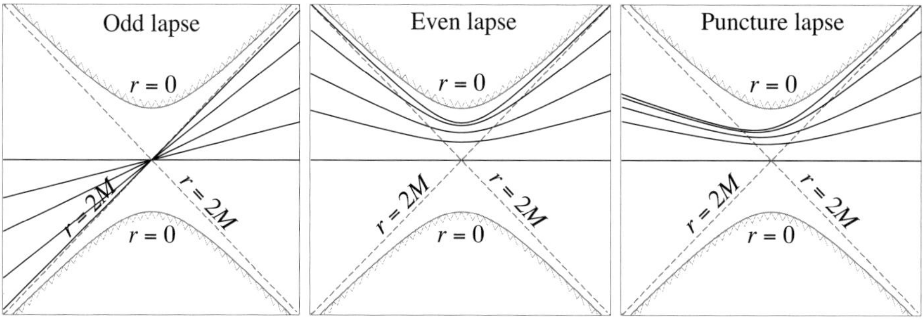

Fig. 4.2: Maximal slicing of Schwarzschild as seen in the Kruskal diagram for odd, even and puncture lapse (plots courtesy of B. Reimann).

studied analytically by Reimann and Bruegmann in [239]. They find that at the puncture the lapse also collapses exponentially with time, but is now such that

$$\alpha_{\tilde{r}=0} \simeq \frac{2\sqrt{2}}{3} \frac{\delta^2}{M^2} \simeq \frac{2\sqrt{2}}{3} \exp\left(2A/\Omega\right) \exp\left(-2\tau/\Omega M\right) , \tag{4.2.47}$$

so the collapse time-scale is twice as fast as at the throat. The limiting value of the lapse at the horizon, however, remains the same as before.

Figure 4.2 shows the maximal slicing of the Schwarzschild spacetime as seen in the Kruskal diagram for the case of the odd, even and puncture lapse. Notice that the odd lapse simply corresponds to the standard Schwarzschild time slices. The even lapse is symmetric across the throat, while the puncture lapse is asymmetric. The plots shown here are the true solution $t = F(r, C)$, with $F(r, C)$ given by (4.2.23). The coordinates $\{t, r\}$ are later transformed to Kruskal–Szekeres coordinates $\{\eta, \xi\}$ as discussed in Section 1.15. It is important to mention that in order to find numerical values for $F(r, C)$, we need first to find the root r_C of the polynomial $P(r)$. Also, the integrand in the expression for $F(r, C)$ has poles at both $r = 2M$ and $r = r_C$, so that the numerical evaluation of this expression is highly non-trivial. A procedure to perform this evaluation accurately has been developed by Thornburg and can be found in Appendix 3 of [285].

4.2.4 *Hyperbolic slicing conditions*

As already mentioned, one disadvantage of using maximal slicing is the fact that, in 3D solving the elliptic equation (4.2.8), every time step requires a lot of computer time. Because of this some alternative slicing conditions have been proposed that share some of the properties of maximal slicing but are at the same time much easier to solve.

Historically, there were two separate routes leading the way to the use of slicing conditions given through an evolution equation for the lapse function. One route was a formal approach to study the well-posedness of the Einstein field equations when viewed as an initial value or *Cauchy* problem. This approach used the so-called harmonic coordinates, which are defined by asking for the wave operator acting on the coordinate functions x^α to vanish:

$$\Box x^\alpha = g^{\mu\nu} \, \nabla_\mu \nabla_\nu \, x^\alpha = 0 \, . \tag{4.2.48}$$

Harmonic coordinate conditions have the important property of allowing the Einstein field equations to be written as a series of wave equations (with non-linear source terms) for the metric coefficients $g_{\mu\nu}$. Because of this, these conditions have been used in many analytical studies of the properties of the Einstein equations, and in particular were used by Choquet-Bruhat to prove the first theorems on the existence and uniqueness of solutions to these equations [84] (we will come back to the issue of the well-posedness of the Einstein evolution equations in Chapter 5).

Expanding the harmonic coordinate condition (4.2.48) in adapted coordinates we find

$$\Gamma^\alpha := g^{\mu\nu} \Gamma^\alpha_{\mu\nu} = 0 \, , \tag{4.2.49}$$

where $\Gamma^\alpha_{\mu\nu}$ are the Christoffel symbols associated with the 4-metric $g_{\mu\nu}$. Concentrating on the condition for the time coordinate $x^0 = t$, and using the expressions for the 4-Christoffel symbols in 3+1 language found in Appendix B, the above equation can be shown to reduce to

$$\frac{d}{dt} \, \alpha \equiv \left(\partial_t - \pounds_{\vec{\beta}} \right) \alpha = -\alpha^2 K \, . \tag{4.2.50}$$

This is known as the *harmonic slicing condition.* Notice that, through the ADM equations, this condition implies that

$$\frac{d}{dt} \, \tilde{\alpha} = 0 \, , \tag{4.2.51}$$

with $\tilde{\alpha} := \alpha / \sqrt{\gamma}$ the densitized lapse introduced in Chapter 2. This implies that, in the case of zero shift, harmonic slicing can also be written in integrated form as $\alpha = h(x^i) \sqrt{\gamma}$, with $h(x^i)$ an arbitrary (but positive) time independent function. It is very important to stress the fact that the integrated relation holds only when moving along the normal direction to the hypersurfaces, and not when moving along the time lines which will differ from the normal direction for any non-zero shift vector. That is, harmonic slicing relates the lapse to the volume elements associated with the Eulerian observers.

A second route to the use of evolution type slicing conditions started with the first three-dimensional evolution codes in the early 1990s. Since solving the maximal slicing condition was very time consuming, some attempts were made to use algebraic slicing conditions, starting from the integrated form of harmonic slicing $\alpha = \sqrt{\gamma}$. However, it was quickly realized that such a slicing condition was not very useful for evolving black hole spacetimes as it approached the singularity very rapidly (we will see below that harmonic slicing is only marginally singularity avoiding). This led to the empirical search for better behaved algebraic slicing conditions [25, 54], and in particular resulted in the discovery that a slicing condition of the form $\alpha = 1 + \ln \gamma$, the so-called *1+log slicing*, was very robust in practice and mimicked the singularity avoiding properties of maximal slicing.

Both routes finally merged with the work on hyperbolic re-formulations of the 3+1 evolution equations of Bona and Masso in the early and mid 1990s [62, 63, 64]. This resulted in the *Bona–Masso* family of slicing conditions [65], which is a generalization of harmonic slicing for which the lapse is chosen to satisfy the following evolution equation

$$\frac{d}{dt} \alpha = -\alpha^2 f(\alpha) \, K \, , \tag{4.2.52}$$

with $f(\alpha)$ a positive but otherwise arbitrary function of α (the reason why $f(\alpha)$ has to be positive will become clear below). Notice that the particular case $f = 1$ reduces to harmonic slicing, while $f = N/\alpha$, with N constant, corresponds (in the case of zero shift) to $\alpha = h(x^i) + \ln \gamma^{N/2}$, so that $N = 2$ reduces to the standard 1+log slicing. The Bona–Masso version of 1+log slicing, *i.e.* equation (4.2.52) with $f = 2/\alpha$, has been found in practice to be extremely robust and well behaved for spacetimes with strong gravitational fields [13, 15, 30], and in recent years has supplanted maximal slicing in most three-dimensional evolution codes dealing with either black holes or neutron stars.

Let us go back to condition (4.2.52). Taking an extra time derivative we find

$$\frac{d^2}{dt^2} \alpha = -\alpha^2 f \left[\frac{d}{dt} K - \alpha (2f + \alpha f') K^2 \right] , \tag{4.2.53}$$

with $f' := df/d\alpha$. Using now the evolution equation for K, equation (4.2.5), we find that (in vacuum)

$$\frac{d^2}{dt^2} \alpha - \alpha^2 f D^2 \alpha = -\alpha^3 f \left[K_{ij} K^{ij} - (2f + \alpha f') \, K^2 \right] . \tag{4.2.54}$$

The last equation shows that the lapse obeys a wave equation with a quadratic source term in K_{ij}. It is because of this that we say that the slicing condition (4.2.52) is a hyperbolic slicing condition: It implies that the lapse evolves with a hyperbolic equation. The wave speed associated with equation (4.2.54) along a specific direction x^i can be easily seen to be

$$v_g = \alpha \sqrt{f \gamma^{ii}} \, . \tag{4.2.55}$$

Notice that this *gauge speed* will only be real if $f(\alpha) \geq 0$, which explains why we asked for $f(\alpha)$ to be positive. In fact, $f(\alpha)$ must be strictly positive because if it were zero we would not have a strongly hyperbolic system (see Chapter 5).

To see how the gauge speed v_g is related to the speed of light consider for a moment a null world-line. It is not difficult to find that such a world-line will have a coordinate speed along the direction x^i given by

$$v_l = \alpha\sqrt{\gamma^{ii}}\,, \tag{4.2.56}$$

so the gauge speed (4.2.55) can be smaller or larger that the speed of light depending on the value of f. In the particular case of harmonic slicing we have $f = 1$, so the gauge speed coincides with the speed of light, but for the 1+log slicing with $f = 2/\alpha$ the gauge speed can easily become superluminal. Having a gauge speed that is larger than the speed of light does not in fact lead to any causality violations, as the superluminal speed is only related with the propagation of the coordinate system, and the coordinate system can be chosen freely. Physical effects, of course, still propagate at the speed of light.

A variation of the Bona–Masso slicing condition has also been proposed that has the property that for static spacetimes it guarantees that the lapse function will not evolve [10, 17, 302]. This condition can easily be obtained by asking for the evolution of the lapse to be such that

$$\partial_t \alpha = \frac{\alpha f(\alpha)}{\gamma^{1/2}}\,\partial_t \gamma^{1/2}\,, \tag{4.2.57}$$

which results in

$$\partial_t \alpha = -\alpha f(\alpha)\left(\alpha K - D_i\beta^i\right)\,. \tag{4.2.58}$$

The modified condition then substitutes the Lie derivative of the lapse with respect to the shift with the divergence of the shift itself (essentially the Lie derivative of γ). This has the consequence that it *will not* result in the same foliation of spacetime for a different shift vector, *i.e.* the foliation of spacetime we obtain will depend on the choice of shift. However, having a slicing condition that is compatible with a static solution might have advantages in some circumstances.

4.2.5 *Singularity avoidance for hyperbolic slicings*

As mentioned in our discussion of maximal slicing, a very important property of slicing conditions is that of singularity avoidance. The singularity avoiding properties of the Bona–Masso slicing condition have been studied in [3, 66]. Following [66], we will start our discussion of this issue by defining a *focusing singularity* as a place where the spatial volume elements $\gamma^{1/2}$ vanish at a bounded rate. Let us assume that such a singularity occurs after a finite proper time τ_s away from our initial time slice (as measured by the normal observers). We will further say that we have a singularity of order m if $\gamma^{1/2}$ approaches zero as

$$\gamma^{1/2} \sim (\tau_s - \tau)^m\,, \tag{4.2.59}$$

with m some constant power (we will see below that the expected order of a focusing singularity is in fact $m = 1$). Notice that we must have $m > 0$ for there to be a singularity at all, and $m \geq 1$ for the singularity to be approached at a bounded rate. From now on we will therefore assume that $m \geq 1$. In the following analysis we will also assume that the shift vector vanishes, so our statements will correspond to the relation between the lapse and the normal volume elements.

In order to study the behavior of the lapse as the singularity is approached, we will start from the fact that the Bona–Masso slicing condition (4.2.52) implies the following relation between α and $\gamma^{1/2}$

$$d \ln \gamma^{1/2} = \frac{d\alpha}{\alpha f(\alpha)} \,, \tag{4.2.60}$$

from which we find

$$\gamma^{1/2} = F(x^i) \exp \left\{ \int \frac{d\alpha}{\alpha f(\alpha)} \right\} \,, \tag{4.2.61}$$

with $F(x^i)$ a time independent function. Also, from the definition of the lapse we see that the elapsed coordinate time at the formation of the singularity will be

$$\Delta t = \int_0^{\tau_s} \frac{d\tau}{\alpha} \,. \tag{4.2.62}$$

Let us now consider what happens to the lapse as the volume elements $\gamma^{1/2}$ approach zero. One possibility would be for α to remain finite as $\gamma^{1/2}$ vanishes, which would clearly mean that the coordinate time remains finite at the singularity, so the singularity would *not* be avoided. Equation (4.2.61), however, implies that if the lapse remains always finite it is impossible for the volume elements to vanish (remember that $f(\alpha)$ is never allowed to be zero). We then conclude that this case can never arise: The Bona–Masso slicing condition (4.2.52) always causes the lapse to collapse when the volume elements approach zero, for any $f(\alpha) > 0$.

Since the lapse can not remain finite at a focusing singularity, we are left with only two other possibilities: Either the lapse becomes zero at the singularity, or it becomes zero before reaching the singularity. The last case implies that the time slices will stop advancing a finite coordinate time before the singularity is reached. We call such behavior *strong singularity avoidance*. If, on the other hand, the lapse becomes zero at the singularity we could still find that the integrated coordinate time is finite or infinite, depending on the speed at which α approaches zero. We will say that a slicing is *marginally singularity avoiding* if the singularity is reached after an infinite coordinate time (but do notice that marginal singularity avoidance implies that we will get arbitrarily close to the singularity after a finite time).

To see under which conditions we can have strong or marginal singularity avoidance we must say something about the form of the function $f(\alpha)$ as we approach the singularity. For this analysis we will assume that as α approaches zero the function $f(\alpha)$ behaves as a power law

$$f(\alpha) = A\alpha^n \,, \tag{4.2.63}$$

with both $A > 0$ and n constants. Such an assumption implies that

$$\int \frac{d\alpha}{\alpha f(\alpha)} = \frac{1}{A} \int \frac{d\alpha}{\alpha^{n+1}} = \begin{cases} \ln \alpha^{1/A} & n = 0 \, , \\ -1/(nA\alpha^n) & n \neq 0 \, . \end{cases} \tag{4.2.64}$$

Consider first the case $n \neq 0$. Equation (4.2.61) can now be integrated to find

$$\gamma^{1/2} \sim \exp\left(-\frac{1}{nA\alpha^n} \right) \, . \tag{4.2.65}$$

We then see that for $n < 0$ the volume elements remain finite as the lapse approaches zero, in other words for this case we have strong singularity avoidance. Notice that 1+log family of slicing conditions $f(\alpha) = N/\alpha$ corresponds to $n = -1$, which implies that it is singularity avoiding in a strong sense, explaining why it has been found to mimic maximal slicing in practice. If, on the other hand, $n > 0$ then both the lapse and the volume elements go to zero at the same time so we can at most have marginal singularity avoidance.

Let us now go back now to the case $n = 0$. Using again (4.2.61) we find that

$$\gamma^{1/2} \sim \alpha^{1/A} \Rightarrow \alpha \sim \gamma^{A/2} \, .$$

It is then clear that in this case α and $\gamma^{1/2}$ also vanish at the same time.

We then find that $n < 0$ guarantees strong singularity avoidance, while for $n \geq 0$ we can have at most marginal singularity avoidance. In order to see if in this last case we reach the singularity in an infinite or a finite coordinate time we need to study the behavior of α as a function of proper time τ as we approach the singularity. Starting from equation (4.2.62) for the elapsed coordinate time we find

$$\Delta t = \int_0^{\tau_s} \frac{d\tau}{\alpha} = \int_{\alpha_0}^0 \frac{d\tau/d\alpha}{\alpha} \, d\alpha \, , \tag{4.2.66}$$

with α_0 the initial lapse. Equation (4.2.66) implies that if $d\tau/d\alpha$ remains different from zero as the lapse collapses then Δt will diverge and we will have marginal singularity avoidance. On the other hand, if $d\tau/d\alpha$ vanishes at the singularity as α^p with $p > 0$ (or faster), then the integral will converge and the singularity will be reached in a finite coordinate time, *i.e.* the singularity will not be avoided. To find the behavior of $d\tau/d\alpha$ as we approach the singularity, we notice that equation (4.2.59) implies

$$\frac{d \ln \gamma^{1/2}}{d\tau} = -\frac{m}{(\tau_s - \tau)} \, . \tag{4.2.67}$$

Using now equation (4.2.60) we find

$$\frac{d\alpha/d\tau}{\alpha f(\alpha)} = -\frac{m}{(\tau_s - \tau)} \, , \tag{4.2.68}$$

which can be integrated to give

$$\tau = \tau_s - \exp\left(\frac{1}{m}\int \frac{d\alpha}{\alpha f(\alpha)}\right) . \tag{4.2.69}$$

Substituting now $f(\alpha) = A\alpha^n$ we obtain

$$\tau = \begin{cases} \tau_s - \alpha^{1/mA} & n = 0 , \\ \tau_s - \exp\left[-1/(nmA\alpha^n)\right] & n > 0 . \end{cases} \tag{4.2.70}$$

Consider first the case $n > 0$. The derivative of τ with respect to α then becomes

$$\frac{d\tau}{d\alpha} = -\frac{1}{mA\alpha^{n+1}} \exp\left(-\frac{1}{nmA\alpha^n}\right) , \tag{4.2.71}$$

from which it is easy to see that as α approaches zero $d\tau/d\alpha$ also approaches zero faster than any power. This means that the singularity is reached in a finite coordinate time and the singularity is not avoided.

For $n = 0$ we have, on the other hand,

$$\frac{d\tau}{d\alpha} = -\frac{1}{mA} \alpha^{(1/mA-1)} . \tag{4.2.72}$$

The behavior of $d\tau/d\alpha$ as the singularity is approached therefore depends on the sign of $(mA - 1)$. We find that for $mA < 1$ the derivative $d\tau/d\alpha$ goes to zero as a positive power of α and the singularity is reached in a finite coordinate time, while in case $mA \geq 1$ the singularity is reached in an infinite coordinate time and we have marginal singularity avoidance.

The final result is that, if $f(\alpha)$ behaves as $f = A\alpha^n$ for small α and we have a focusing singularity of order m, then for $n < 0$ we have strong singularity avoidance, and for $n = 0$ and $mA \geq 1$ we have marginal singularity avoidance. In the cases $n > 0$, and $n = 0$ with $mA < 1$, the singularity is not avoided even though the lapse collapses to zero, *i.e.* the collapse of the lapse does not always imply singularity avoidance.

We still have not addressed the issue of what value of m should be expected, that is, how fast do the volume elements collapse to zero at the singularity. It is in fact not difficult to give an estimate of this. First, notice that from the ADM equations the volume elements behave as

$$\partial_t \ln \gamma^{1/2} = -\alpha K , \tag{4.2.73}$$

so along the normal lines we have ($d\tau = \alpha dt$)

$$\frac{d}{d\tau} \ln \gamma^{1/2} = -K . \tag{4.2.74}$$

Consider now the evolution equation for K (4.2.5). Using the Hamiltonian constraint we find

$$\partial_t K - \beta^i \partial_i K = \alpha \left[R + K^2 + 4\pi \left(S - 3\rho \right) \right] \; . \tag{4.2.75}$$

Now, according to the well known BKL conjecture (Belinskii, Khalatnikov and Lifshitz [53]), in the approach to a singularity the velocity terms can be expected to dominate (the solution is *velocity term dominated* or VTD), and both the Ricci scalar and the matter terms can be ignored compared to the quadratic term K^2, so that along the normal lines we have

$$\frac{d}{d\tau} K \sim K^2 \qquad \Rightarrow \qquad K \sim \frac{1}{\tau_s - \tau} \; . \tag{4.2.76}$$

Substituting this back into the evolution equation for $\gamma^{1/2}$ we finally find

$$\gamma^{1/2} \sim \tau_s - \tau \; . \tag{4.2.77}$$

The expected order of the singularity is therefore $m = 1$. This means that for $n = 1$ we will have marginal singularity avoidance if $A \geq 1$. In other words, the case $f = 1$ corresponding to harmonic slicing marks the limit of the region with marginal singularity avoidance.

There is one final important point to consider in the strongly singularity avoiding case $n < 0$. From the form of the slicing condition (4.2.52) we see that if $n \leq -2$, then as the lapse approaches zero we can not guarantee that $\partial_t \alpha$ will also approach zero. In fact, $\partial_t \alpha$ could remain finite or even become arbitrarily negative; the slices will therefore not only avoid the singularity, but will in fact move back away from it as the lapse function becomes negative. This type of behavior is not desirable as the time slices could easily stop being spacelike. If we want to guarantee that we have strong singularity avoidance without the lapse becoming negative we must therefore choose $f(\alpha)$ such that as α approaches zero n remains in the region $-2 < n < 0$. Since the 1+log family corresponds to $n = -1$ it is just in the middle of this region, making it a very good choice indeed.

4.3 Shift conditions

Considerably less is know about shift conditions than about slicing conditions. The main reason is that simply taking the shift equal to zero works well in many cases. Still, there are indeed many proposals for how to choose the shift vector, and it is known that in some cases taking a zero shift vector is a very poor choice. In particular, evolving black hole spacetimes with a vanishing shift vector causes the black hole horizon to grow rapidly in coordinate space (since the Eulerian observers keep falling in), which means that eventually all the computational domain will be inside the black hole. This implies that for long-term evolutions of black hole spacetimes it is crucial to have an outward pointing shift vector that will prevent the time lines from falling into the black hole. Also, in the past few years it has been realized that for systems with angular momentum (rotating neutron stars or black holes), the dragging of inertial frames can be so severe that a non-zero rotational shift vector is absolutely essential in order to avoid having large shears developing in the spatial metric that will rapidly cause the simulation to crash. This type of situation also occurs in the case of orbiting compact objects, and it is because of this that the study of shift conditions has become a topic of great relevance in the past few years.

Below I will discuss both the classic elliptic shift conditions and the recently proposed hyperbolic-type conditions that have been very successful in the context of simulation of orbiting compact objects.

4.3.1 *Elliptic shift conditions*

The classic elliptic shift conditions where originally proposed by Smarr and York in their work on the so-called *radiation gauge* in the late 1970s [272, 273]. Consider a congruence of timelike curves with unit tangent vector Z^μ. The projection operator normal to Z^μ is given by

$$P_{\mu\nu} = g_{\mu\nu} + Z_\mu Z_\nu \; . \tag{4.3.1}$$

In terms of $P^{\mu\nu}$, the *twist* $\omega_{\mu\nu}$, *strain* $\theta_{\mu\nu}$ and *acceleration* ζ^μ of the congruence are defined as

$$\omega_{\mu\nu} := P^\alpha_\mu P^\beta_\nu \, \nabla_{[\alpha} Z_{\beta]} \; , \tag{4.3.2}$$

$$\theta_{\mu\nu} := P^\alpha_\mu P^\beta_\nu \, \nabla_{(\alpha} Z_{\beta)} \; , \tag{4.3.3}$$

$$\zeta^\mu := Z^\nu \nabla_\nu Z^\mu \; . \tag{4.3.4}$$

The strain tensor $\theta_{\mu\nu}$ can be further decomposed into its trace $\theta = \nabla_\mu Z^\mu$, also know as the *expansion*, and its free part $\sigma_{\mu\nu} = \theta_{\mu\nu} - (\theta/3)P_{\mu\nu}$, known as the *shear*. In terms of the quantities defined above, the covariant derivative of Z^μ can be expressed as

$$\nabla_\mu Z_\nu = \omega_{\mu\nu} + \theta_{\mu\nu} - \zeta_\mu Z_\nu \; . \tag{4.3.5}$$

Assume now that our congruence corresponds to the world-lines of the Eulerian observers. We then have $Z^\mu = n^\mu$, with n^μ the normal vector to the spatial hypersurfaces. This immediately implies that $\omega_{\mu\nu} = 0$, since the congruence is hypersurface orthogonal. We also find $\zeta^\mu = a^\mu$, with a^μ the acceleration of the Eulerian observers, and $\theta_{\mu\nu} = -K_{\mu\nu}$, with $K_{\mu\nu}$ the extrinsic curvature.

The strain tensor $\theta_{\mu\nu} = -K_{\mu\nu} = 1/2 \, \pounds_{\vec{n}} \gamma_{\mu\nu}$ corresponds to motion along the normal lines. In an analogous way, we can define a strain tensor along the time lies t^μ in the following way

$$\Theta_{ij} := \frac{1}{2} \, \pounds_{\vec{t}} \, \gamma_{ij} = -\alpha K_{ij} + \frac{1}{2} \, \pounds_{\vec{\beta}} \, \gamma_{ij} \; , \tag{4.3.6}$$

where we have only considered the spatial components, as the strain tensor is clearly purely spatial.

The strain tensor Θ_{ij} just defined measures the change in both the size *and* form of the volume elements along the time lines. We can then try to use a shift vector to minimize some measure of the strain tensor in order to reduce as much as possible the changes in the spatial metric itself. Smarr and York propose to minimize the non-negative square of the strain tensor $\Theta_{ij}\Theta^{ij}$ in a global sense over the spatial hypersurface. If we integrate $\Theta_{ij}\Theta^{ij}$ over the slice, variation with respect to the shift vector β^i can be shown to result in the equation

$$D_j\Theta^{ij} = 0 \quad \Rightarrow \quad D^j\left(D_i\beta_j + D_j\beta_i\right) = 2D^j\left(\alpha K_{ij}\right) , \tag{4.3.7}$$

which can be rewritten as

$$D^2\beta^i + D^i D_j\beta^j + R_{ij}\beta^j = 2D^j\left(\alpha K_{ij}\right) , \tag{4.3.8}$$

where the Ricci tensor appears when we commute covariant derivatives of the shift. This equation is known as the *minimal strain* shift condition and gives us three elliptic equations that can be used to find the three components of the shift vector given appropriate boundary conditions.

The minimal strain condition minimizes a global measure of the change in the volume elements associated with the time lines. However, as the change in the volume is related to the trace of the extrinsic curvature K, it would seem better to use the shift vector to minimize only the changes in the shape of the volume elements, independently of their size. We then define the shear associated with the time lines as the tracefree part of Θ_{ij}:

$$\Sigma_{ij} := \Theta_{ij} - \frac{1}{3}\gamma_{ij}\Theta . \tag{4.3.9}$$

Smarr and York call Σ_{ij} the *distortion tensor* in order to distinguish it from the shear tensor σ_{ij} associated with the normal lines. From the definition above we find that the distortion tensor is given by

$$\Sigma_{ij} = -\alpha A_{ij} + \frac{1}{2}(\mathbf{L}\beta)_{ij} , \tag{4.3.10}$$

where as before $A_{ij} = K_{ij} - (\gamma_{ij}/3)\, K$ is the tracefree part of the extrinsic curvature, and $(\mathbf{L}\beta)_{ij}$ is the conformal Killing form associated with the shift (see equation (3.2.7)):

$$(\mathbf{L}\beta)_{ij} := D_i\beta_j + D_j\beta_i - \frac{2}{3}\,\gamma_{ij}D_k\beta^k . \tag{4.3.11}$$

We can also rewrite the distortion tensor in a way that is perhaps more illustrative by noticing that (4.3.10) is equivalent to

$$\Sigma_{ij} = \frac{1}{2}\gamma^{1/3}\pounds_{\vec{t}}\,\tilde{\gamma}_{ij} , \tag{4.3.12}$$

with $\tilde{\gamma}_{ij} = \gamma^{-1/3}\gamma_{ij}$ the conformal metric. The distortion tensor is therefore essentially the velocity of the conformal metric.

Minimizing now the integral of $\Sigma_{ij}\Sigma^{ij}$ over the spatial hypersurface with respect to the shift yields the condition

$$D_j \Sigma^{ij} = 0 \; , \tag{4.3.13}$$

which implies

$$\Delta_{\mathbf{L}} \, \beta^i = 2 D_j \left(\alpha A^{ij} \right) \; , \tag{4.3.14}$$

where the operator $\Delta_{\mathbf{L}}$ is defined as

$$\begin{aligned} \Delta_{\mathbf{L}} \, \beta^i &:= D_j (\mathbf{L}\beta)^{ij} = D^2 \beta^i + D_j D^i \beta^j - \frac{2}{3} \, D^i D_j \beta^j \\ &= D^2 \beta^i + \frac{1}{3} \, D^i D_j \beta^j + R^i_j \beta^j \; . \end{aligned} \tag{4.3.15}$$

Equation (4.3.14) is known as the *minimal distortion* shift condition, and again gives us three elliptic equations for the three components of the shift. The minimal distortion condition is a very natural condition that will minimize changes in the shape of volume elements during an evolution. It can also be shown that when the gravitational field is weak, it includes the TT gauge used to study gravitational waves [273]. Still, the fact that it is given through three coupled elliptic equations has meant that it has not been extensively used in three-dimensional numerical simulations.

Notice that, in particular, the expression (4.3.12) for the distortion tensor implies that the minimal distortion condition can also be written as

$$D_j \left(\partial_t \tilde{\gamma}^{ij} \right) = 0 \; . \tag{4.3.16}$$

This is very closely related to a proposal of Dirac for finding an analog of the radiation gauge in general relativity [111, 273], for which he suggested using

$$\partial_j \partial_t \tilde{\gamma}^{ij} = \partial_t \partial_j \tilde{\gamma}^{ij} = 0 \; . \tag{4.3.17}$$

Remarkably, this equation has been recently rediscovered in the context of the BSSNOK formulation of the 3+1 evolution equations. Notice that, from equation (2.8.8), the conformal connection functions are given by $\tilde{\Gamma}^i = -\partial_j \tilde{\gamma}^{ij}$, so the above gauge condition is equivalent to

$$\partial_t \tilde{\Gamma}^i = 0 \; , \tag{4.3.18}$$

This shift condition, known as *Gamma freezing*, has been proposed as a natural shift condition in the context of the BSSNOK formulation, as it freezes three of the independent degrees of freedom. To see the relationship between the minimal distortion and Gamma freezing conditions we notice that the evolution equation for the conformal connection functions (2.8.25) can be written in terms of Σ_{ij} as

$$\partial_t \tilde{\Gamma}^i = 2 \partial_j \left(\gamma^{1/3} \Sigma^{ij} \right) \; . \tag{4.3.19}$$

More explicitly, we have

$$\partial_t \tilde{\Gamma}^i = 2e^{4\phi} \left[D_j \Sigma^{ij} - \tilde{\Gamma}^i_{jk} \Sigma^{jk} - 6\Sigma^{ij} \partial_j \phi \right] . \tag{4.3.20}$$

We then see that the minimal distortion condition $D^j \Sigma_{ij} = 0$, and the Gamma freezing condition $\partial_t \tilde{\Gamma}^i = 0$, are equivalent up to terms involving first spatial derivatives of the conformal metric and the conformal factor. In particular, all terms involving second derivatives of the shift are identical in both cases (but not so terms with first derivatives of the shift which appear in the distortion tensor itself). That the difference between both conditions involves Christoffel symbols should not be surprising since the minimal distortion condition is covariant while the Gamma freezing condition is not.

Very recently there has been an important development in looking for more natural shift conditions. Jantzen and York [168] have proposed a modified form of the minimal distortion shift vector that takes into account the new weighted conformal decomposition of symmetric tensors, and the role that the densitized lapse plays in the ADM evolution equations (see Section 3.2.3). The proposal of Jantzen and York is to minimize the square of a lapse-corrected distortion tensor Σ_{ij}/α (corresponding to the change of the conformal metric with respect to proper time), and use the full spacetime metric determinant as measure in the integral (thus giving an extra factor α in the volume element). This new variation results in the modified condition

$$D_j \left(A^{ij}/\alpha \right) = 0 \quad \Rightarrow \quad D_j \left[(\mathbf{L}\beta)^{ij} /2\alpha - A^{ij} \right] = 0 \, , \tag{4.3.21}$$

The difference between the original minimal distortion condition and the modified version is essentially the position of the factor α inside the divergence. However, the new version has a closer relationship to the initial data problem, in particular in terms of the conformal thin-sandwich approach of Section 3.3. To see this, we first rewrite the modified minimal distortion condition (4.3.21) in terms of conformal quantities using the transformations given in Chapter 3. We find that the new condition is equivalent to

$$\bar{D}_j \left[\left(\bar{\mathbf{L}}\beta \right)^{ij} /2\bar{\alpha} - \bar{A}^{ij} \right] = 0 \, , \tag{4.3.22}$$

Compare now this last equation with the expression for the conformal tracefree extrinsic curvature in terms of the shift and the velocity of the conformal metric, equation (3.3.7). We see that if the shift vector satisfies the new minimal distortion condition, then the time derivative of the conformal metric $\bar{u}_{ij}$ will remain transverse during the evolution, *i.e.* $\bar{D}_j \bar{u}^{ij} = 0$, so that only the transverse-traceless part of the conformal metric evolves ($\bar{u}_{ij}$ is traceless by construction). Thus the new shift condition reduces the evolution of the conformal metric to its "dynamical" part. This new version of minimal distortion is therefore more natural and is closer to the original motivation of finding a gauge in general relativity that separates out the evolution of the physical degrees of freedom from the gauge degrees of freedom.

4.3.2 *Evolution type shift conditions*

Just as in the case of the lapse, solving the elliptic minimal distortion and Gamma freezing shift conditions is very computationally demanding in three dimensions (considerably more so than maximal slicing since there are three coupled equations). Because of this, in the past few years, there has been a search for robust shift conditions that can be implemented as evolution equations for the shift components. One natural approach is based on the idea that in order to solve elliptic conditions numerically we usually replace such equations by parabolic equations in a fictitious time, and then evolve the parabolic equation until its solution relaxes to the solution of the elliptic equation we are looking for. Assume, for example, that we wish to solve Laplace's equation with some given boundary conditions of either Dirichlet or Newmann type:

$$D^2\phi = 0 \,. \tag{4.3.23}$$

In practice we replace this equation with the heat equation

$$\partial_\tau \phi = \epsilon D^2 \phi \,, \tag{4.3.24}$$

where τ is a fictitious time and $\epsilon > 0$ is some constant that fixes the time-scale of relaxation. We can then start with arbitrary initial data and evolve this equation until ϕ relaxes to a stationary solution that satisfies the given boundary conditions. If we reach a stationary situation clearly we have $\partial_\tau \psi = 0$, so we have found the solution of the Laplace equation we wanted. In practice, direct relaxation methods to solve elliptic equations are very slow to converge and are hardly ever used anymore. More powerful methods like multi-grid are the preferred choice nowadays, but at their core they still use the relaxation idea. The main problem has to do with the fact that numerical stability for an equation of parabolic type like the heat equation requires that we take $\Delta\tau \lesssim \epsilon(\Delta x)^2$, so that as we refine the spatial grid the time step rapidly becomes prohibitively small. In principle, instead of using the heat equation we could substitute Laplace's equation with the wave equation in fictitious time,

$$\partial_\tau^2 \phi - v^2 D^2 \phi = 0 \,, \tag{4.3.25}$$

for which the restriction on the time step is much less severe and is given instead by the CFL condition $v\Delta\tau \lesssim \Delta x$, with v the characteristic speed of the system (see Chapter 9). However, this approach does not work well in practice since the wave equation with arbitrary boundary conditions will typically not relax as waves keep getting reflected by the boundaries.[39]

[39]For some special boundary conditions, however, using the wave equation does work and converges much faster than using the heat equation.

An important development in the search for evolution-type gauge conditions was the observation by Balakrishna *et al.* [46] that since in the case of the minimal distortion condition we are actually solving for a gauge function, and the gauge is arbitrary, it does not really matter too much if we have an exact solution to the minimal distortion equation, as an approximate solution might already work extremely well in practice. We can therefore relax the elliptic equation in real time instead of fictitious time. This led to the proposal of the so-called *driver conditions*, of which the first is the parabolic minimal distortion driver:

$$\partial_t \beta^i = \epsilon \left[D^2 \beta^i + \frac{1}{3} D^i D_j \beta^j + R^i_j \beta^j - 2 D_j \left(\alpha A^{ij} \right) \right] , \tag{4.3.26}$$

where the parameter $\epsilon > 0$ is chosen small enough to guarantee numerical stability in the range of resolutions considered, but large enough to allow the shift to react to changes in the geometry.

We can now make a second observation. In practice we find that parabolic drivers like the one introduced above usually do not allow the shift to respond rapidly enough to changes in the geometry unless ϵ is large, but in that case the numerical stability of the whole evolution system rapidly becomes dominated by the parabolic shift condition. One way to fix this is to use instead a *hyperbolic driver* condition of the form

$$\partial_t^2 \beta^i = \alpha^2 \xi \left[D^2 \beta^i + \frac{1}{3} D^i D_j \beta^j + R^i_j \beta^j - 2 D_j \left(\alpha A^{ij} \right) \right] , \tag{4.3.27}$$

where now $\xi > 0$ is an arbitrary positive function of position and/or α that controls the wave speed of the different shift components, and the explicit factor of α^2 is there for later convenience when analyzing this wave speed (we will come back to this issue below). At this point we might worry about the fact mentioned above that relaxing via wave equations usually does not work since waves will reflect from the boundaries for generic boundary conditions. But again we notice that we are solving for a gauge function, so we are also free to choose boundary conditions any way we like. In particular, we can use *outgoing wave boundary conditions* that allow the gauge waves to leave the computational domain with minimal reflections.

Driver conditions have recently become very important in the context of three-dimensional evolutions using the BSSNOK system of equations. As already mentioned, the minimal distortion condition is very closely related to the Gamma freezing condition (4.3.18). This has led to the suggestion of Alcubierre *et al.* of using a *Gamma driver* shift condition of the form [13]

$$\partial_t^2 \beta^i = \alpha^2 \xi \, \partial_t \tilde{\Gamma}^i . \tag{4.3.28}$$

In practice, a damping term is added to this condition to obtain

$$\partial_t^2 \beta^i = \alpha^2 \xi \, \partial_t \tilde{\Gamma}^i - \eta \, \partial_t \beta^i , \tag{4.3.29}$$

where the parameter $\eta > 0$ is used to avoid strong oscillations in the shift (a similar damping term can also be added to the minimal distortion driver).[40]

Both the hyperbolic minimal distortion driver and the hyperbolic Gamma driver provide us with shift conditions that will attempt to control the distortion of the volume elements with some time delay. In practice, the hyperbolic Gamma driver condition has been found to be extremely robust and well behaved in black hole simulations with puncture initial data, controlling both the slice stretching and the shear due to the rotation of the hole. The minimal distortion hyperbolic driver condition has not been used extensively in practice, but we could expect that it would be just as robust. In fact, on purely theoretical grounds the minimal distortion driver condition should probably be preferred as it results in a condition that is 3-covariant. The Gamma driver condition, on the other hand, is clearly not 3-covariant implying that if we change our spatial coordinates (say from Cartesian to spherical), we will obtain a different shift.

Before moving to a more general class of hyperbolic shift conditions, there is an important point related to the characteristic speeds associated with the hyperbolic driver conditions that should be mentioned. A formal description of how to obtain such characteristic speeds will have to wait until Chapter 5, but here we can anticipate some of the results. Notice that the hyperbolic minimal distortion and Gamma driver conditions are in fact identical up to the principal part (*i.e.* all terms with second derivatives are the same in both cases), so that their characteristic structure is the same. Consider first the asymptotically flat region where the covariant derivatives can be substituted by partial derivatives. It is easy to see that the characteristic speeds in that region along a fixed direction x^i are

$$\lambda_l \simeq \pm\alpha\sqrt{4\gamma^{ii}\xi/3}\,, \qquad \lambda_t \simeq \pm\alpha\sqrt{\gamma^{ii}\xi}\,, \tag{4.3.30}$$

where here λ_l refers to the speed associated with the longitudinal shift component and λ_t to the speed associated with the transverse shift components, so that longitudinal and transverse components propagate at different speeds. In numerical simulations it is usual to take $\xi = 3/4$ (or $\xi = \alpha^n 3/4$, with n some integer), so that the longitudinal components propagate at the speed of light asymptotically.[41]

There is yet another important point to make, related to the characteristic speeds. A more formal analysis using the techniques described in Chapter 5 shows that the characteristic cones associated with the shift modes for the driver conditions (4.3.27) and (4.3.29) have rather strange properties when compared with the characteristic cones associated with all other modes. For all other modes

[40] When using the Gamma driver shift, the specific value of the damping parameter η has an important effect in achieving long-lasting evolutions of black hole spacetimes. Notice, however, that in contrast with ξ, the parameter η is not dimensionless, so that it must be scaled with the total mass of the spacetime M. The typical value used in black hole simulations is $\eta \sim 2/M$.

[41] Recently it has been found that taking $n = -2$, which effectively makes the gauge speeds independent of the lapse α, is a good choice in black hole simulations since it allows the shift to evolve even in regions where the lapse has collapsed to zero.

it turns out that the width of the characteristic cones is independent of the shift, and the shift's only effect is to tilt the cones by $-\beta^i$, which is precisely what we would expect from the geometrical interpretation of the shift vector. In the case of the driver conditions as given above we find, however, that the characteristic cones associated with the shift modes are only tilted by $-\beta^i/2$, and also that their width depends on the magnitude of the shift. This unexpected behavior can be easily cured by changing the minimal distortion driver to

$$\partial_0 \beta^i = B^i \, , \tag{4.3.31}$$

$$\partial_0 B^i = \alpha^2 \xi \left[D^2 \beta^i + \frac{1}{3} \, D^i D_j \beta^j + R^i_j \beta^j - 2 D_j \left(\alpha A^{ij} \right) \right] \, , \tag{4.3.32}$$

with $\partial_0 := \partial_t - \beta^i \partial_i$. The Gamma driver condition is also changed in an analogous way to

$$\partial_0 \beta^i = B^i \, , \tag{4.3.33}$$

$$\partial_0 B^i = \alpha^2 \xi \, \partial_0 \tilde{\Gamma}^i - \eta B^i \, . \tag{4.3.34}$$

Apart from a trivial rewriting of the above conditions in first order in time form, all we have actually done is added an advection term to the time derivatives to fix the structure of the light cones. An advection term of this type was in fact considered initially by the authors of [13] but later rejected (and never published) as it brings the shift condition to a form very similar to the simple equation $\partial_t v - v \, \partial_x v = 0$, known as *Burger's equation*, which is well know to result in the formation of shocks. However, recent numerical simulations indicate that, at least in the case of the Gamma driver condition, the advection terms do not give rise to shocks (the shift condition is coupled to all other dynamical equations, so that this superficial resemblance to Burger's equation is misleading). Moreover, in simulations of binary black hole systems with moving punctures [44] (see Chapter 6), it has been found that the inclusion of advection terms is important in order to avoid having a perturbation in the constraints remaining at the initial location of the punctures.[42]

The driver conditions described above have been proposed as a way of finding a shift vector that is close to the solution of the corresponding elliptic equations and at the same time is easy to solve for. As such these conditions, though certainly robust in practice, are rather *ad hoc*. There is, however, a more natural approach to obtaining hyperbolic shift conditions similar to those used for the lapse. In Section 4.2.4 we introduced the idea of generalizing the condition for a harmonic time coordinate to obtain different slicing conditions. This led us

[42] Recently there has been both a formal analysis, and a series of numerical tests involving moving black holes, of variations in the way in which advection terms are be added to the Gamma driver shift condition [156, 293] and their impact on the hyperbolicity of the system and the accuracy of the numerical simulations.

to the Bona–Masso family which has been found to be extremely successful in practice. Recently, it has been suggested to use a similar approach to find hyperbolic shift conditions as generalizations of the condition for harmonic spatial coordinates [16].

Let us start from the harmonic condition for spatial coordinates

$$\Box x^i = 0 \; , \tag{4.3.35}$$

which implies

$$\Gamma^i := g^{\mu\nu} \Gamma^i_{\mu\nu} = 0 \; , \tag{4.3.36}$$

with $\Gamma^\alpha_{\mu\nu}$ the Christoffel symbols associated with the spacetime metric $g_{\mu\nu}$. Using the expressions found in Appendix B, we can easily show that in 3+1 language the above condition reduces to

$$\begin{aligned} \partial_t \beta^i &= \beta^a \partial_a \beta^i - \alpha \partial^i \alpha + \alpha^2 \, {}^{(3)}\Gamma^i \\ &+ \frac{\beta^i}{\alpha} \left(\partial_t \alpha - \beta^a \partial_a \alpha + \alpha^2 K \right) \; , \end{aligned} \tag{4.3.37}$$

where now ${}^{(3)}\Gamma^i$ is defined in terms of the three-dimensional Christoffel symbols ${}^{(3)}\Gamma^i_{jk}$. This condition has in fact been known for a long time (see *e.g.* [140, 305]), though it is usually written down assuming that the lapse is also harmonic so that the term in parenthesis vanishes.[43] Having the evolution equation for the shift depend on the time derivative of the lapse is clearly inconvenient if we want to use harmonic spatial coordinates with a different slicing condition, say maximal slicing. Remarkably, it turns out that if we rewrite the evolution equation for the shift in terms of a rescaled shift vector of the form $\sigma^i = \beta^i / \alpha$, then the spatial harmonic condition decouples completely from the evolution of the lapse so we can work with an arbitrary slicing condition. We find

$$\partial_t \sigma^i = \alpha \sigma^a \partial_a \sigma^i - \partial^i \alpha + \alpha \left(\sigma^i K + {}^{(3)}\Gamma^i \right) \; . \tag{4.3.38}$$

This can now be generalized in a natural way by considering instead the following evolution equation for σ^i

$$\partial_t \sigma^l = \alpha \sigma^m \partial_m \sigma^l - \partial^l \alpha + \alpha h \left(\sigma^l K + {}^{(3)}\Gamma^l \right) \; , \tag{4.3.39}$$

with $h(\alpha) > 0$ an arbitrary function of the lapse [16]. The last condition is known as the *generalized harmonic shift condition*, and is closely related to shift conditions recently proposed by Lindblom and Scheel [190], and by Bona and Palenzuela [67]. It is not difficult to see that by choosing the free parameters in these references appropriately, we can in fact recover the above condition, but only *provided* we also take the lapse to evolve via the Bona–Masso slicing condition (4.2.52) and take $h(\alpha) = f(\alpha)$.

[43] Equation (4.3.37) fixes a sign error in [305], and includes a term missing in [140].

There are several properties of the shift condition (4.3.39) that are important to mention. First, in an analogous way to the Bona–Masso slicing condition, it leads to a characteristic speed associated with the shift modes of the form

$$\lambda = \alpha \left(-\sigma^i \pm \sqrt{h\gamma^{ii}} \right) = -\beta^i \pm \alpha \sqrt{h\gamma^{ii}} \,, \tag{4.3.40}$$

so that we clearly must ask for $h > 0$. In fact, the function h plays exactly the same role for the shift as the function f did in the case of the Bona–Masso slicing condition. Notice also that the generalized harmonic shift condition (4.3.39) has again the Burger's type structure $\partial_t v - v\partial_x v = 0$, but a detailed analysis shows that it does not lead to the formation of shocks (see [16] and also Chapter 5).

There is a final issue to discuss regarding the generalized harmonic shift condition (4.3.39). Notice that, just as was the case with the Gamma driver, this condition is not covariant with respect to changes in the spatial coordinates. That is, starting from exactly the same 3-geometry but with different spatial coordinates we will get a different evolution of the shift vector. In particular, for curvilinear systems of coordinates we could find that even starting from a flat slice of Minkowski spacetime we would still have non-trivial shift evolution driven by the fact that the initial ${}^{(3)}\Gamma^i$ do not vanish. Worse still, in many cases it can happen that the ${}^{(3)}\Gamma^i$ of flat space are not only non-zero but are also singular, as is the case with spherical coordinates for which ${}^{(3)}\Gamma^r$ is of order $1/r$. In response to this it has been suggested in [16] that the generalized harmonic shift condition should be interpreted as always being applied in a coordinate system that is topologically Cartesian.

Of course, we would still like to be able to express the condition in a general curvilinear coordinate system. Let us denote by $\{x^{\bar{a}}\}$ the reference topologically Cartesian coordinates, and by $\{x^i\}$ the general curvilinear coordinates. We find that in the curvilinear coordinate system the shift condition should in fact be replaced with

$$\begin{aligned} \partial_t \sigma^l &= \alpha \sigma^m D_m \sigma^l - D^l \alpha + \alpha h \, \sigma^l K \\ &\quad + \alpha \left(h\gamma^{mn} - \sigma^m \sigma^n \right) \Delta^l_{mn} \,, \end{aligned} \tag{4.3.41}$$

where as before D_i is the three-dimensional covariant derivative and where we have defined

$$\Delta^l_{mn} := {}^{(3)}\Gamma^l_{mn} - F^l_{mn} \,, \tag{4.3.42}$$

with

$$F^i_{jk} := \partial_{\bar{a}} x^i \, \partial_j \partial_k x^{\bar{a}} \,. \tag{4.3.43}$$

The shift condition (4.3.41) is in fact 3-covariant, as we can readily verify that the Δ^l_{mn} transform as the components of a 3-tensor. But the price we have paid is that we have chosen a privileged Cartesian coordinate system to be used as a reference in order to define the F^l_{mn}. A similar idea can also be used to obtain a fully covariant version of the Gamma driver condition (4.3.28).

In practice, we can use the fact that for flat space in Cartesian coordinates the Christoffel symbols vanish, which implies

$$F^l_{mn} = {}^{(3)}\Gamma^l_{mn}\Big|_{\text{flat}} , \tag{4.3.44}$$

so that

$$\Delta^l_{mn} = {}^{(3)}\Gamma^l_{mn} - {}^{(3)}\Gamma^l_{mn}\Big|_{\text{flat}} . \tag{4.3.45}$$

The covariant version of the generalized harmonic shift has been used successfully in evolutions of spherically symmetric systems, but so far has not been tried in the full three-dimensional case where the hyperbolic Gamma driver condition still dominates.

4.3.3 *Corotating coordinates*

Consider a system of two orbiting compact objects (neutron stars or black holes). If the objects are sufficiently far away, the inspiral time-scale of the binary due to emission of gravitational waves will be large compared to the orbital period. In such a case we can in fact use a shift vector to go to a corotating coordinate frame where the metric dynamics will be greatly reduced and numerical errors will become smaller (in particular, numerical conservation of angular momentum will improve in a corotating frame). Corotating coordinates have been used, for example, in [114] for the case of binary neutron stars, and in [11, 80, 110] for the case of binary black holes. Also, a shift vector corresponding to a corotating frame is the expected result of the quasi-equilibrium approach to constructing initial data for orbiting binaries (see Section 3.5.2), as only for such a shift can we expect to have a quasi-stationary evolution.

Let us consider how we can implement a corotating reference frame. A first approach is simply to go to a frame that rotates with some fixed angular velocity Ω. Now let $\{t, x, y, z\}$ be the non-rotating original coordinates and $\{\bar{t}, \bar{x}, \bar{y}, \bar{z}\}$ be the rotating coordinates. Assuming a rotation on the x-y plane, *i.e* $\vec{\Omega} = \Omega\, \vec{e}_z$, the transformation of coordinates is given by

$$t = \bar{t} , \tag{4.3.46}$$

$$z = \bar{z} , \tag{4.3.47}$$

$$x = \bar{x}\cos(\Omega t) - \bar{y}\sin(\Omega t) , \tag{4.3.48}$$

$$y = \bar{x}\sin(\Omega t) + \bar{y}\cos(\Omega t) . \tag{4.3.49}$$

This corresponds to the following gauge transformation

$$\bar{\alpha} = \alpha , \tag{4.3.50}$$

$$\bar{\beta}^i = \beta^i + \left(\vec{\Omega} \times \vec{\rho}\right)^i , \tag{4.3.51}$$

with $\vec{\rho} := (x, y, 0)$. So in order to go to a corotating frame all we need to do is add a rigid rotation to the shift vector. Of course, we must know the correct value of the orbital angular velocity Ω, but this can usually be obtained (at least approximately) from the initial data.[44]

Notice that if we are using an elliptic shift condition like minimal distortion, the rigid rotation term will only fix the boundary condition, and the value of the shift in the interior will be given by the solution of the elliptic equation. If, on the contrary, we use an evolution equation for the shift, then the rigid rotation will fix the initial data for the shift everywhere. There is, however, an important point that should be mentioned. Having an initial shift given by a pure rigid rotation term assumes that we are working on a flat spacetime. In practice we can expect this to work well in the asymptotically flat region far from the sources, but not in the inner regions where the spacetime can be far from flat. To see the type of problem we face let us consider for a moment maximal slicing ($K = 0$) and conformally flat initial data ($\gamma_{ij} = \psi^4 \delta_{ij}$) for an orbiting binary, and an initial shift given by a rigid rotation as above. Now, the ADM evolution equation for the volume elements implies that the conformal factor ψ evolves as

$$\partial_t \psi = -\frac{\psi}{6} \left(\alpha K - \partial_i \beta^i \right) + \beta^i \partial_i \psi \, . \tag{4.3.52}$$

But in this case we have $K = 0$, and also for a rigid rotation it is easy to verify that $\partial_i \beta^i = 0$. The last equation then reduces to

$$\partial_t \psi - \beta^i \partial_i \psi = 0 \, . \tag{4.3.53}$$

This is nothing more than the advection equation. We then find that ψ will keep its original profile and will just advect with a speed given by $-\beta^i$ (at least initially). The curious thing about this result is that naively we could have expected that some value of Ω would bring us to a corotating frame where ψ essentially does not evolve, but we have instead found that given a rigid rotation with *any* value of Ω, the conformal factor ψ will just start rotating the opposite way, so we would have been better off having zero shift. Of course, this is only initially, we would expect that if we have a good value of the angular velocity everything would settle after some initial transient to the true corotating frame.

A extreme example of the situation just mentioned occurs for binary black holes that use puncture type data. In that case the initial conformal factor behaves as $1/r$ near the puncture, so that as long as K and $\partial_i \beta^i$ remain regular we will find that near the puncture the advection term $\beta^i \partial_i \psi$ will dominate so that again ψ will just advect with speed $-\beta^i$. In other words, if the shift is not zero at the puncture then *the puncture will move.* Since the whole point of going to a corotating frame is to have the black holes approximately stationary, we find that the rigid rotation must be modified at the puncture position to ensure

[44] The effective potential method gives an approximate value of the angular velocity as $\Omega = \partial E_b / \partial J$, with E_b the effective potential energy and J the ADM angular momentum. The quasi-equilibrium approach usually gives a much better value of Ω by asking for the Komar and ADM masses to be equal (see Chapter 3).

that the shift becomes zero. The standard way to achieve this in practice is to multiply the initial rigid rotation term with a negative power of ψ to guarantee that the shift will vanish at the puncture [11, 80]:

$$\vec{\beta}_{t=0} = \frac{1}{\psi^3} \, \vec{\Omega} \times \vec{\rho} \, . \qquad (4.3.54)$$

The third power is chosen because this guarantees that the magnitude of the shift will become zero at the puncture in the non-conformal physical metric, and also guarantees that the puncture will not evolve [13].

There is another important issue to consider when dealing with corotating coordinates. Notice that as we move away from the center of rotation, the magnitude of the rigid rotation shift vector increases as $\rho\Omega$. For a given value of Ω there will then be a distance such that the shift will become superluminal – this is known as the *light-cylinder*. We might then worry that outside the light-cylinder causality effects will cause problems with a numerical simulations. In fact, some simple numerical methods do indeed become unstable when the light-cones tip beyond the vertical, but this problem is in fact very easy to overcome. Methods using first-order in time evolution equations can be made to remain stable well beyond the light-cylinder by simply using one-sided differences for the advection terms on the shift (*i.e.* terms of the form $\beta^i \partial_i$). What we do have to worry about numerically is the fact that, as the magnitude of the corotating shift vector increases with distance, the restrictions on numerical stability associated with the CFL condition become more severe and the time step must be taken smaller and smaller.

There is, however, a serious issue related to causality for superluminal shift vectors that has to do with the outer boundary conditions. If we are evolving on a cubical numerical grid and the boundary is located beyond the light-cylinder, then there will be regions of the boundary whose entire causal past is outside the computational domain (see Figure 4.3). We then have absolutely no physical information that can be used to set up boundary conditions at those points. In real-life simulations, standard practice has been to just keep using the same boundary conditions we would use for simulations with a small shift at the boundary and "hope for the best", taking the pragmatic position that as long as the boundaries remain stable and are sufficiently far away we can live with unphysical boundary conditions. But it is well understood that the boundary issue is a serious problem for corotating grids. A much better solution would be to use a cylindrical computational domain.

As already mentioned, although corotating frames seem attractive in some circumstances, they have a series of disadvantages associated mainly with the boundary conditions applied at the edge of the computational domain. Another major disadvantage is the fact that it is more difficult to extract gravitational radiation from a corotating frame, plus this radiation will be seriously contaminated by errors coming from the boundary conditions. Finally, recent advances

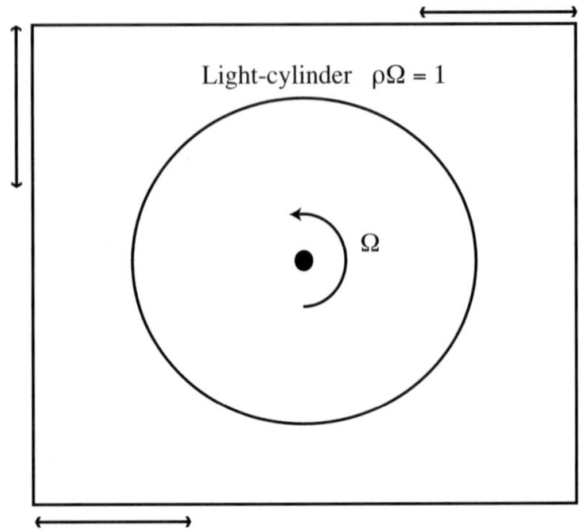

Fig. 4.3: The light-cylinder marks the boundary of the region where the shift becomes superluminal (essentially $\rho\Omega > 1$); outside this region the light cones are tilted beyond the vertical. The arrows at the boundaries indicate those points whose causal past is completely outside the computational domain so that there is no physical information available to set up boundary conditions.

in gauge conditions and formulations of the evolution equations, and in particular the development of new techniques that permit the simulation of binary black hole spacetimes without the need of singularity excision by allowing the punctures to move [44, 93] (see Chapter 6), have shown that the use of corotating coordinates is not necessary to achieve stable and accurate numerical evolutions of binary black hole systems. Because of all this the use of corotating coordinates is rapidly becoming less common, though they are still used in the particular case of numerical codes based on spectral methods and multiple coordinate patches [191].

5

HYPERBOLIC REDUCTIONS OF THE FIELD EQUATIONS

5.1 Introduction

When discussing the ADM equations in Chapter 2 we encountered the fact that the 3+1 evolution equations are not unique, as we can add arbitrary multiples of the constraints to obtain new systems of evolution equations that will have the same physical solutions but will differ in their unphysical (*i.e.* constraint violating) solutions and, more importantly, in their mathematical properties. The original ADM equations differ from the standard version used in numerical relativity (due to York) by precisely the addition of a term proportional to the Hamiltonian constraint to the evolution equation of K_{ij}. Also, the BSSNOK system uses both the Hamiltonian constraint to modify the evolution equation for the trace of the extrinsic curvature K, and the momentum constraints to modify the evolution equations for the conformal connection functions $\tilde{\Gamma}^i$.

There are of course an infinite number of ways in which we can add multiples of the constraints to the evolution equations. The key question is then: Which of these possible systems of equations will be better behaved both mathematically and numerically? Historically, attempts to write down different physically-equivalent forms of the evolution equations for general relativity have followed two different routes: an empirical route that started in the late 1980s and looked for systems of equations that were simply better-behaved numerically than ADM, and a mathematically formal route that looked for well-posed, particularly hyperbolic, reformulations of the evolution equations and that can be traced back to the work of Choquet-Bruhat in the 1950s on the existence and uniqueness of solutions to the Einstein equations [84]. In recent years these two approaches have finally merged, and the traditionally more pragmatic numerical community has realized that well-posedness of the underlying system of evolution equations is essential in order to obtain stable and robust numerical simulations. At the same time, it has been empirically discovered that well-posedness in not enough, as some well-posed formulations have been found to be far more robust in practice than others.

From the early 1990s a large number of alternative formulations of the 3+1 evolution equations have been proposed. We have now reached a point where more formulations exist than there are numerical groups capable of testing them. Because of this, here we will limit ourselves to discussing a small number of formulations, chosen both because they are a representative sample of the different approaches used, and because they correspond to the formulations used by the majority of numerical evolution codes that exist today. The discussion of well-posedness and hyperbolicity presented here is of necessity brief and touches only on the main ideas, a more formal discussion can be found in the book by Kreiss and Lorenz [177] (see also the review paper of Reula [241]).

5.2 Well-posedness

Consider a system of partial differential equations of the form

$$\partial_t u = P(D) u \,, \tag{5.2.1}$$

where u is some n-dimensional vector-valued function of time and space, and $P(D)$ is an $n \times n$ matrix with components that depend smoothly on spatial derivative operators.[45] The *Cauchy* or *initial value* problem for such a system of equations corresponds to finding a solution $u(t, x)$ starting from some known initial data $u(t = 0, x)$.

A crucial property of a system of partial differential equations like the one considered above is that of *well-posedness*, by which we understands that the system is such that its solutions depend continuously on the initial data, or in other words, that small changes in the initial data will correspond to small changes in the solution. More formally, a system of partial differential equations is called well-posed if we can define a norm $|| \cdot ||$ such that

$$||u(t, x)|| \leq k e^{\alpha t} ||u(0, x)|| \,, \tag{5.2.2}$$

with k and α constants that are independent of the initial data. That is, the norm of the solution can be bounded by the same exponential for all initial data.

Most systems of evolution equations we usually find in mathematical physics turn out to be well-posed, which explains why there has been some complacency in the numerical relativity community about this issue. However, it is in fact not difficult to find rather simple examples of evolution systems that are not well-posed. We will consider three such examples here. The easiest example is the inverse heat equation which can be expressed as

$$\partial_t u = -\partial_x^2 u \,. \tag{5.2.3}$$

Assume now that as initial data we take a Fourier mode $u(0, x) = e^{ikx}$. In that case the solution to the last equation can be easily found to be

$$u(x, t) = e^{k^2 t + ikx} \,. \tag{5.2.4}$$

We then see that the solution grows exponentially with time, with an exponent that depends on the frequency of the initial Fourier mode k. It is clear that by

[45] One should not confuse the vectors we are considering here with vectors in the sense of differential geometry. A vector here only represents an ordered collection of independent variables.

increasing k we can increase the rate of growth arbitrarily, so the general solution can not be bounded by an exponential that is independent of the initial data. This also shows that given any arbitrary initial data, we can always add to it a small perturbation of the form ϵe^{ikx}, with $\epsilon \ll 1$ and $k \gg 1$, such that after a finite time the solution can be very different, so there is no continuity of the solutions with respect to the initial data.

A second example is the two-dimensional Laplace equation where one of the two dimensions is taken as representing "time":

$$\partial_t^2 \phi = -\partial_x^2 \phi \, . \tag{5.2.5}$$

This equation can be trivially written in first order form by defining $u_1 := \partial_t \phi$ and $u_2 := \partial_x \phi$. We find

$$\partial_t u_1 = -\partial_x u_2 \, , \tag{5.2.6}$$

$$\partial_t u_2 = +\partial_x u_1 \, , \tag{5.2.7}$$

where the second equation simply states that partial derivatives of ϕ commute. Again, consider a Fourier mode as initial data. The solution is now found to be

$$\phi = \phi_0 e^{kt+ikx} \, , \quad u_1 = k\phi_0 e^{kt+ikx} \, , \quad u_2 = ik\phi_0 e^{kt+ikx} \, . \tag{5.2.8}$$

We again see that the solution grows exponentially with a rate that depends on the frequency of the initial data k, so it can not be bounded in a way that is independent of the initial data. This shows that the Laplace equation is ill-posed when seen as a Cauchy problem, and incidentally explains why numerical algorithms that attempt to solve the Laplace equation by giving data on one boundary and then "evolving" to the opposite boundary are bound to fail (numerical errors will explode exponentially as we march ahead).

The two examples above are rather artificial, as the inverse heat equation is unphysical and the Laplace equation is not really an evolution equation. Our third example of an ill-posed system is more closely related to the problem of the 3+1 evolution equations. Consider the simple system

$$\partial_t u_1 = \partial_x u_1 + \partial_x u_2 \, , \tag{5.2.9}$$

$$\partial_t u_2 = \partial_x u_2 \, . \tag{5.2.10}$$

This system can be rewritten in matrix notation as

$$\partial_t u = M \partial_x u \, , \tag{5.2.11}$$

with $u = (u_1, u_2)$ and M the matrix

$$M = \begin{pmatrix} 1 & 1 \\ 0 & 1 \end{pmatrix} \, . \tag{5.2.12}$$

Again, consider the evolution of a single Fourier mode. The solution of the system of equations can then be easily shown to be

$$u_1 = (ikAt + B)\, e^{ik(t+x)} , \qquad u_2 = Ae^{ik(t+x)} , \tag{5.2.13}$$

with A and B constants. Notice that u_2 is oscillatory in time, so it is clearly bounded. However, u_1 has both an oscillatory part and a linear growth in time with a coefficient that depends on the initial data. Again, it is impossible to bound the growth in u_1 with an exponential that is independent of the initial data, as for any time t we can always choose k large enough to surpass any such bound. The system is therefore ill-posed.

Systems of this last type in fact often appear in reformulations of the 3+1 evolution equations (particularly in the ADM formulation). The problem can be traced back to the form of the matrix M above. Such a matrix is called a *Jordan block* (of order 2 in this case), and has two identical real eigenvalues but can not be diagonalized.

In the following Section we will consider a special type of system of partial differential equations called *hyperbolic* that can be shown to be well-posed under very general conditions.

5.3 The concept of hyperbolicity

Consider a first order system of evolution equations of the form

$$\partial_t u + M^i \partial_i u = s(u) , \tag{5.3.1}$$

where M^i are $n \times n$ matrices, with the index i running over the spatial dimensions, and $s(u)$ is a source vector that may depend on the u's but not on their derivatives. In fact, if the source term is linear in the u's we can show that the full system will be well-posed provided that the system without sources is well-posed. We will therefore ignore the source term from now on. Also, we will assume for the moment that the coefficients of the matrices M^i are constant.

There are several different ways of introducing the concept of *hyperbolicity* of a system of first order equations like (5.3.1).[46] Intuitively, the concept of hyperbolicity is associated with systems of evolution equations that behave as generalizations of the simple wave equation. Such systems are, first of all, well-posed, but they also should have the property of having a finite speed of propagation of signals, or in other words, they should have a finite past domain of dependence.

We will start by defining the notion of hyperbolicity based on the properties of the matrices M^i, also called the *characteristic matrices.* Consider an arbitrary unit vector n_i, and construct the matrix $P(n_i) := M^i n_i$, also known as the

[46]One can in fact also define hyperbolicity for systems of second order equations (see for example [154, 155, 212]), but here we will limit ourselves to first order systems as we can always write the 3+1 evolution equations in this form.

principal symbol of the system of equations (one often finds that P is multiplied with the imaginary unit i, but here we will assume that the coefficients of the M^i are real so we will not need to do this). We then say that the system (5.3.1) is *strongly hyperbolic* if the principal symbol has real eigenvalues and a complete set of eigenvectors for all n_i. If, on the other hand, P has real eigenvalues for all n_i but does not have a complete set of eigenvectors then the system is said to be only *weakly hyperbolic* (an example of a weakly hyperbolic system is precisely the Jordan block considered in the previous Section). For a strongly hyperbolic system we can always find a positive definite Hermitian (*i.e.* symmetric in the purely real case) matrix $H(n_i)$ such that

$$HP - P^T H^T = HP - P^T H = 0 \; , \qquad (5.3.2)$$

where the superindex T represents the transposed matrix. In other words, the new matrix HP is also symmetric, and H is called the *symmetrizer*. The symmetrizer is in fact easy to find. By definition, if the system is strongly hyperbolic the symbol P will have a complete set of eigenvectors e_a such that (here the index a runs over the dimensions of the space of solutions u)

$$Pe_a = \lambda_a e_a \; , \qquad (5.3.3)$$

with λ_a the corresponding eigenvalues. Define now R as the matrix of column eigenvectors. The matrix R can clearly be inverted since all the eigenvectors are linearly independent. The symmetrizer is then given by

$$H = (R^{-1})^T R^{-1} \; , \qquad (5.3.4)$$

which is clearly Hermitian and positive definite. To see that HP is indeed symmetric notice first that

$$R^{-1} P R = \Lambda \; , \qquad (5.3.5)$$

with $\Lambda = \mathrm{diag}(\lambda_a)$ (this is just a similarity transformation of P into the basis of its own eigenvectors). We can then easily see that

$$HP = (R^{-1})^T R^{-1} P = (R^{-1})^T \Lambda R^{-1} \; . \qquad (5.3.6)$$

But Λ is diagonal so that $\Lambda^T = \Lambda$, which immediately implies that $(R^{-1})^T \Lambda R^{-1}$ is symmetric. Of course, since the eigenvectors e_a are only defined up to an arbitrary scale factor and are therefore not unique, the matrix R and the symmetrizer H are not unique either.

We furthermore say that the system of equations is *symmetric hyperbolic* if all the M^i are symmetric, or more generally if the symmetrizer H is independent of n_i. Symmetric hyperbolic systems are therefore also strongly hyperbolic, but not all strongly hyperbolic systems are symmetric. Notice also that in the case of one spatial dimension any strongly hyperbolic system can be symmetrized, so the distinction between symmetric and strongly hyperbolic systems does not arise. We can also define a *strictly hyperbolic* system as one for which the eigenvalues of the principal symbol P are not only real but are also distinct for all n_i. Of course, this immediately implies that the symbol can be diagonalized, so strictly hyperbolic systems are automatically strongly hyperbolic. This last concept, however, is of little use in physics where we often find that the eigenvalues of P are degenerate, particularly in the case of many dimensions.

The importance of the symmetrizer H is related to the fact that we can use it to construct an inner product and norm for the solutions of the differential equation in the following way

$$\langle u, v \rangle := u^\dagger H v \,, \tag{5.3.7}$$

$$||u||^2 := \langle u, u \rangle = u^\dagger H u \,, \tag{5.3.8}$$

where $u^\dagger$ is the adjunct of u, *i.e.* its complex-conjugate transpose (we will allow complex solutions in order to use Fourier modes in the analysis). In geometric terms the matrix H plays the role of the metric tensor in the space of solutions. The norm defined above is usually called an *energy norm* since in some simple cases it coincides with the physical energy.

We can now use the evolution equations to estimate the growth in the energy norm. Consider a Fourier mode of the form

$$u(x,t) = \tilde{u}(t) e^{ik\vec{x}\cdot\vec{n}} \,. \tag{5.3.9}$$

We will then have

$$\begin{aligned} \partial_t ||u||^2 &= \partial_t \left(u^\dagger H u \right) = \partial_t (u^\dagger) H u + u^\dagger H \partial_t (u) \\ &= ik\tilde{u}^T P^T H \tilde{u} - ik\tilde{u}^T H P \tilde{u} \\ &= ik\tilde{u}^T \left(P^T H - H P \right) \tilde{u} = 0 \,, \end{aligned} \tag{5.3.10}$$

where on the second line we have used the evolution equation (assuming $s = 0$). We then see that the energy norm remains constant in time. This shows that strongly and symmetric hyperbolic systems are well-posed. We can in fact show that hyperbolicity and the existence of a conserved energy norm are equivalent, so instead of analyzing the principal symbol P we can look directly for the existence of a conserved energy to show that a system is hyperbolic. Notice that for symmetric hyperbolic systems the energy norm will be independent of the vector n_i, but for systems that are only strongly hyperbolic the norm will in general depend on n_i.

Now, for a strongly hyperbolic system we have by definition a complete set of eigenvectors and we can construct the matrix of eigenvectors R. We will use this matrix to define the *eigenfunctions* w_i (also called *eigenfields*) as

$$u = R\,w \quad \Rightarrow \quad w = R^{-1}\,u \,. \tag{5.3.11}$$

Notice that, just as was the case with the eigenvectors, the eigenfields are only defined up to an arbitrary scale factor. Consider now the case of a single spatial dimension x. By multiplying equation (5.3.1) with R^{-1} on the left we find that

$$\partial_t w + \Lambda \, \partial_x w = 0 \; , \tag{5.3.12}$$

so that the evolution equations for the eigenfields decouple. We then have a set of independent advection equations, each with a speed of propagation given by the corresponding eigenvalue λ_a. This is the mathematical expression of the notion that associates a hyperbolic system with having independent "wave fronts" propagating at (possibly different) finite speeds. Of course, in the multidimensional case the full system will generally not decouple even for symmetric hyperbolic systems, as the eigenfunctions will depend on the vector n_i.

We can in fact use the eigenfunctions also to study the hyperbolicity of a system; the idea here would be to construct a complete set of linearly independent eigenfunctions w_a that evolve via simple advection equations starting from the original variables u_a. If this is possible then the system will be strongly hyperbolic. For systems with a large number of variables this method is often simpler than constructing the eigenvectors of the principal symbol directly, as finding eigenfunctions can often be done by inspection (this is in fact the method we will use in the following Sections to study the hyperbolicity of the different 3+1 evolution systems).

Up until now we have assumed that the characteristic matrices M^i have constant coefficients, and also that the source term $s(u)$ vanishes. In the more general case when $s(u) \neq 0$ and $M^i = M^i(t,x,u)$ we can still define hyperbolicity in the same way by linearizing around a background solution $\hat{u}(t,x)$ and considering the local form of the matrices M^i, and we can also show that strong and symmetric hyperbolicity implies well-posedness. The main difference is that now we can only show that solutions exist locally in time, as after a finite time singularities in the solution may develop (*e.g.* shock waves in hydrodynamics, or spacetime singularities in relativity). Also, the energy norm does not remain constant in time but rather grows at a rate that can be bounded independently of the initial data. A particularly important sub-case is that of *quasi-linear* systems of equations where we have two different sets of variables u and v such that derivatives in both space and time of the u's can always be expressed as (possibly non-linear) combinations of v's, and the v's evolve though equations of the form $\partial_t v + M^i(u)\, \partial_i v = s(u,v)$, with the matrices M^i functions only of the u's. In such a case we can bring the u's freely in and out of derivatives in the evolution equations of the v without changing the principal part by replacing all derivatives of u's in terms of v's, and all the theory presented here can be applied directly. As we will see later, the Einstein field equations have precisely this property, with the u's representing the metric coefficients (lapse, shift and spatial metric) and the v's representing both components of the extrinsic curvature and spatial derivatives of the metric.

First order systems of equations of type (5.3.1) are often written instead as

$$\partial_t u + \partial_i F^i(u) = s(u) \; , \tag{5.3.13}$$

where F^i are vector valued functions of the u's (and possibly the spacetime coordinates), but not of their derivatives. The vectors F^i are called the *flux vectors*, and it terms of them the characteristic matrices are simply given by

$$M^i_{ab} := \frac{\partial F^i_a}{\partial u_b} , \tag{5.3.14}$$

that is, the matrices M^i are simply the Jacobian matrices associated with the fluxes F^i (notice that here $i \in (1, 2, 3)$ runs over the spatial dimensions, while $(a, b) \in (1, ..., n)$ run over the dimensions of the space of solutions). A system of the form (5.3.13) is called a *balance law*, since the change in u within a small volume element is given by a balance between the fluxes entering (or leaving) the volume element and the sources. In the special case where the sources vanish, the system is called instead a *conservation law*. This is because in such a case we can use the divergence theorem to show that if u has compact support then the integral of u over a volume outside such support is independent of time, *i.e.* it is conserved.

The simplest example of a strongly hyperbolic equation is of course the one-dimensional advection equation itself

$$\partial_t u + v\, \partial_x u = 0 , \tag{5.3.15}$$

with v a real constant. The solution of this equation simply propagates the initial data with a speed v without changing its original profile. In other words if $u(t = 0, x) = f(x)$, then $u(t, x) = f(x - vt)$. Advection terms like this appear all the time in the 3+1 evolution equations, as the Lie derivative with respect to the shift introduces terms of precisely the type $\beta^i \partial_i$.

The next interesting example is the one-dimensional wave equation

$$\partial_t^2 \phi - v^2 \nabla^2 \phi = 0 , \tag{5.3.16}$$

with v the wave speed. This equation can written in first order form by defining

$$\Pi := \partial_t \phi , \tag{5.3.17}$$

$$\Psi_i := v\, \partial_i \phi . \tag{5.3.18}$$

The wave equation then turns into the system

$$\partial_t \Pi - v \sum_i \partial_i \Psi_i = 0 , \tag{5.3.19}$$

$$\partial_t \Psi_i - v\, \partial_i \Pi = 0 . \tag{5.3.20}$$

Choose now a vector n_i. The principal symbol is clearly

$$P(n_i) = v \begin{pmatrix} 0 & -n_x & -n_y & -n_z \\ -n_x & 0 & 0 & 0 \\ -n_y & 0 & 0 & 0 \\ -n_z & 0 & 0 & 0 \end{pmatrix} . \tag{5.3.21}$$

Notice that the symbol is already symmetric, so the symmetrizer H is unity and the system is symmetric hyperbolic. Remembering now that the vector n_i is unitary (*i.e.* $n_x^2 + n_y^2 + n_z^2 = 1$), we find that the eigenvalues of the principal symbol are

$$\lambda_1 = +v\,, \quad \lambda_2 = -v\,, \quad \lambda_3 = \lambda_4 = 0\,. \tag{5.3.22}$$

So we have two degenerate eigenvalues and the system is not strictly hyperbolic. The matrix of column eigenvectors is given by

$$R(n_i) = \begin{pmatrix} 1 & 1 & 0 & 0 \\ -n_x & n_x & -n_z & -n_y \\ -n_y & n_y & 0 & n_x \\ -n_z & n_z & n_x & 0 \end{pmatrix}, \tag{5.3.23}$$

and its inverse is

$$R^{-1}(n_i) = \begin{pmatrix} 1/2 & -n_x/2 & -n_y/2 & -n_z/2 \\ 1/2 & n_x/2 & n_y/2 & n_z/2 \\ 0 & -n_z & -n_y n_z/n_x & (n_x^2 + n_y^2)/n_x \\ 0 & -n_y & (n_x^2 + n_z^2)/n_x & -n_y n_z/n_x \end{pmatrix}. \tag{5.3.24}$$

The eigenfields then turn out to be (we have rescaled them for convenience)

$$w_1 = \Pi - \sum_i n_i \Psi_i\,, \tag{5.3.25}$$

$$w_2 = \Pi + \sum_i n_i \Psi_i\,, \tag{5.3.26}$$

$$w_3 = \sum_i p_i \Psi_i\,, \tag{5.3.27}$$

$$w_4 = \sum_i q_i \Psi_i\,, \tag{5.3.28}$$

with $p = (-n_x n_z, -n_y n_z, n_x^2 + n_y^2)$ and $q = (-n_y n_x, n_x^2 + n_z^2, -n_y n_z)$. Notice that both p_i and q_i are orthogonal to n_i. So we see that the longitudinal modes (parallel to n_i) propagate at speeds $\pm v$, and the transverse modes (orthogonal to n_i) do not propagate. This example clearly shows that for symmetric hyperbolic systems the eigenfields w_a in general depend on the vector n_i, even if the symmetrizer does not.

Consider now the energy norm. Since the symmetrizer for the wave equation is unity (the principal symbol is already symmetric), our energy norm is simply

$$||u||^2 = \Pi^2 + \sum_i (\Psi_i)^2 = (\partial_t \phi)^2 + v^2 \sum_i (\partial_i \phi)^2\,, \tag{5.3.29}$$

which in this case coincides with the physical energy density of the wave. Notice that we should really understand this energy in an integrated sense:

$$\mathcal{E} = \int ||u||^2 dV = \int \left(\Pi^2 + \sum_i (\Psi_i)^2 \right) dV \, . \tag{5.3.30}$$

The change in time of the energy will then be

$$\begin{aligned} \frac{d}{dt}\mathcal{E} &= \frac{d}{dt} \int \left(\Pi^2 + \sum_i (\Psi_i)^2 \right) dV = 2 \int \left(\Pi \, \partial_t \Pi + \sum_i \Psi_i \, \partial_t \Psi_i \right) dV \\ &= 2v \sum_i \int \left(\Pi \, \partial_i \Psi_i + \Psi_i \, \partial_i \Pi \right) dV = 2v \sum_i \int \partial_i \left(\Pi \Psi_i \right) dV \, . \end{aligned} \tag{5.3.31}$$

The divergence theorem now implies that for initial data with compact support $d\mathcal{E}/dt = 0$.

As a final comment we should mention the fact there is one more equation that we have left out of the above analysis, namely $\partial_t \phi = \Pi$. But this equation has no fluxes and evolves only through a source term. If we included it in the analysis we would simply have to add one new row and column to the matrix P with zeroes as entries. The result would be one more eigenfunction that does not propagate – namely ϕ itself. Since this does not add any new information we will generally leave out of the analysis the evolution equations of lower order quantities such as ϕ.

5.4 Hyperbolicity of the ADM equations

We will now return to the 3+1 evolution equations and study under which conditions they are well-posed. In order to do this we will rewrite these equations as a system of the form (5.3.13), and study the structure of its characteristic matrices, *i.e.* its *characteristic structure*. There are many different ways in which we can do this, the direct way would be to write down the principal symbols and find its eigenvalues and eigenvectors. In some cases, however, it turns out that we can in fact discover the eigenfields by inspection. This is the method we will use here and in the following Sections.

We will start by rewriting the standard ADM equations (2.3.11) and (2.5.5) described in Chapter 2

$$\partial_t \gamma_{ij} - \pounds_{\vec{\beta}} \gamma_{ij} = -2\alpha K_{ij} \, , \tag{5.4.1}$$

$$\partial_t K_{ij} - \pounds_{\vec{\beta}} K_{ij} = -D_i D_j \alpha + \alpha \left(R_{ij} + K K_{ij} - 2 K_{ik} K^k_j \right) \, , \tag{5.4.2}$$

where for simplicity we have assumed that we are in vacuum (the matter terms can in any case be considered as sources). This system is first order in time but second order in space, as the Ricci tensor R_{ij} contains second derivatives of the spatial metric γ_{ij}, and also there are second derivatives of the lapse function α in the evolution equation for K_{ij}. In order to have a purely first order system we introduce the quantities

$$a_i := \partial_i \ln \alpha \, , \qquad d_{ijk} := \frac{1}{2} \, \partial_i \gamma_{jk} \, . \tag{5.4.3}$$

We will also assume that the shift vector β^i is a known (*i.e. a priori* given) function of space and time. The lapse, on the other hand, will be considered a dynamical quantity that evolves through an equation of the form

$$\partial_t \alpha - \beta^i \partial_i \alpha = -\alpha^2 Q \,, \tag{5.4.4}$$

with the explicit form of the gauge source function Q to be fixed later.

Since the characteristic structure is only related to the principal part, from now on we will ignore all source terms, that is all terms that do not contain derivatives of a_i, d_{ijk}, and K_{ij}. In terms of the d_{ijk}, the Ricci tensor can be written as

$$\begin{aligned} R_{ij} &\simeq -\frac{1}{2}\gamma^{lm}\partial_l\partial_m\gamma_{ij} + \gamma_{k(i}\partial_{j)}\Gamma^k \\ &\simeq -\partial_m {d^m}_{ij} + 2\,\partial_{(i}{d^m}_{mj)} - \partial_{(i}{d_{j)m}}^m \,, \end{aligned} \tag{5.4.5}$$

where here the symbol $\simeq$ denotes "equal up to principal part". There is in fact an ordering ambiguity in the last expression, as the definition of d_{ijk} implies the constraint

$$\partial_i d_{jmn} = \partial_j d_{imn} \,, \tag{5.4.6}$$

so that we can write the Ricci tensor in different ways that are only equivalent if the constraint above is satisfied. However, here we will ignore this issue and use the expression for the Ricci tensor given above (known as the *De Donder–Fock decomposition*). We are now in a position to write down the evolution equations for a_i, d_{ijk} and K_{ij}:

$$\partial_0\, a_i \simeq -\alpha\, \partial_i Q \,, \tag{5.4.7}$$

$$\partial_0\, d_{ijk} \simeq -\alpha\, \partial_i K_{jk} \,, \tag{5.4.8}$$

$$\partial_0\, K_{ij} \simeq -\alpha\, \partial_k \Lambda^k_{ij} \,, \tag{5.4.9}$$

where $\partial_0 := \partial_t - \beta^k \partial_k$, and where we have defined (following [65])

$$\Lambda^k_{ij} := {d^k}_{ij} + \delta^k_{(i}\left(a_{j)} + {d_{j)m}}^m - 2{d^m}_{mj)}\right) \,. \tag{5.4.10}$$

Notice that here we have another ordering ambiguity with the second derivatives of α, but this ambiguity can be resolved in a unique way by taking the explicitly symmetric expression: $\partial_i\partial_j\alpha = [\partial_i(\alpha a_j) + \partial_j(\alpha a_i)]/2$.

As mentioned in the previous Section, for the characteristic analysis we will ignore the evolution of the lower order quantities α and γ_{ij}. We then have a system of 27 equations to study corresponding to the three components of a_i, the 18 independent components of d_{ijk}, and the six independent components of K_{ij}.

To proceed with the characteristic analysis we will choose a specific direction x and ignore derivatives along the other directions; in effect we will only be analyzing the matrix M^x. The reason why we can do this instead of analyzing the full symbol $P = n_i M^i$ is that the tensor structure of the equations makes all spatial directions have precisely the same structure, so it is enough to analyze just one of them. The idea is then to find 27 independent eigenfunctions that will allow us to recover the 27 original quantities, where by eigenfunctions here we mean linear combinations of the original quantities $u = (a_i, d_{ijk}, K_{ij})$ of the form $w_a = \sum_b C_{ab} u_b$, that up to principal part evolve as $\partial_t w_a + \lambda_a \partial_x w_a \simeq 0$, with λ_a the corresponding eigenspeeds. Notice that even if the coefficients C_{ab} should not depend on the u's, they can in fact be functions of the lapse α and the spatial metric γ_{ij} (in order to avoid confusion we will use the Latin indices at the beginning of the alphabet $\{a, b, c, ...\}$ to identify the different fields, and those starting from $\{i, j, k, ...\}$ to denote tensor indices).

Then taking into account only derivatives along the x direction we immediately see that there is a set of fields that propagate along the time lines with velocity $\lambda^0 = -\beta^x$. These fields are

$$a_q \, , d_{qij} \qquad (q \neq x) \; . \tag{5.4.11}$$

We have therefore already found 14 out of a possible 27 characteristic fields.

Consider now the evolution equation for the Λ^x_{pq}, with $p, q \neq x$. Ignoring again derivatives along directions different from x we find

$$\partial_0 \, \Lambda^x_{pq} = \partial_0 \, d^x{}_{pq} \simeq -\alpha \gamma^{xx} \, \partial_x K_{pq} \; , \tag{5.4.12}$$

Comparing with (5.4.9) we see that we have found another set of characteristic fields that propagate along the light-cones with speeds[47]

$$\lambda^{\text{light}}_{\pm} = -\beta^x \pm \alpha \sqrt{\gamma^{xx}} \; . \tag{5.4.13}$$

These eigenfields are

$$\sqrt{\gamma^{xx}} \, K_{pq} \mp \Lambda^x_{pq} \; . \tag{5.4.14}$$

This gives us six new characteristic fields, so we now have a total of 20 out of 27.

From this point on things become somewhat more complicated. Notice that we still need to find characteristic fields that will allow us to recover the seven fields a_x, d_{xxi}, and K_{xi}. We start by writing down the evolution equation for Λ^x_{xq}:

$$\begin{aligned} \partial_0 \, \Lambda^x_{xq} &= \partial_0 \, d^x{}_{xq} + \frac{1}{2} \, \partial_0 \left(a_q + d_{qm}{}^m - 2 d^m{}_{mq} \right) \\ &\simeq -\alpha \gamma^{xx} \, \partial_x K_{xq} + \alpha \, \partial_x K^x_q \\ &\simeq \alpha \gamma^{xp} \, \partial_x K_{pq} \; . \end{aligned} \tag{5.4.15}$$

[47]Here we have used the observation that a pair of functions (u_1, u_2) that evolve through equations of the form

$$\partial_t u_1 = a \, \partial_x u_2 \; , \qquad \partial_t u_2 = b \, \partial_x u_1 \; ,$$

have the structure of the simple one-dimensional wave equation. The speeds of propagation for this system are simply $\pm\sqrt{ab}$, and the corresponding eigenfields are $w_{\pm} = u_1 \mp (a/b)^{1/2} u_2$.

This equation presents us with a serious problem: It shows that while K_{xq} evolves through derivatives of Λ^x_{xq}, the evolution of Λ^x_{xq} is independent of K_{xq} and only involves derivatives of K_{pq}. This subsystem can therefore not be diagonalized. Notice in particular that if the metric γ_{ij} is diagonal then Λ^x_{xq} will remain constant up to principal part, which implies that K_{xq} will grow linearly unless Λ^x_{xq} vanishes initially (of course, the presence of source terms makes things more complicated). This shows that the ADM system is not strongly hyperbolic; it is only weakly hyperbolic as we can in fact show that the eigenvalues of the principal symbol are all real.

We have then found that we can not recover the four quantities K_{xq} and Λ^x_{xq} (and hence d_{xxq}) from characteristic fields. What about the quantities a_x, d_{xxx}, and K_{xx}? Even though we already know that the full system is not strongly hyperbolic, there is still something very important to learn from these last three quantities. As it turns out, it is in fact easier to work with the traces K and $\Lambda^x := \gamma^{mn}\Lambda^x_{mn}$. The evolution equation for K clearly has the form

$$\partial_0 K \simeq -\alpha\, \partial_x \Lambda^x \,. \tag{5.4.16}$$

The evolution equation for Λ^x, on the other hand, becomes

$$\begin{aligned} \partial_0 \Lambda^x &= \partial_0 a^x + 2\, \partial_0 \left(d^{xm}{}_m - d_m{}^{mx} \right) \\ &\simeq -\alpha\gamma^{xx}\, \partial_x Q - 2\alpha \left(\gamma^{xx}\, \partial_x K - \partial_x K^{xx} \right) \,. \end{aligned} \tag{5.4.17}$$

Notice first that this is the only equation that involves the gauge source function Q, so that the explicit form of Q only affects the evolution of K and Λ^x (and a_x of course). Before discussing the form of Q let us for a moment recall the explicit form of the momentum constraints:

$$D_j \left(K^{ij} - \gamma^{ij} K \right) = 8\pi j^i \,, \tag{5.4.18}$$

which up to principal part becomes

$$\partial_j K^{ij} - \gamma^{ij} \partial_j K \simeq 0 \,. \tag{5.4.19}$$

Considering now only derivatives along x we see that this implies, in particular,

$$\partial_x K^{xx} \simeq \gamma^{xx} \partial_x K \,, \tag{5.4.20}$$

$$\partial_x K^x_q \simeq 0 \,. \tag{5.4.21}$$

This means that if the momentum constraints are satisfied identically the evolution equations for Λ^x_{xq} and Λ^x, equations (5.4.15) and (5.4.17), become

$$\partial_0 \Lambda^x_{xq} \simeq -\alpha\gamma^{xx}\, \partial_x K_{xq} \tag{5.4.22}$$

$$\partial_0 \Lambda^x \simeq -\alpha\gamma^{xx}\, \partial_x Q \tag{5.4.23}$$

This immediately solves our problem with the pair K_{xq} and Λ^x_{xq}, as now they behave exactly like K_{pq} and Λ^x_{pq}, so this subsystem would also be strongly hyperbolic with the eigenfunctions $\sqrt{\gamma^{xx}}\, K_{xq} \mp \Lambda^x_{xq}$ propagating at the speed of light.

On the other hand, for K and Λ^x we still need to say something about the form of Q. The simplest choice is to say that, just like the shift, the lapse is also an *a priori* known function of spacetime, in which case Q is just a source term and can be ignored. We can now clearly see, however, that this is a very bad idea as we would then have Λ^x constant up to principal part and K evolving through derivatives of Λ^x, which is precisely the same type of problem we had identified before. So having a prescribed lapse will result in a weakly hyperbolic system *even if* the momentum constraints are identically satisfied. Notice, however, that if we were to take instead the densitized lapse $\tilde{\alpha} = \alpha/\sqrt{\gamma}$ as a prescribed function of spacetime, then we would have

$$\begin{aligned}\left(\partial_t - \pounds_{\vec{\beta}}\right)\alpha &= \frac{\tilde{\alpha}}{2\sqrt{\gamma}}\left(\partial_t - \pounds_{\vec{\beta}}\right)\gamma + \sqrt{\gamma}\left(\partial_t - \pounds_{\vec{\beta}}\right)\tilde{\alpha} \\ &= -\alpha^2 K + \sqrt{\gamma}\, F(x,t) , \end{aligned} \tag{5.4.24}$$

with $F(x,t) := (\partial_t - \pounds_{\vec{\beta}})\,\tilde{\alpha}$. This essentially means that $Q = K$ (the second term can be ignored as it contributes only to a source term). The equations for K and Λ^x would now become

$$\partial_0 K \simeq -\alpha\, \partial_x \Lambda^x , \tag{5.4.25}$$

$$\partial_0 \Lambda^x \simeq -\alpha\gamma^{xx}\, \partial_x K , \tag{5.4.26}$$

so that we have another pair of modes propagating at the speed of light, namely $\sqrt{\gamma^{xx}}\, K \mp \Lambda^x$. More generally, we can take a slicing condition of the Bona–Masso family (4.2.52), which implies $Q = f(\alpha)K$. The gauge eigenfields would then be $\sqrt{f\gamma^{xx}}\, K \mp \Lambda^x$, and they would propagate with the gauge speeds

$$\lambda_{\pm}^{\text{gauge}} = -\beta^x \pm \alpha\sqrt{f\gamma^{xx}} . \tag{5.4.27}$$

These two gauge eigenfields will allow us to recover two of the three quantities a_x, d_{xxx}, and K_{xx}. The last quantity can be recovered by noticing that

$$\partial_0 \left(a_x - f d_{xm}{}^m\right) \simeq -\alpha\, \partial_x \left(Q - fK\right) = 0 , \tag{5.4.28}$$

so that we have another independent eigenfield propagating along the time lines.

We have then found that, *provided* that 1) the momentum constraints can be guaranteed to be identically satisfied, and 2) either the densitized lapse $\tilde{\alpha}$ is assumed to be a known function of spacetime (but not the lapse itself), or we use a slicing condition of the Bona–Masso family, then the ADM system would be strongly hyperbolic.[48]

Taking a slicing condition of the Bona–Masso type is simple enough, but guaranteeing that the momentum constraints are identically satisfied is altogether a different matter. First, numerically the constraints will inevitably be

[48] In fact, we could also use maximal slicing since in that case $K = 0$ and the a_i are obtained through an elliptic equation. We would then have a coupled elliptic-hyperbolic system.

violated, and second, even at the continuum level strong hyperbolicity would only be guaranteed for a very specific type of initial data, so the system would not be well-posed as such. We are then forced to conclude that ADM is in fact only weakly hyperbolic. Still, we have learned two important lessons here: The lapse itself can not be assumed to be a known function of spacetime, and the momentum constraints must play a crucial role.

As a final comment, notice that here we have only considered the standard (York) version of the ADM equations. The original ADM evolution equations add a multiple of the Hamiltonian constraint which makes the analysis somewhat more difficult. Still, we will come back to the original ADM system in the next Section.

5.5 The Bona–Masso and NOR formulations

In the last Section we showed that the standard ADM evolution system is only weakly hyperbolic and is therefore not well-posed. It is no wonder then that the first fully three-dimensional codes written in the early and mid 1990s based on the ADM equations had serious stability problems (as we will see in Chapter 9, numerical stability can be seen essentially as a discrete version of well-posedness). Well-posed versions of the Einstein equations have in fact been known since the 1950s and where used to prove the first theorems on the existence and uniqueness of solutions [84], but they were not based on a 3+1 decomposition and required the use of fully harmonic spacetime coordinates, so they were by and large ignored by the numerical community (but not completely, see *e.g.* [231] for a very recent and highly successful application of such techniques). Efforts to find well-posed versions of the 3+1 evolution equations had to wait until the late 1980s and early 1990s.

I will concentrate first on the Bona–Masso formulation, which is arguably the simplest way to write down a strongly hyperbolic reformulation of the 3+1 evolution equations [62, 63, 64, 65, 66].[49] The Bona–Masso formulation starts by defining the three auxiliary variables

$$V_i := d_{im}{}^m - d^m{}_{mi} \,. \tag{5.5.1}$$

In terms of the V_i, the Ricci tensor can be written up to principal part as

$$R_{ij} \simeq -\partial_m d^m{}_{ij} - \partial_i \left(V_j - \frac{1}{2}\, d_{jm}{}^m \right) - \partial_j \left(V_i - \frac{1}{2}\, d_{im}{}^m \right) \,. \tag{5.5.2}$$

We then write the evolution equations for a_i, d_{ijk} and K_{ij} just as before as

[49] One often finds some confusion between the Bona–Masso family of slicing conditions described in Chapter 4 (and also used here), and the Bona–Masso formulation of the 3+1 evolution equations. The Bona–Masso formulation uses their slicing condition, but the slicing condition itself is in fact quite general and can be used with any form of the evolution equations.

$$\partial_0 \, a_i \simeq -\alpha \, \partial_i Q \; , \tag{5.5.3}$$

$$\partial_0 \, d_{ijk} \simeq -\alpha \, \partial_i K_{jk} \; , \tag{5.5.4}$$

$$\partial_0 \, K_{ij} \simeq -\alpha \, \partial_k \Lambda^k_{ij} \; , \tag{5.5.5}$$

where now

$$\Lambda^k_{ij} := d^k{}_{ij} + \delta^k_{(i} \left(a_{j)} + 2V_{j)} - d_{j)m}{}^m \right) \; . \tag{5.5.6}$$

Up to this point we have done nothing more than introduce a short-hand for a certain combination of d's, but now we will take a big leap and consider the V_i to be independent dynamical quantities in their own right and take their definition (5.5.1) as a constraint. If the V_i are independent quantities we will need an evolution equation for them, which can be obtained directly from their definition. In order to write these evolution equations we first notice that

$$V^i = \frac{1}{2} \left(d^{im}{}_m - \Gamma^i \right) \; , \tag{5.5.7}$$

with $\Gamma^i := \gamma^{lm} \Gamma^i_{lm}$ the contracted Christoffel symbols. We can show, after some algebra, that

$$\partial_t \Gamma^i = \pounds_{\vec{\beta}} \Gamma^i + \gamma^{lm} \partial_l \partial_m \beta^i - D_l \left[\alpha \left(2K^{il} - \gamma^{il} K \right) \right] + 2\alpha K^{lm} \Gamma^i_{lm} \; , \tag{5.5.8}$$

where the Lie derivative of Γ^i is to be understood as that of an ordinary vector. The fact that the Γ^i are not components of a true vector is what gives rise to the term with the flat Laplacian of the shift. An interesting observation at this point is that we can rewrite the flat Laplacian of the shift as

$$\gamma^{lm} \partial_l \partial_m \beta^i = D^2 \beta^i - \pounds_{\vec{\beta}} \Gamma^i - 2\Gamma^i_{lm} D^m \beta^n + R^i_m \beta_m \; , \tag{5.5.9}$$

so that the evolution equation for Γ^i can be written in a more covariant-looking way as

$$\partial_t \Gamma^i = D^2 \beta^i + R^i_m \beta_m - D_l \left[\alpha \left(2K^{il} - \gamma^{il} K \right) \right] + 2 \left[\alpha K^{lm} - D^l \beta^m \right] \Gamma^i_{lm} \; . \tag{5.5.10}$$

Using now the evolution equation for Γ^i we find that the V_i evolve up to principal part as

$$\partial_0 V_i \simeq \alpha \left(\partial_j K^j_i - \partial_i K \right) \; . \tag{5.5.11}$$

This is a very beautiful result, as it shows that the principal part of the evolution equation for V_i is precisely the same as the principal part of the momentum constraints.

Assume now that we add $2\alpha M^i$ to the evolution equation for Γ^i, where $M^i := D_j \left(K^{ij} - \gamma^{ij} K \right) - 8\pi j^i = 0$ are the momentum constraints. We are of course perfectly free to do this as the physical solutions will remain the same. We would then have

$$\partial_t \Gamma^i = \pounds_{\vec{\beta}} \Gamma^i + \gamma^{lm} \partial_l \partial_m \beta^i - \alpha \gamma^{il} \partial_l K \\ + \alpha a_l \left(2K^{il} - \gamma^{il} K \right) + 2\alpha K^{lm} \Gamma^i_{lm} - 16\pi j^i \,, \tag{5.5.12}$$

with $a_i = \partial_i \ln \alpha$ as before. The evolution equation for the V_i up to principal part now becomes

$$\partial_0 V_i \simeq 0 \,, \tag{5.5.13}$$

so that V_i evolves along the time lines.

We can now repeat the analysis we did before for ADM, but now for the 30 independent quantities $u = (a_i, d_{ijk}, K_{ij}, V_i)$. Considering as before only derivatives along the x direction we immediately see that there are now 17 fields that propagate along the time lines, namely

$$a_q \,, d_{qij} \,, V_i \qquad (q \neq x) \,. \tag{5.5.14}$$

For Λ^x_{pq} and K_{pq} with $p, q \neq x$, we find the same thing as in the ADM case, as independently of the introduction of the V_i we still find that $\Lambda^x_{pq} = d^x{}_{pq}$, so that again we have the six eigenfields

$$\sqrt{\gamma^{xx}}\, K_{pq} \mp \Lambda^x_{pq} \,, \tag{5.5.15}$$

propagating along the light-cones.

The difference with ADM is related to the remaining seven fields a_x, d_{xxi}, and K_{xi}. For the evolution of Λ^x_{xq} we now find

$$\partial_0 \, \Lambda^x_{xq} = \partial_0 \, d^x{}_{xq} + \frac{1}{2} \, \partial_0 \left(a_q + 2V_q - d_{qm}{}^m \right) \\ \simeq -\alpha \gamma^{xx} \, \partial_x K_{xq} \,, \tag{5.5.16}$$

so that the four eigenfields

$$\sqrt{\gamma^{xx}}\, K_{xq} \mp \Lambda^x_{xq} \,, \tag{5.5.17}$$

also propagate along the light-cones. Moreover, for the evolution of the trace $\Lambda^x := \gamma^{mn} \Lambda^x_{mn}$ we now have

$$\partial_0 \, \Lambda^x = \partial_0 a^x + 2 \, \partial_0 V^x \\ \simeq -\alpha \gamma^{xx} \, \partial_x Q \,. \tag{5.5.18}$$

If we again take a slicing condition of the Bona–Masso type so that $Q = f(\alpha) K$, we find the two gauge eigenfields

$$\sqrt{f \gamma^{xx}}\, K \mp \Lambda^x \,, \tag{5.5.19}$$

that propagate with the gauge speeds

$$\lambda^{\text{gauge}}_{\pm} = -\beta^x \pm \alpha \sqrt{f \gamma^{xx}} \,. \tag{5.5.20}$$

The last eigenfield is again $a_x - f d_{xm}{}^m$ which also propagates along the time lines. In summary, we have 18 eigenfields propagating along the time lines, namely a_q, d_{qij}, V_i, $a_x - f d_{xm}{}^m$, 10 more eigenfields propagating along the light cones, $w^l_{iq\pm} = \sqrt{\gamma^{xx}}\, K_{iq} \mp \Lambda^x_{iq}$, and two gauge eigenfields $w^f_\pm = \sqrt{f\gamma^{xx}}\, K \mp \Lambda^x$. The Bona–Masso formulation is therefore strongly hyperbolic as long as the evolution equation for the V_i is modified using the momentum constraints to make its principal part vanish.

Instead of the V_i, we can choose to evolve the Γ^i as auxiliary quantities using (5.5.12), and write the Ricci tensor R_{ij} as

$$R_{ij} \simeq -\partial_m d^m{}_{ij} + \partial_{(i}\Gamma_{j)} \, , \tag{5.5.21}$$

which results in the Λ^k_{ij} being given by

$$\Lambda^k_{ij} = d^k{}_{ij} + \delta^k_{(i}\left(a_{j)} - \Gamma_{j)}\right) \, . \tag{5.5.22}$$

This is entirely equivalent since changing the V_i for the Γ^i only corresponds to a simple change of basis in the characteristic matrices, and does not alter their eigenvalues or eigenvectors.

The Bona–Masso formulation just presented can in fact be easily generalized by introducing two parameters that corresponds to adding an arbitrary multiple of the momentum constraints to the evolution equation for the Γ^i (or equivalently the V_i), and a multiple of the Hamiltonian constraints to the evolution equation for K_{ij}. The resulting equations are known as the Nagy–Ortiz–Reula (NOR) formulation [212]. The NOR formulation modifies the evolution equations for K_{ij} in the following way

$$\partial_t K_{ij} - \pounds_{\vec{\beta}} K_{ij} = -D_i D_j \alpha + \alpha\left(R_{ij} + K K_{ij} - 2K_{ik}K^k_j\right) + \alpha\eta\gamma_{ij} H \, , \tag{5.5.23}$$

with $H := R + K^2 - K_{ij}K^{ij} - 16\pi\rho = 0$ the Hamiltonian constraint and η a constant parameter. At the same time, the evolution equation for Γ^i is taken as

$$\partial_t \Gamma^i = \pounds_{\vec{\beta}}\Gamma^i + \gamma^{lm}\partial_l\partial_m\beta^i - D_l\left[\alpha\left(2K^{il} - \gamma^{il}K\right)\right] + 2\alpha K^{lm}\Gamma^i_{lm} + \alpha\xi M^i \, , \tag{5.5.24}$$

where as before M^i are the momentum constraints and ξ is also a constant parameter. Notice that the particular case $\eta = 0$, $\xi = 2$ reduces to the Bona–Masso formulation.[50] Also, the case $\eta = \xi = 0$ is essentially equivalent to the standard ADM formulation (*à la* York), while $\xi = 0$ and $\eta = -1/4$ corresponds to the original ADM formulation. The fact that the Γ^i are introduced as independent quantities has no effect if we take $\xi = 0$, since in that case their evolution equations are not modified and the characteristic structure of the system is identical to that of ADM with the addition of three extra eigenfields that propagate along the time lines, namely $\Gamma^i - 2d_m{}^{mi} + d^{im}{}_m$.

[50] In fact, the Bona–Masso formulation also considered non-zero values of η, with the special cases $\eta = 0$ and $\eta = -1/4$ called the "Ricci" and "Einstein" systems respectively.

The hyperbolicity analysis for the NOR system can be done in a number of different ways. For example, Nagy, Ortiz, and Reula use a method based on pseudo-differential operators in order to analyze directly the second order system without the need to introduce the first order quantities d_{ijk} [212]. For simplicity, however, here we will keep with our first order approach, but will not go into a detailed analysis since it proceeds in just the same way as before. If we again use a slicing condition of the Bona–Masso type[51], and consider only derivatives along the x direction, we find that there are 18 fields that propagate along the time lines with speed $-\beta^x$; they are

$$a_q , \; d_{qij} , \; a_x - f d_{xm}{}^m , \; \Gamma^i + (\xi - 2)\, d_m{}^{mi} + (1 - \xi)\, d^{im}{}_m \, , \tag{5.5.25}$$

with $q \neq x$. For the rest of the characteristic fields it turns out to be convenient to define the projected metric onto the two-dimensional surface of constant x as $h_{ij} := \gamma_{ij} - (s^x)_i (s^x)_j$, with $\vec{s}^{\,x}$ the normal vector to the surface:

$$(s^x)_i = \delta^x_i / \sqrt{\gamma^{xx}} \qquad (s^x)^i = \gamma^{xi} / \sqrt{\gamma^{xx}} \, . \tag{5.5.26}$$

We can now use this projected metric to define the "surface-trace" of K_{ij} as

$$\hat{K} := h^{ij} K_{ij} = K - K^{xx} / \gamma^{xx} \, , \tag{5.5.27}$$

with an analogous definition for $\hat{\Lambda}^x$. Using this we find that the following characteristic fields

$$\sqrt{\gamma^{xx}} \left(K_{pq} - \frac{h_{pq}}{2} \hat{K} \right) \mp \left(\Lambda^x_{pq} - \frac{h_{pq}}{2} \hat{\Lambda}^x \right) , \qquad (p, q) \neq x \, , \tag{5.5.28}$$

propagate along the light-cones with speeds

$$\lambda^{\text{light}}_{\pm} = -\beta^x \pm \alpha \sqrt{\gamma^{xx}} \, . \tag{5.5.29}$$

Notice that these are only four independent eigenfields, since the combination $K_{pq} - h_{pq} \hat{K}/2$ is both symmetric and surface-traceless. This result is to be expected, as these fields represent the transverse-traceless part associated with the gravitational waves. The four fields correspond to having two independent polarizations that can travel in either the positive or negative x directions. The traces $\hat{K}$ and $\hat{\Lambda}^x$ also form an eigenfield pair given by

$$\left\{ \gamma^{xx} \left[1 + 2\eta \, (2 - \xi) \right] \right\}^{1/2} \hat{K} \mp \hat{\Lambda}^x \, , \tag{5.5.30}$$

propagating with the speeds

$$\lambda^{\text{trace}}_{\pm} = -\beta^x \pm \alpha \left\{ \gamma^{xx} \left[1 + 2\eta \, (2 - \xi) \right] \right\}^{1/2} \, . \tag{5.5.31}$$

Notice that for $\xi = 2$ these two fields also propagate at the speed of light.

[51] The original NOR system in fact does not use the Bona–Masso slicing condition, but rather takes a general densitized lapse of the form $Q = \alpha \gamma^{-\sigma/2}$, with $\sigma =$constant, as a fixed function of spacetime. The results, however, are entirely equivalent to the ones discussed here if we take $\sigma = f$.

Another set of four eigenfields, corresponding to the longitudinal components of the extrinsic curvature, turns out to be given by

$$\sqrt{\gamma^{xx}(\xi/2)}\, K^x_q \mp \Lambda^{xx}{}_q \,, \tag{5.5.32}$$

with a characteristic speed of

$$\lambda^{\rm long}_{\pm} = -\beta^x \pm \alpha\sqrt{\gamma^{xx}(\xi/2)} \,. \tag{5.5.33}$$

Again, for $\xi = 2$ these fields propagate along the light-cones.

The last two eigenfields are again related to the gauge choice, and propagate with the gauge speeds

$$\lambda^{\rm gauge}_{\pm} = -\beta^x \pm \alpha\sqrt{f\gamma^{xx}} \,. \tag{5.5.34}$$

These fields are

$$\sqrt{f\gamma^{xx}}\left(K + F\hat{K}\right) \mp \left(\Lambda^x + F\hat{\Lambda}^x\right) \,, \tag{5.5.35}$$

with $F := (1+3\eta)(2-\xi)/[f-1-2\eta(2-\xi)]$.

Collecting all these results we find that the NOR system is strongly hyperbolic if the function $f(\alpha)$ and the parameters η and ξ fall into any of the three cases:

1. $\eta = 0$, $\xi = 2$ and $f > 0$.
2. $\eta = 0$, $\xi > 0$, $f > 0$, and $f \neq 1$.
3. $\eta \neq 0$, $\xi > 0$, $f > 0$, and $\eta\,(2-\xi) > -1/2$.

We see that if we take $\eta = 0$ and $\xi = 2$, then the system is strongly hyperbolic for any $f > 0$, but this we already knew as it corresponds to the Bona–Masso system. More generally, if we take $\eta = 0$ and any $\xi > 0$, the system remains strongly hyperbolic as long as $f > 0$ and $f \neq 1$. This case, however, is probably not very useful as it explicitly excludes harmonic slicing $f = 1$. More interesting is the last case that shows that we can take any $f > 0$ and any $\xi > 0$, provided that $\eta \neq 0$ and $\eta\,(2-\xi) > -1/2$. Notice that standard ADM corresponds to $\eta = \xi = 0$, so that it fails to be strongly hyperbolic since in all cases the parameter ξ must be strictly positive. The original ADM system, with $\eta = -1/4$ and $\xi = 0$, is in fact worse as it would in principle fall into the third case but it fails to satisfy two of the necessary inequalities, since we now have both $\xi = 0$ and $\eta\,(2-\xi) = -1/2$. By "worse" here we mean that there will be a smaller number of independent eigenfunctions, but in any case both versions of the ADM system are still weakly hyperbolic since they violate the inequalities only marginally and the eigenspeeds are still real (satisfying the reversed inequalities would result in systems that are not even weakly hyperbolic as the eigenspeeds would become complex). Taking $\xi = 2$, as in the Bona–Masso case, seems to be an optimal choice as in that case all fields other than those related to the slicing condition propagate either along the time lines or with the speed of light. Also, in that case the value of η plays no role in any of the eigenfields or eigenspeeds, so we are free to choose an arbitrary value without affecting the characteristic structure in any way (using different values of η in this case might in fact improve the numerical stability as even though the characteristic structure in unaffected, the source terms will certainly change, but this issue has not been studied in any detail).

5.6 Hyperbolicity of BSSNOK

As already mentioned in Chapter 2, in the past few years the BSSNOK formulation has become the most widely used in three-dimensional numerical codes based on the 3+1 decomposition.[52] The reason for this is that this formulation has been found to be very robust in practice in a large class of systems with strong and dynamical gravitational fields, both with and without matter. In the original paper of Baumgarte and Shapiro [50], the success of the new formulation was attributed to the fact that it had a "more hyperbolic flavor". This statement was later put on somewhat ground by Alcubierre *et al.* in [6]. Finally, Sarbach *et al.* showed that the BSSNOK formulation is in fact strongly hyperbolic [249], so it should come as no surprise that it behaves better than ADM which, as we have seen, is only weakly hyperbolic. Hyperbolicity, however, is clearly not the only important property of this formulation, because BSSNOK has also been found empirically to out-perform other strongly hyperbolic formulations, such as the Bona–Masso formulation presented in the previous Section. Since the main difference between these two formulations is the conformal decomposition of BSSNOK, such a decomposition clearly seems to be important, though at this point it is still not understood why this should be so.

The BSSNOK formulation takes as independent variables the conformal metric $\tilde{\gamma}_{ij}$, the conformal factor ϕ, the trace of the extrinsic curvature K, the conformally rescaled tracefree extrinsic curvature $\tilde{A}_{ij}$, and the contracted conformal Christoffel symbols $\tilde{\Gamma}^i := \tilde{\gamma}^{mn}\tilde{\Gamma}^i_{mn}$, plus the lapse α and shift vector β^i. In order to study its characteristic structure we will again introduce the spatial derivatives of α, $\tilde{\gamma}_{ij}$, and ϕ as independent quantities,

$$a_i := \partial_i \ln \alpha \, , \qquad \tilde{d}_{ijk} := \frac{1}{2}\, \partial_i \tilde{\gamma}_{jk} \, , \qquad \Phi_i := \partial_i \phi \, , \tag{5.6.1}$$

and assume the shift to be a given function of spacetime. For the hyperbolicity analysis we now have to consider the 30 quantities $u = (a_i, \Phi_i, \tilde{d}_{ijk}, K, \tilde{A}_{ij}, \tilde{\Gamma}^i)$. Notice that these are in fact only 30 quantities and not 34, as $\tilde{A}_{ij}$ is traceless by definition, and also $\tilde{\gamma}^{jk}\tilde{d}_{ijk} = 0$ as a consequence of the fact that $\tilde{\gamma}_{ij}$ has unit determinant. The evolution equations guarantee that if these constraints are satisfied initially they will remain satisfied during evolution (numerically, however, we finds that this does not hold exactly and the trace of $\tilde{A}_{ij}$ drifts, so that the constraint $\mathrm{tr}\tilde{A}_{ij} = 0$ must be actively enforced during a simulation).

[52]At the time of writing this text, BSSNOK has become the dominant formulation used in 3D numerical relativity, and is used in one way or another in practically all 3+1 based codes. The only real contenders to BSSNOK are hyperbolic formulations that evolve the full four-dimensional spacetime metric and are not directly based on the 3+1 formalism [191, 231].

Starting from the BSSNOK evolution equations of Section 2.8, we find that up to principal part the evolution equations for our dynamical quantities become

$$\partial_0 a_i \simeq -\alpha\, \partial_i Q \,, \tag{5.6.2}$$

$$\partial_0 \Phi_i \simeq -\frac{1}{6}\, \alpha\, \partial_i K \,, \tag{5.6.3}$$

$$\partial_0 \tilde{d}_{ijk} \simeq -\alpha\, \partial_i \tilde{A}_{jk} \,, \tag{5.6.4}$$

$$\partial_0 K \simeq -\alpha e^{-4\phi} \tilde{\gamma}^{mn} \partial_m a_n \tag{5.6.5}$$

$$\partial_0 \tilde{A}_{ij} \simeq -\alpha e^{-4\phi} \partial_k \tilde{\Lambda}^k_{ij} \,, \tag{5.6.6}$$

$$\partial_0 \tilde{\Gamma}^i \simeq \alpha \partial_k \left[(\xi - 2)\, \tilde{A}^{ik} - \frac{2}{3}\, \xi \tilde{\gamma}^{ik} K \right] , \tag{5.6.7}$$

where we have now defined

$$\tilde{\Lambda}^k_{ij} := \left[\tilde{d}^k_{ij} + \delta^k_{(i} \left(a_{j)} - \tilde{\Gamma}_{j)} + 2\Phi_{j)} \right) \right]^{TF} , \tag{5.6.8}$$

and where we have allowed for an arbitrary multiple of the momentum constraint to be added to the evolution equation for $\tilde{\Gamma}^i$ (standard BSSNOK corresponds to choosing $\xi = 2$). In the following we will again assume that we are using a slicing condition of the Bona–Masso family so that $Q = fK$.

For the hyperbolicity analysis we again consider only derivatives along the x direction. Again we immediately find that there are 18 fields that propagate along the time lines with speed $-\beta^x$:

$$a_q \,, \Phi_q \,, \tilde{d}_{qij} \,, a_x - 6f\Phi_x \,, \tilde{\Gamma}^i + (\xi - 2)\, \tilde{d}_m{}^{mi} - 4\xi\, \tilde{\gamma}^{ik} \Phi_k \,, \tag{5.6.9}$$

with $q \neq x$ (again, these are only 18 fields since $\tilde{d}_{qij}$ is traceless). Now, the fact that in the BSSNOK formulation the trace of the extrinsic curvature K has been promoted to an independent variable and the Hamiltonian constraint has been used to simplify its evolution equation, makes the gauge eigenfields very easy to identify. The two gauge eigenfields are

$$e^{-2\phi} \sqrt{f \tilde{\gamma}^{xx}} K \mp a^x \,, \tag{5.6.10}$$

which propagate with the gauge speeds

$$\lambda^{\text{gauge}}_{\pm} = -\beta^x \pm \alpha e^{-2\phi} \sqrt{f \tilde{\gamma}^{xx}} \,. \tag{5.6.11}$$

There is another set of four eigenfields that again correspond to longitudinal components and are be given by

$$e^{2\phi} \sqrt{\tilde{\gamma}^{xx} \left(\xi/2 \right)}\, \tilde{A}^x_q \mp \tilde{\Lambda}^{xx}{}_q \,, \tag{5.6.12}$$

with corresponding characteristic speeds

$$\lambda^{\text{long}}_{\pm} = -\beta^x \pm \alpha e^{-2\phi} \sqrt{\tilde{\gamma}^{xx} \left(\xi/2 \right)} \,. \tag{5.6.13}$$

These eigenfields in fact correspond to the fields (5.5.32) of the NOR formulation discussed in the previous Section. Notice that for standard BSSNOK we have

$\xi = 2$, so that these eigenfields propagate along the light-cones. In any case, we see that we must have $\xi > 0$ for the characteristic speeds to be real.

The transverse-traceless part is now represented by the eigenfields

$$e^{2\phi}\sqrt{\tilde{\gamma}^{xx}}\left(\tilde{A}_{pq} + \frac{\tilde{\gamma}_{pq}}{2\tilde{\gamma}^{xx}}\tilde{A}^{xx}\right) \mp \left(\tilde{\Lambda}^{x}_{pq} + \frac{\tilde{\gamma}_{pq}}{2\tilde{\gamma}^{xx}}\tilde{\Lambda}^{xxx}\right), \tag{5.6.14}$$

with $p, q \neq x$, which propagate with speeds

$$\lambda^{\text{light}}_{\pm} = -\beta^x \pm \alpha e^{-2\phi}\sqrt{\tilde{\gamma}^{xx}}\,, \tag{5.6.15}$$

i.e. along the light-cones. There are in fact only four independent eigenfields here, because we can easily check that

$$\tilde{A}_{pq} + \frac{\tilde{\gamma}_{pq}}{2\tilde{\gamma}^{xx}}\tilde{A}^{xx} = e^{-4\phi}\left(K_{pq} - \frac{h_{pq}}{2}\hat{K}\right), \tag{5.6.16}$$

with h_{ij} the metric tensor projected onto the surface of constant x introduced in the previous Section and $\hat{K} = h^{ij}K_{ij}$. These eigenfields are therefore symmetric and surface-traceless, so that they have only four independent components.

The final two characteristic fields represent the trace of the transverse components and are given by

$$e^{2\phi}\sqrt{\tilde{\gamma}^{xx}(2\xi - 1)/3}\left(\tilde{A}^{xx} - \frac{2}{3}\tilde{\gamma}^{xx}K\right) \mp \left(\tilde{\Lambda}^{xxx} - \frac{2}{3}\tilde{\gamma}^{xx}\,\tilde{a}^x\right), \tag{5.6.17}$$

where $\tilde{a}^x := \tilde{\gamma}^{xm}a_m$. These eigenfields propagate with the characteristic speeds

$$\lambda^{\text{trace}} = -\beta^x + \alpha e^{-2\phi}\sqrt{\tilde{\gamma}^{xx}(2\xi - 1)/3}\,, \tag{5.6.18}$$

which will only be real if

$$2\xi - 1 > 0 \quad \Rightarrow \quad \xi > 1/2\,. \tag{5.6.19}$$

We then conclude that the generalized BSSNOK system is strongly hyperbolic only for $\xi > 1/2$. The standard BSSNOK system corresponding to $\xi = 2$ is again an optimal choice, as in that case all eigenfields other than those associated with the gauge propagate either along the time lines or along the light-cones. It is important to observe that the BSSNOK system with no modification of the evolution equation for the $\tilde{\Gamma}^i$, *i.e.* $\xi = 0$, is not even weakly hyperbolic as the eigenspeeds (5.6.18) become complex, so this choice would be far worse than ADM.

As a final observation it should be mentioned that Sarbach *et al.* [249] have also shown that in some special cases the generalized BSSNOK system can be made to be symmetric hyperbolic. In order to see this, we will attempt to write down a conserved energy norm for this system that is independent of any particular direction. By "conserved" what we really mean here is that up to principal part its time derivative can be written as a full divergence, so that its integral in space vanishes for data of compact support.

The energy norm must be positive definite, and must include all the 30 independent fields we are considering. In order to construct this energy norm we will start by noticing that the terms corresponding to fields that propagate along the time lines are trivial. For example,

$$\begin{aligned} \partial_t \left(a_n - 6f\Phi_n\right)^2 &\simeq 2\tilde{\gamma}^{mn} \left(a_m - 6f\Phi_m\right) \partial_t \left(a_n - 6f\Phi_n\right) \\ &= -2\alpha\tilde{\gamma}^{mn} \left(a_m - 6f\Phi_m\right) \left(f - f\right) \partial_n K = 0 \,, \end{aligned} \tag{5.6.20}$$

where again we are only considering principal terms. Here and in what follows, the square of an object with indices will be understood as the norm calculated using the conformal metric, *i.e.* $(T_n)^2 \equiv \tilde{\gamma}^{mn} T_m T_n$ and so on. In a similar way we also find that

$$\partial_t \left(\tilde{\Gamma}_n + (\xi - 2)\tilde{d}^m{}_{mn} - 4\Phi_n\right)^2 \simeq 0 \,. \tag{5.6.21}$$

So we already have two terms in the energy norm. The gauge sector is also easily found, as we have

$$\partial_t K^2 \simeq -2\alpha K e^{-4\phi} \tilde{\gamma}^{mn} \partial_m a^n \,, \tag{5.6.22}$$

$$\partial_t \left(a_n\right)^2 \simeq -2\alpha f \tilde{\gamma}^{mn} a_n \partial_m K \,, \tag{5.6.23}$$

so that

$$\partial_t \left[K^2 + \frac{e^{-4\phi}}{f} \left(a_n\right)^2\right] \simeq -2\partial_m \left(\alpha \tilde{\gamma}^{mn} a_n K\right) \,, \tag{5.6.24}$$

that is, we have a complete divergence.

For the energy terms involving $\tilde{A}_{mn}$ and $\tilde{d}_{kmn}$ we notice that

$$\partial_t \left(\tilde{d}_{kmn}\right)^2 \simeq -2\alpha\, \tilde{d}^{kmn} \partial_k \tilde{A}_{mn} \,, \tag{5.6.25}$$

$$\partial_t \left(\tilde{A}_{mn}\right)^2 \simeq -2\alpha\, e^{-4\phi} \tilde{A}^{mn} \left[\partial_k \tilde{d}^k{}_{mn} + \partial_m \left(a_n - \tilde{\Gamma}_n + 2\Phi_n\right)\right] . \tag{5.6.26}$$

The time derivative of $(\tilde{A}_{mn})^2$ involves terms other than the divergence of $\tilde{d}_{kmn}$, so we need to study how these other terms evolve. Consider then

$$\partial_t \left(a_n - \tilde{\Gamma}_n + 2\Phi_n\right)^2 \simeq 2\tilde{\gamma}^{mn} \left(a_n - \tilde{\Gamma}_n + 2\Phi_n\right) \partial_t \left(a_m - \tilde{\Gamma}_m + 2\Phi_m\right) \tag{5.6.27}$$

$$\simeq -2\alpha \left(a_n - \tilde{\Gamma}_n + 2\Phi_n\right) \left[(\xi - 2)\, \partial_l \tilde{A}^{lm} - \left(\frac{2\xi - 1}{3} - f\right) \tilde{\partial}^m K\right] . \tag{5.6.28}$$

The first term involves a divergence of $\tilde{A}_{mn}$ so that we can in principle use it together with the time derivative of $(\tilde{A}_{mn})^2$ to form a total divergence, but the second term involving the gradient of K spoils this. In order to build our total divergence we must then ask for

$$\frac{2\xi - 1}{3} - f = 0 \quad \Rightarrow \quad \xi = \frac{1+3f}{2} \,. \tag{5.6.29}$$

This means that if we want to have a symmetric hyperbolic system, the amount of momentum constraint added to the evolution equation for $\tilde{\Gamma}^i$ must be determined by the gauge source function f. Doing this we find that

$$\partial_t \left[\left(\tilde{A}_{mn} \right)^2 + e^{-4\phi} \left(\tilde{d}_{kmn} \right)^2 + \frac{e^{-4\phi}}{\xi - 2} \left(a_n - \tilde{\Gamma}_n + 2\Phi_n \right)^2 \right] \tag{5.6.30}$$

$$\simeq -2\alpha e^{-4\phi} \left\{ \partial_k \left(\tilde{A}^{mn} \tilde{d}^k{}_{mn} \right) + \partial_k \left[\tilde{A}^{km} \left(a_n - \tilde{\Gamma}_n + 2\Phi_n \right) \right] \right\} , \tag{5.6.31}$$

so we have another total divergence. Collecting terms we obtain the following final expression for the conserved energy norm

$$\mathcal{E} = (a_n - 6f\Phi_n)^2 + \left(\tilde{\Gamma}_n + (\xi - 2)\tilde{d}^m{}_{mn} - 4\Phi_n \right)^2 + K^2 + \frac{e^{-4\phi}}{f} \left(a_n \right)^2$$
$$+ \left(\tilde{A}_{nm} \right)^2 + e^{-4\phi} \left(\tilde{d}_{kmn} \right)^2 + \frac{e^{-4\phi}}{\xi - 2} \left(a_n - \tilde{\Gamma}_n + 2\Phi_n \right)^2 \,. \tag{5.6.32}$$

Notice that since the energy norm must be positive definite, we must also ask for $\xi > 2$, as otherwise the last term would be negative. In particular, standard BSSNOK with $\xi = 2$ is not symmetric hyperbolic. The restriction $\xi > 2$, together with $f = (2\xi - 1)/3$, implies that we must also have $f > 1$ (*i.e.* the gauge speed must be superluminal), so that for the BSSNOK family we cannot have a symmetric hyperbolic system with harmonic slicing. However, for the case of 1+log slicing with $f = 2/\alpha$, the above restriction implies $\alpha < 2$ which will typically always be satisfied (the lapse usually collapses below 1 in the presence of a gravitational field – it does not grow), so that we could in principle use a symmetric hyperbolic version of BSSNOK with 1+log slicing.[53]

5.7 The Kidder–Scheel–Teukolsky family

As mentioned in the previous Sections, the Bona–Masso, NOR and BSSNOK formulations are closely related to each other and in particular follow the same route to obtain a strongly hyperbolic system. The main idea in those formulations is to introduce three new independent quantities related to contractions of the Christoffel symbols associated with either the physical or the conformal metric, and then to modify the evolution equations for these quantities using the

[53]Gundlach and Martín-García have in fact found a different conserved energy norm for BSSNOK (*i.e.* a different symmetrizer) for which all characteristic speeds can remain causal, and that, in particular, allows us to have a symmetric hyperbolic system with harmonic slicing [154]. Such an energy norm, however, involves mixed terms of the form $\tilde{\gamma}^{nl}\tilde{d}^m{}_{mn}(a_l - \tilde{\Gamma}_l + 2\Phi_l)$, so that showing that it is positive definite becomes considerably more difficult.

momentum constraints. These formulations also remain second order in space, though they can trivially be rewritten as fully first order formulations by introducing the first derivatives of the spatial metric d_{ijk} as independent quantities and considering their definitions $d_{ijk} := \partial_i \gamma_{jk}/2$ as new constraints (which is what we have done in order to study their hyperbolicity properties). An important point is that in these first order versions the evolution equations for the d_{ijk} are obtained directly from their definition and are left unmodified. This has the consequence that the derivative constraints remain trivially satisfied if they were satisfied initially (even at the numerical level it is possible to arrange things so that these constraints remain satisfied to the level of machine round-off error).

There is a different approach that can be used to obtain hyperbolic formulations of the 3+1 evolution equations based on the idea of going to a fully first order system and modifying directly the evolution equations for the d_{ijk} using the momentum constraints, but *without* introducing extra independent quantities like the V^i of Bona–Masso or the $\tilde{\Gamma}^i$ of BSSNOK. This approach can be traced back to the work of Frittelli and Reula [135], and Anderson and York [21] (the so-called Einstein–Christoffel formulation) from the mid and late 1990s. The most general version of these type of formulations is known as the Kidder–Scheel–Teukolsky (KST) family and has twelve free parameters [172] (there have been some further generalizations of this idea that have resulted in formulations with even more free parameters, see *e.g.* [250]). Here, however, we will limit ourselves to presenting a very simplified version of KST with only three free parameters that nevertheless includes the main features of this formulation.

The KST system starts by introducing the derivatives of the spatial metric as independent quantities

$$d_{ijk} := \frac{1}{2}\, \partial_i \gamma_{jk} \; . \tag{5.7.1}$$

The factor 1/2 above in fact does not appear in the original KST formulation, but here we introduce it in order to have a notation consistent with that of the previous Sections.

For the Ricci tensor we use again (5.5.2), which we rewrite here for completeness

$$R_{ij} \simeq -\partial_m d^m{}_{ij} - \partial_{(i} \left(2V_{j)} - d_{j)m}{}^m \right) \; , \tag{5.7.2}$$

with the V_i given by

$$V_i := d_{im}{}^m - d^m{}_{mi} \; . \tag{5.7.3}$$

Notice that these are the same quantities used in the Bona–Masso formulation, but here we will not promote them to independent variables and will instead consider them just a shorthand for the given combination of d's. It is important to mention the fact that in the standard KST formulation a parameter is added to the above expression for the Ricci tensor R_{ij} to allow for different ways to resolve the ordering ambiguity of the second derivatives of the metric. However, in order to simplify the system, here we have chosen the same specific ordering used in previous Sections.

We will now proceed to modify the evolution equations for the extrinsic curvature K_{ij} and the metric derivatives d_{ijk} by adding multiples of the constraints in the following way

$$\partial_t d_{ijk} = (\cdots) + \alpha\xi\gamma_{i(j}M_{k)} + \alpha\chi\gamma_{jk}M_i \,, \tag{5.7.4}$$
$$\partial_t K_{ij} = (\cdots) + \alpha\eta\gamma_{ij}H \,, \tag{5.7.5}$$

with (ξ, χ, η) constant parameters, and where $(\cdots)$ denotes the original ADM values for the right hand side prior to adding multiples of the constraints, and where, as before, H and M^i are the Hamiltonian and momentum constraints, which up to principal part become

$$H \simeq R \simeq -2\partial_m V^m \,, \tag{5.7.6}$$
$$M_i \simeq \partial_m K^m_i - \partial_i K \,. \tag{5.7.7}$$

For the analysis of this system we will again use a Bona–Masso type slicing condition $\partial_t\alpha = -\alpha^2 f K$, and introduce the derivatives of the lapse $a_i := \partial_i \ln\alpha$. There is, however, an important point related to the slicing condition that should be mentioned. Just as is done in the NOR system, the standard KST formulation uses a fixed densitized lapse of the form $Q = \alpha\gamma^{-\sigma/2}$ instead of the Bona–Masso slicing condition we will use here. When discussing the NOR system in Section 5.5 we mentioned the fact that these two approaches were actually equivalent; however, in the case of KST this is not true anymore. The reason for this is that by densitizing the lapse we are transforming derivatives of a_i into derivatives of ${d_{im}}^m$. Now, as long as the evolution equations for the d's are not modified we find that ${d_{im}}^m$ evolves only through K, so using a Bona–Masso slicing condition is completely equivalent to densitizing the lapse. But if the evolution equations for the d's are modified using the momentum constraints, as is done in KST, then the evolution of ${d_{im}}^m$ involves other components of K_{ij}, and the two approaches are not equivalent anymore. Still, the use of a Bona–Masso slicing condition instead of a densitized lapse results in simpler equations so we will follow this approach here.

Going back to our evolution equations we find that up to principal part they take the form

$$\partial_0 a_i \simeq -\alpha f \partial_i K \,, \tag{5.7.8}$$
$$\partial_0 d_{ijk} \simeq -\alpha\partial_i K_{jk} + \alpha\xi\gamma_{i(j}M_{k)} + \alpha\chi\gamma_{jk}M_i \,, \tag{5.7.9}$$
$$\partial_0 K_{ij} \simeq -\alpha\partial_m\Lambda^m_{ij} \,, \tag{5.7.10}$$

where as before $\partial_0 = \partial_t - \beta^i\partial_i$, and where we have now defined

$$\Lambda^k_{ij} := {d^k}_{ij} + \delta_{(i}\left(a_{j)} + 2V_{j)} - {d_{j)m}}^m\right) + 2\eta\gamma_{ij}V^k \,. \tag{5.7.11}$$

Notice that in this case we have 27 independent dynamical variables, given by the three a_i, the six K_{ij}, and the 18 d_{ijk}. The hyperbolicity analysis is simplified considerably by noticing that

$$\partial_0 V_i \simeq \alpha \, (1 - \xi + 2\chi) \, M_i \; , \tag{5.7.12}$$

so that the combination of d's represented by V_i evolves only through the momentum constraints. However, and in contrast with what happened in the Bona–Masso formulation, it is now not a good idea to choose the parameters in such a way that the V_i remain constant up to principal part, for a reason that will become clear below.

The analysis now proceeds in much the same way as before. If we consider only derivatives along the x direction we find that there are now 15 fields that propagate along the time lines, namely ($q \neq x$)

$$a_q \; , \; d_{qij} - \frac{1}{1 - \xi + 2\chi} \left(\xi \gamma_{q(i} V_{j)} + \chi \gamma_{ij} V_q \right) \; , \tag{5.7.13}$$

$$a_x - f \left(d_{xm}{}^m - \frac{\xi + 3\chi}{1 - \xi + 2\chi} \, V_x \right) \; . \tag{5.7.14}$$

From the above expressions it is clear that we must ask for

$$1 - \xi + 2\chi \neq 0 \; , \tag{5.7.15}$$

as otherwise these fields will degenerate. In other words, we need certain combination of d's, namely the V_i, to evolve through the momentum constraints to be able to cancel out terms in the evolution of d_{qij} and $d_{xm}{}^m$ that will otherwise make it impossible to diagonalize the system. It is therefore not possible to arrange things in such a way that the V_i remain constant up to principal part and still obtain a strongly hyperbolic system.

For the remaining 12 fields we start by noticing that the longitudinal components again result in four eigenfields given by

$$\sqrt{\gamma^{xx} \left(\xi - \chi/2 \right)} \, K^x_q \mp \Lambda^{xx}{}_q \; , \tag{5.7.16}$$

with corresponding characteristic speeds

$$\lambda^{\text{long}}_{\pm} = -\beta^x \pm \alpha \sqrt{\gamma^{xx} \left(\xi - \chi/2 \right)} \; , \tag{5.7.17}$$

so that we clearly must ask for $\xi - \chi/2 > 0$.

In order to proceed we again introduce the surface-trace of K_{ij} as

$$\hat{K} := h^{ij} K_{ij} = K - K^{xx}/\gamma^{xx} \; , \tag{5.7.18}$$

with h_{ij} the projected metric onto the surface of constant x (with an analogous definition for $\hat{\Lambda}^x$). Doing this we find the four transverse-traceless eigenfields

$$\sqrt{\gamma^{xx}} \left(K_{pq} - \frac{h_{pq}}{2} \hat{K} \right) \mp \left(\Lambda^x_{pq} - \frac{h_{pq}}{2} \hat{\Lambda}^x \right) \; , \qquad (p, q) \neq x \; , \tag{5.7.19}$$

propagating along the light-cones with speeds

$$\lambda_{\pm}^{\text{light}} = -\beta^x \pm \alpha\sqrt{\gamma^{xx}} \,. \tag{5.7.20}$$

The surface-traces also result in two eigenfields given by

$$\left\{\gamma^{xx}\left[1 + 2\chi + 4\eta\left(1 - \xi + 2\chi\right)\right]\right\}^{1/2} \hat{K} \mp \hat{\Lambda}^x \,, \tag{5.7.21}$$

and propagating with the speeds

$$\lambda_{\pm}^{\text{trace}} = -\beta^x \pm \alpha\left\{\gamma^{xx}\left[1 + 2\chi + 4\eta\left(1 - \xi + 2\chi\right)\right]\right\}^{1/2} \,. \tag{5.7.22}$$

Finally, the two gauge modes are

$$\sqrt{f\gamma^{xx}}\left(K + F\hat{K}\right) \mp \left(\Lambda^x + F\hat{\Lambda}^x\right) \,, \tag{5.7.23}$$

with $F = 2(1+3\eta)(1-\xi+2\chi)/[f-(1+2\chi+4\eta(1-\xi+2\chi)]$, and they propagate with the gauge speeds

$$\lambda_{\pm}^{\text{gauge}} = -\beta^x \pm \alpha\sqrt{f\gamma^{xx}} \,. \tag{5.7.24}$$

We then conclude that the system will be strongly hyperbolic as long as

$$f > 0 \,, \tag{5.7.25}$$

$$\xi - \chi/2 > 0 \,, \tag{5.7.26}$$

$$1 + 2\chi + 4\eta\left(1 - \xi + 2\chi\right) > 0 \,, \tag{5.7.27}$$

$$1 - \xi + 2\chi \neq 0 \,. \tag{5.7.28}$$

In particular we see that $\xi = \chi = 0$ is not allowed, so the evolution equation for the d_{ijk} *must* be modified. An optimal choice seems to be

$$\eta = -1/3 \,, \qquad \xi = 1 + \chi/2 \,, \qquad \chi \neq 0 \,, \tag{5.7.29}$$

in which case $F = 0$ and all fields other than those associated with the gauge propagate either along the time lines or the light-cones.

As we have seen, the KST formulation has the advantage of providing us with a strongly hyperbolic system without the need to introduce extra quantities like the V^i or $\tilde{\Gamma}^i$ of the Bona–Masso and BSSNOK formulations. However, by modifying the evolution equations for the metric derivatives d_{ijk} it forces us to go to a fully first order form of the equations. This is in contrast with formulations like Bona–Masso or BSSNOK for which we can keep working at second order without the need to introduce all the d_{ijk} as independent variables, so in practical applications KST has a larger number of independent variables.

5.8 Other hyperbolic formulations

All the formulations discussed in the previous Sections are based directly in the 3+1 evolution equations, and differ from ADM in the fact that they introduce either all the first derivatives of the spatial metric γ_{ij}, or some special combinations of them, as new independent quantities whose evolution equations are then modified using the momentum constraints. There are, however, other routes to constructing hyperbolic systems of evolution equations for relativity. Here we will briefly introduce two such routes, one based on going to higher order derivatives and another one that in fact starts by modifying the four-dimensional Einstein equations themselves. Below we will present the basic ideas behind these approaches, but will not go into a detailed analysis.

It is important to mention the fact that there are still other forms of obtaining hyperbolic systems of evolution equations for general relativity. In particular, we can mention the formulations of Friedrich that involve the Bianchi identities [131] and make use of the conformal Weyl tensor and its decomposition into "electric" and "magnetic" parts (see Chapter 8). These formulations are very elegant, but so far have not been used in practical applications, so here we will not discuss them further and will refer the reader to [131].

5.8.1 *Higher derivative formulations*

Higher derivative formulations start by calculating the time derivative of the spatial Ricci tensor R_{ij} using the ADM equations. Consider the change in the Ricci tensor under an infinitesimal variation of the spatial metric $\delta\gamma_{ij}$. Up to principal part we find

$$\delta R_{ij} \simeq \frac{\gamma^{mn}}{2} \left(-\partial_m \partial_n \delta\gamma_{ij} - \partial_i \partial_j \delta\gamma_{mn} + 2\partial_{(i} \partial_m \delta\gamma_{j)n} \right) , \tag{5.8.1}$$

which implies that

$$\partial_0 R_{ij} \simeq \alpha \left(\gamma^{mn} \partial_m \partial_n K_{ij} + \partial_i \partial_j K - 2\partial_{(i} \partial_m K^m_{j)} \right) , \tag{5.8.2}$$

where as before $\partial_0 := \partial_t - \beta^i \partial_i$. From the ADM equations we find that this implies, in particular that

$$\partial_0^2 K_{ij} \simeq -\partial_i \partial_j \left(\partial_0 \alpha \right) + \alpha^2 \left(\gamma^{mn} \partial_m \partial_n K_{ij} + \partial_i \partial_j K - 2\partial_{(i} \partial_m K^m_{j)} \right) . \tag{5.8.3}$$

If we now assume that the lapse evolves via a Bona–Masso type slicing condition we can transform the last expression into

$$\begin{aligned} \partial_0^2 K_{ij} &\simeq \alpha^2 f \, \partial_i \partial_j K + \alpha^2 \left(\gamma^{mn} \partial_m \partial_n K_{ij} + \partial_i \partial_j K - 2\partial_{(i} \partial_m K^m_{j)} \right) \\ &\simeq \alpha^2 \left[\gamma^{mn} \partial_m \partial_n K_{ij} + (f-1) \, \partial_i \partial_j K - 2\partial_{(i} M_{j)} \right] , \end{aligned} \tag{5.8.4}$$

where $M_i \simeq \partial_m K^m_i - \partial_i K \simeq 0$ are the momentum constraints. Using now the fact that the momentum constraints vanish we finally find that

$$\partial_0^2 K_{ij} \simeq \alpha^2 \left[\gamma^{mn} \partial_m \partial_n K_{ij} + (f-1) \, \partial_i \partial_j K \right] . \tag{5.8.5}$$

Assume now that we have harmonic slicing, *i.e.* $f = 1$ (or equivalently, take the densitized lapse $\tilde{\alpha} = \alpha\gamma^{-1/2}$ as a known function of spacetime). The K_{ij} now satisfy the second order equation

$$\partial_0^2 K_{ij} \simeq \alpha^2 \gamma^{mn} \partial_m \partial_n K_{ij} \,, \tag{5.8.6}$$

which is nothing more than the wave equation with a wave speed given by the physical speed of light. We now have a system of six simple wave equations for the independent components of K_{ij} coupled only through source terms, so the systems is clearly not only strongly hyperbolic, but even symmetric hyperbolic. The price we have paid is that now we have moved to a system that essentially involves third derivatives in time of the spatial metric. A system of this type has been proposed by Choquet-Bruhat and York in [100], and also by Abrahams *et al.* in [1]. We can in fact relax the gauge choice and take any $f > 0$ to obtain a system that, though no longer symmetric, is still strongly hyperbolic. These results shouldn't be surprising since we can see them as essentially a time derivative of the Bona–Masso or NOR systems. The interesting feature of these third order formulations is the fact that by taking an extra time derivative there is no need to introduce auxiliary quantities.

5.8.2 *The Z4 formulation*

A very different approach to constructing a hyperbolic system of evolution equations has been proposed by Bona *et al.* [59, 60] (see also [68]). The main idea of this so-called Z4 system is to modify the Einstein field equations at the four-dimensional level by rewriting them as (cf. (1.13.4))

$$R_{\mu\nu} + \nabla_\mu Z_\nu + \nabla_\nu Z_\mu = 8\pi \left(T_{\mu\nu} - \frac{1}{2} g_{\mu\nu} T \right) \,, \tag{5.8.7}$$

where an extra 4-vector Z_μ has been introduced that is assumed to vanish for physical solutions (Z_μ is a "zero" 4-vector). That is, we introduce the extra constraints

$$Z_\mu = 0 \,. \tag{5.8.8}$$

By taking the trace of (5.8.7) we find the following useful relationship

$$\nabla_\mu Z^\mu = -\frac{1}{2} R - 4\pi T \,. \tag{5.8.9}$$

Taking now a divergence of (5.8.7), and using the conservation of the Einstein tensor $G_{\mu\nu} = R_{\mu\nu} - g_{\mu\nu} R/2$ and of the stress-energy tensor $T_{\mu\nu}$, we find that the vector Z_μ must evolve via

$$\nabla^\nu \nabla_\nu Z_\mu + R_{\mu\nu} Z^\nu = 0 \,. \tag{5.8.10}$$

The four-dimensional equations (5.8.7) can now be written in 3+1 form as (in vacuum)

$$\left(\partial_t - \pounds_{\vec{\beta}}\right) \gamma_{ij} = -2\alpha K_{ij} \; , \tag{5.8.11}$$

$$\left(\partial_t - \pounds_{\vec{\beta}}\right) K_{ij} = -D_i\alpha_j + \alpha \left[R_{ij} + D_i Z_j + D_j Z_i \right.$$
$$\left. -2K_{im}K^m_j + (K - 2\Theta) K_{ij}\right] \; , \tag{5.8.12}$$

$$\left(\partial_t - \pounds_{\vec{\beta}}\right) \Theta = \frac{\alpha}{2} \left[R + (K - 2\Theta) K - K_{mn}K^{mn} \right.$$
$$\left. +2D_m Z^m - 2Z^m \partial_m \ln \alpha\right] \; , \tag{5.8.13}$$

$$\left(\partial_t - \pounds_{\vec{\beta}}\right) Z_i = \alpha \left[D_m K^m_i - D_i K \right.$$
$$\left. +\partial_i \Theta - 2K^m_i Z_m - \Theta \, \partial_i \ln \alpha\right] \; , \tag{5.8.14}$$

where for simplicity we have assumed that we are in vacuum, and where we have introduced the scalar quantity Θ defined as the projection of Z_μ along the normal direction to the spatial hypersurfaces

$$\Theta := n_\mu Z^\mu = \alpha Z^0 \; . \tag{5.8.15}$$

Notice that the equations above represent all 10 field equations, in other words, we have 10 evolution equations and no independent constraint equations. The evolution equations for Θ and Z_i now play the role of the Hamiltonian and momentum constraints, and they clearly reduce to them in the case when Θ and Z_i vanish. The crucial point is that we now have no elliptic constraints, but rather 10 evolution equations and four algebraic constraints, namely

$$\Theta = 0 \; , \qquad Z_i = 0 \; . \tag{5.8.16}$$

One can now study the hyperbolicity properties of the Z4 system. We will not do a detailed analysis here, but just mention the main points. One key observation is the fact that in order to obtain a hyperbolic system we must now use a slicing condition of the form

$$\left(\partial_t - \pounds_{\vec{\beta}}\right) \alpha = -\alpha^2 f(\alpha) \left(K - m\Theta\right) \; , \tag{5.8.17}$$

which is a generalization of the standard Bona–Masso slicing condition, and in fact reduces to it for physical solutions since in that case Θ vanishes. The motivation for this change in the slicing condition can be traced back to the fact that the principal part of the fully covariant field equations (5.8.7) now takes the form

$$\Box g_{\mu\nu} - \partial_\mu \left(\Gamma_\nu + 2Z_\nu\right) - \partial_\nu \left(\Gamma_\mu + 2Z_\mu\right) \simeq 0 \; , \tag{5.8.18}$$

with $\Gamma^\mu := g^{\nu\lambda}\Gamma^\mu_{\nu\lambda}$, so that these equations will transform into wave equations for the 4-metric if we ask for

$$\Gamma^\mu = -2Z^\mu \; , \tag{5.8.19}$$

which for the standard Einstein equations reduces to the harmonic condition $\Gamma^\mu = 0$. This means, in particular, that the condition for harmonic slicing must be modified to

$$\Gamma^0 = -2Z^0 = -2\frac{\Theta}{\alpha} \,, \tag{5.8.20}$$

which, using the expressions of Appendix B, becomes

$$\left(\partial_t - \pounds_{\vec{\beta}}\right) \alpha = -\alpha^2 f(\alpha) \left(K - 2\Theta\right) \,. \tag{5.8.21}$$

The final result is that, when using a slicing condition of the form (5.8.17), the Z4 formulation is strongly hyperbolic for $f > 0$, but we must take $m = 2$ in the particular case when $f = 1$, so that $m = 2$ is the preferred choice.

5.9 Boundary conditions

When we do a numerical simulation of a gravitational system (or for that matter any other physical system with no boundaries), we have to deal with the fact that the memory of the computer is necessarily finite, while space is infinite in extent. We therefore can only cover a finite region of space and must impose some artificial boundary condition on a timelike boundary at a finite distance. There are, in fact, some alternatives to this approach. We can, for example, use a transformation of coordinates to bring spatial infinity to a finite distance, and impose boundary conditions there, though this is generally not a good idea as waves traveling outwards become less and less resolved in the computational grid causing the waves to be "back-scattered" by the numerical grid (nevertheless, this approach has been used recently by Pretorius with good results [231]). Another approach is to use hyperboloidal slices that reach null infinity instead, and then bring null infinity to a finite distance (see [132, 165]) . Yet another possibility is to use a 3+1 interior region and a characteristic exterior region where again null infinity is brought to a finite distance, a technique known as Cauchy-characteristic matching (see *e.g.* [57]). However, for Cauchy-characteristic matching to work we still need to be able to identify which degrees of freedom we are allowed to "inject" at the boundary from one region to the other between the Cauchy interior and characteristic exterior.

In the case of the 3+1 formalism, there are in fact two separate issues related to the choice of boundary condition. The first has to do with the well-posedness of the resulting initial-boundary value problem. We can find examples of well-posed systems of evolution equations that become ill-posed when the "wrong" boundary condition is used. Fortunately, there exist results that show that for the case of symmetric hyperbolic systems we can impose boundary conditions in such a way that the full initial-boundary value problem will remain well-posed. The second issue has to do with the constraint equations, as we can have perfectly well-posed boundary conditions that violate the constraints and are therefore inconsistent with the Einstein field equations. This second issue is a matter of current research and has not yet been fully resolved in the general case, though significant progress has been made in some special cases.[54]

[54] For systems of equations that have no elliptic constraints, like the Z4 system of Section 5.8, this issue is less important, a property that makes such formulations particularly attractive.

In practice, most codes today attempt to solve the boundary problem by using some simple *ad hoc* boundary condition and pushing the boundaries as far away as possible, ideally sufficiently far away so that they will remain causally disconnected from the interior regions during the dynamical time-scale of the problem under consideration. This approach, however, is extremely expensive in terms of computational resources (though the use of mesh refinement can alleviate the problem). Also, if we are solving a coupled elliptic-hyperbolic problem (for example when using elliptic gauge conditions), boundary effects will propagate at an infinite speed and will affect the interior regardless of how far away the boundaries are. Because of this, the correct treatment of the boundaries is rapidly becoming a very important issue in numerical relativity.

5.9.1 *Radiative boundary conditions*

Before going into a more formal discussion of the boundary conditions, it is interesting to describe what many existing numerical codes based on the BSSNOK formulation use in practice as a simple "naive" boundary condition. Since most of these codes do not decompose the dynamical fields into eigenfields at the boundaries, they use an *ad hoc* boundary condition based on the idea that far away all fields behave as spherical waves traveling outward, that is $f \sim f0 + u(r - vt)/r$, which results in the so-called *radiative boundary condition*:

$$\partial_t f + v\, \partial_r f + v\, (f - f_0)/r \sim 0\,, \tag{5.9.1}$$

where f_0 is an asymptotic value for the specific variable considered (typically 1 for the lapse and diagonal metric components and 0 for everything else), and v is some wave speed. The radiative boundary condition above is a boundary condition of Sommerfeld type, that is, a condition that assumes outgoing radiation at the boundaries.

If the boundary is sufficiently far away we can assume that the spacetime is close to being flat so that the coordinate speed of light is approximately 1. We then take $v = 1$ for most fields. However, the gauge variables can easily propagate with a different speed implying a different value of v (for example, for Bona–Masso slicing we would take $v = \sqrt{f}$ for the lapse itself and the trace of the extrinsic curvature). Boundary conditions of this type work well in practice in the sense that they remain stable. They depend on three basic assumptions: 1) The spacetime is asymptotically flat, 2) the sources of the gravitational field are localized in a small region so that far away we have a spherical front of gravitational waves, and 3) the shift is small far away so that its effect on the characteristic speeds can be ignored (this is why standard radiative boundary conditions fail for corotating coordinate systems for which the shift is typically very large at the boundaries). Of course, since these radiative conditions completely ignore the constraints we would expect that some constraint violation will be introduced by the boundary conditions that will subsequently propagate inward at essentially the speed of light. Because of this it is important to place the boundaries as far away as possible.[55]

In practical applications, condition (5.9.1) is modified in a number of different ways since even though it models reasonably well the radiative behavior, it is very poor at modeling the behavior of the non-radiative Coulomb part of the gravitational field (see for example [13]).

Quite apart from the issue of the constraint violation introduced by the radiative boundary condition, we might also worry about it being well-posed. Intuitively, it would seem that we are imposing two many conditions, since even though we can model incoming fields in any way we like, outgoing fields must be left alone. In order to see what is the effect of the radiative boundary conditions, let us for a moment consider the case of simple wave equation in flat space

$$\partial_t^2 \varphi - v^2 \nabla^2 \varphi = 0 \; , \tag{5.9.2}$$

where as before the operator ∇^2 refers to the standard three-dimensional flat space Laplacian. Now, in spherical coordinates the Laplacian takes the form

$$\begin{aligned} \nabla^2 \varphi &= \frac{1}{r^2}\, \partial_r \left(r^2 \partial_r \varphi \right) + \frac{1}{r^2 \sin\theta}\, \partial_\theta \left(\sin\theta \partial_\theta \varphi \right) + \frac{1}{r^2 \sin^2\theta}\, \partial_\phi^2 \varphi \\ &= \partial_r^2 \varphi + \frac{2}{r}\, \partial_r \varphi + \frac{1}{r^2 \sin\theta}\, \partial_\theta \left(\sin\theta \partial_\theta \varphi \right) + \frac{1}{r^2 \sin^2\theta}\, \partial_\phi^2 \varphi \; . \end{aligned} \tag{5.9.3}$$

Notice that, far away, the first two terms dominate, so that we can ignore the angular dependency. Far away we can therefore take the wave equation to be

$$\partial_t^2 \varphi - v^2 \left(\partial_r^2 \varphi + \frac{2}{r}\, \partial_r \varphi \right) = 0 \; , \tag{5.9.4}$$

It is now clear that

$$\varphi = \varphi_0 + u(r - vt)/r \tag{5.9.5}$$

is an exact solution of the last equation for any arbitrary function u and constant φ_0. This solution also satisfies

$$\partial_t \varphi + v\, \partial_r \varphi + v\, (\varphi - \varphi_0)/r = 0 \; , \tag{5.9.6}$$

which justifies the radiative boundary condition given above.

Of course, the Einstein equations are not simple wave equations of this type for at least two reasons. First, we have source terms that in principle should be

[55] We typically measure distances in terms of the total ADM mass of the system (see Appendix A). In practice, we find that "sufficiently far away" means that the boundaries should be at least at a distance of $\sim 50M$ in order to have violations of the constraints at the boundaries of the order of a few percent. Recently, however, simulations using different forms of mesh refinement have placed the boundaries at $\sim 500M$ or more in order to make sure that the boundaries remain causally disconnected from the central regions during the dynamical time-scale of the systems under study.

taken into account, but these are typically quadratic in quantities that become small far away, so that ignoring them is a reasonable approximation. Second, in the case of the Einstein equations not all fields propagate with the same speed, so that we must be careful with the value of v chosen. However, as we have seen in previous sections, the standard BSSNOK formulation has only eigenfields that propagate along the time lines, at the speed of light which far away is essentially 1, or at the gauge speed which far away is $\sqrt{f}$. We can therefore take all speeds equal to 1 except for the fields associated with the gauge, namely the lapse itself, the conformal factor ϕ, and the trace of the extrinsic curvature K. What about the fields that propagate along the time lines? It turns out that due to the presence of source terms, these fields also behave as pulses moving outward, but since the sources have contributions from many different fields, choosing one fixed value for the wave speed might introduce numerical reflections from the boundaries.[56]

Let us now rewrite the wave equation in a more "ADM-like" form by defining $\Pi := \partial_t \varphi$. We then have

$$\partial_t \varphi = \Pi \,, \tag{5.9.7}$$

$$\partial_t \Pi = v^2 \left(\partial_r^2 \varphi + \frac{2}{r} \, \partial_r \varphi \right) \,. \tag{5.9.8}$$

Notice that if we have a solution of the form $\varphi = u(r - vt)/r$, then the behavior of Π will be

$$\Pi = -vu'(r - vt)/r \,, \tag{5.9.9}$$

which is precisely of the form $\Pi = u_0(r - vt)/r$, with $u_0 = -vu'$. This justifies the fact that in relativity we can use a radiative boundary condition for the extrinsic curvature as well as for the metric components. What about the spatial derivative? Defining $\Psi := v\partial_r \varphi$, we now find that if $\varphi = u(r - vt)/r$ then

$$\Psi = v \left(u'(r - vt)/r - u/r^2 \right) \,, \tag{5.9.10}$$

which is *not* of the same form as φ. We therefore should not use the radiative boundary condition for spatial derivatives.

Notice that in the case of the BSSNOK formulation, as long as we keep the evolution equations second order in space, we will in fact be using the radiative boundary condition for the six components of the extrinsic curvature, and the three $\tilde{\Gamma}^i$ (plus the quantities $\{\alpha, \phi, \tilde{\gamma}_{ij}\}$). Now, since we know that we have three quantities associated with the $\tilde{\Gamma}^i$ that propagate along the time lines, namely $\tilde{\Gamma}^i - 8\,\tilde{\gamma}^{ik}\partial_k \phi$ (cf. equation (5.6.9)), giving boundary data for the $\tilde{\Gamma}^i$ is clearly inconsistent, but probably not too bad since the fields are not outgoing (a much

[56] A good idea might be to choose the gauge function f such that far away it goes to 1. Harmonic slicing corresponds precisely to $f = 1$, but as we saw in Chapter 4 it does not have good singularity avoiding properties. The standard 1+log slicing takes $f = 2/\alpha$, so that far away $f \sim 2$, which will cause precisely the boundary reflections mentioned here.

better alternative would be to evolve the quantity $\tilde{V}^i := \tilde{\Gamma}^i - 8\,\tilde{\gamma}^{ik}\partial_k\phi$, instead of the $\tilde{\Gamma}^i$, because it propagates along time lines and would therefore require no boundary condition in the case of zero shift). Far worse, however, would be to impose a radiative boundary condition on the spatial derivatives of the $\tilde{\gamma}_{ij}$, since in that case we would be giving boundary data for outgoing fields as well. But as already mentioned, this is not done when we works with evolution equations that are second order in space.

5.9.2 *Maximally dissipative boundary conditions*

Let us now go back to the issue of finding well-posed boundary conditions for a symmetric hyperbolic system of equations. The restriction to symmetric hyperbolic systems is important in order to be able to prove well-posedness. For systems that are only strongly hyperbolic but not symmetric, like BSSNOK, no rigorous results exist about the well-posedness of the initial-boundary value problem. We then start by considering, as before, an evolution system of the form

$$\partial_t u + M^i \partial_i u = 0 \,, \tag{5.9.11}$$

where the matrices M^i are constant, and the domain of dependence of the solution is restricted to the region $\vec{x} \in \Omega$. Let us now construct the principal symbol $P(n_i) := M^i n_i$, with n_i an arbitrary unit vector. We will assume that the system is symmetric hyperbolic, which implies that there exists a symmetrizer H, independent of the vector n_i, that is a Hermitian matrix such that $HP - P^T H = 0$. Consider now the energy norm

$$E(t) = \int_\Omega u^\dagger H u \, dV \,. \tag{5.9.12}$$

Taking a time derivative of this energy we find

$$\begin{aligned} \frac{dE}{dt} &= - \int_\Omega \left[(\partial_i u^\dagger) {M^i}^T H u + u^\dagger H M^i (\partial_i u) \right] dV \\ &= - \int_\Omega \left[(\partial_i u^\dagger) H M^i u + u^\dagger H M^i (\partial_i u) \right] dV \\ &= - \int_\Omega \partial_i \left(u^\dagger H M^i u \right) dV \,, \end{aligned} \tag{5.9.13}$$

where in the second line we used the fact that H is the *same symmetrizer* for all n_i, which in particular means that all three matrices HM^i are symmetric. This is precisely the place where the assumption that we have a symmetric hyperbolic system becomes essential. Using now the divergence theorem we finally find

$$\frac{dE}{dt} = - \int_{\partial\Omega} \left(u^\dagger H M^i u \right) n_i \, dA = - \int_{\partial\Omega} \left(u^\dagger H P(\vec{n}) u \right) dA \,, \tag{5.9.14}$$

where $\partial\Omega$ is the boundary of Ω, $\vec{n}$ and dA are the normal vector to the boundary and its corresponding area element, and $P(\vec{n}) = M^i n_i$ is the symbol associated with $\vec{n}$. In contrast to what we have done before, we will now not assume that the surface integral above vanishes. Instead, we now make use of equation (5.3.6):

$$HP = (R^{-1})^T R^{-1} P = (R^{-1})^T \Lambda R^{-1} \; , \tag{5.9.15}$$

with R the matrix of column eigenvectors of $P(\vec{n})$ and $\Lambda = \mathrm{diag}(\lambda_i)$ the matrix of corresponding eigenvalues. We can therefore rewrite the change in the energy norm as

$$\frac{dE}{dt} = - \int_{\partial\Omega} \left(u^\dagger (R^{-1})^T \Lambda R^{-1} u \right) dA = - \int_{\partial\Omega} \left(w^\dagger \Lambda w \right) dA \; , \tag{5.9.16}$$

where $w := R^{-1} u$ are the eigenfields. Let w_+, w_-, w_0 now denote the eigenfields corresponding to eigenvalues of $P(\vec{n})$ that are positive, negative and zero respectively, *i.e.* eigenfields that propagate outward, inward, and tangential to the boundary respectively.[57] We then find

$$\begin{aligned} \frac{dE}{dt} &= - \int_{\partial\Omega} \left(w_+^\dagger \Lambda_+ w_+ \right) dA - \int_{\partial\Omega} \left(w_-^\dagger \Lambda_- w_- \right) dA \\ &= \int_{\partial\Omega} \left(w_-^\dagger \left| \Lambda_- \right| w_- \right) dA - \int_{\partial\Omega} \left(w_+^\dagger \left| \Lambda_+ \right| w_+ \right) dA \; , \end{aligned} \tag{5.9.17}$$

with Λ_+ and Λ_- the sub-matrices of positive and negative eigenvalues. We clearly see that the first term in the last expression is always positive, while the second term is always negative. This shows that outward propagating fields (those with positive speed) reduce the energy norm since they are leaving the region Ω, while inward propagating modes (those with negative speed) increase it since they are coming in from the outside.

Assume that we now impose a boundary condition of the following form

$$w_-|_{\partial\Omega} = S \, w_+|_{\partial\Omega} \; , \tag{5.9.18}$$

with S some matrix that relates incoming fields at the boundary to outgoing ones. We then have

$$\begin{aligned} \frac{dE}{dt} &= \int_{\partial\Omega} \left(w_+^\dagger S^T \left| \Lambda_- \right| S \, w_+ \right) dA - \int_{\partial\Omega} \left(w_+^\dagger \left| \Lambda_+ \right| w_+ \right) dA \\ &= \int_{\partial\Omega} \left[w_+^\dagger \left(S^T \left| \Lambda_- \right| S - \left| \Lambda_+ \right| \right) w_+ \right] dA \; . \end{aligned} \tag{5.9.19}$$

From this we clearly see that if we take S to be "small enough" in the sense that $w_+^\dagger S^T \left| \Lambda_- \right| S \, w_+ \leq w_+^\dagger \left| \Lambda_+ \right| w_+$, then the energy norm will not increase

[57] We should be careful with the interpretation of w_+ and w_-, because in many references we find their meaning reversed. This comes from the fact that we often find the evolution system written as $\partial_t u = M^i \partial_i u$ instead of the form $\partial_t u + M^i \partial_i u = 0$ used here, which of course reverses the signs of all the matrices and in particular of the matrix of eigenvalues Λ.

with time and the full system including the boundaries will remain well-posed. Boundary conditions of this form are known as *maximally dissipative* [185]. The particular case $S = 0$ corresponds to saying that the incoming fields vanish, and this results in a Sommerfeld-type boundary condition. This might seem the most natural condition, but it is in fact not always a good idea, as we might find that in order to reproduce the physics correctly (*e.g.* to satisfy the constraints) we might need to have some non-zero incoming fields at the boundary.

We can in fact generalize the above boundary condition somewhat to allow for free data to enter the domain. We can then take a boundary condition of the form

$$w_-|_{\partial\Omega} = S\, w_+|_{\partial\Omega} + g(t) \,, \tag{5.9.20}$$

where $g(t)$ is some function of time that represents incoming radiation at the boundary, and where as before we ask for S to be small. In this case we are allowing the energy norm to grow with time, but in a way that is bounded by the integral of $|g(t)|$ over the boundary, so the system remains well-posed. In the same way we can also allow for the presence of source terms on the right hand side of the evolution system (5.9.11).

As a simple example of the above results we will consider again the wave equation in spherical symmetry

$$\partial_t^2 \varphi - v^2 \left(\partial_r^2 \varphi + \frac{2}{r}\, \partial_r \varphi \right) = 0 \,. \tag{5.9.21}$$

Introducing the first order variables $\Pi := \partial_t \varphi$ and $\Psi := v\partial_r \varphi$, the wave equation can be reduced to the system

$$\partial_t \varphi = \Pi \,, \tag{5.9.22}$$

$$\partial_t \Pi = v \partial_r \Psi + \frac{2v}{r} \Pi \,, \tag{5.9.23}$$

$$\partial_t \Psi = v \partial_i \Pi \,. \tag{5.9.24}$$

The system above is clearly symmetric hyperbolic, with eigenspeeds $\{0, \pm v\}$ and corresponding eigenfields

$$w_0 = \varphi \,, \qquad w_\pm = \Pi \mp \Psi \,. \tag{5.9.25}$$

Let us consider now the maximally dissipative boundary conditions at a sphere of radius $r = R$. These boundary conditions have the form

$$w_- = S w_+ + g(t) \,. \tag{5.9.26}$$

The requirement for S to be small now reduces simply to $S^2 \leq 1$. We will consider three particular cases:

- $S = -1$. This implies $\Pi + \Psi = -(\Pi - \Psi) + g(t)$, or in other words $\Pi = g(t)/2$. Since $\Pi = \partial_t \varphi$, this boundary condition fixes the evolution of φ at the boundary, so it corresponds to a boundary condition of Dirichlet type. The particular case $g = 0$ results in a standard reflective boundary condition, where the sign of φ changes as it reflects from the boundary.
- $S = +1$. This now implies $\Psi = g(t)/2$, which fixes the evolution of the spatial derivative of the wave function φ and corresponds to a boundary condition of Newmann type. Again, the case $g = 0$ corresponds to reflection, but preserving the sign of φ.
- $S = 0$. In this case we have $\Pi + \Psi = g(t)$, or in terms of the wave function $\partial_t \varphi + v \partial_r \varphi = g(t)$. This is therefore a boundary condition of Sommerfeld type.

From the expressions for dE/dt given above, it is easy to see that the choices $S = \pm 1$ with $g = 0$ imply that the energy norm is preserved (all the energy that leaves the domain through the outgoing modes comes back in through the incoming modes), so the wave is reflected at the boundary. Notice also that in the Sommerfeld case $S = 0$ we have not quite recovered the radiative boundary condition of the previous Section. But this is not a serious problem and only reflects the fact that we have excluded the source terms from all of our analysis. However, it does show that in many cases we need to consider a more general boundary condition of the form

$$w_-|_{\partial\Omega} = (S_+ w_+ + S_0 w_0)|_{\partial\Omega} + g(t) \ . \tag{5.9.27}$$

5.9.3 *Constraint preserving boundary conditions*

As discussed in the previous Section, the use of maximally dissipative boundary conditions for a symmetric hyperbolic system is crucial if we wish to have a well-posed initial-boundary value problem. However, this is not enough in the case of the 3+1 evolution equations since well-posed boundary conditions can still introduce a violation of the constraints that will then propagate into the computational domain at essentially the speed of light (the specific speed will depend on the form of the evolution equations used). We then have to worry about finding boundary conditions that are not only well-posed, but at the same time are compatible with the constraints. In a seminal work [133], Friedrich and Nagy have shown for the first time that it is possible to find a well-posed initial-boundary value formulation for the Einstein field equations that preserves the constraints. Their formulation, however, is based on the use of an orthonormal tetrad and takes as dynamical variables the components of the connection and the Weyl curvature tensor, so it is very different from most 3+1 formulations that evolve the metric and extrinsic curvature directly. It is therefore not clear how to apply their results to these standard "metric" formulations.

In the past few years, there have been numerous investigations related to the issue of finding well-posed constraint preserving boundary conditions [40, 61, 87, 88, 89, 137, 138, 154, 174, 191, 251, 278, 279, 280]. Here we will just present the basic idea behind these approaches without considering any specific formulation of the evolution equations. In order to simplify the discussion further, we will also assume that the shift vector vanishes at the boundary (the results can be easily generalized to the case when the shift vector is tangential to the boundary, but the presence of a component of the shift normal to the boundary complicates the analysis considerably). The discussion presented here follows closely that of [154].

Consider an arbitrary strongly or symmetrically hyperbolic formulation of the 3+1 Einstein evolution equations. The system therefore has the form

$$\partial_t u + M^i \partial_i u = s(u) , \tag{5.9.28}$$

with M^i the characteristic matrices, $s(u)$ some source terms, and u the vector of main evolution variables, essentially the extrinsic curvature K_{ij} and the spatial derivatives of the metric $d_{ijk} = \partial_i \gamma_{jk}/2$, plus possibly some combinations of these (such as the Γ^i of the NOR formulation, or the $\tilde{\Gamma}^i$ of the BSSNOK formulation). Given such a formulation we will have a series of constraints that involve spatial derivatives of the main evolution variables u. These constraints will include both the Hamiltonian and the momentum constraints, plus constraints related to the definition of some first order quantities of the form $C_{ijk} := d_{ijk} - \partial_i \gamma_{jk}/2 = 0$. Let us collectively denote the constraints as c_a, where the index a runs over all the constraints. Using now the evolution equations plus the (twice contracted) Bianchi identities, it is always possible to derive a system of evolution equations for the constraints. Since the constraints are compatible with the evolution equations, this system will have the form

$$\partial_t c + N^i \partial_i c = q(c) , \tag{5.9.29}$$

with N^i some characteristic matrices and $q(c)$ some source terms. This system is clearly closed in the sense that if all constraints are initially zero they will remain zero during the subsequent evolution.

Now, if the main evolution system (5.9.28) is strongly or symmetrically hyperbolic, the constraint evolution system will inherit this property. This comes about because the constraints must be compatible with the evolution equations. Let us denote by w the eigenfields of the main evolution system, and by W those of the constraint system. Consider now a boundary surface with unit normal vector n^i. The compatibility of the constraints with the main evolution system implies, in particular, that for every pair of constraint characteristic variables $W_\pm$ that propagate with speeds $\pm\lambda$ along the normal direction, there will be an associated pair of main characteristic variables $w_\pm$ that propagate with the same speeds.[58] Since the constraints are given in terms of spatial derivatives of

[58] The inverse of this statement is clearly not true as there are in general more main variables than constraints. In particular, both gauge and physical characteristic modes (those associated with the gravitational waves) can have no counterpart in the constraint system.

the main variables, it will always be possible to choose the normalization such that these associated pairs of variables are related through

$$W_\pm = \partial_n w_\pm + \dots \,, \tag{5.9.30}$$

where $\partial_n := n^i \partial_i$ is the normal derivative and the dots indicate transversal derivatives and lower order terms.

If we want the constraint evolution system to be well-posed we would need to impose maximally dissipative boundary conditions of the form (5.9.20)

$$W_- - S W_+ = 0 \,. \tag{5.9.31}$$

However, in practice it is not possible to impose this condition as we are not evolving the constraints directly. The above condition must therefore be rewritten in terms of the main variables as

$$\partial_n w_- - S \partial_n w_+ \simeq 0 \,, \tag{5.9.32}$$

where we are ignoring transverse derivatives, and as before $\simeq$ denotes equal up to principal part. Unfortunately, the last condition is not maximally dissipative as it does not involve the characteristic variables directly but rather their normal derivatives. We can, however, use the evolution equations for the $w_\pm$, namely $\partial_t w_\pm \pm \lambda \partial_n w_\pm \simeq 0$, to trade normal derivatives for time derivatives. The boundary condition then becomes

$$\partial_t w_- + S \partial_t w_+ \simeq 0 \,. \tag{5.9.33}$$

Define now a new variable $X := w_- + S w_+$ which is restricted to the boundary. The boundary condition then implies that X evolves through

$$\partial_t X \simeq 0 \,, \tag{5.9.34}$$

i.e. X is constant up to principal part. We can then go back and impose the following boundary condition on the main system

$$w_- + S w_+ = X \,. \tag{5.9.35}$$

This gives us boundary conditions for those characteristic variables of the main evolution system that are associated with constraint modes. For the remaining characteristic variables (those associated with the gauge and physical degrees of freedom) we are of course free to choose any maximally dissipative boundary conditions we desire.

If the quantities X were truly constant on the boundary, then the above boundary conditions would in fact be maximally dissipative conditions. However, in general X is only constant up to principal part, and its evolution is coupled to the other variables through both source terms and tangential derivatives, so that we do not have a true maximally dissipative boundary condition.

There are in fact some interesting exceptions to this where, for some specific evolution systems, we can find special choices of the matrix S for which the boundary system becomes closed in the sense that the evolution of the different X's is given only in terms of the X's themselves (the boundary system decouples from the "bulk" system). In such a case we can evolve the boundary system first, and then treat the X's as *a priori* given functions so that boundary conditions for the w_- become truly maximally dissipative. Such special cases can be shown to yield boundary conditions of the Dirichlet or Newmann types, as they reduce to imposing boundary data either on the components of the extrinsic curvature or on the components of the metric derivatives (for details see [154] and references therein). However, these special cases are very restrictive and will result in reflections of the constraint violations at the boundaries. A much more interesting case would be that of a Sommerfeld type boundary condition but this is still a matter for further research.

The requirement of having maximally dissipative boundary conditions might in fact be too strong, and there are some current efforts to find conditions for the well-posedness of initial-boundary value problems with more general differential boundary conditions (see *e.g.* [251]). At the time of writing this book, the problem of having well-posed and constraint-preserving boundary conditions for metric formulations of the 3+1 evolution equation is still not entirely solved and promises to be a very active area of research in the near future.

6

EVOLVING BLACK HOLE SPACETIMES

6.1 Introduction

The study of black hole spacetimes has been a central theme in numerical relativity since its origins. Black holes are important for a variety of reasons. First they correspond to the final state of the gravitational collapse of compact objects and are therefore expected to form even in situations that start from completely regular initial data, such as supernova core collapse or the collision of neutron stars. At the same time, spacetimes containing black holes can be seen as the simplest representation of gravitating bodies in general relativity. Indeed, being vacuum solutions, black holes avoid the need to consider non-trivial matter distributions with their corresponding complications, *i.e.* dealing with hydrodynamics, thermodynamics, micro-physics, electromagnetic fields, *etc.*[59] Also, we should remember the fact that general relativity is a self-consistent theory where energy and momentum conservation, and as a consequence the dynamics of matter, are intimately linked to the Einstein field equations. Because of this, in relativity it is inconsistent to think of distributions of matter whose motion is determined "from the outside". For example, trying to solve for the gravitational field associated with two solid spheres at rest is not entirely consistent in general relativity since we would immediately need to ask which forces are responsible both for the solid nature of the spheres and for keeping them at rest, and what is the stress-energy tensor associated with those forces, with is corresponding effect on the gravitational field itself (solutions of this type typically contain artifacts such as singularity "struts" connecting the two bodies to cancel the gravitational attraction that would otherwise move them toward each other). Considerations of this type imply that black holes are in fact the cleanest way to approach the two body problem in general relativity. Black holes, however, are clearly not the simple point particles of Newtonian dynamics and they bring extra complications of their own associated with the presence of horizons and singularities, as well as the possibility of non-trivial internal topologies.

Historically, the numerical study of binary black hole spacetimes started in 1964 with the work of Hahn and Lindquist [158] (though at that time the term "black hole" had not yet been coined). The binary black hole problem has continued as a major area of research in numerical relativity for the past 40 years, both because of purely theoretical considerations, and because of the fact that

[59]We must remember that from the point of view of general relativity, anything other than the gravitational field itself is seen as "matter", including physical entities such as scalar fields and electromagnetic fields.

binary black hole inspiral collisions are considered one of the most promising sources of gravitational waves, and as such are potentially observable in the next few years by the large interferometric gravitational wave observatories that are only just coming on line. Indeed, the binary black hole problem has been considered for a long time as the "holy grail" of numerical relativity. The problem has proved a difficult one to solve, hitting serious stumbling blocks and taking numerous detours over the years. In the past two years, however, enormous progress has been made following two different routes, first by Pretorius using an approach based on evolving the full 4-metric using generalized harmonic coordinates [232, 231, 233], and more recently by the Brownsville and Goddard groups (independently of each other) using a more traditional 3+1 approach based on the BSSNOK formulation with advanced hyperbolic gauge conditions and the crucial new ingredient of allowing for moving punctures [44, 45, 92, 93].

In the following Sections we will consider the different issues associated with the numerical evolution of black hole spacetimes. These issues can roughly be separated into three areas: 1) How to evolve black holes successfully and in particular how to deal with the presence of singularities, 2) how to locate the black hole horizons in a numerically generated spacetime, and 3) how to measure physical quantities such as mass and angular momentum associated with a black hole. There are a series of good references where we can look at all of these issues. In particular, for locating event and apparent horizons there is an excellent review by Thornburg [287].

6.2 Isometries and throat adapted coordinates

When we want to evolve black hole initial data, the first problem that needs to be considered is the fact that black holes have both singularities and horizons associated with them. Perhaps surprisingly, the presence of horizons is not really a serious problem in black hole evolutions as the geometry is locally well behaved at a horizon. The problem with standard Schwarzschild coordinates at the black hole horizon is a consequence of insisting on having a time independent metric with a vanishing shift vector, which by necessity implies that the coordinates can not penetrate the horizon. This problem is in fact very easy to fix by going to *e.g.* isotropic coordinates (see equation (1.15.28)). We only need to worry about the causal structure inside a horizon if we wish to use horizon penetrating coordinates that keep the horizon at an approximately fixed coordinate location, in which case we need to use a superluminal shift vector which might cause some simple numerical schemes to become unstable.

The more serious problem when evolving black hole spacetimes is the presence of singularities where the geometric quantities become infinite. Since we can not deal with infinite quantities directly in a numerical simulation, it is clear that we must somehow hide the infinity from the numerics. There are in fact two different types of singularities associated with black holes. On the one hand we have the real physical singularity where the gravitational field becomes infinite, and on the other hand there are also possible coordinate singularities where the spacetime is regular but the metric components are nevertheless ill-behaved. Fortunately, as discussed in Chapter 3, we can construct initial data for multiple black holes for which the physical singularity is in the future, so it does not represent a problem during the first stages of an evolution (though we still have to worry about not reaching it as time moves forward – see Section 6.4 below). The price to pay is that these types of initial data have non-trivial topologies, with multiple universes connected to each other through a series of wormholes. We can, of course, also construct initial data that do contain the physical singularity, as in the Kerr–Schild type initial data, but for the moment we will assume that this is not the case.

Let us then consider the evolution of initial data of Brill–Lindquist or Misner type that does not contain the physical singularity, but has instead a series of wormholes connecting different asymptotically flat regions. As already seen in Chapter 3, such initial data can be represented by a conformally flat metric with a conformal factor ψ that is infinite at a series of points inside each black hole horizon. If we insist on representing such initial data in $\Re^3$, then we find that at each of these points the metric is ill-defined, resulting in a series of coordinate singularities. For Brill–Lindquist type data, there is just one such coordinate singularity for each hole (the so-called "puncture"), but for Misner type data there is in fact an infinite number of coordinate singularities for each hole (cf. equations (3.4.12) and (3.4.15)).

We can think of two different strategies for dealing with such coordinate singularities, one better adapted to Brill–Lindquist type puncture data and the other to Misner type data. Since the number of singularities for Brill–Lindquist data is small, we can try to factor out the singular part from the metric and evolve the regular and singular parts separately. This is the basic idea of the *puncture evolution method*, and we will have more to say about it in the following Sections. For Misner data, on the other hand, there is an infinite number of singular points so this strategy is doomed to failure. Still, Misner type data has one key property that we can exploit in a numerical simulation. Since it is constructed precisely so that it represents two completely isometric universes, we can in fact do an evolution in $\Re^3$ minus a series of spheres and impose isometry boundary conditions on each of these spheres.

To see how this idea works let us start by considering a single Schwarzschild black hole in isotropic coordinates, for which the metric has the form (1.15.28):

$$ds^2 = -\left(\frac{1 - M/2r}{1 + M/2r}\right)^2 dt^2 + \psi^4 \left(dr^2 + r^2 d\Omega^2\right) , \tag{6.2.1}$$

with the conformal factor given by $\psi = 1 + M/2r$. The isotropic radial coordinate r is related to the standard "areal" Schwarzschild radius r_{Schwar} by

$$r_{\text{Schwar}} := r\psi^2 = r\left(1 + M/2r\right)^2 . \tag{6.2.2}$$

Let us for the moment ignore the time component of the metric and concentrate on the spatial metric

$$dl^2 = \psi^4 \left(dr^2 + r^2 d\Omega^2 \right) . \tag{6.2.3}$$

As already mentioned in Chapter 1, for this metric the horizon is located at $r_0 = M/2$ and coincides with the throat of the wormhole (the Einstein–Rosen bridge).[60] This metric has an isometry with respect to the transformation

$$r \to M^2/4r = r_0^2/r . \tag{6.2.4}$$

To evolve the initial data (6.2.3) we now place a boundary at the throat $r = r_0$, evolve the exterior metric, and impose isometry boundary conditions at the throat. In order to find the form that these isometry boundary conditions should take, let us consider a map $\bar{\mathbf{x}} = \mathbf{J}(\mathbf{x})$ from one side of the wormhole to the other that has the following form in spherical coordinates

$$\bar{r} = r_0/r , \quad \bar{\theta} = \theta , \quad \bar{\phi} = \phi , \tag{6.2.5}$$

The isometry now implies that any field should remain identical under the above map. For example, for a scalar field Φ this implies

$$\Phi\left(\mathbf{x}\right) = \pm \Phi\left(\mathbf{J}(\mathbf{x})\right) . \tag{6.2.6}$$

In the previous expression we allow the possibility of a minus sign since the square of the map $\bar{\mathbf{x}}$ is clearly the identity, plus the fact that considering the antisymmetric case has some important applications (see below). For a vector v^i and a one-form w_i, the isometry condition will take the form

$$v^i\left(\mathbf{x}\right) = \pm \left(\Lambda^i_m\right)^{-1} v^m\left(\mathbf{J}(\mathbf{x})\right) , \qquad w_i\left(\mathbf{x}\right) = \pm \Lambda^m_i \, w_m\left(\mathbf{J}(\mathbf{x})\right) , \tag{6.2.7}$$

where

$$\Lambda^i_j := \frac{\partial J^i}{\partial x^j} . \tag{6.2.8}$$

The transformation above looks very much like a standard coordinate transformation, the main difference being that here we assume that the components v^i and w_i have *the same functional form* before and after the transformation.

Finally, for a tensor T_{ij} the isometry condition becomes

$$T_{ij}\left(\mathbf{x}\right) = \pm \Lambda^m_i \Lambda^n_j \, T_{mn}\left(\mathbf{J}(\mathbf{x})\right) . \tag{6.2.9}$$

For the spatial metric γ_{ij} the minus sign is clearly unphysical, but this is not necessarily so for other tensors, such as for example the extrinsic curvature.

If we now differentiate the above expressions and evaluate them at the fixed point of the isometry we will obtain boundary conditions for the different geometric quantities at the throat. For a scalar field the antisymmetric case must

[60] For a single Schwarzschild black hole Brill–Lindquist and Misner data are in fact identical.

clearly satisfy $\Phi|_{\rm throat} = 0$, while the boundary condition for the symmetric case takes the form

$$(\partial_i \Phi)|_{\rm throat} = (\Lambda_i^m \partial_m \Phi)|_{\rm throat} \,, \tag{6.2.10}$$

which for the transformation (6.2.5) reduces to $\partial_r \Phi|_{r_0} = 0$. For vectors and one-forms the boundary conditions become

$$\partial_i v^j \big|_{r_0} = \pm \left(v^m \partial_i \left(\Lambda_m^j \right)^{-1} + \left(\Lambda_m^j \right)^{-1} \Lambda_i^n \, \partial_n v^m \right) \Big|_{r_0} \,, \tag{6.2.11}$$

$$\partial_i w_j \big|_{r_0} = \pm \left(w_m \partial_i \Lambda_j^m + \Lambda_j^m \Lambda_i^n \, \partial_n w_j \right) \big|_{r_0} \,, \tag{6.2.12}$$

while for tensors they take the form

$$\partial_i T_{jk} \big|_{r_0} = \pm \left(T_{mn} \partial_i \left(\Lambda_j^m \Lambda_k^n \right) + \Lambda_j^m \Lambda_k^n \Lambda_i^p \partial_p T_{mn} \right) \big|_{r_0} \,. \tag{6.2.13}$$

In the particular case of the transformation (6.2.5), these boundary conditions reduce, in the symmetric case, to

$$v^r \big|_{r_0} = 0 \,, \qquad w_r \big|_{r_0} = 0 \,, \qquad \partial_r T_{rr} \big|_{r_0} = -2 \, \frac{T_{rr}|_{r_0}}{r_0} \,, \tag{6.2.14}$$

and in the antisymmetric case to,

$$\partial_r v^r \big|_{r_0} = \frac{v^r|_{r_0}}{r_0} \,, \qquad \partial_r w_r \big|_{r_0} = - \frac{w_r|_{r_0}}{r_0} \,, \qquad T_{rr} \big|_{r_0} = 0 \,. \tag{6.2.15}$$

Notice that different fields can have different symmetry properties, for example we can have a symmetric spatial metric and an antisymmetric extrinsic curvature.

Antisymmetric fields are interesting if we wish, for example, to use a lapse function that vanishes at the throat. In that case we can take the lapse to be an antisymmetric scalar function, and use an antisymmetric extrinsic curvature to guarantee that the spatial metric itself remains symmetric. This is precisely what happens if we choose the lapse associated with the full isotropic metric (6.2.1), which takes the form

$$\alpha = \frac{1 - M/2r}{1 + M/2r} \,. \tag{6.2.16}$$

In this case time will move in opposite directions on the different sides of the wormhole and the throat will not evolve, which implies that the spatial slices will not penetrate inside the horizon.

Notice also that the conformal factor ψ is *not* a scalar function but rather a scalar density. Its boundary conditions are therefore inherited from those for the spatial metric, and for the symmetric case take the form

$$\partial_r \psi \big|_{r_0} = -\psi(r_0)/2r_0 \,. \tag{6.2.17}$$

It is easy to see that the conformal factor $\psi = 1 + r_0/r$ satisfies this condition.

The isometry boundary conditions simplify considerably is instead of r we use a logarithmic radial coordinate defined as

$$\eta := \ln\left(r/r_0\right) . \tag{6.2.18}$$

Notice that η now goes from $-\infty$ to ∞, with the throat located at $\eta_0 = 0$ and the isometry map taking the simple form $\bar{\eta} = -\eta$. The boundary conditions then become, in the symmetric case

$$\partial_\eta \Phi|_0 = v^\eta|_0 = w_\eta|_0 = \partial_\eta T_{\eta\eta}|_0 = 0 , \tag{6.2.19}$$

and in the antisymmetric case

$$\Phi|_0 = \partial_\eta v^\eta|_0 = \partial_\eta w_\eta|_0 = T_{\eta\eta}|_0 = 0 , \tag{6.2.20}$$

Because of the simplicity of these boundary conditions, the evolution of isometric single black hole spacetimes is best done using the logarithmic coordinate η.

The isometry boundary conditions described above can also be used in the case of two or more black holes, the main difference being that we now have a series of isometry maps, one for each throat. There is, however, one important issue that must be addressed: The isometry boundary conditions become very difficult to implement if we use a coordinate system that is not adapted to the throats of the individual black holes. Ideally we should then use a coordinate system such that the individual throats correspond to surfaces where some "radial" coordinate remains constant, and with the lines associated with the "angular" coordinates perpendicular to this surface.

In order to construct such throat adapted coordinates, it turns out to be convenient to think of a coordinate transformation starting from the standard cylindrical coordinates $(z, \rho = \sqrt{x^2 + y^2})$ as a map from the complex plane onto itself (see [271]). In order to do this we define the complex variable $\zeta = z + i\rho$ and consider a complex function $\chi(\zeta)$. The new coordinates (η, ξ) will correspond to the real and imaginary parts of χ:

$$\eta = \mathrm{Re}(\chi) , \qquad \xi = \mathrm{Im}(\chi) . \tag{6.2.21}$$

The simplest example corresponds to the function

$$\chi = \ln \zeta . \tag{6.2.22}$$

If we rewrite ζ as $\zeta = re^{i\theta}$, with $r = \sqrt{\rho^2 + z^2}$ and $\theta = \arctan(\rho/z)$, we find immediately find that

$$\chi = \ln r + i\theta , \tag{6.2.23}$$

which implies

$$\eta = \ln r , \qquad \xi = \theta . \tag{6.2.24}$$

We then see that this transformation corresponds to polar coordinates on the (ρ, z) plane, with η a logarithmic radial coordinate and ξ just the standard angular coordinate.

A more interesting example corresponds to *bipolar coordinates* (also called bispherical), which are obtained by considering the sum of two opposite poles at $z = \pm Z$:

$$\chi = \ln(\zeta + Z) - \ln(\zeta - Z) = \ln\left[\frac{\zeta + Z}{\zeta - Z}\right] . \tag{6.2.25}$$

In this case we find after some algebra that

$$\coth\eta = \left(\frac{\rho^2 + z^2 + Z^2}{2zZ}\right) , \tag{6.2.26}$$

$$\cot\xi = \left(\frac{\rho^2 + z^2 - Z^2}{2\rho Z}\right) , \tag{6.2.27}$$

where we have in fact taken ξ with the opposite sign as before in order to recover the standard convention for bipolar coordinates. The inverse transformation is

$$z = \frac{Z\sinh\eta}{\cosh\eta - \cos\xi} , \tag{6.2.28}$$

$$\rho = \frac{Z\sin\xi}{\cosh\eta - \cos\xi} . \tag{6.2.29}$$

Figure 6.1 shows a representation of bipolar coordinates in the (ρ, z) plane. Notice that this coordinate system is in fact singular at $\eta = \xi = 0$, which corresponds to points at infinity. The coordinate η goes to $\pm\infty$ at the poles $(z = \pm Z, \rho = 0)$, and behaves as a logarithmic radial coordinate close to each pole. The coordinate ξ, on the other hand, goes from 0 to π and behaves as a standard angular coordinate close to the poles (measured from the z axis). The value $\xi = \pi$ corresponds to the line joining the two poles. Notice also that both the η and ξ coordinate lines correspond to circles on the (ρ, z) plane. In particular, the lines of constant η are circles centered at the points $z_0 = Z\coth\eta$, with radii $a = Z/\sinh\eta$.

As discussed in Chapter 3, binary black hole initial data of the Misner type are given in terms of a parameter μ that is related to the coordinate center of the black hole throats and their coordinate radius through $z_0 = \pm\coth\mu$ and $a = 1/\sinh\mu$ (cf. equation (3.4.14)). We then see that if we take $Z = 1$, the black hole horizons will correspond to the surfaces $\eta = \pm\mu$. Bipolar coordinates were in fact used by Hahn and Lindquist in their pioneering work on black hole collisions [158], and also by Misner [204] for his work on initial data.

Another common choice for throat-adapted coordinates are the so-called *Čadež coordinates*. Here we choose the function $\chi(\zeta)$ as the sum of two equal sign logarithmic poles plus a series of higher multipoles:

$$\chi = \frac{1}{2}\left[\ln(\zeta + Z) + (\zeta - Z)\right] + \sum_{n=1}^{\infty} C_n \left(\frac{1}{(\zeta + Z)^n} + \frac{1}{(\zeta - Z)^n}\right) . \tag{6.2.30}$$

The coefficients C_n need to be chosen by a least-squares method in such a way as to guarantee that the black hole throats lie on an $\eta =$ constant line. Čadež

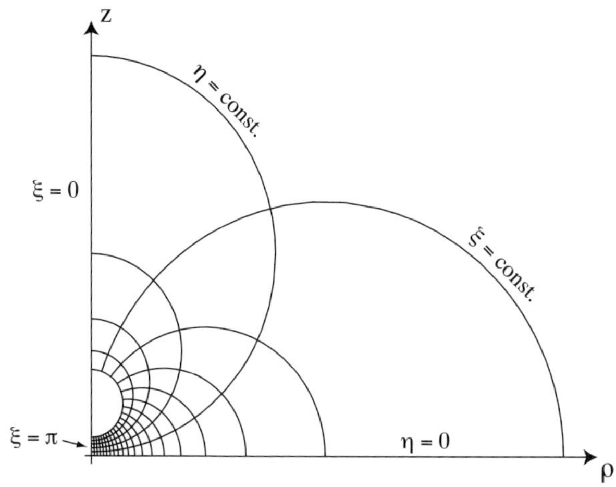

Fig. 6.1: Bipolar coordinates in the (ρ, z) plane.

coordinates must therefore be constructed numerically for each value of the parameter μ. Figure 6.2 shows a representation of the Čadež coordinate grid in the (ρ, z) plane for the case when $\mu = 2.5$.

The advantage of using Čadež coordinates is that these coordinates approach spherical coordinates close to the black hole throats and also far away from both holes. The coordinate η goes to 0 at the center of both throats, while ξ goes from 0 to $\pi/2$ as we go half way around the throat close to a black hole, or a quarter of the way around the origin in the faraway region. The coordinates are clearly singular at the saddle point $\rho = z = 0$ between both holes, where the coordinate line $\xi = \pi/2$ has a kink. Čadež coordinates have been used extensively for the simulation of black hole collisions using Misner initial data [27, 28, 85, 123, 270, 271].

Notice that since both for bipolar and Čadež coordinates η behaves as a logarithmic radial coordinate close to the throats, the isometry boundary conditions simplify just as they did in the case of a single black hole.

There are yet other throat adapted coordinate systems that have been proposed in the literature (see *e.g.* [24]), but we will not discuss them here. Finally, we can mention another approach to adapt our coordinates to the black hole throats: Instead of using a global coordinate system that is adapted to *both* black hole throats, we can use multiple coordinate patches with ordinary spherical coordinates close to each hole and a Cartesian patch everywhere else. This approach has been recently advocated and is being actively pursued by a number of groups (particularly in the context of black hole excision).

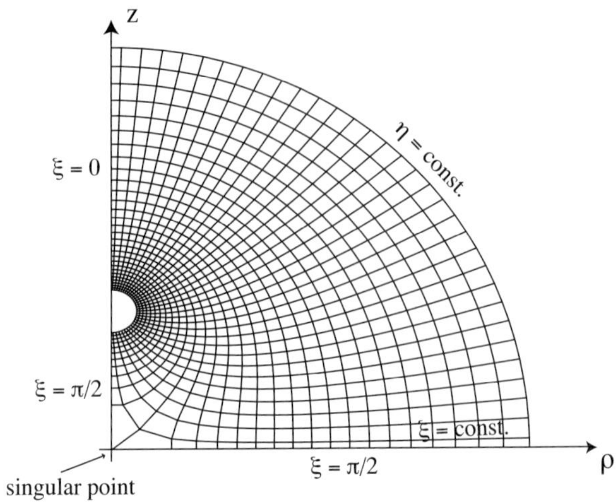

Fig. 6.2: Čadež coordinates. These coordinates are constructed numerically for each value of the parameter μ associated with Misner initial data. Here we show the particular case $\mu = 2.5$ (the data for this figure is courtesy of S. Brandt).

6.3 Static puncture evolution

As already mentioned in the previous Section, the use of throat adapted coordinates coupled with isometry boundary conditions is particularly well suited for Misner type initial data. For Brill–Lindquist puncture data, on the other hand, both the fact that the location of the throats must be determined numerically plus the lack of an isometry between the different asymptotically flat regions, imply that such an approach can not be used. However, since puncture data has only one singular point for each black hole, we can think instead of factoring out the singular terms from the metric and evolving the regular part separately. This idea is commonly known as the *puncture evolution method*. When we use a formulation that explicitly separates the volume elements using a conformal factor, as in the case of the BSSNOK formulation, such an approach becomes even more attractive as the singular terms are absorbed by the conformal factor itself while the conformal metric remains regular.

The idea of factoring out singular terms from the metric goes all the way back to the work of Hahn and Lindquist [158]. In that case, the use of bipolar coordinates (η, ξ) resulted in a coordinate singularity at the point $\eta = \xi = 0$, which was factored out explicitly from the metric. Similarly, in their work on head-on black hole collisions, Smarr *et al.* [271] also factored out the conformal factor and only evolved the conformal metric in order to improve the accuracy of their results, even though the conformal factor remained finite in their domain of integration.

Puncture evolution in a more modern sense, that is the idea of factoring out analytically the singular conformal factor from Brill–Lindquist type data and evolving only the regular conformal metric in *all* $\Re^3$, was first used by Anninos *et al.* in 1994 [25] for a single Schwarzschild black hole. In that reference the puncture idea was arrived at empirically by using a code that was originally designed to follow the approach of Smarr *et al.* that consisted of imposing an isometry condition at the throat *and* at the same time factoring out the conformal factor, and then simply turning off the isometry boundary condition and evolving the conformal metric everywhere. In 1997 Brügmann [82] proposed puncture evolutions as a general method for the evolution of black hole spacetimes constructed with a conformally flat metric and Bowen–York extrinsic curvature (*i.e.* puncture initial data). The idea was later described in detail by Alcubierre *et al.* in [13], both for the ADM and BSSNOK formulations of the evolution equations. Since the BSSNOK formulation already separates out the volume elements from the conformal metric, it is perhaps more natural to discuss puncture evolutions in this context.

Starting from the BSSNOK rescaling of the spatial metric:

$$\tilde{\gamma}_{ij} := e^{-4\phi}\gamma_{ij} \; , \tag{6.3.1}$$

we further split the conformal factor as

$$\phi = \hat{\phi} + \ln \psi_{\mathrm{BL}} \; , \tag{6.3.2}$$

where ψ_{BL} is the singular Brill–Lindquist conformal factor. For pure Brill–Lindquist type data initially we will have $\hat{\phi}(t=0)=0$, but for more general puncture data the initial conformal factor has an extra regular piece so that $\hat{\phi}(t=0) \neq 0$ (see Chapter 3). During the subsequent evolution, the regular part of the conformal factor $\hat{\phi}$ is evolved, but the singular part is kept constant in time. Terms in the evolution equations that make reference to the singular piece of the conformal factor or its derivatives are calculated analytically. Since the singular piece of the conformal factor is kept constant in time, it is perhaps better to refer to this method as the *static puncture evolution method.*

We can argue that it is possible to obtain regular evolutions using this approach by examining the evolution equations for the geometric quantities and the gauge conditions close to a puncture. For simplicity, let us assume that we have a single puncture located at the origin with zero linear momentum and spin, and consider the limit $r \to 0$ (in the more general case the analysis is more complicated but the results are similar). For the conformal factor we have $\psi_{\mathrm{BL}} = O(1/r)$ initially, and the different BSSNOK quantities will have the following initial behavior and assumed subsequent evolution:

$$\hat{\phi} = O(1) \to O(1) \; , \tag{6.3.3}$$

$$\tilde{\gamma}_{rr} = O(1) \to O(1) \; , \tag{6.3.4}$$

$$K = 0 \to O(r^2) \; , \tag{6.3.5}$$

$$\tilde{A}_{rr} = 0 \to O(r^2) \; , \tag{6.3.6}$$

$$\tilde{\Gamma}^r = 0 \to O(r) \; . \tag{6.3.7}$$

Assume furthermore that the initial lapse is $\alpha = O(1)$, and the shift is purely radial at the punctures with $\beta^r = O(r)$. Notice also that first derivatives of $O(1)$ quantities are $O(r)$ and second derivatives are again order $O(1)$. The BSSNOK evolution equations (2.8.9), (2.8.10), (2.8.11), (2.8.12), and (2.8.25) can then be shown to imply the following behavior near the puncture

$$\partial_t \hat{\phi} = O(1) + O(r^2) \,, \tag{6.3.8}$$

$$\partial_t \tilde{\gamma}_{rr} = O(1) + O(r^2) \,, \tag{6.3.9}$$

$$\partial_t K = O(r^2) + O(r^4)(O(1) + O(r^2) + r\, O(\partial_r \phi)) \,, \tag{6.3.10}$$

$$\partial_t \tilde{A}_{rr} = O(r^2) + O(r^4)\left(O(1) + O(r^2) + r\, O(\partial_r \phi) + O(\partial_r^2 \phi) + O(\partial_r \phi)^2\right) \,, \tag{6.3.11}$$

$$\partial_t \tilde{\Gamma}^r = O(r) + O(r^3) + O(r^2) O(\partial_r \phi) \,, \tag{6.3.12}$$

where we have kept the explicit dependence on derivatives of ϕ. If the assumption about the behavior of the different quantities during evolution is to hold for all times, then we must clearly say something about the derivatives of the conformal factor ϕ. Since we are assuming that $\hat{\phi}$ is regular, we will have

$$\phi = O(\ln r) \,, \quad \partial_r \phi = O(1/r) \,, \quad \partial_r^2 \phi = O(1/r^2) \,. \tag{6.3.13}$$

Substituting this into the above expressions we see that it is compatible with our assumptions, so that to leading order we have

$$\partial_t \hat{\phi} = O(1) \,, \tag{6.3.14}$$

$$\partial_t \tilde{\gamma}_{rr} = O(1) \,, \tag{6.3.15}$$

$$\partial_t K = O(r^2) \,, \tag{6.3.16}$$

$$\partial_t \tilde{A}_{rr} = O(r^2) \,, \tag{6.3.17}$$

$$\partial_t \tilde{\Gamma}^r = O(r) \,. \tag{6.3.18}$$

Notice, in particular, that K, $\tilde{A}_{ij}$ and $\tilde{\Gamma}^i$ will remain zero at the puncture. In contrast, $\hat{\phi}$ and the conformal metric $\tilde{\gamma}_{ij}$ will evolve at the puncture for non-zero shift, but will remain at their initial values if the shift vanishes or if it behaves as $O(r^3)$ instead of $O(r)$.

In the previous analysis we have assumed that the initial lapse is $\alpha = O(1)$, but we can start instead the evolution with a lapse of the form $\alpha = O(r^n)$ with n an even positive number (a "pre-collapsed" lapse). In that case the above counting arguments will change for vanishing shift, but they will remain the same if $\beta^r = O(r)$ because the shift terms will dominate.

The static puncture evolution method allows us to deal with the infinities in the spatial metric in a clean and efficient way. Nevertheless, it has a series of disadvantages. From a purely numerical point of view, the infinity still exists in the static conformal factor, so we must be careful to avoid placing a grid point directly at the puncture. In practice, the numerical grid is set up in such a way that the puncture lies between grid points, *i.e.* the puncture is staggered. Still, as resolution is increased it is clear that we will have to deal with larger and larger values of the conformal factor, so that convergence near the puncture will not be achieved.[61] Other more serious effects are related to the fact that since the singular conformal factor is static, we are connected at all times with the other asymptotic infinities.

Perhaps the worst disadvantage of this approach, however, is the fact that having a static singular term implies that the punctures can not move by construction, so that the black hole horizons must always surround the original position of the puncture, even when the individual black holes have non-zero linear momentum. This is really a serious drawback, as it forces us to use coordinate systems that follow the motion of the individual black holes so that they remain at an approximately fixed position in coordinate space. For binary black holes in orbiting configurations, for example, this implies that we must use a coordinate system that is corotating with the black holes. Still, the static puncture approach has been used successfully for the simulation of the grazing collision of two black holes [7, 82], and also of orbiting black holes with corotating coordinates [12, 110].

The static puncture method has been recently superseded by an approach that allows the punctures to move, resulting in much more accurate and stable simulations of black hole spacetimes. We will discuss this moving puncture method in Section 6.6 below.

6.4 Singularity avoidance and slice stretching

When we use initial data of Misner or Brill–Lindquist type, the singularities present are purely coordinate singularities. The physical singularity associated with the black holes is not present in the initial data. During evolution, however, the physical singularity will be rapidly approached unless our slicing condition is chosen in such a way as to avoid the singularity. The issue of singularity avoidance was already mentioned in Chapter 4 when we discussed different slicing conditions, so here we will just summarize the main results.

The simplest possible slicing condition, namely geodesic slicing which corresponds to choosing $\alpha = 1$ during the whole evolution, has the problem that it will rapidly encounter the physical singularity of the black hole. For example, in the case of a single Schwarzschild black hole, the isotropic initial data touches the horizon but does not penetrate inside it. The time to reach the singularity when using geodesic slicing will then coincide with the free-fall time for an observer initially at rest at the Schwarzschild horizon, which can be easily found to be $t = \pi M$.

[61] Empirical experience shows, however, that this lack of convergence does not leak out of the black hole horizon. In fact, it usually does not extend more than a few grid points away from the puncture itself.

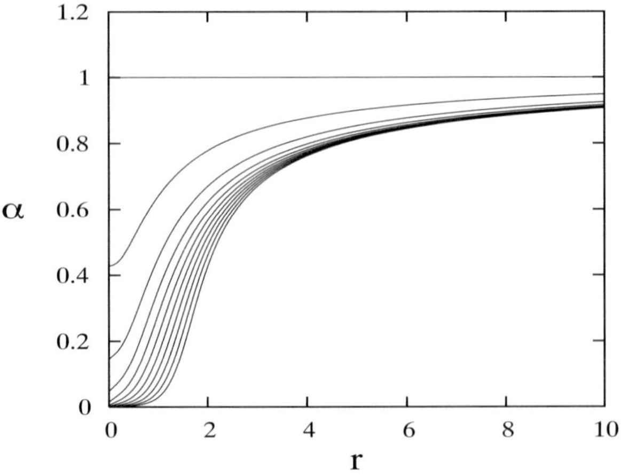

Fig. 6.3: Evolution of the lapse function α for Schwarzschild using maximal slicing. The value of α is shown every $t = 1M$. The collapse of the lapse is clearly visible.

Other slicing conditions are much better at avoiding the singularity by slowing down the evolution in the regions approaching the singularity. The classical example of a singularity avoiding slicing condition is maximal slicing which was discussed in detail in Section 4.2.2, and corresponds to asking for the trace of the extrinsic curvature to remain zero during the evolution, $K = \partial_t K = 0$. This requirement results in the following elliptic condition on the lapse function

$$D^2\alpha = \alpha \left[K_{ij} K^{ij} + 4\pi \left(\rho + S \right) \right] . \tag{6.4.1}$$

For Schwarzschild we can in fact show analytically that the maximal slices approach a limiting surface given by $r_{\text{Schwar}} = 3M/2$, with r_{Schwar} the standard Schwarzschild areal radius, so that the singularity is avoided. Because of its singularity avoiding properties, for many years maximal slicing was the standard slicing condition for evolving black hole spacetimes.

As already mentioned in Chapter 4, when using maximal slicing with puncture initial data, a specific boundary condition arises naturally at the punctures that corresponds to the lapse having zero gradient at the puncture, the so-called puncture lapse. Figure 6.3 shows an example of the evolution of the lapse function for a numerical simulation of the Schwarzschild spacetime starting from isotropic (puncture) initial data and using maximal slicing. The collapse of the lapse is evident.

Maximal slicing is not the only singularity avoiding slicing condition. Another common choice for a singularity avoiding slicing condition is the 1+log slicing, which is a member of the Bona–Masso family of slicing conditions (4.2.52) and has the form

$$\partial_t \alpha - \beta^i \partial_i \alpha = -2\alpha K . \tag{6.4.2}$$

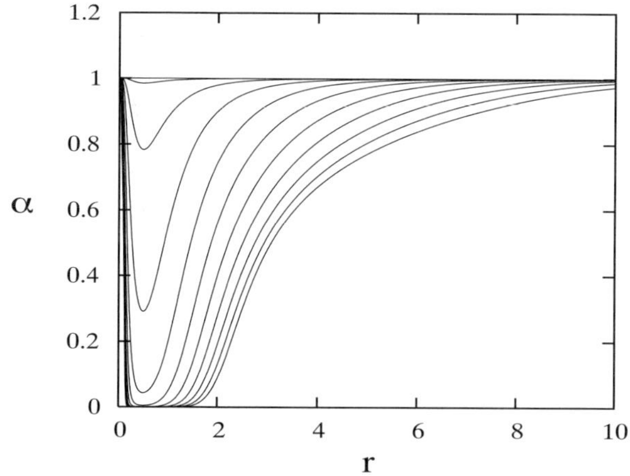

Fig. 6.4: Evolution of the lapse function α for Schwarzschild using 1+log slicing. The value of α is shown every $t = 1M$. Notice how the lapse remains equal to one at the puncture and large gradients develop.

The 1+log slicing condition has been found to be very robust in black hole evolutions. It is also much easier to solve than maximal slicing, and because of this it has supplanted maximal slicing in recent years. There is, however, an important point to keep in mind when using 1+log slicing with puncture data. Since 1+log slicing is a hyperbolic slicing condition, it has a finite speed of propagation given by $v = \alpha\sqrt{2\gamma^{rr}} = \alpha\sqrt{2\tilde{\gamma}^{rr}}/\psi^2$. At the puncture ψ diverges, so the speed of propagation vanishes (this reflects the fact that the puncture is an infinite proper distance away). As a result of this, for standard 1+log slicing the lapse will collapse in the region around the throat of the wormhole, but will remain equal to one both at infinity and at the puncture location. Having the lapse remain equal to one at the puncture will cause extremely large gradients to develop close to the puncture that can cause the numerical simulation to fail.[62] This phenomenon can be clearly seen in Figure 6.4, which shows the evolution of the lapse function for a numerical simulation of the Schwarzschild spacetime starting from puncture initial data and using the 1+log slicing.

This problem of 1+log slicing with puncture data can be cured in a number of different ways. First, we can simply ignore the region close to the puncture and cut it from the simulation using so-called *excision techniques* (see Section 6.5). Another possibility is to modify the 1+log condition so that the speed of propagation of signals becomes infinite in physical space, while remaining finite in coordinate space. This can be done by changing the 1+log condition to

[62] Depending on the numerical method used, we can find instead that the numerical dissipation is so strong that these large gradients force the lapse to collapse at the puncture. But this is a purely numerical artifact.

$$\partial_t \alpha = \beta^i \partial_i \alpha - 2\alpha\psi^m K , \tag{6.4.3}$$

with $m > 0$. A natural choice is $m = 4$, in which case the speed of propagation becomes finite at the puncture. Yet another possibility is to keep the original form of the 1+log slicing condition, but start the simulation with a "pre-collapsed" lapse that is already zero at the puncture. We can choose, for example, $\alpha(t = 0) = 1/\psi^n$ with $n \geq 2$, which will result in an initial lapse that is zero at the puncture and also has zero gradient. Choosing $n = 2$ has the advantage that the initial lapse has the same limit for $r \to \infty$ as the lapse of the static Schwarzschild metric in isotropic coordinates $\alpha = (1 - M/2r)/(1 + M/2r)$.

The use of a pre-collapsed initial lapse has recently become the standard in black hole simulations. In particular, the choice $n = 4$ seems to be required in the case where we have spinning punctures, for which the initial extrinsic curvature diverges as $1/r^3$.

There is a phenomenon related to singularity avoidance that can also cause problems for black hole simulations. Since time freezes inside the black hole but keeps advancing outside, the spatial slices become distorted, giving rise to the phenomenon of *slice stretching* which results in a rapid growth of the radial metric component and the development of large gradients near the throat of the black hole. We have already discussed this phenomenon in Chapter 4, where we mentioned that it is in fact a combination of two separate effects, one of them related to the differential infall of coordinate observers close to the black hole, and the other the result of the slices "wrapping" around the singularity. Figure 6.5 shows an example of the slice stretching effect for a numerical simulation of the Schwarzschild spacetime using maximal slicing. The figure shows the evolution of the conformal radial metric component $\tilde{\gamma}_{rr} = \gamma_{rr}/\psi^4$. From the figure we can clearly see how the metric grows rapidly in the region around the horizon of the black hole.

Historically, there have been two main strategies to avoid slice stretching. One strategy is again to simply ignore what happens inside the horizon by excising the black hole interior. The second strategy has involved the use of a shift vector to counter the effects of slice stretching. For some time there was the hope that shift conditions such as minimal distortion (4.3.14) would suffice to cure slice stretching, but early numerical simulations of Schwarzschild seemed to indicated that this was not the case [55]. In those simulations, the evolutions with no shift resulted in the familiar growth of the radial metric component γ_{rr} in the region around the horizon of the black hole. Evolutions using a minimal distortion shift, on the other hand, did indeed eliminate the growth in γ_{rr} close to the horizon, but at the price of producing an even larger growth in the region close to the throat. With hindsight, it is clear why in such simulations the shift condition failed to cure slice stretching. We must bear in mind the fact that all the shift vector can do is move the coordinate lines around, but these coordinate lines must come from somewhere. For a black hole evolved in isotropic coordinates

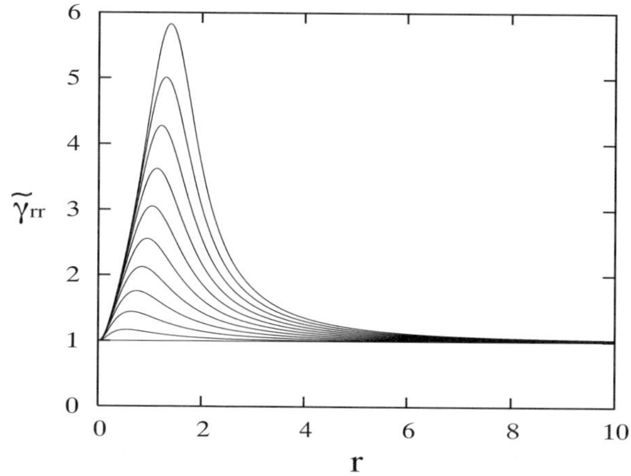

Fig. 6.5: Evolution of the conformal radial metric $\tilde{\gamma}_{rr} = \gamma_{rr}/\psi^4$ for Schwarzschild using maximal slicing. The value of the metric is shown every $t = 1M$. The slice stretching effect is evident.

with an isometry boundary condition, there is a large growth of the radial metric close to the horizon as coordinate observers get separated from each other. This growth can be cured by pushing coordinate lines from the region around the throat to the region around the horizon. But this clearly will result in a growth of the radial metric close to the throat as this region has now been depleted of coordinate lines. Notice, however, that since the region close to the throat is fast approaching the limiting surface $r = 3M/2$, there is very little distortion of the volume elements there. The depletion of coordinate lines in that region will only result in an increase of the volume elements with essentially no distortion, to which the minimal distortion condition is blind.

Because of the negative early results with the use of a shift vector, excision was considered for a long time as the only viable way to avoid the effects of grid stretching (though excision also requires a good shift choice in order to be successful). Recently, however, it has been shown that modern hyperbolic shift conditions such as the Gamma driver condition (4.3.34) can indeed cure slice stretching when used in conjunction with the puncture evolution method. In this type of simulations the shift is initially zero, but as the slice stretching develops the shift reacts by pulling out points from the inner asymptotically flat region near the punctures (so the isometry is clearly broken). The region around the puncture is therefore providing the necessary coordinate lines to counter slice stretching close to the horizon. The result is that with these new techniques the slice stretching phenomenon can be controlled and the evolution of the radial metric component can be virtually frozen [13].

The development of modern gauge conditions like the 1+log slicing and the Gamma driver shift, coupled with the puncture evolution method, has finally allowed the use of singularity avoiding slicings without the negative effects of slice stretching, and this has resulted in the possibility of having long-lived and accurate evolutions of black hole spacetimes.

6.5 Black hole excision

When evolving black hole spacetimes we are mostly interested in studying physical effects outside of the black hole itself, such as for example the orbital dynamics of the black holes and the gravitational wave signal emitted to infinity. Since the black hole horizon is a causal boundary, it is clear that whatever happens inside the black hole can not affect the physics outside. This observation gave rise to the idea of cutting or excising the black hole interior from the simulations altogether, thus eliminating the need to deal with the singularities present inside the black hole, either physical or coordinate, and also cleanly solving the problems associated with punctures, singularity avoidance and slice stretching. Black hole excision was first attempted successfully by Seidel and Suen in spherical symmetry in 1992 [261], and was later studied in more detail by Anninos *et al.* [26]. However, the original idea has been attributed by Thornburg [284, 285] to a suggestion by Unruh from 1984.[63]

Black hole excision in fact consists of two separate ingredients: First, we place a boundary inside the black hole and excise its interior from the computational domain. Second, we use a non-zero shift vector that keeps the horizon roughly in the same coordinate location during the evolution. Since no information can leave the interior of the black hole, excision should have no effect on the physics outside. Ideally, we would like to know the position of the event horizon which marks the true causal boundary, but the global character of its definition means that in principle we can only locate it once the whole evolution of the spacetime is known. The apparent horizon, on the other hand, can be located on every time slice and is guaranteed to be inside the event horizon (assuming cosmic censorship holds). In practice we therefore needs to find the apparent horizon and excise a region inside it.

Over the years, a number of different excision techniques have been developed by different researchers, with limited degrees of success. Early on it was realized that in order for excision to work, and for the black hole horizon to remain with an approximately fixed size in coordinate space, a superluminal shift vector was required. It is in fact easy to see why this should be so. Consider for a moment a point with fixed spatial coordinates x^i directly on the black hole horizon. If the shift is zero, this point will move on a timelike trajectory (the normal to the hypersurfaces), so at any later time it must be well inside the black hole, but since it still has the same spatial coordinates this means that the horizon must have moved outward in coordinate space (in other words, the black hole swallows the coordinate lines). So for the horizon to remain fixed in coordinate space we

[63] In the early literature the name "apparent horizon boundary condition" is used instead of the more modern name of "black hole excision" [26, 261]. This older name is now considered inadequate as no boundary condition is placed at the apparent horizon directly.

need an outward pointing shift vector to counter this effect. In fact, the shift vector must be such that the coordinate lines move out at the speed of light at the horizon, and faster than light inside the horizon. This is, of course, no cause for alarm as we are only talking about the "motion" of the coordinates and not any physical effect.

As it turns out, some simple numerical methods become unstable when the shift is superluminal, so in the early 1990s numerical techniques known as "causal differencing" [261] and "causal reconnection" [20] where developed that tried to tilt the computational molecules in order for them to follow the physical light-cones. However, later on it was realized that when using hyperbolic formulations of the evolution equations this was unnecessary, and quite standard numerical techniques should be able to cope with superluminal shifts as long as the CFL stability condition remains satisfied (see Chapter 9).[64] Moreover, when using hyperbolic formulations of the evolution equations we find that inside the black hole all characteristics move inward, so there is no need for a boundary condition on the excision surface as all the necessary causal information is in fact known. In more geometric terms, the excision surface is spacelike, so it is not a boundary as such, but rather a "time" at which we simply stop integrating – see Figure 6.6 (a possible exception to this would be the existence of gauge modes that move faster than light and that could have incoming characteristics at the excision surface, but we can always choose a gauge that does not behave in this way).

Though black hole excision has been successful in spherical symmetry [26, 196, 173, 151, 253, 254, 256, 261], it has been very difficult to implement with a 3+1 approach in three dimensions [25, 81, 104, 107, 286, 296] (black hole excision using a characteristic formulation, on the other hand, has been very successful in 3D, allowing stable evolutions of perturbed black holes for essentially unlimited times [147]). The main problem is related to the fact that the excision surface is typically of spheroidal shape and this is hard to represent on a Cartesian grid. It is therefore difficult to even define at a given point on the excision surface what is the "outward" direction. This inevitably leads to the use of extrapolations in several directions in order to be able to do the finite differencing, and this extrapolation is typically very unstable. One important development in this area was the introduction of so-called *simple excision* techniques [10]. The original idea behind such techniques was to excise a simple surface in Cartesian coordinates (*e.g.* a cube), and then to apply very simple boundary conditions on that surface, in particular simply copying the values of all dynamical variables onto the excision cube from the point directly adjacent to it. Such techniques have have later been generalized to the case of more complicated excision surfaces (so-called "lego-spheres"), and have allowed the simulation of black hole collisions for some limited times [12, 15]. Still, excision techniques in Cartesian grids have

[64] A common practice is to use one-sided differences in the shift advection terms (terms involving $\beta^i \partial_i$) to improve accuracy and stability, this being the only place where any consideration about causality enters the numerical code (*i.e.* the direction of the tilt of the light-cones).

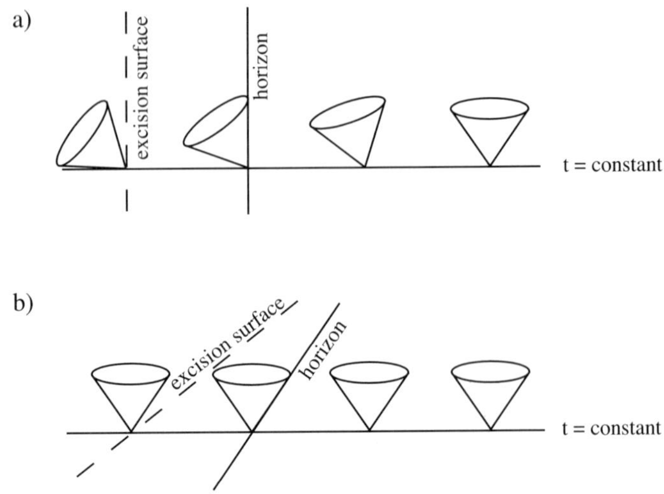

Fig. 6.6: Two different views of excision: a) Excision as seen in coordinate space. The excision surface is inside the horizon, where the light-cones are titled beyond the vertical and all characteristics are outgoing so no boundary condition is needed. b) Excision as seen locally by the normal observers. Here the light-cones are not distorted and the horizon is a null surface moving at the speed of light. The excision surface is spacelike, and again we can see that no boundary condition is needed as we simply stop integrating when this surface is reached (do notice that this diagram is valid only locally – in particular the horizon and excision surface do not really intersect since spacetime is not flat).

had a very difficult time evolving binary black hole spacetimes for more than about one orbit in the best of cases.

The problems faced by excision algorithms using Cartesian grids suggest another approach. Ideally, the excision surface should correspond to a coordinate surface where some "radial" coordinate remains fixed, so the use of spherical coordinates would seem ideal. In such a case defining the outgoing direction becomes trivial, so that when used in combination with a hyperbolic formulation of the evolution equations the excision algorithm should become easy to implement and can be guaranteed to remain stable. When dealing with more than one black hole, however, the use of a Cartesian grid is desirable to describe the space between the black holes. This has led to algorithms that use a series of overlapping coordinate patches: spherical coordinates near each black hole, Cartesian in between, and possibly spherical again far away. Several such multi-patch codes are currently under active development [255, 257].

Before finishing this Section it is important to mention that the use of excision algorithms has lately become less common as there has recently been a crucial breakthrough in puncture evolution techniques that has allowed the long-term simulation of binary black hole spacetimes without the need to excise the black hole interior (see the following Section).

6.6 Moving punctures

In Section 6.3 we described the basic ideas behind the static puncture evolution method. This method allows us to handle the infinities associated with puncture initial data in a clean way, but it has one major disadvantage, namely that by factoring out the infinities analytically and keeping them static it forces the punctures to remain at a fixed coordinate location even for black holes that have non-zero linear momentum.

Very recently there has been a true breakthrough in black hole simulations by taking some of the basic ideas of the puncture evolution method while at the same time allowing the punctures to evolve. This simple idea has finally allowed the accurate simulation of binary black holes for multiple orbits, with the black holes moving through the grid as we would expect, instead of remaining glued in a fixed position as the static puncture evolution method required. The idea of evolving the punctures was arrived at independently, using slightly different but equivalent methods, by Campanelli *et al.* [92, 93] and Baker *et al.* [44, 45].[65]

6.6.1 *How to move the punctures*

The moving puncture idea starts from the BSSNOK formulation, as in the static puncture approach, but now the singular part of the conformal factor is *not* factored out. The dynamical conformal factor $\phi = \ln \psi$ then has a logarithmic singularity which is directly evolved. The approaches of Campanelli *et al.* and Baker *et al.* differ on how this singularity is evolved in practice.

The approach of Baker *et al.* is to simply evolve ϕ directly. Notice that as long as we make sure that initially the point $r = 0$ does not correspond to a grid point (the puncture is staggered) no actual infinities will be present on the numerical grid. Of course, we could worry about the effect of taking numerical derivatives across the logarithmic singularity in ϕ. Empirically we find that such derivatives, though certainly introducing large numerical errors, nevertheless result in stable evolutions. Moreover, the numerical errors associated with taking derivatives across the puncture do not seem to leak out of the black hole and in fact only affect the region immediately adjacent to the puncture.

In contrast, Campanelli *et al.* define a new function χ through

$$\chi = \psi^{-4} = e^{-4\phi} \,. \tag{6.6.1}$$

[65] One should mention the fact that the first simulation of binary black holes through several orbits with the black holes moving through the grid was in fact done by Pretorius using an approach based on evolving the full 4-metric of the spacetime in generalized harmonic coordinates [231, 232, 233]. This method seems to be extremely robust and powerful and it is actively being used by several numerical relativity groups, but as it is not directly based on a 3+1 approach we will not discuss it here. Both the approach of Pretorius and the moving puncture method where developed within a year of each other in 2004–05, showing the tremendous amount of progress that the field of black hole simulations has seen within the last few years.

The evolution equation for χ follows from that for ϕ and has the form

$$\partial_t \chi - \beta^i \partial_i \chi = \frac{2}{3} \chi \left(\alpha K - \partial_i \beta^i \right) . \tag{6.6.2}$$

Notice that while ψ has a $1/r$ pole at the puncture, χ is instead $O(r^4)$ at the puncture. We then do not have to differentiate an infinite quantity at the puncture, but rather a C^4 quantity. Still, since there will be explicit divisions by χ in the evolution equations, we must ensure that χ is never exactly zero (in practice we can set the value of χ to some very small positive number whenever it either gets too close to zero or becomes negative due to numerical error).

Of course, evolving the puncture directly does not immediately imply that it will move across the grid. From the discussion in Section 6.3 we can see that for vanishing shift the punctures will simply not evolve. We therefore need to have a good choice of gauge conditions. Moving puncture approaches typically use a 1+log slicing condition

$$\partial_t \alpha - \beta^i \partial_i \alpha = -2 \alpha K , \tag{6.6.3}$$

together with a shift vector of the Gamma driver family (4.3.34) of the form

$$\partial_0 \beta^i = B^i , \qquad \partial_0 B^i = \frac{3}{4} \, \partial_0 \tilde{\Gamma}^i - \eta B^i . \tag{6.6.4}$$

Notice that no power of the lapse is present in the coefficient in front of $\partial_0 \tilde{\Gamma}^i$, as we want the shift to evolve at the puncture where the lapse will collapse to zero. The operator ∂_0 is taken to be equal to ∂_t in some approaches and to $\partial_t - \beta^i \partial_i$ in others (and even different combinations are used in the different terms). Notice that having a non-zero shift at the position of the puncture is precisely what allows it to move. The position of the puncture x_p^i can be tracked by integrating the equation of motion[66]

$$\frac{dx_p^i}{dt} = -\beta^i \left(x_p^j \right) . \tag{6.6.5}$$

The shift is typically chosen to vanish initially, but the Gamma driver condition rapidly causes the shift to evolve in such a way that it not only counteracts the longitudinal slice stretching effect, but for orbiting black holes it also automatically acquires a tangential component that allows the punctures to orbit one another. Also, because of the problem of the lack of collapse of the 1+log lapse at the puncture discussed in the previous Section, the initial lapse is usually taken to be of a pre-collapsed type $\alpha(t = 0) = \psi^{-n}$, with n equal to 2 or 4.

Both the ϕ-method and the χ-method have been shown empirically to lead to robust and stable simulations of black hole spacetimes, allowing us to follow binary black holes for several orbits through the merger and ring-down, and to

[66] In practice, this equation does not need to be integrated very accurately, since with the shift conditions typically used (*i.e.* the Gamma driver shift) the position of the puncture behaves as an attractor.

extract accurate gravitational waves, with evolution times that can last for many hundreds of M's. A recent review of the moving puncture approach can be found in [83].[67]

6.6.2 *Why does evolving the punctures work?*

The fact that moving puncture evolutions work can seem at first sight to be quite surprising. After all, we are "moving" through the numerical grid a point that corresponds to another infinitely far away, asymptotically flat region, and at which the conformal factor diverges. The sole idea of taking finite differences across this singular point should be enough to send any experienced numerical analyst into a panic attack. Clearly, we would expect numerical errors to get worse close, the puncture as the resolution is increased, destroying any chance of convergence. These errors should then quickly contaminate the entire domain of integration, rendering the whole numerical solution useless. Indeed, it was precisely this fear of dealing numerically with infinities that gave rise to the ideas of black hole excision and static puncture evolutions in the first place. But numerical experiments have shown that the fears were exaggerated, and that simulations involving moving punctures not only do converge but can also be highly accurate. Of course, empirical evidence aside, we would like to understand the reason why evolving the punctures works. The problem can in fact be separated in two distinct issues: First, from the point of view of numerical analysis, why is it that the formally infinite errors near the puncture do not contaminate the whole solution? And second, from a point of view of general relativity, how is it possible that near each puncture the lapse and shift conditions seem to drive the solution into a stationary situation?

The first question has in fact not been tackled formally yet, but there is a heuristic argument as to why numerical errors do not contaminate the whole solution. First, we must remember that the punctures are inside the black hole horizons, so that the effects of the numerical errors would have to propagate faster than light to contaminate the exterior solution. As puncture evolutions are typically done using hyperbolic formulations it is clear that errors will propagate at a finite speed. If this speed is smaller than the speed of light the errors should not be able to "escape" from the black hole. Moreover, as we will see below, it turns out that at late times the puncture is in fact not at asymptotic infinity, but rather truly inside the black hole, so that all characteristics near the puncture point inwards, and the errors can be expected to remain "trapped" close to the puncture. Of course, this heuristic explanation is not sufficient as it is clear that when using a centered numerical scheme (as is usually done) numerical errors *must* propagate to the whole grid regardless of the sign of the physical characteristics. So what is really needed is a formal proof that shows that as

[67]Systematic studies have also shown that extremely high numerical resolution is required to evolve the black holes accurately. Because of this, modern algorithms use fourth order differencing in both space and time, plus some form of mesh refinement, to allow high resolution close to each of the black holes while keeping the outer boundaries as far away as possible.

resolution is increased, the effect of these errors leaking out of the black hole become smaller and smaller so that the simulation converges. At the time of writing this book such an analysis has not yet been done, but in some simplified systems it might not be so difficult to address.

There has been more progress on the second issue regarding the final stationary state reached near each puncture. Recent work by Hannam *et al.* [159] has shown that, at least in the case of Schwarzschild, the 1+log slicing condition together with a Gamma driver shift condition does indeed allow us to find an explicitly time independent form of the Schwarzschild metric that penetrates the horizon and corresponds to the final state of the "evolving puncture" simulations to within numerical error. The surprising result has been that as the simulation proceeds the numerical puncture ceases to correspond to another asymptotically flat region and moves to a minimal surface *inside* the black hole. Also, the singularity in the conformal factor changes character. Without going into the details of their argument, Hannam *et al.* show that once the stationary state is reached, the lapse and conformal factor behave near the puncture as

$$\alpha \sim r \,, \qquad \psi \sim 1/\sqrt{r} \,, \tag{6.6.6}$$

which is in contrast with the initial conformal factor of the isometric initial data $\psi \sim 1/r$, and the initial pre-collapsed lapse typically chosen for these simulations $\alpha \sim r^2$. More interesting is the fact that since the area of spheres close to the puncture goes as $r^2\psi^4$, we now find that at the puncture itself this area remains finite, so that the puncture no longer corresponds to asymptotic infinity. For standard 1+log slicing we can in fact show that the puncture corresponds to a cylinder with areal radius $R_0 \sim 1.31M < 2M$, *i.e.* well inside the black hole horizon (the puncture, however, is still an infinite proper distance away from the horizon).

We can ask how it is possible that a slice that initially reaches the other asymptotic infinity ends up instead reaching a cylinder inside the black hole. What is happening is that, as the evolution proceeds, the slicing condition causes the throat of the Einstein-Rosen bridge to collapse toward the limiting cylinder with radius R_0, while at the same time the shift condition makes the throat approach the puncture exponentially in coordinate space. After a relatively short time, the throat is less that one grid point away from the puncture, so the whole space between the throat and the inner asymptotic infinity is squeezed into a region that can simply not be resolved numerically, and in practice the puncture can now be assumed to correspond to the throat itself.

As a final comment, the fact that we start with two asymptotically flat regions but end up with a single region that reaches an inner throat at the puncture immediately raises the possibility of constructing initial data that already has this final topology, and using it instead of the standard Brill–Lindquist type data. This is an issue that will no doubt receive much attention in the near future.

6.7 Apparent horizons

Being able to evolve a black hole spacetime in a stable and consistent way, though clearly crucial, is not enough. We would also like to be able to extract physical information from the numerically generated spacetime. Such physical information includes both far field studies such as the extraction of the emitted gravitational waves (see Chapter 8), and near field studies related to the dynamics of the black hole horizons and the extraction of physical quantities such as the black hole mass and spin. Locating black hole horizons is also crucial if we are using an approach based on singularity excision, since in that case we need to be sure that the region excised from the computational domain is safely inside the black hole.

There are in fact two different types of horizon associated with black holes. On the one hand, we define the event horizon as the boundary between null lines that "escape" to infinity and those that fall back into the black hole and hit the singularity instead. Since an event horizon is defined in a global way we need in principle to know the full evolution of the spacetime (or a significant part of it) in order to locate it, which means that it can not be used as an indicator of the presence and location of a black hole during an evolution. Apparent horizons, on the other hand, are defined locally as the outermost marginally trapped surface on a given spatial hypersurface, that is, a closed two-dimensional surface S such that the expansion of outgoing null geodesics orthogonal to S is zero everywhere. An important property of apparent horizons is the fact that, if the cosmic censorship conjecture holds and the null energy condition is satisfied, then an apparent horizon implies the existence of an event horizon exterior to it, or coincident with it in the stationary case.

Apparent horizons can be located during an evolution on each spatial hypersurface. They are thus ideal indicators of the presence of a black hole, and can be used both to locate the black holes in order to use *e.g.* black hole excision techniques, and also to measure physical quantities associated with each black hole such as mass and angular momentum. In order to locate apparent horizons on a given spatial hypersurface we must first derive an equation that characterizes them.

Consider then a closed two-dimensional surface S immersed in our three-dimensional spatial hypersurface Σ. Let $\vec{s}$ be the spacelike unit outward-pointing normal vector to S, and $\vec{n}$ the timelike unit future-pointing normal vector to Σ. We can now use these vectors to construct the outgoing null vector $\vec{l}$ as[68]

$$\vec{l} = \vec{n} + \vec{s} \,. \tag{6.7.1}$$

The expansion of the null lines is essentially the change in the area elements of S along $\vec{l}$. If we denote by $h_{\mu\nu}$ the two-dimensional metric on S induced by the spacetime metric $g_{\mu\nu}$ we will have

[68]The normalization of the null vector $\vec{l}$ is in fact arbitrary. Throughout most of the book we will usually take $\vec{l} = (\vec{n} + \vec{s})/\sqrt{2}$, but here we drop the factor $1/\sqrt{2}$ to simplify the notation.

$$h_{\mu\nu} = g_{\mu\nu} + n_\mu n_\nu - s_\mu s_\nu \; . \tag{6.7.2}$$

The expansion H of the outgoing null geodesics will then be

$$H = -\frac{1}{2}\, h^{\mu\nu} \pounds_{\vec{l}} h_{\mu\nu} = -\frac{1}{2}\, h^{\mu\nu} \left(\pounds_{\vec{s}}\, h_{\mu\nu} + \pounds_{\vec{n}}\, h_{\mu\nu} \right) \; . \tag{6.7.3}$$

But $\pounds_{\vec{s}}\, h_{\mu\nu} = -2X_{\mu\nu}$, with $X_{\mu\nu}$ the extrinsic curvature of S, so that

$$h^{\mu\nu} \pounds_{\vec{s}}\, h_{\mu\nu} = -2\, h^{\mu\nu} X_{\mu\nu} = -2X = -2D_m s^m \; , \tag{6.7.4}$$

with D_i the three-dimensional covariant derivative in Σ. On the other hand

$$\begin{aligned} h^{\mu\nu} \pounds_{\vec{n}}\, h_{\mu\nu} &= h^{\mu\nu} \left[\pounds_{\vec{n}}\, g_{\mu\nu} + \pounds_{\vec{n}}\, n_\mu n_\nu - \pounds_{\vec{n}}\, s_\mu s_\nu \right] \\ &= -2h^{\mu\nu} K_{\mu\nu} + h^{\mu\nu} \left[\pounds_{\vec{n}}\, n_\mu n_\nu - \pounds_{\vec{n}}\, s_\mu s_\nu \right] \; , \end{aligned} \tag{6.7.5}$$

where $K_{\mu\nu}$ is the extrinsic curvature of Σ. Using now the fact that $n_\alpha n^\alpha = -1$, $s_\alpha s^\alpha = 1$ and $n_\alpha s^\alpha = 0$ we find

$$h^{\mu\nu} \pounds_{\vec{n}}\, n_\mu n_\nu = h^{\mu\nu} \pounds_{\vec{n}}\, s_\mu s_\nu = 0 \; , \tag{6.7.6}$$

so that

$$h^{\mu\nu} \pounds_{\vec{n}}\, h_{\mu\nu} = -2\, h^{\mu\nu} K_{\mu\nu} = -2K + 2K_{mn} s^m s^n \; , \tag{6.7.7}$$

with K the trace of $K_{\mu\nu}$. Collecting results we finally find

$$\begin{aligned} H &= D_m s^m - K + K_{mn} s^m s^n \\ &= \left(\gamma^{mn} - s^m s^n \right) \left(D_m s_n - K_{mn} \right) \; , \end{aligned} \tag{6.7.8}$$

where in the latter equality we used the fact that $s^m s_m = 1$ implies $s^m D_n s_m = 0$.

The condition for having an apparent horizon is by definition $H = 0$, so that the equation that characterizes the apparent horizon takes the final form

$$H = D_m s^m - K + K_{mn} s^m s^n = \left(\gamma^{mn} - s^m s^n \right) \left(D_m s_n - K_{mn} \right) = 0 \; . \tag{6.7.9}$$

Notice that the condition for having a minimal (or rather extremal) surface, like the throat of a wormhole for example, is simply

$$D_m s^m = 0 \; , \tag{6.7.10}$$

so that for time symmetric data with $K_{ij} = 0$ the apparent horizon will coincide (at least initially) with the minimal surface, *i.e.* with the throat of the Einstein–Rosen bridge.

It is often convenient to characterize an apparent horizon by assuming that the surface has been parameterized as a level set of some scalar function F:

$$F(x^i) = 0 \; . \tag{6.7.11}$$

It is then straightforward to rewrite H in terms of the function F and its derivatives. First, we write the unit normal vector s^i as

$$s^i = D^i F / u \; , \tag{6.7.12}$$

with $u := |DF| = (\gamma^{mn} D_m F D_n F)^{1/2}$. Substituting this in equation (6.7.9) results, after some algebra, in

$$H = \left(\gamma^{mn} - \frac{D^m F D^n F}{u^2} \right) \left(\frac{D_m D_n F}{u} - K_{mn} \right) = 0 \; . \tag{6.7.13}$$

Having found the equation that must be satisfied by an apparent horizon, the next step is to develop an algorithm to solve this equation. For this we will consider the cases of spherical and axial symmetry separately since in those cases the problem simplifies considerably. A very complete discussion of the different algorithms for finding apparent horizons can be found in the recent review by Thornburg [287].

As a final comment, it is important to mention the fact that apparent horizons do not evolve in a smooth way during gravitational collapse. On the contrary, we typically find that an apparent horizon can jump discontinuously when a new larger marginally trapped surface appears outside the earlier apparent horizon. This is typical of black hole collisions where a common apparent horizon suddenly appears outside the individual apparent horizons.

6.7.1 *Apparent horizons in spherical symmetry*

Let us consider first a spherically symmetric spacetime. We can write the spatial metric quite generally in terms of spherical coordinates $\{r, \theta, \phi\}$ as

$$dl^2 = A(r,t) dr^2 + r^2 B(r,t) \, d\Omega^2 \; , \tag{6.7.14}$$

with A and B positive, and where $d\Omega^2 = d\theta^2 + \sin^2\theta d\phi^2$ is the standard solid angle element. In order to obtain an expression for the expansion of the outward-pointing null geodesics we can take $\vec{s}$ as the unit radial vector: $\vec{s} = (1/\sqrt{A}, 0, 0)$. After some simple algebra the expansion H can be found to be

$$H = \frac{1}{\sqrt{A}} \left(\frac{2}{r} + \frac{\partial_r B}{B} \right) - 2 K^\theta_\theta \; , \tag{6.7.15}$$

so that the condition for an apparent horizon reduces to

$$\frac{1}{\sqrt{A}} \left(\frac{2}{r} + \frac{\partial_r B}{B} \right) - 2 K^\theta_\theta = 0 \; . \tag{6.7.16}$$

Notice that this should not be understood as a differential equation for the metric coefficient B, but rather as a condition that indicates the presence of an apparent horizon at some value of the radial coordinate r.

As an example, consider the case of a single black hole in Schwarzschild coordinates for which we have $K_\theta^\theta = 0$, $A = 1/(1-2M/r)$ and $B = 1$, so that the horizon condition becomes

$$\frac{2}{r}\sqrt{1-2M/r} = 0 \quad \Rightarrow \quad r = 2M \; . \tag{6.7.17}$$

On the other hand, in isotropic coordinates we have $K_\theta^\theta = 0$ and $A = B = \psi^4$, with $\psi = 1 + M/2r$ (cf. equation (1.15.28)), so that the apparent horizon condition takes the form

$$\frac{1}{\psi^2}\left(\frac{2}{r} + \frac{4\,\partial_r\psi}{\psi}\right) = 0 \quad \Rightarrow \quad r = M/2 \; . \tag{6.7.18}$$

Here we must remember that the radial coordinate r in each case has a different meaning, with the isotropic and Schwarzschild radii $r_{\rm iso}$ and $r_{\rm Schwar}$ related through $r_{\rm Schwar} = r_{\rm iso}\left(1 + M/2r_{\rm iso}\right)^2$, so that $r_{\rm iso} = M/2$ implies $r_{\rm Schwar} = 2M$, in agreement with the above results.

In a more general case all we need to do in order to locate an apparent horizon in spherical symmetry is to simply evaluate the expansion (6.7.15) over the whole computational domain and look for places where it vanishes (accuracy higher than the grid spacing can be obtained by using linear or quadratic interpolation). The outermost zero will indicate the apparent horizon, while regions with $H < 0$ will correspond to trapped surfaces. In some cases we can find that H vanishes at more than one place, implying the presence of other marginally trapped surfaces, so-called *inner horizons*. This situation is typical of black holes formed through gravitational collapse, where a single marginally trapped surface forms first as H becomes zero at a single point while remaining positive everywhere else. This initial apparent horizon later splits into an inner and an outer horizon, with a trapped region in between. The inner horizon then moves toward a smaller radius as the physical singularity is approached.

When dealing with black hole spacetimes represented through wormholes, however, there is a final subtlety that should be kept in mind. Although outside the wormhole it is clear that the direction of increasing r points outward, inside the wormhole this is not true and the "outward" direction corresponds in fact to that of decreasing r (we are going out on the other side). This means that if we are trying to locate an apparent horizon on the other side of the wormhole the normal vector should instead be $\vec{s} = (-1/\sqrt{A}, 0, 0)$, and the expression for the expansion H must be changed to

$$H = -\frac{1}{\sqrt{A}}\left(\frac{2}{r} + \frac{\partial_r B}{B}\right) - 2K_\theta^\theta \; . \tag{6.7.19}$$

6.7.2 *Apparent horizons in axial symmetry*

Consider next the case of an axisymmetric spacetime. In this case the apparent horizon will be a surface of revolution around the axis of symmetry. It can then be described in spherical coordinates as a level surface of the form

a) b)

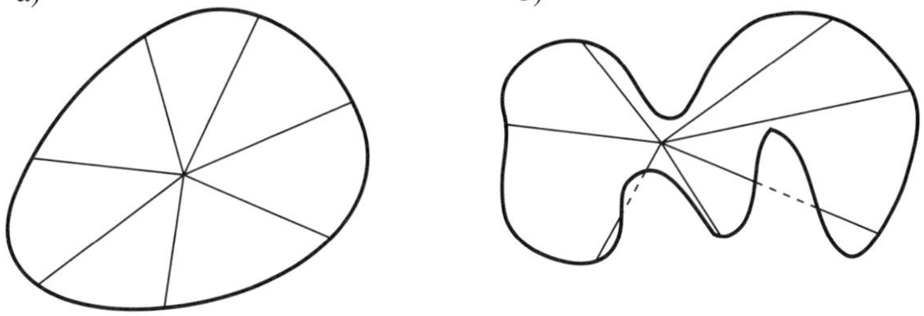

Fig. 6.7: a) A strahlkörper or ray-body: The rays from the "center" intersect the surface only once. b) A closed surface with spherical topology that is nevertheless not a ray-body: There are rays that intersect the surface more than once.

$$F(r,\theta) = r - h(\theta) \; . \tag{6.7.20}$$

The parameterization above is not in fact completely general, and it does imply a restriction in the allowed form of the apparent horizon. Namely, the horizon is assumed to be a so-called *strahlkörper* or *ray-body*, which is defined as a surface with spherical topology with the property that we can always find a "center" inside it such that all rays leaving this center intersect the surface once and only once (see Figure 6.7).[69] This simplifying assumption is typical for horizon finding algorithms, and implies that horizons with very complicated shapes will not be found.

Substituting now the form of F given above into equation (6.7.13), and assuming that we are working in spherical coordinates (r,θ,φ), we obtain a non-linear second order differential equation for h of the form

$$\frac{d^2h}{d\theta^2} = \frac{-1}{\gamma^{rr}\gamma^{\theta\theta} - (\gamma^{r\theta})^2} \left(u^2 \gamma^{ij} - \partial^i F \, \partial^j F \right) \left(\Gamma^k_{ij} \, \partial_k F + u K_{ij} \right) \; , \tag{6.7.21}$$

with Γ^k_{ij} the Christoffel symbols associated with the spatial metric γ_{ij}, and where

$$\partial_i F = \left(1, -\frac{dh}{d\theta}, 0 \right) \; , \qquad \partial^i F = \gamma^{im} \partial_m F = \gamma^{ir} - \gamma^{i\theta} \left(\frac{dh}{d\theta} \right) \; , \tag{6.7.22}$$

and

$$u = \left[\partial_i F \, \partial^i F \right]^{1/2} = \left[\gamma^{rr} - 2\gamma^{r\theta} \left(\frac{dh}{d\theta} \right) + \gamma^{\theta\theta} \left(\frac{dh}{d\theta} \right)^2 \right]^{1/2} \; . \tag{6.7.23}$$

The crucial observation here is the fact that this is an ordinary differential equation (ODE) for h, which can be solved by giving some appropriate boundary conditions and using any standard ODE integrator (*e.g.* Runge–Kutta).

[69]The original definition of a strahlkörper is due to Minkowski.

The boundary conditions can be derived from the fact that the apparent horizon should be a smooth surface. If we integrate from $\theta = 0$ to $\theta = \pi$, smoothness requires that $h(\theta)$ should have zero derivative at both boundaries, that is

$$\partial_\theta h = 0 \,, \qquad \theta = 0, \pi \,. \tag{6.7.24}$$

In some cases we have equatorial symmetry as well as axial symmetry, in which case we can stop the integration at $\theta = \pi/2$ and impose the boundary condition $\partial_\theta h = 0$ there.

The way to find a horizon in practice is to start at some arbitrary point $r = r_0$ on the symmetry axis (say, a few grid points away from the origin), and integrate the second order differential equation (6.7.21) with the boundary conditions $h = r_0$ and $\partial_\theta h = 0$ at $\theta = 0$. When we reach $\theta = \pi$ (or $\theta = \pi/2$ for equatorial symmetry), we check whether the boundary condition $\partial_\theta h = 0$ is satisfied at that point to some numerical tolerance. If it is, we have found an apparent horizon, but if it isn't we move up along the axis and start again. If no horizon has been found as we reach the end of the axis, then no horizon exists in our spacetime. This method is usually called a *shooting method* in the ODE literature, because we start with some initial guess at one boundary and "shoots" to the other boundary, adjusting the initial guess (we "aim better") until we obtain the desired boundary condition on the other side (we "hit the target").

6.7.3 *Apparent horizons in three dimensions*

As we have seen in the previous Sections, finding apparent horizons in spherical symmetry reduces to locating the zeros of a one-dimensional function, while in axial symmetry it reduces to integrating an ordinary differential equation subject to certain boundary conditions. In three dimensions, however, the problem is considerably more complicated. Nevertheless, a number of different algorithms designed for finding apparent horizons in three dimensions have been proposed over the years. Here I will concentrate on three such algorithms and mention only the main ideas associated with each of them. For a more detailed description of the algorithms the reader is directed to the recent review by Thornburg [287] (this review includes an extensive list of references of working implementations of the different algorithms).

6.7.3.1 *Minimization algorithms.* Minimization algorithms for finding apparent horizons were among the first methods ever tried and were in fact the original methods used by Brill and Lindquist [79] and by Eppley [124]. The basic idea behind a minimization algorithm is to expand the parameterization function $F(x^i)$ in terms of some set of basis functions, and then minimize the surface integral of the square of the expansion H^2 over the space of these basis functions. At an apparent horizon this integral will vanish and we will have a global minimum. Numerically, of course, the integral will never vanish exactly so we set a tolerance level below which a horizon is assumed to have been found.

A typical minimization apparent horizon finder uses a parameterization of the horizon of the form

$$F(r,\theta,\phi) = r - h(\theta,\phi) \; , \tag{6.7.25}$$

again, assuming the horizon is a ray-body. The function $h(\theta,\phi)$ in then expanded in terms of spherical harmonics as

$$h(\theta,\phi) = \sum_{l=0}^{l_{\max}} \sum_{m=-l}^{l} a_{l,m} Y^{l,m}(\theta,\phi) \; , \tag{6.7.26}$$

with $a_{l,m}$ some constant coefficients.

Given a trial function h (*i.e.* a trial set of $a_{l,m}$), we reconstruct $F = r - h$, find the expansion H, and calculate the surface integral of H^2. A standard minimization algorithm can then be used to determine the values of the coefficients $a_{l,m}$ for which the integral reaches a minimum. Notice that we can in principle calculate H over the entire three-dimensional grid by using finite differences in Cartesian coordinates, and later interpolate it into the surface $r = h(\theta,\phi)$ in order to do the integral. This procedure is computationally expensive but very simple to code. Alternatively, we can calculate H directly on the surface by using finite differencing in the angular coordinates.

Even though the idea behind minimization algorithms is straightforward, they have at least two serious drawbacks. First, the algorithm can very easily settle on a local minimum for which the expansion is not zero, so a good initial guess is often required. Also, minimization algorithms tend to be very slow: If N is the total number of terms in the spectral decomposition, a minimization algorithm requires a few times N^2 evaluations of the surface integrals. Finally, minimization algorithms are only as accurate as the tolerance given, and in practice the tolerance can not be too small since we run the risk of not finding the horizon at all. Because of this, minimization algorithms are now rarely used in practice.

6.7.3.2 *Flow algorithms.* Flow algorithms work by evolving an initial trial surface in some unphysical time λ in such a way that the trial surface approaches the apparent horizon asymptotically. The basic idea behind a flow algorithm is the following: We starts with a spherical trial surface that is considerably further out than the expected apparent horizon, and calculate the expansion H over the whole surface which, being well outside the horizon, should be positive everywhere. At each point, the trial surface is now moved inward a distance proportional to the value of H at that point. If we have a point on the trial surface with coordinates x^i, it will then move according to the equation

$$\frac{dx^i}{d\lambda} = -H s^i \; , \tag{6.7.27}$$

where s^i is the unit normal to the surface at that point. A flow algorithm of this type for finding apparent horizons was first proposed by Tod in 1991 [291]. In the case when the extrinsic curvature vanishes we have $H = \nabla_i s^i$, the apparent horizon then reduces to a minimal surface and the flow method is guaranteed to converge (the method is then known as *mean curvature flow*). In the more general case no such proof exists, but in practice flow algorithms do seem to converge to the apparent horizon.

If we parameterize our family of surfaces as $F(x^i, \lambda) = 0$, we then find

$$\frac{dF}{d\lambda} = \partial_\lambda F + \frac{dx^i}{d\lambda} \nabla_i F = 0 \,, \tag{6.7.28}$$

which using the flow equation can be reduced to

$$\partial_\lambda F = H s^i \nabla_i F = |\nabla F| \, H \,, \tag{6.7.29}$$

where we have used the fact that the normal vector is given in terms of F as $s^i = \nabla^i F / |\nabla F|$. Taking now as usual $F = r - h(\theta, \phi)$, the flow equation becomes

$$\partial_\lambda h = - |\nabla (r - h)| \, H \,. \tag{6.7.30}$$

We can now use this equation directly to evolve the function h in the unphysical time λ until the right hand side vanishes and the evolution reaches a stationary state, at which point we have found the apparent horizon.

The direct flow algorithm just described has one serious disadvantage: When we write the expansion H explicitly, it turns out to be a second order elliptic operator on the function $h(\theta, \phi)$. The flow equation then has the structure of a non-linear heat equation, so that numerical stability demands a very small time-step (the stability condition takes the form $\Delta\lambda \sim (\Delta x)^2$, with x any of the angular coordinates). For a reasonably high angular resolution the algorithm can then become too slow to use more than a few times during an evolution.

In order to improve on its speed, the basic flow algorithm can be generalized in a number of different ways. For example, we can use implicit integrators in the unphysical time λ (see for example [269]). Gundlach, however, has suggested a very different way of improving flow algorithms by considering a generalized flow equation of the form [152]

$$\partial_\lambda h = -A \left(1 - B L^2\right)^{-1} \rho H \,, \tag{6.7.31}$$

with ρ some scalar function constructed from $\{K_{ij}, g_{ij}, s^i\}$, and where L^2 denotes the angular part of the flat-space Laplace operator:

$$L^2 f := \frac{1}{\sin\theta} \, \partial_\theta \left(\sin\theta \, \partial_\theta\right) + \frac{1}{\sin^2\theta} \, \partial^2_\varphi f \,. \tag{6.7.32}$$

The expression $\left(1 - BL^2\right)^{-1}$ denotes the inverse of the operator $\left(1 - BL^2\right)$. We can compute this inverse explicitly by expanding the function h in terms of spherical harmonics as in (6.7.26). Gundlach's flow equation then becomes

$$\partial_\lambda a_{l,m} = -\frac{A}{1 + Bl(l+1)} \left(\rho H\right)_{l,m} \; . \tag{6.7.33}$$

This can now be solved by forward differencing to obtain the following iterative procedure

$$a_{l,m}^{(n+1)} = a_{l,m}^{(n)} - \frac{A}{1 + Bl(l+1)} \left(\rho H\right)_{l,m}^{(n)} \; . \tag{6.7.34}$$

Notice that if we take $\rho = |\nabla F|$, $A > 0$ and $B = 0$ we recover the standard flow algorithm. Gundlach shows that for values of B different from zero, and for other choices of ρ, we can obtain flows that are significantly faster (essentially because of the smoothing property of the $(1 - BL^2)^{-1}$ operator), giving rise to the term *fast flow* algorithms. In particular, the well known algorithm of Nakamura, Kojima and Oohara of 1984 [214] can be rewritten as a fast flow for a specific form of ρ in the limit when $A = B \to \infty$.

Flow algorithms have one important strength and one main drawback. Their strength is related to the fact that they are global algorithms is the sense that they do not need a good initial guess in order to find the apparent horizon. Their main disadvantage is the fact that, even in their faster versions, they are still considerably slower than the direct solvers we will discuss below (there is at least one major exception in the flow algorithm of Meztger which uses finite elements and is reported to be very fast [203]). However, flow algorithms are still used by a number of different groups and their conceptual simplicity will probably guarantee that they will remain in use for the foreseeable future.

6.7.3.3 *Direct elliptic solvers.* This family of algorithms takes the standard parameterization of the level surface $H(r, \theta, \phi) = r - h(\theta, \phi)$, and interprets the horizon equation (6.7.13) as a non-linear second order elliptic differential equation for the function h in the space of the two angular coordinates. The idea is then to solve this equation using standard techniques for solving non-linear elliptic problems. Typically, we can calculate all derivatives using finite differences and later use Newton's method to solve the resulting system of non-linear algebraic equations.

Many different versions of these direct elliptic solvers have been used to locate apparent horizons in three dimensions, and we will not go into the details here here (but see Thornburg's review for a more detailed description [287]). These types of finder have the advantage that they can be very fast indeed (for example, Thornburg reports that his direct elliptic solver finder can be over 30 times faster than a horizon finder based on Gundlach's fast-flow algorithm on the same data sets). The main disadvantage is that they need a good initial guess to converge rapidly, or even to converge at all (Newton's method can in fact *diverge* if the initial guess is too far off). Still, in many cases we usually have a reasonably good idea of the region where an apparent horizon is expected, so this is not such a serious drawback (in particular, during an evolution we can always use the last known location of the apparent horizon as an initial guess).

6.8 Event horizons

In contrast with apparent horizons, the event horizon is defined globally as the boundary between those light rays that reach infinity, and those that fall back into the singularity. The event horizon is therefore the true boundary of the black hole, so that locating it in a numerical simulation becomes important in order to decide when a black hole has really formed. The global nature of its definition, however, means that in principle we can only locate an event horizon if we know the entire evolution of the spacetime, though in practice it is sufficient to know the evolution of the spacetime up to the time when the final black hole has settled down.

The first studies of event horizons in numerical simulations go back to the work of Shapiro and Teukolsky in the early 1980s [262, 263, 265, 266]. In this pioneering work, the event horizon was located approximately by continuously shooting null geodesics outward and tracking them during the simulation. At the end of the simulation one would look back to check which of these null rays went back into the black hole interior and which moved outward, thus getting a rough idea of the position of the event horizon. Since this technique requires us to follow a representative number of the outgoing null geodesics, it is better adapted to systems with a high degree of symmetry where the possible directions in which light rays can move is limited. Tracking outgoing null rays in this way, however, has the serious disadvantage that the event horizon "repels" null lines in the sense that if we start slightly outside of it the corresponding light ray will quickly move outward, while if we start slightly inside the event horizon the light ray will rapidly fall inward toward the singularity.

Fortunately, we can use the fact that light rays rapidly diverge from the event horizon to our advantage by noticing that if we reverse the flow of time then the light rays will converge on the event horizon; in other words, the event horizon behaves as an *attractor* for null geodesics propagating backward in time. This was first realized by Anninos *et al.* [23] and Libson *et al.* [188], who proposed saving the data for a full numerical run and then integrating the outgoing null geodesics *backward in time* in order to locate the event horizon. As the event horizon attracts light rays when we go back in time, this technique is much more stable than the forward integration of null geodesics.

The backward in time integration of null geodesics, though a major improvement in itself, nevertheless has one drawback. This has to do with the fact that if we are not in spherical symmetry it is not immediately obvious what are the correct "inward" and "outward" directions, and any tangential motion of the null geodesics with respect to the event horizon can cause large inaccuracies to develop. For example, the different null geodesics can cross, or worse, they can leave the vicinity of the event horizon entirely. The problem, together with its solution, was already pointed out by Libson *et al.* in [188] (see also [296]). The idea is to consider the evolution not of independent null geodesics, but rather of a complete *null surface*, where by a null surface we mean a three-dimensional surface in spacetime such that its normal vectors $\vec{l}$ are null everywhere.

Just as we did when looking for apparent horizons, let us assume that we want to parameterize our null surface as a level set of some function $F(x^\mu)$ of the form $F = 0$, the main difference being that $F(x^\mu)$ is now a function of all four spacetime dimensions so that its level sets are in fact three-dimensional. Since we are looking for a horizon, we will also assume that F is constructed in such a way that the intersections of the different level sets of F with the spatial hypersurfaces are always closed two-dimensional surfaces. The condition for the surface to be null will just be

$$g^{\mu\nu}\partial_\mu F\,\partial_\nu F = 0\;, \tag{6.8.1}$$

where $g_{\mu\nu}$ is the full spacetime metric. Since we want to integrate F in time, let us explicitly separate its time and space derivatives:

$$g^{tt}\left(\partial_t F\right)^2 + 2g^{it}\partial_t F\,\partial_i F + g^{ij}\partial_i F\,\partial_j F = 0\;. \tag{6.8.2}$$

This is a quadratic equation for $\partial_t F$ which can be easily solved to find

$$\partial_t F = \frac{1}{g^{tt}}\left\{-g^{it}\partial_i F \pm \left[\left(g^{it}\partial_i F\right)^2 - g^{tt}g^{ij}\partial_i F\,\partial_j F\right]^{1/2}\right\}\;. \tag{6.8.3}$$

The positive sign in the above equation corresponds to outgoing light rays and the negative sign to ingoing ones. Since the event horizon is associated with outgoing light rays, from now on we will keep only the positive sign. In terms of 3+1 quantities, the null surface equation takes the simple form

$$\partial_t F = \beta^i\partial_i F - \alpha\left(\gamma^{ij}\partial_i F\partial_j F\right)^{1/2}\;. \tag{6.8.4}$$

In order to locate the horizon we then start with a given initial profile for the function $F(t=0)$ that grows monotonically outward, and then integrates the above equation back in time. Since the event horizon behaves as an attractor, the initial surface $F(t=0)=0$ can in principle correspond to any arbitrary closed surface. However, convergence will be faster if we choose F such that $F(t=0)=0$ is already close to the expected position of the event horizon. The location of the apparent horizon on the last slice of our numerical spacetime is a good initial guess, as for a black hole that is almost stationary it will be very close to the true event horizon (the apparent horizon will be somewhat inside the event horizon).

As a simple example, assume that we are in Minkowski spacetime. The above equation will then reduce to

$$\partial_t F = -\partial_r F\;. \tag{6.8.5}$$

If at $t=0$ we take $F = r - r0$, so that $F=0$ corresponds to a sphere of radius $r = r_0$, then the solution of this equation will be $F = r - r_0 - t$. Our null surface will then be given by $r = r_0 + t$. It is clear that this solution does not settle on any stationary surface as the light rays just keep moving out, which is expected since Minkowski has no event horizon (notice that since we have an exact solution there is no need to look at the evolution of the surface backward in time).

It is important to notice that, in contrast with the geodesic equation needed to integrate null rays, the above equation for the evolution of a null surface makes no reference to derivatives of the metric (no Christoffel symbols are present). This means that the numerical integration of this equation will be more accurate than the integration of individual geodesics. Another important property of this method is the fact that, even if we are focusing on the particular level set $F = 0$, all the different level sets of F will correspond to a separate null surface. This implies that by evolving the single scalar function F we are actually tracking a whole sequence of concentric light-fronts as they propagate through our spacetime.

The idea of tracking null surfaces backward in time in order to locate the event horizon is extremely robust, but there are two important issues that must be taken into account. The first has to do with the fact that, since the event horizon behaves as an attractor as we go back in time, light-fronts that are already close to it will approach it very slowly, while those further away will approach it much faster. This has the effect that close to the event horizon the function F will rapidly become very steep as the different level sets (*i.e.* different light-fronts) approach each other, resulting in large gradients that can cause the numerical integration to lose accuracy.

To see how this happens consider the case of a single Schwarzschild black hole in Kerr–Schild coordinates described by the metric (1.15.20), which corresponds to the 3+1 quantities

$$\alpha = \frac{1}{(1+2M/r)^{1/2}} \, , \qquad \beta^r = \frac{2M/r}{1+2M/r} \, , \qquad \gamma_{rr} = 1 + 2M/r \, . \tag{6.8.6}$$

The evolution equation for the null surface then becomes

$$\partial_t F = \left[\beta^r - \alpha \left(\gamma^{rr} \right)^{1/2} \right] \partial_r F = \frac{2M/r - 1}{1 + 2M/r} \, \partial_r F \, . \tag{6.8.7}$$

Remember now that we want to evolve this back in time, so we must in fact take

$$\partial_{\bar{t}} F = \frac{1 - 2M/r}{1 + 2M/r} \, \partial_r F \, , \tag{6.8.8}$$

with $\bar{t} = -t$. Figure 6.8 shows the evolution of the function F for a black hole of mass $M = 1$, starting from $F(\bar{t} = 0) = r - 2$ and evolving up to $\bar{t} = 1$. Since in this case the horizon position is precisely at $r = 2$, we expect the level set $F = 0$ to remain fixed at the horizon location. We can clearly see from the figure how this is indeed the case. Notice also how the other level sets rapidly approach the horizon from both sides causing the function F to become steeper very rapidly.

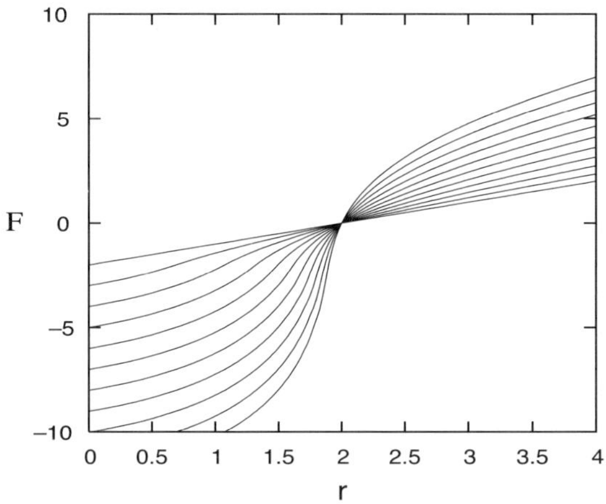

Fig. 6.8: Evolution of the function F backward in time for a Schwarzschild black hole with $M = 1$ in Kerr–Schild coordinates. The evolution is shown for a total time of $\bar{t} = 1$, with snapshots every $\Delta\bar{t} = 0.1$.

The problem of the steepening of the function F close to the event horizon is well known and can be solved by regularly re-initializing the function F without changing its zero level set. For example, in [109] Diener suggests re-initializing F whenever its gradient becomes too large by evolving the following equation in the unphysical time λ until a steady state is achieved:

$$\frac{dF}{d\lambda} = -\frac{F}{\sqrt{F^2+1}} \left(|\nabla F| - 1 \right) , \tag{6.8.9}$$

with $|\nabla F| := \sqrt{\gamma^{ij}\partial_i F \partial_j F}$. The last equation will have the effect of driving the magnitude of the gradient of F to 1 everywhere. The factor $F/\sqrt{F^2+1}$ is there to guarantee that the level set $F = 0$ does not change position during re-initialization, while at the same time limiting the size of the coefficient in front of $|\nabla F|$ since too large a coefficient would require a very small $\Delta\lambda$ to maintain numerical stability (this is related to the CFL stability condition, see Chapter 9). Notice, however, that as soon as we start re-initializing F, the level sets different from $F = 0$ will stop corresponding to null surfaces, so that in the end we will only be tracking the single null surface $F = 0$.

Another potential problem with the algorithm is related to the possibility of the event horizon changing topology as black holes merge. As we are integrating backward in time, what we will in fact see is a single horizon that splits into two separate ones. Simple geometry indicates that at the point where the split happens the scalar function F will necessarily develop a saddle point where the spatial gradient vanishes, so that according to equation (6.8.4) it will stop evolving (at the saddle point it is simply impossible to decide which direction is "outward"). Luckily, this seems to be one of those situations where the inaccuracies of numerical approximations come to our help: Numerically, we will never be *exactly* on the saddle point of F, so that the gradient will never vanish completely. Empirically we find that the surface simply evolves right through a topology change without any problem.

The discussion presented here has been very brief. For a much more detailed explanation of the null surface algorithm the reader should look at the recent paper by Diener on a general-purpose three-dimensional event horizon finder [109].

6.9 Isolated and dynamical horizons

As we have already mentioned, the formal definition of a black hole is based on the notion of an event horizon, which marks the boundary between those light rays that escape to infinity and those that do not. The global nature of this definition, however, means that it can not be used to characterize a black hole locally during an evolution. Recently, however, a new theoretical paradigm has emerged to describe black holes locally, based on the notions of *isolated* and *dynamical* horizons.[70] These notions allow us, for example, to define the mass and angular momentum associated with a black hole during a dynamical evolution, and have rapidly been adopted in numerical relativity simulations [113, 181]. The theory behind these new concepts is still being actively developed, and we are now at the point where numerical relativity is inspiring, and perhaps even driving, further theoretical developments.

There is already an extensive literature devoted to isolated and dynamical horizons [32, 33, 34, 38, 35, 36, 39, 148, 149]. Here we will just consider some of the basic notions and their application to numerical relativity. We will start by recalling the definition of a trapped surface: A closed two-dimensional surface S is said to be an *outer trapped surface* if the expansion of the outgoing null vectors normal to S is everywhere negative. In the special case when this expansion is zero the surface is called *marginally trapped.* An apparent horizon is then defined as the *outermost* marginally trapped surface.

We can prove that provided the weak energy condition holds (and in particular it does in vacuum), then the existence of a trapped surface implies the formation of a singularity. This means, in particular, that if cosmic censorship is true then a trapped surface implies the existence of an event horizon, *i.e.* a black hole (see *e.g.* [161]).

We now define a *non-expanding horizon* as a null three-dimensional hypersurface H, with topology $S^2 \times \Re$, that is foliated by marginally trapped surfaces (see Figure 6.9). Notice that if we have a stationary black hole, then necessarily the world-tube formed by simply stacking the apparent horizons at each spatial hypersurface will form a non-expanding horizon.

[70]One might argue that black holes are by definition non-local. However, the new concepts apply to geometric objects that are more closely related to the *astrophysical* notion of a black hole: something that forms as a result of gravitational collapse and has a horizon around it.

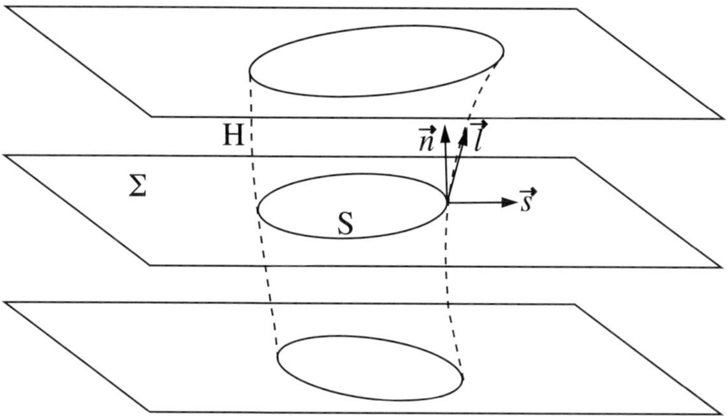

Fig. 6.9: On each spatial slice Σ the intersection of the hypersurface H is a marginally trapped surface S. If $\vec{s}$ is the unit normal to S in Σ, and $\vec{n}$ is the timelike unit normal to Σ, then $\vec{l} = (\vec{n} + \vec{s})/\sqrt{2}$ is an outgoing null vector. For a non-expanding horizon, $\vec{l}$ must be tangential to H.

Inspired by the notion of a non-expanding horizon, several authors have recently introduced different local concepts of "black hole". For example, in 1999 Ashtekar *et al.* introduced the concept of an *isolated horizon* which is defined as a non-expanding horizon H with the extra requirement that its intrinsic geometry (induced from the four dimensional metric) is not evolving along the null generators [33]. The distinction between non-expanding horizons and isolated horizons is rather technical, but it allows us to use a Hamiltonian formulation to define the mass and angular momentum for the horizon H. It turns out, however, that the explicit expressions for the mass and angular momentum are in fact independent of the extra structure of an isolated horizon, and hold true for a non-expanding horizon as well.

Without going into the details of the derivation, we will write here the final expressions of the mass and angular momentum for an isolated horizon. For the definition of the angular momentum we require that the surface S is axisymmetric, or in other words that there exists a Killing vector field $\vec{\varphi}$ on the horizon. If we take $\vec{s}$ to be the unit outward pointing normal to S, then the Killing condition implies that

$$\pounds_{\vec{\varphi}} \, q_{ij} = 0 \; , \tag{6.9.1}$$

where $q_{ij} := \gamma_{ij} - s_i s_j$ is the induced metric on the horizon. Notice that $\vec{\varphi}$ only needs to be defined on the horizon itself.

Once we have a Killing field $\vec{\varphi}$, the magnitude of the angular momentum on the horizon can be written in 3+1 terms as

$$J_H = \frac{1}{8\pi} \oint_S \varphi^l s^m K_{lm} dA \; , \tag{6.9.2}$$

with K_{ij} the extrinsic curvature of the tri-dimensional spatial hypersurfaces Σ and dA the area element on S. It is instructive to compare the expression for J_H above with the expression for the ADM angular momentum discussed in Appendix A, equation (A.8). Both expressions are essentially identical with the exception that one is calculated at the horizon and the other at infinity. The Killing field $\vec{\varphi}$ is needed in equation (6.9.2) in order to identify the correct angular coordinate around the horizon. In [113], Dreyer *at al.* describe an algorithm for finding an approximate Killing field $\vec{\varphi}$ on a closed surface S (which may or may not be an apparent horizon). This algorithm is now being used by a large number of groups doing numerical simulations of black holes spacetimes.

Having found the angular momentum, the horizon mass M_H turns out to be given by

$$M_H^2 = \frac{A_H}{16\pi} + \frac{4\pi J_H^2}{A_H} , \tag{6.9.3}$$

with A_H the area of the surface S. This is nothing more than the standard relation for a rotating black hole, equation (1.16.10). Notice, however, that this is not just a definition, but the result of the Hamiltonian approach.

The definition of isolated horizons, though extremely useful, has nevertheless the drawback that it only applies to black holes that are stationary. Because of this, theoretical work has now shifted to the study of the so-called *dynamical horizons*. In 1994 Hayward introduced the notion of a *future outer trapping horizon*, which is essentially a three-dimensional hypersurface foliated by marginally trapped surfaces [162]. This definition, though very similar to that of a non-expanding horizon, is nevertheless more general as it does not ask for the hypersurface to be null. Notice that in a dynamical situation we in fact expect the world-tube formed by the apparent horizons to be spacelike. This last statement can in fact be proven formally, but for our purposes it is enough to notice that if a black hole is dynamical, the area of the apparent horizon must grow (this is the so-called second law of black hole dynamics, see *e.g.* [160]), while the area of a null surface that momentarily coincides with the apparent horizon remains constant by definition (the expansion is zero). The apparent horizon must therefore move outwards faster than light, *i.e.* the world-tube of apparent horizons is spacelike. This result motivates the definition of a *dynamical horizon* as a spacelike three-dimensional hypersurface H with topology $S^2 \times \Re$ that is foliated by marginally trapped surfaces [37]. Notice that this definition is almost the same as that of a non-expanding horizon given above, with the important difference that now the hypersurface is not null, but is instead spacelike.

We can use the notion of a dynamical horizon to define fluxes and balance laws for mass and angular momentum on a horizon, and we can also define the angular momentum of the dynamical horizon itself [37, 72, 148]. In this case we still need to find a "rotation" vector $\vec{\varphi}$ which, however, does not need to be a Killing field anymore. It is enough now for the vector $\vec{\varphi}$ to be such that it has closed orbits on the horizon and zero divergence. This definition, unfortunately, is somewhat ambiguous since in general we can have many different vector fields on the horizon that satisfy these conditions, and each will give us a different value for the angular momentum.

As a final comment, it should be mentioned that the need to study the evolution of the direction as well as the magnitude of the angular momentum of a black hole during a dynamical simulation has recently led to the introduction of the so-called *coordinate angular momenta*, for which we simply take the standard Euclidean rotational vector fields around the coordinate axis, $\vec{\varphi}_x = (0, -z, y)$, $\vec{\varphi}_y = (z, 0, -x)$ and $\vec{\varphi}_z = (-y, x, 0)$, to define the three components of the angular momentum of the horizon [94]. These coordinate angular momenta are not well defined in any formal way, the main problem being that it is difficult to associate a "vector" with the horizon, because, in differential geometry vectors are only defined at points, and a single vector cannot be related clearly with a surface. Still, we would like to be able to recover the notion from classical mechanics that requires three numbers to completely define the angular momentum of a physical object. Keeping this in mind, the informal definition of coordinate angular momenta has already been found to provide useful physical information during the simulation of black hole collisions. We are now in a situation where current numerical simulations are demanding further theoretical developments. These are indeed exciting times.

7

RELATIVISTIC HYDRODYNAMICS

7.1 Introduction

In the previous Chapter we considered the simulation of black hole spacetimes, which corresponds to the study of the Einstein field equations in their purest form, *i.e.* in vacuum. It is clear, however, that most relativistic astrophysical systems involve matter sources: stars, accretion flows, jets, gravitational collapse, *etc.* Since many of these systems involve gases it is natural to model them using the theory of fluid dynamics. The fluid approximation describes matter as a continuum, this means that when we consider an "infinitesimal" fluid element we are in effect assuming that the fluid element is small compared to the system as a whole, but still big enough to contain a very large number of particles. In order for the fluid model to be a good approximation there are two basic conditions that must be satisfied: Both the mean distance between the particles *and* the mean free path between collisions must be much smaller than the macroscopic characteristic length of the system. Fluids are therefore particularly bad at modeling collisionless particles, unless their paths are not expected to cross, in which case we can use a fluid with zero pressure (known as "dust").

The state of the fluid is usually described in terms of the velocity field of the fluid elements v^i, the mass-energy density ρ, and the pressure p. There are two distinct approaches that can be used to follow the fluid motion. One possibility is to consider a fixed coordinate system and study the motion of the fluid as seen by observers at rest in this coordinate system – such an approach is called *Eulerian* (as in the Eulerian observers of the 3+1 formulation). The other possibility is to tie the coordinate system to the fluid elements themselves, corresponding to the so-called *Lagrangian* approach. The Lagrangian point of view has some advantages in the case of simple fluid motion (*e.g.* spherical collapse), as conservation of mass can be trivially guaranteed. However, a Lagrangian scheme becomes inadequate when the fluid motion has large shears since in that case the coordinate system can easily become entangled. Because of this, in the presentation given here we will always consider a Eulerian approach.

In the following Sections we will derive the dynamical equations governing the motion of a perfect fluid, known as the *Euler equations*, both in the special and general relativistic cases, and will also consider some simple equations of state. We will later discuss the hyperbolicity properties of the hydrodynamic system of evolution equations, and consider the concept of weak solutions and shock waves. Finally, we will discuss how to generalize the Euler equations to the case of imperfect fluids with heat conduction and viscosity. The description presented here will be very brief. There exists, however, a large literature dedicated exclusively to the study of fluid mechanics, most notably the beautiful book of Landau and Lifshitz [184]. There are also many books that discuss the numerical treatment of the hydrodynamic equations, particularly in the non-relativistic case. In the case of relativity there is the recent book by Wilson and Mathews on relativistic numerical hydrodynamics [300], and the review papers by Marti and Müller [200], and Font [129].

7.2 Special relativistic hydrodynamics

The starting point to the study of relativistic hydrodynamics is the stress-energy tensor for a perfect fluid, *i.e.* a fluid with zero viscocity and no heat conduction. Such a stress-energy tensor has already been introduced in equation (1.12.4) of Chapter 1, and has the form

$$T_{\mu\nu} = (\rho + p)\, u_\mu u_\nu + p\, \eta_{\mu\nu} \; , \tag{7.2.1}$$

where u^μ is the 4-velocity of the fluid elements (the average 4-velocity of the particles), ρ and p are the energy density and pressure as measured in the fluid's rest frame, and where, for the moment, we have assumed that we are in special relativity so the metric is given by the Minkowski tensor $g_{\mu\nu} = \eta_{\mu\nu}$.

The stress-energy tensor above is usually written in a simplified form by first separating the total energy density ρ into contributions coming from the *rest mass energy density* ρ_0 and the internal energy:

$$\rho = \rho_0 \,(1 + \epsilon) \; , \tag{7.2.2}$$

where ϵ is the *specific internal energy* (internal energy per unit mass) of the fluid. Let us now introduce the so-called *specific enthalpy* of the fluid defined as[71]

$$h := 1 + \epsilon + \frac{p}{\rho_0} \; . \tag{7.2.3}$$

In terms of h, the stress-energy tensor then takes the simple form

$$T_{\mu\nu} = \rho_0 h\, u_\mu u_\nu + p\, \eta_{\mu\nu} \; . \tag{7.2.4}$$

The rest mass energy density is also often written in terms of the particle number density n as

$$\rho_0 = nm \; , \tag{7.2.5}$$

with m the rest mass of the fluid particles.

[71] In thermodynamics the enthalpy H is defined as the sum of the internal energy U plus the pressure times the volume, $H = U + pV$. In other words, the enthalpy represents the total energy in the system capable of doing mechanical work. In relativity we also add the rest mass energy M to the definition of enthalpy, so that $H = M + U + pV = M(1+\epsilon) + pV$. The *specific enthalpy* is then defined as the enthalpy per unit mass: $h = H/M = 1 + \epsilon + p/\rho_0$.

A note about the interpretation of the fluid variables is important at this point. In the case of general relativity, the 3+1 evolution equations and the Hamiltonian constraint involve the energy density of matter as a source of the gravitational field. However, this energy density is assumed to be measured in the Eulerian reference frame (the one associated with the observers whose 4-velocity is normal to the spatial hypersurface), which will in general differ from the fluid's frame of reference (the Lagrangian frame). In order to avoid confusion we will from now on denote the energy density that appears in the ADM equations by $\rho_{\rm ADM}$. We can derive the relationship between the different energy densities by starting from the definition of $\rho_{\rm ADM}$:

$$\rho_{\rm ADM} := n^\mu n^\nu T_{\mu\nu} \,, \tag{7.2.6}$$

with n^μ the unit normal to the spacelike hypersurfaces. Substituting the above stress-energy tensor here and using the fact that $n_\mu n^\mu = -1$ we find

$$\rho_{\rm ADM} = \rho_0 h \left(u_\mu n^\mu\right)^2 - p = \rho_0 h W^2 - p \,, \tag{7.2.7}$$

where we have defined

$$W := -u^\mu n_\mu = u^0 \,. \tag{7.2.8}$$

The last equality follows from the fact that, in special relativity, we have $n_\mu = (-1, 0)$. Finally, using the fact that $u_\mu u^\mu = -1$ we obtain

$$W = \left(1 + \sum_i u^i\right)^{1/2} \,. \tag{7.2.9}$$

But this is nothing more than the Lorentz factor $1/\sqrt{1-v^2}$, since the standard three-dimensional speed of the fluid is given by

$$v^i = u^i/u^0 = \left(1 - v^2\right)^{1/2} u^i \,, \tag{7.2.10}$$

which can be shown to imply

$$W = 1/\sqrt{1-v^2} \,. \tag{7.2.11}$$

In the particular case when the local coordinates follow the fluid element we have $W = 1$, and the energy densities become equal:

$$\rho_{\rm ADM} = \rho_0 h W^2 - p = \rho_0 h - p = \rho_0 (1 + \epsilon) = \rho \,. \tag{7.2.12}$$

Notice, however, that if the flow is non-uniform we can not adapt the coordinate system to follow the fluid elements (the Lagrangian approach) without being forced to replace the Minkowski metric $\eta_{\mu\nu}$ with a more general metric $g_{\alpha\beta}$, since the fluid motion will in general deform the volume elements.

The state of the fluid at any given time is given in terms of the six variables $(\rho_0, \epsilon, p, v^i)$, which from now on will be called the *primitive variables.* The evolution equations for the fluid now follow from the conservation laws. We have in fact two sets of conservation laws, namely the conservation of particles and the conservation of energy-momentum:

$$\partial_\mu \left(\rho_0 u^\mu\right) = 0 \; , \tag{7.2.13}$$

$$\partial_\mu T^{\mu\nu} = 0 \; . \tag{7.2.14}$$

Notice that these conservation laws provide us with five equations. In order to close the system we therefore need an equation of state which can be assumed to be of the form

$$p = p\left(\rho_0, \epsilon\right) \; . \tag{7.2.15}$$

To proceed let us now introduce the quantity

$$D := \rho_0 W \; , \tag{7.2.16}$$

which is nothing more than the rest mass density as seen in the Eulerian frame. The conservation of particles now implies

$$\partial_t D + \partial_k \left(D v^k\right) = 0 \; . \tag{7.2.17}$$

This is known as the continuity equation and has exactly the same form as in the Newtonian case, but now D includes the relativistic correction coming from the Lorentz factor W. The continuity equation can be interpreted as an evolution equation for D.

For the conservation of momentum we first define the quantities

$$S^\mu := \rho_0 h W u^\mu \; . \tag{7.2.18}$$

Notice that since $u^i = W v^i$, the spatial components $S^i = \rho_0 h W^2 v^i$ are nothing more that the momentum density as seen in the Eulerian frame, with the correct Lorentz factors (the fact that the enthalpy h appears in the expression for S^i shows that in relativity the pressure contributes to the momentum density). In terms of S^μ, the mixed components of the stress-energy tensor become

$$T^\mu_\nu = \frac{S_\nu u^\mu}{W} + p\,\delta^\mu_\nu \; , \tag{7.2.19}$$

The conservation of momentum then takes the form

$$\begin{aligned} &\partial_\mu \left(S_i u^\mu / W + p\,\delta^\mu_i\right) = 0 \\ \Rightarrow \quad &\partial_t S_i + \partial_k \left(S_i v^k\right) + \partial_i p = 0 \; . \end{aligned} \tag{7.2.20}$$

These are the evolution equations for the momentum density and are known as the *Euler equations.* Notice that they have a structure similar to that of the continuity equation, but now there is an extra term given by the gradient of the pressure. The momentum density can then change both because of the flow of momentum out of the volume element represented by the term $\partial_j \left(S_i v^j\right)$, and because of the existence of a force given by the gradient of the pressure $\partial_i p$. The Euler equations above have again exactly the same form as in the Newtonian case, but now the definition of the momentum density S_i includes the relativistic corrections.

We are still missing an evolution equation for the energy density. Such an equation can be obtained in a number of different ways. Experience has shown that it is in fact convenient to subtract the rest mass energy density in order to have higher accuracy, since for systems that are not too relativistic the rest mass can dominate the total energy density. However, there are several non-equivalent ways to do this. As a first approach, consider the internal energy density as measured in the Eulerian frame:

$$E = \rho_0 \epsilon W \; . \tag{7.2.21}$$

Notice that there is only one Lorentz factor W coming from the Lorentz contraction of the volume elements, since the specific internal energy ϵ can be considered a scalar (this is by definition the internal energy per particle in the fluid's frame). In order to derive an evolution equation for E we first notice that the conservation equations imply that

$$\partial_\mu \left(u_\nu T^{\mu\nu}\right) = T^{\mu\nu} \partial_\mu u_\nu \; . \tag{7.2.22}$$

Substituting here the expression for the stress-energy tensor, and remembering that $u_\mu u^\mu = -1$ implies $u_\mu \partial_\nu u^\mu = 0$, we find

$$\partial_\mu \left(u_\nu T^{\mu\nu}\right) = p \, \partial_\mu u^\mu \; . \tag{7.2.23}$$

On the other hand

$$u_\nu T^{\mu\nu} = -\rho_0 \left(1 + \epsilon\right) u^\mu \; , \tag{7.2.24}$$

and using now the conservation of particles, this implies

$$\partial_\mu \left(u_\nu T^{\mu\nu}\right) = -\partial_\mu \left(\rho_0 \epsilon u^\mu\right) \; . \tag{7.2.25}$$

Collecting results we obtain

$$\partial_\mu \left(\rho_0 \epsilon u^\mu\right) + p \, \partial_\mu u^\mu = 0 \; , \tag{7.2.26}$$

which can be rewritten as

$$\partial_t E + \partial_k \left(E v^k\right) + p \left[\partial_t W + \partial_k \left(W v^k\right)\right] = 0 \; . \tag{7.2.27}$$

This equation has been used successfully by Wilson and collaborators to evolve relativistic fluids (see *e.g.* [300]). However, as an evolution equation for E it has one serious drawback, namely that it also involves the time derivative of W, so that it can not be written as a balance law, which in particular makes it impossible to use for analyzing the characteristic structure of the system.

Fortunately, there exists an alternative way of subtracting the rest mass energy from the system that does yield an equation in balance law form. We can simply decide to evolve instead the difference between the total energy density and the mass energy density as measured in the Eulerian frame:

$$\mathcal{E} := \rho_{\mathrm{ADM}} - \rho_0 W = \rho_{\mathrm{ADM}} - D = \rho_0 h W^2 - p - D \; . \tag{7.2.28}$$

Notice that the energies E and $\mathcal{E}$ differ since E does not include contributions from the kinetic energy while $\mathcal{E}$ does. To find the evolution equation for $\mathcal{E}$ we first notice that from the definition of S^μ we have $S^0 = \rho_0 h W^2$. The conservation of energy then takes the form

$$0 = \partial_\mu T^{0\mu} = \partial_\mu \left(S^0 u^\mu / W + p \eta^{0\mu} \right) \; , \tag{7.2.29}$$

which immediately yields

$$\partial_t S^0 + \partial_k \left(S^0 v^k \right) - \partial_t p = 0 \; . \tag{7.2.30}$$

Using now the evolution equation for D we finally find

$$\partial_t \mathcal{E} + \partial_k \left[(\mathcal{E} + p) \, v^k \right] = 0 \; , \tag{7.2.31}$$

where we used the fact that $S^0 = \mathcal{E} + D + p$.

Our set of evolution equations then becomes a system of conservation laws of the form[72]

$$\partial_t D + \partial_k \left(D v^k \right) = 0 \; , \tag{7.2.32}$$

$$\partial_t S_i + \partial_k \left(S_i v^k + p \, \delta^k_i \right) = 0 \; , \tag{7.2.33}$$

$$\partial_t \mathcal{E} + \partial_k \left[(\mathcal{E} + p) \, v^k \right] = 0 \; , \tag{7.2.34}$$

with the conserved quantities $(D, S_i, \mathcal{E})$ given in terms of the primitive quantities $(\rho_0, \epsilon, p, v^i)$ as

$$D = \rho_0 W \; , \qquad S_i = \rho_0 h W^2 v_i \; , \qquad \mathcal{E} = \rho_0 h W^2 - p - \rho_0 W \; . \tag{7.2.35}$$

Note that the Euler equations are frequently written for the speed v_i instead of the flux S_i and have the form (see *e.g.* [297])

$$\partial_t v_i + v^k \partial_k v_i = - \left[\partial_i p + v_i \partial_t p \right] / (\rho_0 h W^2) \; . \tag{7.2.36}$$

These equations can be easily derived by combining the evolution equations for D, S_i and $\mathcal{E}$. However, they are not as convenient as the evolution equations for

[72]The hydrodynamic equations in the conservative form given here were first derived by Marti, Ibañez, and Miralles at the University of Valencia in Spain and are often called the *Valencia formulation* of relativistic hydrodynamics [130, 198] (see also [129, 200]).

S_i since they are not written as conservation laws, and in particular involve the time derivative of the pressure.

There is another important consequence of the conservation equations. Consider the contraction

$$u_\mu \partial_\nu T^{\mu\nu} = 0 \, . \tag{7.2.37}$$

Substituting the expression for $T^{\mu\nu}$, and using again the fact that $u_\mu \partial_\nu u^\mu = 0$, we find

$$u^\mu \partial_\mu p - \partial_\nu \left(\rho_0 h u^\mu \right) = 0 \, . \tag{7.2.38}$$

This can be further simplified with the help of the equation for conservation of particles to

$$u^\mu \partial_\mu p - \rho_0 u^\mu \partial_\mu h = 0 \, , \tag{7.2.39}$$

and using now the expression for the specific enthalpy h we finally obtain

$$\frac{d\epsilon}{d\tau} + p \, \frac{d}{d\tau} \left(\frac{1}{\rho_0} \right) = 0 \, , \tag{7.2.40}$$

where $d/d\tau := u^\mu \partial_\mu$ is the derivative along the trajectory of the fluid elements. This equation is in fact nothing more than the local version of the first law of thermodynamics. To see this, consider a fluid element with rest mass M, internal energy U, and volume V. We then have in general that

$$\rho_0 = \frac{M}{V} \quad \Rightarrow \quad dV = M d \left(\frac{1}{\rho_0} \right) \, , \tag{7.2.41}$$

and similarly

$$\epsilon = \frac{U}{M} \quad \Rightarrow \quad dU = M d\epsilon \, . \tag{7.2.42}$$

The first law of thermodynamics then implies that

$$dQ = dU + p dV = M \left[d\epsilon + p \, d \left(\frac{1}{\rho_0} \right) \right] \, . \tag{7.2.43}$$

This shows that (7.2.40) is precisely the first law of thermodynamics for a fluid element for which $dQ = 0$ (this is to be expected since by definition a perfect fluid has no heat conduction). And since in general $dQ = TdS$, with T the temperature and S the entropy of the fluid, we see that a perfect fluid behaves in such a way that entropy is preserved along flow lines.

Let us now go back and consider the relation between the primitive and conserved variables (7.2.35). In the Newtonian limit these relations reduce to $D = \rho_0$, $S_i = \rho_0 v_i$ and $\mathcal{E} = \rho_0(\epsilon + v^2/2)$, so that they are very easy to invert. In the relativistic case, however, inverting the relations becomes much more difficult since first W involves v^2, and also the pressure appears explicitly in the expression for

$\mathcal{E}$, so the equation of state is needed in order to recover the primitive variables. Unfortunately, the evolution equations for the conserved quantities involve the primitive variables directly, so that these must be recovered every time step. This requires that an algorithm for recovering such variables is implemented. Such an algorithm starts by choosing some trial value of the pressure p^* (for example the old value at the corresponding grid cell). Then, from the expressions for D and $\mathcal{E}$ we can recover the speed v^i in the following way

$$v_i(p^*) = \frac{S_i}{\rho_0 h W^2} = \frac{S_i}{\mathcal{E} + D + p^*} \,. \tag{7.2.44}$$

Having found v_i we then compute W as

$$W(p^*) = \frac{1}{\sqrt{1 - v^2(p^*)}} \,. \tag{7.2.45}$$

This allows us to find the density ρ_0 as

$$\rho_0(p^*) = \frac{D}{W(p^*)} \,. \tag{7.2.46}$$

Finally, we find the specific internal energy ϵ through the definition of h

$$\begin{aligned} \epsilon(p^*) &= \frac{\mathcal{E} + \rho_0 W\,(1 - W) + p^*\left(1 - W^2\right)}{\rho_0 W^2} \\ &= \frac{\mathcal{E} + D\,(1 - W(p^*)) + p^*\left(1 - W^2(p^*)\right)}{D W(p^*)} \end{aligned} \tag{7.2.47}$$

Of course, the chosen value of p^* will almost certainly not satisfy the equation of state $p = p(\rho_0, \epsilon)$, so we must now evaluate the residual $r(p^*)$ defined as

$$r(p^*) := p(\rho_0(p^*), \epsilon(p^*)) - p^* \,, \tag{7.2.48}$$

and change the value of p^* until this residual vanishes. This can typically be accomplished by standard non-linear root-finding techniques (*e.g.* one-dimensional Newton–Raphson). For some simple equations of state, such as that of an ideal gas discussed in Section 7.5, the whole procedure can in fact be done analytically and involves finding the physically admissible root of a high order polynomial (a fourth order polynomial in the case of an ideal gas). However, this is typically more computationally expensive than using the non-linear root finder directly.

7.3 General relativistic hydrodynamics

The generalization of the evolution equations (7.2.32)–(7.2.34) to the case of a non-trivial gravitational field is rather straightforward. We again start from the stress-energy tensor for a perfect fluid, but now for an arbitrary metric $g_{\mu\nu}$:

$$T_{\mu\nu} = \rho_0 h u_\mu u_\nu + p\, g_{\mu\nu} \,, \tag{7.3.1}$$

where as before u^μ is the 4-velocity of the fluid elements, ρ_0 is the rest mass energy density measured in the fluid's rest frame, p is the pressure and h is the specific enthalpy

$$h := 1 + \epsilon + \frac{p}{\rho_0} \, , \tag{7.3.2}$$

with ϵ the specific internal energy. The evolution equations for the fluid again follow from the conservation laws, which however now take the form

$$\nabla_\mu \left(\rho_0 u^\mu \right) = 0 \, , \tag{7.3.3}$$

$$\nabla_\mu T^{\mu\nu} = 0 \, . \tag{7.3.4}$$

Using the fact that the divergence of a vector can be written in general as

$$\nabla_\mu \xi^\mu = \frac{1}{\sqrt{-g}} \partial_\mu \left(\sqrt{-g} \, \xi^\mu \right) \, , \tag{7.3.5}$$

with g the determinant of the metric tensor $g_{\mu\nu}$, we can immediately rewrite the conservation of particles as

$$\partial_\mu \left(\sqrt{-g} \, \rho_0 u^\mu \right) = 0 \, , \tag{7.3.6}$$

and the conservation of energy and momentum as

$$\partial_\mu \left(\sqrt{-g} \, T^\mu_\nu \right) = \sqrt{-g} \, \Gamma^\alpha_{\mu\nu} T^\mu_\alpha \, , \tag{7.3.7}$$

with $\Gamma^\alpha_{\mu\nu}$ the Christoffel symbols associated with the metric $g_{\mu\nu}$.

We now assume that we are using a standard 3+1 decomposition of spacetime, in which case we find

$$g = -\alpha^2 \gamma \quad \Rightarrow \quad \sqrt{-g} = \alpha \sqrt{\gamma} \, , \tag{7.3.8}$$

with α the lapse function and γ the determinant of the spatial metric γ_{ij}.

Just as we did in special relativity, let us define the scalar parameter W as

$$W := -u^\mu n_\mu \, , \tag{7.3.9}$$

with n^μ the unit normal to the spatial hypersurfaces. In this case we have $n_\mu = (-\alpha, 0)$, so that

$$W = \alpha u^0 \, . \tag{7.3.10}$$

Define now

$$v^i := \frac{u^i}{\alpha u^0} + \frac{\beta^i}{\alpha} \, , \tag{7.3.11}$$

with β^i the shift vector. With this definition v^i is precisely the speed of the fluid elements as seen by the Eulerian observers. To see this notice that u^i/u^0 is the coordinate speed of the fluid elements, so we first need to add the shift to go to the Eulerian reference frame and then divide by the lapse to use the proper time of the Eulerian observers instead of coordinate time. Notice also that since u^μ is a 4-vector while v^i is only a 3-vector, when we lower the indices we have

$$\begin{aligned} u_i &= g_{i\mu} u^\mu = \beta_i u^0 + \gamma_{ik} u^k \\ &= \beta_i u^0 + \gamma_{ik} u^0 \left(\alpha v^k - \beta^k\right) = \alpha u^0 v_i = W v_i \ . \end{aligned} \tag{7.3.12}$$

Using again the fact that $u_\mu u^\mu = -1$, we find that W takes the simple form

$$W = 1/\sqrt{1 - v^2} \ , \tag{7.3.13}$$

where now $v^2 := \gamma_{ij} v^i v^j$, *i.e.* W is again the Lorentz factor as seen by the Eulerian observers. Define again D as the rest mass density measured by the Eulerian observers

$$D := \rho_0 W \ , \tag{7.3.14}$$

we can then rewrite the conservation of particles as

$$\partial_t \left(\sqrt{\gamma}\, D\right) + \partial_k \left[\sqrt{\gamma}\, D \left(\alpha v^k - \beta^k\right)\right] = 0 \ . \tag{7.3.15}$$

This is again a conservation law for D, but in contrast to the case of special relativity it involves the lapse α, the shift vector β^k, and the determinant of the spatial metric γ (compare with equation (7.2.32)).

For the conservation of momentum we again introduce the quantities

$$S^\mu := \rho_0 h W u^\mu \ , \tag{7.3.16}$$

and rewrite the stress-energy tensor as

$$T^\mu_\nu = \frac{u^\mu S_\nu}{W} + p\, \delta^\mu_\nu \ . \tag{7.3.17}$$

The conservation of momentum then becomes

$$\partial_t \left(\sqrt{\gamma} S_i\right) + \partial_k \left\{\sqrt{\gamma} \left[S_i \left(\alpha v^k - \beta^k\right) + \alpha p\, \delta^k_i\right]\right\} = \alpha \sqrt{\gamma}\, \Gamma^\mu_{\nu i} T^\nu_\mu \ , \tag{7.3.18}$$

where $S_i = \rho_0 h W u_i = \rho_0 h W^2 v_i$. The last equations are the general relativistic version of the Euler equations. When we compare them with their special relativistic version (7.2.33) we see that in the case of general relativity, apart from the lapse and shift factors that correct for the motion of the Eulerian observers, we don't have strict conservation of momentum anymore since there is a source term on the right hand side that represents the gravitational forces.

Finally, for the conservation of energy we again start by defining

$$\mathcal{E} = \rho_0 h W^2 - p - D \ . \tag{7.3.19}$$

The conservation of energy then takes the form

$$\partial_\mu \left(\alpha \sqrt{\gamma}\, T^{0\mu}\right) = -\alpha \sqrt{\gamma}\, \Gamma^0_{\mu\nu} T^{\mu\nu} \ . \tag{7.3.20}$$

It is in fact convenient to rewrite the term on the left hand side as

$$\partial_\mu \left(\alpha\sqrt{\gamma}\, T^{0\mu}\right) = \frac{1}{\alpha}\partial_\mu \left(\alpha^2\sqrt{\gamma}\, T^{0\mu}\right) - \sqrt{\gamma}\, T^{0\mu}\partial_\mu\alpha \; , \tag{7.3.21}$$

so that the conservation of energy becomes

$$\partial_\mu \left(\alpha^2\sqrt{\gamma}\, T^{0\mu}\right) = \alpha^2\sqrt{\gamma}\left(T^{0\mu}\partial_\mu \ln\alpha - \Gamma^0_{\mu\nu}T^{\mu\nu}\right) \, . \tag{7.3.22}$$

Using now the expression for $T^{0\mu}$ we obtain, after some algebra,

$$\begin{aligned}\partial_\mu \left(\alpha^2\sqrt{\gamma}\, T^{0\mu}\right) &= \partial_t \left[\sqrt{\gamma}\left(\rho_0 h W^2 - p\right)\right] \\ &\quad + \partial_k \left\{\sqrt{\gamma}\left[\rho_0 h W^2\left(\alpha v^k - \beta^k\right) + p\beta^k\right]\right\} \, ,\end{aligned} \tag{7.3.23}$$

and using the evolution equation for D we find that the final expression for the conservation of energy takes the form

$$\begin{aligned}\partial_t \left(\sqrt{\gamma}\,\mathcal{E}\right) &+ \partial_k \left\{\sqrt{\gamma}\left[\mathcal{E}\left(\alpha v^k - \beta^k\right) + \alpha p v^k\right]\right\} \\ &= \alpha^2\sqrt{\gamma}\left(T^{0\mu}\partial_\mu \ln\alpha - \Gamma^0_{\mu\nu}T^{\mu\nu}\right) \, .\end{aligned}$$

The final set of evolution equations is then

$$\partial_t \left(\sqrt{\gamma}\, D\right) + \partial_k \left[\sqrt{\gamma}\, D\left(\alpha v^k - \beta^k\right)\right] = 0 \; , \tag{7.3.24}$$

$$\partial_t \left(\sqrt{\gamma}\, S_i\right) + \partial_k \left\{\sqrt{\gamma}\left[S_i\left(\alpha v^k - \beta^k\right) + \alpha p \delta^k_i\right]\right\} = \alpha\sqrt{\gamma}\,\Gamma^\mu_{\nu i}T^\nu_\mu \; , \tag{7.3.25}$$

$$\begin{aligned}\partial_t \left(\sqrt{\gamma}\,\mathcal{E}\right) + \partial_k \left\{\sqrt{\gamma}\left[\mathcal{E}\left(\alpha v^k - \beta^k\right) + \alpha p v^k\right]\right\} &= \alpha^2\sqrt{\gamma}\left(T^{0\mu}\partial_\mu \ln\alpha\right. \\ &\quad \left. - \Gamma^0_{\mu\nu}T^{\mu\nu}\right) \, ,\end{aligned} \tag{7.3.26}$$

where the conserved quantities $(D, S_i, \mathcal{E})$ and primitive variables $(\rho_0, \epsilon, p, v^i)$ are related through

$$D = \rho_0 W \; , \qquad S_i = \rho_0 h W^2 v_i \; , \qquad \mathcal{E} = \rho_0 h W^2 - p - D \, . \tag{7.3.27}$$

Notice that the system of equations (7.3.24)–(7.3.26) reduces to the special relativistic counterpart (7.2.32)-(7.2.34) when we take $\alpha = 1$, $\beta^i = 0$ and $\gamma_{ij} = \delta_{ij}$, in which case $\Gamma^\alpha_{\mu\nu} = 0$ and the system is truly conservative. The presence of a non-trivial gravitational field implies that there is no longer true conservation of energy and momentum, but the equations are still in the form of balance laws: $\partial_t u + \partial_k F^k(u) = s(u)$.

Before concluding this Section, it is important to write down the relation between the quantities $(D, S_i, \mathcal{E})$ and the matter terms measured by the Eulerian observers that appear in the ADM equations, namely the energy density $\rho_{\rm ADM}$, the momentum density $j^i{}_{\rm ADM}$ and the stress tensor $S^{ij}{}_{\rm ADM}$. Using the expression for $T^{\mu\nu}$ we find

$$\rho_{\rm ADM} := n^\mu n^\nu T_{\mu\mu} = \rho_0 h W^2 - p = \mathcal{E} + D \; , \tag{7.3.28}$$

$$j^i{}_{\rm ADM} := -n^\mu P^{\nu i} T_{\mu\nu} = \rho_0 h W^2 v^i = S^i \; , \tag{7.3.29}$$

$$S^{ij}{}_{\rm ADM} := P^{\mu i} P^{\nu j} T_{\mu\nu} = \rho_0 h W^2 v^i v^j + \gamma^{ij} p \; , \tag{7.3.30}$$

where $P^{\mu\nu} = g^{\mu\nu} + n^\mu n^\nu$ is the standard projection operator onto the spatial hypersurfaces.

7.4 3+1 form of the hydrodynamic equations

The general relativistic hydrodynamic evolution equations (7.3.24)–(7.3.26) derived in the previous Section have been written as a set of balance laws, which has some advantages from a numerical point of view. However, from the perspective of the 3+1 formulation of general relativity they are not written in the most convenient form as they are not manifestly 3-covariant. In this Section we will rewrite these equations as tensor equations in 3+1 language.

Let us start from the evolution equation for the rest mass density D. Since this is just the particle density as seen by the Eulerian observers, times the rest mass of the individual particles, it is in fact a scalar in 3+1 terms. The original evolution equation has the form

$$\partial_t \left(\sqrt{\gamma}\, D\right) + \partial_k \left[\sqrt{\gamma}\, D \left(\alpha v^k - \beta^k\right)\right] = 0 \; . \tag{7.4.1}$$

Notice first that, for any three-dimensional vector w^i we have

$$\frac{1}{\sqrt{\gamma}}\, \partial_k \left(\sqrt{\gamma} w^k\right) = D_k w^k \; , \tag{7.4.2}$$

with D_k the covariant derivative associated with the spatial metric γ_{ij}. This implies that

$$\frac{1}{\sqrt{\gamma}}\, \partial_k \left[\sqrt{\gamma}\, D \left(\alpha v^k - \beta^k\right)\right] = D_k \left(\alpha D v^k\right) - \left(D_k \beta^k\right) D - \beta^k \partial_k D \; . \tag{7.4.3}$$

On the other hand

$$\frac{1}{\sqrt{\gamma}}\, \partial_t \left(\sqrt{\gamma}\, D\right) = \partial_t D + \frac{D}{2}\, \partial_t \ln \gamma \; . \tag{7.4.4}$$

Now, from the ADM evolution equations for the spatial metric (2.3.12), we can easily find that

$$\partial_t \ln \gamma = -2\alpha K + 2 D_k \beta^k \; , \tag{7.4.5}$$

with K the trace of the extrinsic curvature K_{ij}. Collecting results we find that the evolution equation for D in 3+1 language takes the final form

$$\partial_t D - \beta^k \partial_k D + D_k \left(\alpha D v^k\right) = \alpha K D \; . \tag{7.4.6}$$

This equation is clearly a scalar equation. The different terms are easy to interpret: The shift appears only in the advection term, as it should, since the only role of the shift is to move the coordinate lines. The last term on the left hand side shows that the change in D along the normal lines is given essentially by the divergence of the flux of particles. Finally, the source term shows that the density of particles D can also change because of an overall change in the spatial volume elements. For example, in the case of cosmology the so-called cosmological fluid is co-moving with the Eulerian observers so that $v^i = \beta^i = 0$, but the density of particles still becomes smaller with time because of the overall expansion of the Universe ($K < 0$).

It is also interesting to note that in the last equation the shift appears as an advection term that is not in flux-conservative form (the shift is outside the spatial derivatives). The flux conservative form (7.3.24) comes about because as we bring the shift vector into the spatial derivative we pick up a term with the divergence of the shift. This term is canceled by a corresponding term coming from the time derivative of the volume element. This shows that, quite generally, advection terms on the shift can be transformed into flux conservative type terms by bringing a $\sqrt{\gamma}$ factor into the time derivative.

Consider next the evolution equation for $\mathcal{E}$. Again, since this is by definition the energy density measured by the Eulerian observers minus the rest mass density, $\mathcal{E} = \rho_{\text{ADM}} - D = \rho_0 h W^2 - p - D$, it is clearly a scalar quantity in 3+1 terms. Its original evolution equation is

$$\begin{aligned}\partial_t \left(\sqrt{\gamma}\, \mathcal{E}\right) + \partial_k \left\{\sqrt{\gamma} \left[\mathcal{E}\left(\alpha v^k - \beta^k\right) + \alpha p v^k\right]\right\} \\ = \alpha^2 \sqrt{\gamma} \left(T^{0\mu} \partial_\mu \ln \alpha - \Gamma^0_{\mu\nu} T^{\mu\nu}\right) .\end{aligned}$$

Let us first look at the source term. Using the expression for the stress-energy tensor (7.3.1), the definition of v^i (7.3.11), and the expressions for the 4-Christoffel symbols in 3+1 language found in Appendix B, it is not difficult to show that

$$\begin{aligned}\alpha^2 \left(T^{0\mu} \partial_\mu \ln \alpha - \Gamma^0_{\mu\nu} T^{\mu\nu}\right) &= \rho_0 h W^2 \left(\alpha v^m v^n K_{mn} - v^m \partial_m \alpha\right) + \alpha p K \\ &= \left(\mathcal{E} + p + D\right) \left(\alpha v^m v^n K_{mn} - v^m \partial_m \alpha\right) + \alpha p K . \qquad (7.4.7)\end{aligned}$$

We can now rewrite the left hand side of the evolution equation for $\mathcal{E}$ in exactly the same way as the evolution equation for D. We then find the following evolution equation for $\mathcal{E}$ in 3+1 form

$$\begin{aligned}\partial_t \mathcal{E} - \beta^k \partial_k \mathcal{E} + D_k \left[\alpha v^k \left(\mathcal{E} + p\right)\right] &= \left(\mathcal{E} + p + D\right) \left(\alpha v^m v^n K_{mn} - v^m \partial_m \alpha\right) \\ &\quad + \alpha K \left(\mathcal{E} + p\right) . \qquad (7.4.8)\end{aligned}$$

The last term on the right hand side is interesting. Assume that we have a fluid that is co-moving with the Eulerian observers so that $v^k = \beta^k = 0$, we then find that $\partial_t \mathcal{E} = \alpha K (\mathcal{E} + p)$. This shows that the internal energy density changes both as a reflection of a simple change in the volume elements ($\alpha K \mathcal{E}$), and because of the existence of a non-zero pressure ($\alpha p K$). But we know that $\alpha K = -\partial_t \ln \sqrt{\gamma}$, so that $\alpha p K = -p\, \partial_t \ln \sqrt{\gamma}$, which is nothing more than the work done by the fluid as space expands. That is, the term $\alpha p K$ in the source term is there in accordance with the first law of thermodynamics.

Finally, let us consider the evolution equation for the momentum density S_i. Since we have $S_i = \rho_0 h W^2 v_i$, then we can consider S_i a vector with respect to the 3-geometry. Its original evolution equation is

$$\partial_t \left(\sqrt{\gamma} S_i\right) + \partial_k \left\{\sqrt{\gamma} \left[S_i \left(\alpha v^k - \beta^k\right) + \alpha p\, \delta^k_i\right]\right\} = \alpha \sqrt{\gamma}\, \Gamma^\mu_{\nu i} T^\nu_\mu , \qquad (7.4.9)$$

From the expression for the stress-energy tensor, the right hand side of this equation can easily be shown to be

$$\alpha\sqrt{\gamma}\,\Gamma^{\mu}_{\nu i}T^{\nu}_{\mu} = p\,\partial_i\left(\alpha\sqrt{\gamma}\right) + \alpha\sqrt{\gamma}\,\frac{\rho_0 h}{2}\,u^{\mu}u^{\nu}\partial_i g_{\mu\nu}\;. \tag{7.4.10}$$

Substituting now the components of the 4-metric $g_{\mu\nu}$ in terms of 3+1 quantities we find, after some algebra, that

$$\frac{\alpha}{2}\,u^{\mu}u^{\nu}\partial_i g_{\mu\nu} = W^2\left[-\partial_i\alpha + v^k D_i\beta_k + {}^{(3)}\Gamma^m_{ik}\,\frac{u^k v_m}{u^0}\right]\;, \tag{7.4.11}$$

where ${}^{(3)}\Gamma^m_{ik}$ are the Christoffel symbols associated with the 3-geometry. On the other hand we have

$$\frac{1}{\sqrt{\gamma}}\,\partial_t\left(\sqrt{\gamma}S_i\right) = \partial_t S_i - \alpha K S_i + S_i D_k\beta^k\;, \tag{7.4.12}$$

and

$$\begin{aligned}\frac{1}{\sqrt{\gamma}}\,\partial_k\left[\sqrt{\gamma}S_i\left(\alpha v^k - \beta^k\right)\right] &= D_k\left[\alpha S_i\left(v^k - \beta^k\right)\right]\\ &\quad + S_m\left(\alpha v^n - \beta^n\right){}^{(3)}\Gamma^m_{in}\;.\end{aligned} \tag{7.4.13}$$

Collecting results we find that the evolution equation for S_i becomes

$$\begin{aligned}\partial_t S_i - \alpha K S_i + D_k\left(\alpha S_i v^k\right) - \beta^k D_k S_i + S_m\left(\alpha v^k - \beta^k\right){}^{(3)}\Gamma^m_{ik} + \frac{1}{\sqrt{\gamma}}\,\partial_i\left(\alpha\sqrt{\gamma}\,p\right)\\ = \frac{p}{\sqrt{\gamma}}\,\partial_i\left(\alpha\sqrt{\gamma}\right) + \rho_0 h W^2\left[-\partial_i\alpha + v^k D_i\beta_k + {}^{(3)}\Gamma^m_{ik}\frac{u^k v_m}{u^0}\right],\end{aligned} \tag{7.4.14}$$

which can be simplified to

$$\partial_t S_i - \pounds_{\vec{\beta}} S_i + D_k\left(\alpha S_i v^k\right) + \partial_i\left(\alpha p\right) = -\left(\mathcal{E} + D\right)\partial_i\alpha + \alpha K S_i\;. \tag{7.4.15}$$

Notice that the shift vector again only appears in the Lie derivative term, as expected.

The full set of hydrodynamic equations in 3+1 form can then be written as

$$\partial_t D - \beta^k\partial_k D + D_k\left(\alpha D v^k\right) = \alpha K D\;, \tag{7.4.16}$$

$$\partial_t S^i - \pounds_{\vec{\beta}} S^i + D_k\left[\alpha\left(S^i v^k + \gamma^{ik} p\right)\right] = -\left(\mathcal{E} + D\right)D^i\alpha + \alpha K S^i\;, \tag{7.4.17}$$

$$\begin{aligned}\partial_t\mathcal{E} - \beta^k\partial_k\mathcal{E} + D_k\left[\alpha v^k\left(\mathcal{E} + p\right)\right] &= \left(\mathcal{E} + p + D\right)\left(\alpha v^m v^n K_{mn} - v^m\partial_m\alpha\right)\\ &\quad + \alpha K\left(\mathcal{E} + p\right)\;.\end{aligned} \tag{7.4.18}$$

The above equations are now manifestly 3-covariant when we consider $(D, \mathcal{E}, p)$ as scalars and S_i as a 3-vector.[73] The 3+1 equations just derived can also be used

[73] These 3+1 hydrodynamic equations have also been derived previously by Salgado using a somewhat different notation in [248].

in the case of special relativity with curvilinear coordinate systems, in which case we only need to take $\alpha = 1$, $\beta^i = 0$ and $K_{ij} = 0$.

7.5 Equations of state: dust, ideal gases and polytropes

As we have mentioned, the hydrodynamic evolution equations by themselves are not enough as we only have five equations for six unknowns, namely $(\rho_0, \epsilon, p, v^i)$. In order to close the system we still need an equation of state of the form $p = p(\rho_0, \epsilon)$. Ideally, an equation of state should be obtained from the microphysics of the fluid we are considering, but there are some particularly simple equations of state that are widely used in numerical simulations.

The simplest choice we can make for an equation of state is to say that the pressure vanishes:

$$p = 0 \,. \tag{7.5.1}$$

Such a fluid is known as *dust* in relativity, and though it is not particularly realistic it is nevertheless a good approximation in some special cases. Since there is no pressure, the motion of the fluid elements in this case is just along geodesics of the spacetime. In terms of particles, we can think of dust as a collection of particles all of which have exactly the same speed locally, so that there is no random component (*i.e.* zero internal energy). We should be careful with the use of the term "dust", as it does not correspond directly with what we usually have in mind when talking about dust particles. In relativity dust is a pressureless fluid, but a continuous fluid nonetheless. In particular, dust can be used to model a collection of cold collisionless particles, such as a uniform matter distribution in cosmology, the structure of rotating rings or disks of particles, or the collapse of a shell of matter initially at rest (known as Oppenheimer–Snyder collapse). However, the dust approximation has a serious drawback in the fact that for a flow that is initially non-uniform there is nothing that can prevent the fluid elements from "colliding" when their trajectories cross, resulting in the so-called *shell-crossing singularities*. Shell-crossings are a particularly simple example of shock formation, though they are not usually seen as such since the dust model is just too simplistic (we will have more to say about shocks in Section 7.7). As soon as a shell-crossing singularity occurs the dust approximation breaks down completely, and because of this, dust is only interesting in situations where we have collisionless particles whose trajectories *are not expected to cross.*[74]

A much more realistic choice is the equation of state for an *ideal gas* which has the form[75]

[74] If we wish to study a system of collisionless particles in cases where the particle paths may cross then we can't think in terms of a continuous fluid anymore and need to go instead to a description based on kinetic theory and the Boltzmann equation.

[75] The terms "perfect fluid" and "ideal gas", though similar, refer in fact to very different things. A perfect fluid is defined as one with no viscosity and no heat conduction, but the equation of state can still be very general. An ideal gas, on the other hand, refers to a very specific equation of state.

$$p = (\gamma - 1)\, \rho_0 \epsilon \,, \tag{7.5.2}$$

with γ a constant known as the *adiabatic index* (not to be confused with the determinant of the spatial metric of the previous Section). To see how this equation comes about consider the standard equation of state for an ideal gas:

$$pV = nkT \,, \tag{7.5.3}$$

where V is the volume, n the number of particles, T the temperature and k the Boltzmann constant. Define now the *specific heat* of the gas as the amount of heat Q per unit mass needed to raise the temperature by one degree. The specific heat is in fact different if we keep the volume or the pressure constant, so we define the specific heats at constant volume and at constant pressure as

$$c_V = \frac{1}{M} \left(\frac{dQ}{dT} \right)_V \,, \qquad c_p = \frac{1}{M} \left(\frac{dQ}{dT} \right)_p \,, \tag{7.5.4}$$

with $M = nm$ the total mass and m the mass of the individual particles. The adiabatic index γ of the gas is defined as the ratio of these two quantities:

$$\gamma := \frac{c_p}{c_V} \,. \tag{7.5.5}$$

Notice that in general we expect to find $c_p > c_V$, so that $\gamma > 1$. This is because it takes more heat to increase the temperature at constant pressure than at constant volume since in the first case part of the heat produces work (pdV), while in the second case no work is allowed ($dV = 0$).

Now, the first law of thermodynamics states that $dQ = dU + pdV$. This implies that, at constant volume, $dQ = dU$, from which we find

$$c_V = \frac{1}{M} \frac{dU}{dT} \quad \Rightarrow \quad dU = M c_V dT \,. \tag{7.5.6}$$

If we now assume that c_V is constant, this relation can be integrated to yield

$$U = M c_V T \,. \tag{7.5.7}$$

On the other hand, at constant pressure the equation of state implies

$$pdV = nkdT \,, \tag{7.5.8}$$

and the first law takes the form

$$dQ = dU + pdV = M c_V dT + nkdT \,, \tag{7.5.9}$$

so that

$$c_p = \frac{1}{M} \left(\frac{dQ}{dT} \right)_p = \frac{M c_V dT + nkdT}{M \, dT} = c_V + k/m \,. \tag{7.5.10}$$

The adiabatic index then becomes

$$\gamma = \frac{c_p}{c_V} = 1 + \frac{k}{mc_V} \, . \tag{7.5.11}$$

Going back to the equation of state, the relation $U = Mc_V T$ allows us to rewrite it as

$$pV = \frac{kU}{mc_V} = (\gamma - 1)\, U \, , \tag{7.5.12}$$

and dividing by the volume we finally find

$$p = (\gamma - 1)\, \frac{U}{V} = (\gamma - 1)\, \rho_0 \epsilon \, , \tag{7.5.13}$$

with ρ_0 the mass density and ϵ the internal energy per unit mass.

In the description of the ideal gas that uses the equation of state (7.5.2) instead of (7.5.3) the temperature is not explicit, but it can be recovered from $T = U/(Mc_V) = (\gamma - 1)U/nk$ which, using the fact that $U = nm\epsilon$, reduces to

$$T = \frac{m}{k}\, (\gamma - 1)\epsilon \, . \tag{7.5.14}$$

Notice that even if in the previous derivation we have assumed that the specific heat c_V is constant, in the general case it can be expected to be a (typically slowly varying) function of temperature. For example, in the particular case of a mono-atomic gas we find from the equipartition theorem that in the non-relativistic limit corresponding to low temperatures $c_V = 3k/2m$ which implies $\gamma = 5/3$, while in the ultra-relativistic regime corresponding to very high temperatures we have instead $c_V = 3k/m$, so that $\gamma = 4/3$ (this is also the value of γ for a gas of photons).[76]

Another very common choice that is closely related to the ideal gas case is the so-called *polytropic* equation of state that has the form

$$p = K\rho_0^\Gamma \equiv K\rho_0^{1+1/N} \, , \tag{7.5.15}$$

where K is a constant and N is called the *polytropic index*. The parameter $\Gamma := 1 + 1/N$ is usually called the adiabatic index of the polytrope, but as we will see below we should be careful before identifying it with the true adiabatic index γ introduced above.[77]

[76] The factor of two difference in the values of c_V for the non-relativistic and ultra-relativistic case comes from the relativistic relation between energy E and momentum p, $E^2 = m^2 + \mathrm{p}^2$, which at low speeds reduces to $E \simeq m + \mathrm{p}^2/2m$ implying $\rho_0\epsilon \simeq 3p/2$, while at high speeds becomes instead instead $E \simeq \mathrm{p}$ which implies $\rho_0\epsilon \simeq 3p$.

[77] In some references we find that, somewhat confusingly, the parameter Γ is called the polytropic index instead of N. Also, we sometimes find that the polytropic equation of state is defined as $p = K\rho^\Gamma = K\rho_0^\Gamma(1+\epsilon)^\Gamma$, which is *not* equivalent to the standard version discussed here.

To understand the origin of the polytropic equation of state consider an adiabatic process for an ideal gas, *i.e.* a process with no heat transfer. The first law (7.2.43) then becomes

$$0 = d\epsilon + p\, d\left(\frac{1}{\rho_0}\right) = \frac{1}{\gamma - 1}\, d\left(\frac{p}{\rho_0}\right) + p\, d\left(\frac{1}{\rho_0}\right) , \tag{7.5.16}$$

which implies

$$\frac{dp}{p} = \gamma\, \frac{d\rho_0}{\rho_0} . \tag{7.5.17}$$

This can now be easily integrated to find

$$p = K \rho_0^\gamma , \tag{7.5.18}$$

with K some constant. Comparing this with equation (7.5.15) we see that Γ plays the role of γ. However, we must remember that the thermodynamic relation (7.5.18) above is only valid for an adiabatic process involving an ideal gas, while equation (7.5.15) is often promoted to an equation of state valid even when there is heat exchange, so they are not entirely equivalent and in the general case Γ will not correspond to the true adiabatic index.

For a polytrope we also finds the following relation between the specific internal energy and the rest mass density

$$\epsilon = \frac{K}{\gamma - 1}\, \rho_0^{\gamma - 1} . \tag{7.5.19}$$

This relation can be easily obtained by substituting the polytropic relation $p = K\rho_0^\gamma$ into the equation of state for an ideal gas $p = (\gamma - 1)\rho_0 \epsilon$. Again, the relation only holds for an adiabatic situation.

In the particular case of a perfect fluid there is no heat exchange by definition, so using the polytropic relation (7.5.18) is equivalent to using the equation of state of an ideal gas (7.5.2). In fact, in this case we can show that the evolution equation for $\mathcal{E}$ can be derived from the evolution equations for D and S_i, together with the equation of state and the polytropic relation $p = K\rho_0^\gamma$, so that we can ignore $\mathcal{E}$ completely (remember that the evolution equation for the energy density $\mathcal{E}$ is essentially the local version of the first law of thermodynamics, and the polytropic relation is an integral of this law). There is, however, one final subtlety to this argument. The hydrodynamic equations only imply the first law of thermodynamics along the trajectory of the fluid elements (cf. equation (7.2.40)), so that the constant K in the integrated polytropic relation (7.5.18) can in principle be different along different flow lines. The value of this constant can in fact be shown to be related to the entropy of the fluid element. To see this, notice that in the situation where there is heat transfer the first law of thermodynamics has the general form

$$T d\sigma = d\epsilon + p\, d\left(\frac{1}{\rho_0}\right) = d\epsilon - \frac{p}{\rho_0^2}\, d\rho_0 , \tag{7.5.20}$$

where T is the temperature and σ the specific entropy (entropy per unit mass) of the gas. Dividing by T and using the expression for the temperature for an ideal gas (7.5.14) we find

$$\frac{m}{k}\, d\sigma = \frac{d\epsilon}{\epsilon(\gamma-1)} - \frac{d\rho_0}{\rho_0} = d\ln\left(\frac{\epsilon^{1/(\gamma-1)}}{\rho_0}\right) , \tag{7.5.21}$$

so that

$$\sigma = \frac{k}{m}\left[C + \ln\left(\frac{\epsilon^{1/(\gamma-1)}}{\rho_0}\right)\right] , \tag{7.5.22}$$

with C some integration constant. For an adiabatic process we can substitute the polytropic relation (7.5.19) to find

$$\sigma = \frac{k}{m}\left[C + \frac{1}{\gamma-1}\ln\left(\frac{K}{\gamma-1}\right)\right] . \tag{7.5.23}$$

Choosing the integration constant such that $C = \ln(\gamma-1)/(\gamma-1)$ we find

$$\sigma = \frac{k}{m(\gamma-1)} \ln K = c_V \ln K . \tag{7.5.24}$$

This can now be inverted to give

$$K = e^{\sigma/c_V} = e^{m(\gamma-1)\sigma/k} . \tag{7.5.25}$$

We then see that taking K to be the same constant everywhere implies that σ is not only constant along flow lines, but is also uniform. Because of this a fluid with K constant everywhere is called *isentropic*. Notice finally that even though for a perfect fluid we expect the flow to be adiabatic, this is only true as long as the flow remains smooth. When shocks develop, kinetic energy is transformed into internal energy (the so-called *shock heating*), so there is heat transfer and the polytropic relation is no longer equivalent to the ideal gas equation of state.

The origin of the word polytrope, which comes from the Greek for "many turns", is in the study of adiabatic processes in convective gases. Polytropic models where first studied by R. Emden in 1907 [122] and they are particularly useful for the study of hydrostatic equilibrium in the Earth's atmosphere. They have also been extensively used in astrophysics for studying stellar structure.

In the particular case of cold compact objects such as white dwarfs and neutron stars that are supported mainly by the Pauli exclusion principle (for electrons in the first case and neutrons in the second), we can not use the ideal gas equation of state discussed above since it only applies to classical gases and does not describe correctly a highly degenerate Fermi gas. However, we can derive from first principles an equation of state for an ideal Fermi gas at zero temperature and it turns out that this equation reduces to a polytropic form both in the non-relativistic and extremely relativistic limits, with $\gamma = 5/3$ and $4/3$ respectively. More generally, polytropic equations of state with an adiabatic index in the range $1 < \gamma < 3$ can be used as very simple models of neutron stars made of "non-ideal" Fermi gases. High values of γ result in stiff equations of state, and low values in soft equations of state (the words "soft" and "stiff" relate to the speed of sound in the fluid – see Section 7.6).

The true equation of state for neutron stars is still largely unknown owing to our lack of knowledge of the interactions of nuclear matter at very high densities. There are, however, a number of proposed equations of state that are considerably more realistic than a simple polytrope, but we will not discuss them here (the interested reader can see *e.g.* [264, 275]).

7.6 Hyperbolicity and the speed of sound

Just as in the case of the evolution equations for the gravitational field, it is also important to analyze the well-posedness of the hydrodynamical evolution equations. As we will see below, it turns out that the equations for hydrodynamics are strongly hyperbolic, but the analysis itself is interesting since it allows us to introduce some physical concepts, in particular the speed of sound. For clarity, we will do the analysis first in the Newtonian case and only later consider the relativistic case.

7.6.1 *Newtonian case*

In order to study the characteristic structure of the hydrodynamic equations, we will start by considering the Newtonian limit since in that case the analysis is considerably simpler. The Newtonian hydrodynamic equations can be written as

$$\partial_t \rho_0 + \partial_k \left(\rho_0 v^k \right) = 0 \,, \tag{7.6.1}$$

$$\partial_t S_i + \partial_k \left(S_i v^k + p \, \delta_i^k \right) = 0 \,, \tag{7.6.2}$$

$$\partial_t \mathcal{E} + \partial_k \left[(\mathcal{E} + p) \, v^k \right] = 0 \,, \tag{7.6.3}$$

where now $S_i = \rho_0 v_i$ and $\mathcal{E} = \rho_0(\epsilon + v^2/2)$. For the hyperbolicity analysis we will concentrate on the x direction and ignore derivatives along y and z. The first step is to solve for the primitive variables (ρ_0, ϵ, v_i) in terms of the conserved quantities:

$$u_1 := \rho_0, \quad u_2 := S_x, \quad u_3 := S_y, \quad u_4 := S_z, \quad u_5 := \mathcal{E} \,. \tag{7.6.4}$$

Inverting these relations we find

$$\rho_0 = u_1, \quad \epsilon = \frac{u_5}{u_1} - \frac{u_2^2 + u_3^2 + u_4^2}{2u_1^2}, \quad v_x = \frac{u_2}{u_1}, \quad v_y = \frac{u_3}{u_1}, \quad v_z = \frac{u_4}{u_1}. \tag{7.6.5}$$

The fluxes can then be written as

$$F_1 = \rho_0 v_x = u_2 \,, \tag{7.6.6}$$

$$F_2 = S_x v_x + p = \frac{u_2^2}{u_1} + p \,, \tag{7.6.7}$$

$$F_3 = S_y v_x = \frac{u_2 u_3}{u_1} \,, \tag{7.6.8}$$

$$F_4 = S_z v_x = \frac{u_2 u_4}{u_1} \,, \tag{7.6.9}$$

$$F_5 = (\mathcal{E} + p)\, v_x = \frac{(u_5 + p)\, u_2}{u_1} \,. \tag{7.6.10}$$

Here we must remember that $p = p(\rho_0, \epsilon)$, so that in general we have

$$\frac{\partial p}{\partial u_i} = \frac{\partial p}{\partial \rho_0} \frac{\partial \rho_0}{\partial u_i} + \frac{\partial p}{\partial \epsilon} \frac{\partial \epsilon}{\partial u_i} \,. \tag{7.6.11}$$

The characteristic matrix is now defined as the Jacobian of the fluxes, *i.e.* $M_{ij} = \partial F_i / \partial u_j$. When we construct this matrix explicitly we find that it has five real eigenvalues and a complete set of eigenvectors, so the system is strongly hyperbolic. The characteristic speeds (eigenvalues) are

$$\lambda_0 = \frac{u_2}{u_1} = v_x \,, \qquad \text{(with multiplicity 3)} \tag{7.6.12}$$

$$\lambda_\pm = \frac{u_2}{u_1} \pm \sqrt{\chi + \frac{p}{u_1^2}\,\kappa} = v_x \pm \sqrt{\chi + \frac{p}{\rho_0^2}\,\kappa} \,, \tag{7.6.13}$$

where we have defined

$$\chi := \frac{\partial p}{\partial \rho_0} \,, \qquad \kappa := \frac{\partial p}{\partial \epsilon} \,. \tag{7.6.14}$$

We clearly see that λ_0 is nothing more than the fluid's flow speed, while $\lambda_\pm$ are symmetric around the flow lines. We can rewrite $\lambda_\pm$ as

$$\lambda_\pm = v_x \pm c_s \,, \tag{7.6.15}$$

where c_s is given by

$$c_s^2 := \chi + \frac{p}{\rho_0^2}\,\kappa \,. \tag{7.6.16}$$

The quantity c_s is known as the *local speed of sound*, and measures the speed at which density perturbations travel as seen in the fluid's reference frame. The speed of sound c_s can also be rewritten in a different way by remembering that for an adiabatic process the first law of thermodynamics has the form

$$0 = d\epsilon + p\, d\left(\frac{1}{\rho_0}\right) = d\epsilon - \frac{p}{\rho_0^2}\, d\rho_0 \,, \tag{7.6.17}$$

which implies

$$d\epsilon = \frac{p}{\rho_0^2}\, d\rho_0 \,. \tag{7.6.18}$$

This shows that for an adiabatic process, ϵ can be considered a function of ρ_0. The change in pressure p given a change in ρ_0 is then

$$dp = \frac{\partial p}{\partial \rho_0}\, d\rho_0 + \frac{\partial p}{\partial \epsilon}\, d\epsilon = \left(\chi + \frac{p}{\rho_0^2}\, \kappa \right) d\rho_0 = c_s^2 d\rho_0 \,, \tag{7.6.19}$$

which can be rewritten as

$$c_s^2 = \left. \frac{dp}{d\rho_0} \right|_\sigma \,, \tag{7.6.20}$$

where as before $\sigma = \mathcal{S}/M$ is the specific entropy that must remain constant for an adiabatic process.

In the particular case of an ideal gas, we know from Section 7.5 that for an adiabatic process the pressure is given in terms of the mass density through the polytropic relation $p = K\rho_0^\gamma$. In that case the local speed of sound becomes

$$c_s^2 = \gamma K \rho_0^{\gamma-1} = \gamma \frac{p}{\rho_0} \,. \tag{7.6.21}$$

This shows that, for a fixed ratio of pressure to density p/ρ_0, large values of the adiabatic index γ correspond to a large speed of sound (a "stiff" fluid), while low values of γ correspond to a low speed of sound (a "soft" fluid).

We can easily construct three linearly independent eigenvectors corresponding to the eigenvalues λ_0. One convenient such set takes the form

$$\vec{e}_1 = \left[1, v_x, v_y, v_z, \epsilon + v^2/2 - \rho_0 \chi/\kappa \right] \,, \tag{7.6.22}$$

$$\vec{e}_2 = \left[v_y, v_y v_x, v_y^2 - v^2/2, v_y v_z, v_y \left(\epsilon - \rho_0 \chi/\kappa \right) \right] \,, \tag{7.6.23}$$

$$\vec{e}_3 = \left[v_z, v_z v_x, v_z v_y, v_z^2 - v^2/2, v_z \left(\epsilon - \rho_0 \chi/\kappa \right) \right] \,, \tag{7.6.24}$$

where $v^2 := v_x^2 + v_y^2 + v_z^2$. Also, the eigenvectors corresponding to $\lambda_\pm$ can be expressed as

$$\vec{e}_\pm = \left[1, v_x \pm c_s, v_y, v_z \epsilon + v^2/2 + p/\rho_0 \pm v_x c_s \right] \,. \tag{7.6.25}$$

As we will see in Chapter 9, knowing the form of the eigenvectors can be very useful for developing numerical methods that can adequately handle shocks. On the other hand, the concept of eigenfunction is not particularly useful in this case since the equations are nonlinear, so that we can't move the matrix of eigenvectors in and out of derivatives without changing the principal part of the system.[78] In other words, there is in general no set of functions that simply propagate along characteristic lines. Of course, we can always linearize the solution around a given background state $(\hat{\rho}_0, \hat{S}^i, \hat{\mathcal{E}})$. In this case the eigenfunctions can

[78]In the case of the Einstein field equations we can define eigenfunctions meaningfully since even though those equations are also non-linear, they are nevertheless quasi-linear and the nonlinearities appear in the source terms.

be defined in a meaningful way and one does find that small perturbations do travel along characteristic lines. However, even if we do this explicitly the form of the eigenfunctions is somewhat complicated and not particularly illuminating.

There is, however, at least one simple case where looking at the eigenfunctions is interesting. Consider a perturbation around a state in which the fluid is at rest with uniform density and internal energy. In that case we have:

$$\rho_0 = \hat{\rho}_0 + \delta\rho_0 \; , \qquad S_i = \delta S_i \; , \qquad \mathcal{E} = \hat{\mathcal{E}} + \delta\mathcal{E} \; , \tag{7.6.26}$$

with $\hat{\rho}_0$ and $\hat{\mathcal{E}}$ constant. We then find that the eigenvalues are $\lambda_0 = 0$ with multiplicity three (since the background speed vanishes), and $\lambda_\pm = \pm\hat{c}_s$, with $\hat{c}_s$ the speed of sound on the background. The three eigenfunctions associated with λ_0 turn out to be

$$w_1 = \delta S_y \; , \qquad w_2 = \delta S_z \; , \qquad w_3 = \delta\mathcal{E} - \left(\hat{\epsilon} + \frac{\hat{p}}{\hat{\rho}_0}\right)\delta\rho_0 \; . \tag{7.6.27}$$

In other words, δS_y and δS_z remain at their initial values, and $\delta\mathcal{E} - (\hat{\epsilon} + \hat{p}/\hat{\rho}_0)\,\delta\rho_0$ also remains constant. On the other hand, the eigenfunctions associated with $\lambda_\pm$ are found to be

$$w_\pm = \left(\hat{\chi} - \frac{\hat{\kappa}\hat{\mathcal{E}}}{\hat{\rho}_0^2}\right)\delta\rho_0 + \frac{\hat{\kappa}}{\hat{\rho}_0}\,\delta\mathcal{E} \pm \hat{c}_s\delta S_x \; . \tag{7.6.28}$$

Notice now that since $w_\pm$ obey the evolution equations

$$\partial_t w_\pm \pm \hat{c}_s \partial_x w_\pm = 0 \; , \tag{7.6.29}$$

then we can take an extra time derivative to find that

$$\partial_t^2 w_\pm - \hat{c}_s^2\,\partial_x^2 w_\pm = 0 \; . \tag{7.6.30}$$

That is, $w_\pm$ obey a simple scalar wave equation with speed $\hat{c}_s$. And since both obey the same wave equation, it is clear that their sum and difference also do. Finally, since (7.6.27) implies that $\delta\rho_0$ and $\delta\mathcal{E}$ are linearly related to each other with constant coefficients, then we can easily deduce that all three perturbations $\delta\rho_0$, $\delta\mathcal{E}$, and δS_x obey the same scalar wave equation.

7.6.2 *Relativistic case*

Let us now consider the relativistic hydrodynamic equations. We will start from the 3+1 version of these equations which for concreteness we rewrite here:

$$\partial_t D - \beta^k\partial_k D + D_k\left(\alpha D v^k\right) = \alpha K D \; , \tag{7.6.31}$$

$$\partial_t S^i - \pounds_{\vec{\beta}} S^i + D_k\left[\alpha\left(S^i v^k + \gamma^{ik} p\right)\right] = -\left(\mathcal{E} + D\right) D^i\alpha + \alpha K S^i \; , \tag{7.6.32}$$

$$\begin{aligned}\partial_t\mathcal{E} - \beta^k\partial_k\mathcal{E} + D_k\left[\alpha v^k\left(\mathcal{E} + p\right)\right] &= \left(\mathcal{E} + p + D\right)\left(\alpha v^m v^n K_{mn} - v^m\partial_m\alpha\right)\\ &\quad + \alpha K\left(\mathcal{E} + p\right) \; ,\end{aligned} \tag{7.6.33}$$

with $D = \rho_0 W$, $S^i = \rho_0 h W^2 v^i$, $\mathcal{E} = \rho_0 h W^2 - p - D$, and where $h = 1 + \epsilon + p/\rho_0$ is the enthalpy and $W = 1/\sqrt{1 - \gamma_{ij}v^i v^j}$ is the Lorentz factor. As before, for the hyperbolicity analysis we will concentrate on the x direction and ignore derivatives along the y and z directions.

The next task would be to find the Jacobian matrix, but in contrast with the Newtonian case, here we face a serious problem. The definitions of the conserved quantities $(D, S^i, \mathcal{E})$ are now considerably more complex since first they involve the Lorentz factor W, and much worse they also involve the pressure p. This means that we can't invert these relations if we don't assume a specific form for the equation of state, and even for simple equations of state the procedure would require finding roots of high order polynomials. And if we can't solve for the primitive variables (ρ_0, ϵ, v^i) in terms of the conserved quantities then we can't find the Jacobian matrix and we can't continue the analysis.

Fortunately, there exists a generalization of the concept of hyperbolicity for systems of this type due to Friedrichs [134]. The idea is to write the system of equations in the following form

$$\partial_\mu F^\mu(u) = s(u) \,, \qquad (7.6.34)$$

where u is the vector of main variables (in our case the primitive variables), each F^μ is a vector of fluxes, and s is a source vector that does not involve derivatives. Construct now the Jacobian matrices $A^\mu_{ij} = \partial F^\mu_i / \partial u_j$. The system will be strongly hyperbolic if given any arbitrary pair of vectors ξ^μ and ζ^μ such that[79]

$$\xi_\mu \xi^\mu = -1 \,, \qquad \zeta_\mu \zeta^\mu = 1 \,, \qquad \xi_\mu \zeta^\mu = 0 \,, \qquad (7.6.35)$$

then the matrix $A^\mu \xi_\mu$ is invertible (its determinant is different from zero), and the characteristic matrix $M := (A^\mu \xi_\mu)^{-1}(A^\mu \zeta_\mu)$ has real eigenvalues and a complete set of eigenvectors. Notice that the matrix M just defined plays the role of the principal symbol in the standard definition of hyperbolicity.

In the case of the relativistic hydrodynamic equations we will take as main variables $u = (\rho_0, v^x, v^y, v^z, \epsilon)$ (in that order). Notice that now it is important to distinguish between contra-variant and co-variant indices in the speed v^i, and for the main variables we have chosen to use the contra-variant components. The fluxes along the time direction are just the definition of the conserved quantities themselves:

$$F^0_1 := D \,, \quad F^0_2 := S^x \,, \quad F^0_3 := S^y \,, \quad F^0_4 := S^z \,, \quad F^0_5 := \mathcal{E} \,, \qquad (7.6.36)$$

while the fluxes along the direction x are

[79]The notation used here would seem to suggest that F^μ is a 4-vector in the language of differential geometry. However, we are only interested in the fact that it is a collection of four flux vectors, one of which is associated with the time direction. Also, the norm of the vectors ξ^μ and ζ^μ is constructed using the Minkowski metric, as all that really matters is that ξ^μ is timelike and ζ^μ spacelike.

$$F_1^x := (\alpha v^x - \beta^x)\, D\,, \tag{7.6.37}$$
$$F_2^x := (\alpha v^x - \beta^x)\, S^x + \alpha\gamma^{xx} p\,, \tag{7.6.38}$$
$$F_3^x := (\alpha v^x - \beta^x)\, S^y + \alpha\gamma^{xy} p\,, \tag{7.6.39}$$
$$F_4^x := (\alpha v^x - \beta^x)\, S^z + \alpha\gamma^{xz} p\,, \tag{7.6.40}$$
$$F_5^x := (\alpha v^x - \beta^x)\, \mathcal{E} + \alpha p v^x\,. \tag{7.6.41}$$

To continue the analysis we construct the Jacobian matrices $A^0 = \partial F^0/\partial u$ and $A^x = \partial F^x/\partial u$, choose $\xi_\mu = (1,0,0,0)$ and $\zeta_\mu = (0,1,0,0)$, and construct the characteristic matrix which in this case becomes $M = (A^0)^{-1}A^x$. After a long algebra we find that M has five real eigenvalues and a complete set of eigenvectors, so the evolution system is strongly hyperbolic. The eigenvalues have the form

$$\lambda_0 = -\beta^x + \alpha v^x\,, \qquad \text{(multiplicity 3)} \tag{7.6.42}$$
$$\lambda_\pm = -\beta^x + \frac{\alpha}{1 - v^2 c_s^2}\left\{ v^x\left(1 - c_s^2\right)\right.$$
$$\left. \pm\, c_s \sqrt{\left(1 - v^2\right)\left[\gamma^{xx}\left(1 - v^2 c_s^2\right) - (v^x)^2\left(1 - c_s^2\right)\right]}\right\}\,, \tag{7.6.43}$$

where $v^2 := \gamma_{ij} v^i v^j$, and where now the local speed of sound c_s is defined as

$$c_s^2 = \frac{1}{h}\left(\chi + \frac{p}{\rho_0^2}\kappa\right)\,, \tag{7.6.44}$$

with $\chi = \partial p/\partial\rho_0$ and $\kappa = \partial p/\partial\epsilon$ as before. Notice that, in contrast to the Newtonian case, the local speed of sound c_s is now divided by the enthalpy (compare with equation (7.6.16)).

The eigenvalues $\lambda_\pm$ have the interesting property that even though they represent the speed of modes propagating along the x direction, they nevertheless involve the tangential speeds v^y and v^z through the combination v^2 (this is a consequence of the relativistic length contraction). One interesting limit is obtained when $v^y = v^z = 0$ and the spacetime metric is flat ($\alpha = 1$, $\beta^i = 0$, $\gamma_{ij} = \delta_{ij}$), corresponding to the case of one-dimensional special relativity. In that case the eigenvalues reduce to

$$\lambda_0 = v^x\,, \qquad \lambda_\pm = \frac{v^x \pm c_s}{1 \pm v^x c_s}\,. \tag{7.6.45}$$

We then see that $\lambda_\pm$ is nothing more than the standard expression for the relativistic composition of the velocities v^x and c_s.

Another special case corresponds to the situation where the fluid is at rest as seen by the Eulerian observers so that $v^x = v^y = v^z = 0$, but the metric is still quite general. In this case the eigenvalues take the simple form

$$\lambda_0 = -\beta^x\,, \qquad \lambda_\pm = -\beta^x \pm \alpha c_s \sqrt{\gamma^{xx}}\,. \tag{7.6.46}$$

The local speed of sound c_s is now just corrected by the geometric factor $\alpha\sqrt{\gamma^{xx}}$ and shifted by $-\beta^x$.

Let us again look at the expression for the speed of sound c_s for an adiabatic process. As we found in the Newtonian case, for an adiabatic process we have

$$d\epsilon = \frac{p}{\rho_0^2}\, d\rho_0 \, , \tag{7.6.47}$$

$$dp = \left(\chi + \frac{p}{\rho_0^2}\right) d\rho_0 \tag{7.6.48}$$

which now implies

$$c_s^2 = \frac{1}{h} \frac{dp}{d\rho_0}\bigg|_\sigma \, . \tag{7.6.49}$$

On the other hand, if we remember that the total energy density measured in the fluid's reference frame is given by $\rho = \rho_0(1+\epsilon)$, then the previous relations imply that for an adiabatic process we have

$$d\rho = (1+\epsilon)\, d\rho_0 + \rho_0 d\epsilon = h\, d\rho_0 \, , \tag{7.6.50}$$

so that the speed of sound can be written in the final form

$$c_s^2 = \frac{dp}{d\rho}\bigg|_\sigma \, . \tag{7.6.51}$$

This is identical to the Newtonian expression (7.6.21), except for the fact that the relativistic expression involves the total energy density and not just the rest mass energy density. For the particular case of an ideal gas we find again from the polytropic relation

$$c_s^2 = \frac{\gamma p}{\rho_0 h} = \frac{\gamma p}{\rho + p} \, . \tag{7.6.52}$$

In contrast with the Newtonian case, we now need to worry about the fact that at high temperature the speed of sound might be larger than the speed of light for some values of the adiabatic index γ, so in general we should ask for the value of γ to be such that $c_s < 1$. For an ultra-relativistic mono-atomic gas we have $\gamma = 4/3$ so that $p = (\gamma - 1)\rho_0\epsilon = \rho_0\epsilon/3$, and also $\rho \simeq \rho_0\epsilon$. The speed of sound then becomes $c_s \simeq 1/\sqrt{3} \simeq 0.58$ which is still lower than the speed of light but already quite large. More generally, inserting the equation of state for an ideal gas into the above expression for c_s^2 we find that

$$c_s^2 = \frac{\gamma\,(\gamma - 1)\,\epsilon}{1 + \gamma\epsilon} \, . \tag{7.6.53}$$

In the limit of very high temperatures ($\epsilon >> 1$) this reduces to $c_s^2 \simeq \gamma - 1 < 1$, so that we must ask for $1 < \gamma < 2$. Of course, if we choose γ to be a slowly varying function of temperature, we can still have values larger than 2 at low temperatures (this is the reason why we can sometimes consider models for neutron stars at $T = 0$ that have values of γ as high as 3).

7.7 Weak solutions and the Riemann problem

When dealing with linear systems of hyperbolic equations we can show that smooth initial data will remain smooth during evolution. Discontinuities can in fact be present in the initial data, but they are rather innocuous as they will simply propagate along characteristic lines. However, in the case of nonlinear equations like those of hydrodynamics this is no longer true as we can easily see that discontinuities can arise even starting from smooth initial data. The simplest equation in which we can see an example of this phenomenon is *Burgers' equation*, which is a scalar hyperbolic equation of the form

$$\partial_t u + u \partial_x u = 0 \; . \tag{7.7.1}$$

Notice that this is essentially an advection equation where the wave speed is now the unknown function itself. Burgers' equation is also the limit of the Euler equations for the case of a pressureless fluid (dust), as can be seen from equation (7.2.36).

The behavior of the solutions to Burgers' equation can be easily understood by noticing that since the characteristic speed is u itself, then the change in u along a characteristic line will be given by

$$\frac{d}{dt} u = \partial_t u + u \partial_x u = 0 \; . \tag{7.7.2}$$

In other words, u remains constant along characteristics lines, but since u is precisely the characteristic speed then those characteristics will be straight lines. If we now have initial data such that in a given region $u > 0$ and $\partial_x u < 0$, then the speed of the characteristics on the left will be larger than the speed of those on the right and the characteristics will inevitably cross (since they are straight lines). If we start with a smooth wavefront the "crest" of the wave will move faster than the front, and the wave will inevitably "break", at which point the solution u will develop an infinite slope. This crossing of characteristic lines is called a *shock*.

Once a shock develops the solution becomes discontinuous and the differential equation stops making sense. Of course, this does not mean that for a real physical fluid we will have a true discontinuity; it rather shows that the assumption of vanishing viscosity that goes into the Euler equations is too simplistic. In reality, when a shock is about to form, viscosity becomes very important and instead of a discontinuity we find a very steep but finite gradient. Still, we can decide to keep using the inviscid equations as long as we know how to deal with the discontinuities in a way that still makes physical sense. This can be done by remembering that the differential evolution equations are just the local version of the conservation laws, so that at a discontinuity we can go back to the more fundamental integral form of the conservation law.

We can use the integral form of the conservation law to define the so-called *weak solutions* of the conservation law.[80] To see how this works, assume then that we have a conservation law of the form

$$\partial_t u + \partial_x F(u) = 0 \; . \tag{7.7.3}$$

We now multiply this equation with a smooth test function $\phi(x,t)$ with compact support (both in space and time), and integrate over space and time

$$\int_0^\infty \int_{-\infty}^\infty \phi \left(\partial_t u + \partial_x F \right) dx dt = 0 \; . \tag{7.7.4}$$

Integrating by parts the first term in time and the second term in space, and using the fact the ϕ has compact support, we find

$$\int_0^\infty \int_{-\infty}^\infty \left(u \partial_t \phi + F \partial_x \phi \right) dx dt = - \int_{-\infty}^\infty \phi(x,0) u(x,0) dx \; . \tag{7.7.5}$$

We now say that u is a weak solution of the conservation law if the last identity holds for all test functions ϕ.

To understand how weak solutions behave we must consider the so-called *Riemann problem*, which is the solution corresponding to piecewise constant initial data with a single discontinuity. As a simple example, assume that we have initial data for Burgers' equation of the form

$$u(x,0) = \begin{cases} u_l & x < 0 \; , \\ u_r & x > 0 \; . \end{cases} \tag{7.7.6}$$

We can consider two distinct possibilities. Assume first that $u_l > u_r$. In this case there is a unique weak solution corresponding to the discontinuity propagating with a speed $s = (u_l + u_r)/2$. To see that this is the case assume that we have a discontinuity propagating at a speed s. We then have

$$\frac{d}{dt} \int_{x_l}^{x_r} u \, dx = \int_{x_l}^{x_r} \partial_t u \, dx = - \int_{x_l}^{x_r} \partial_x F(u) \, dx = F(u_l) - F(u_r) \; , \tag{7.7.7}$$

where x_l and x_r are such that the discontinuity remains in the interior region during the time considered, and $F(u)$ is the flux function. On the other hand, for a discontinuity propagating with speed s we clearly have

$$\frac{d}{dt} \int_{x_l}^{x_r} u \, dx = \frac{d}{dt} \left[\left(x_r - x_l \right) u_r + st \left(u_l - u_r \right) \right] = s \left(u_l - u_r \right) \; , \tag{7.7.8}$$

where without loss of generality we have assumed that at $t = 0$ the discontinuity is at $x = x_l$. This shows that for a traveling discontinuity we should generally have

[80]The discussion that follows is based on that of [187].

Fig. 7.1: Shocks and rarefaction waves. The left panel shows a shock wave, *i.e.* a traveling discontinuity for which the speed on the left is larger than the speed on the right. The discontinuity travels at a speed given by the Rankine–Hugoinot jump condition. The right panel shows a rarefaction wave corresponding to the case where the speed on the left is lower than that on the right. The solution is then an interpolating line between states of constant u.

$$F(u_l) - F(u_r) = s\,(u_l - u_r) \quad \Rightarrow \quad s = \frac{F(u_r) - F(u_l)}{u_r - u_l} = \frac{[F]}{[u]}\,, \tag{7.7.9}$$

where $[\cdot]$ denotes the jump of a given quantity across the discontinuity. The last expression is known as the *Rankine–Hugoinot jump condition* and governs the behavior of conservation laws across discontinuities.

In the particular case of Burgers' equation we can easily see that the flux is given by $F(u) = u^2/2$. The jump condition then implies that the speed of propagation of the discontinuity must be $s = (u_r^2/2 - u_l^2/2)/(u_r - u_l) = (u_r + u_l)/2$.

We have so far assumed that $u_l > u_r$. However, nothing in the derivation of the jump condition used this assumption, so that we can expect that a propagating discontinuity with speed $s = (u_r + u_l)/2$ would also be a weak solution to Burgers' equation when $u_l < u_r$. This is indeed true, but in that case the weak solution turns out not to be unique. In fact, it is not even stable as we can show that by adding even a small amount of viscosity the solution will change completely. The stable weak solution in this case is in fact very different and has the form

$$u(x,t) = \begin{cases} u_l & x < u_l t\,, \\ x/t & u_l t \leq x \leq u_r t\,, \\ u_r & x > u_r t\,. \end{cases} \tag{7.7.10}$$

This solution interpolates two regions of constant u with a straight line whose slope decreases with time.

We then have two different types of weak solutions depending on the relative sizes of u_r and u_l: A traveling discontinuity for $u_l > u_r$ known as a *shock wave*, and an interpolating solution for $u_r > u_l$ known as a *rarefaction wave* (see Figure 7.1).

In the case of a rarefaction wave, we have already mentioned that the weak solution is in fact not unique. Physically, the way to choose the correct weak solution is to consider the equation with non-vanishing viscosity, by adding a term of the form $\epsilon\,\partial_x^2 u$ to the right hand side of the conservation law (with $\epsilon > 0$ the *viscosity coefficient*), and then taking the limit of the solution when $\epsilon \to 0$. In practice, however, this is very difficult to do so that some simpler physical criteria are required to choose among the different weak solutions. Such simpler criteria are commonly known as *entropy conditions*, since they are based on an analogy with the hydrodynamic equations where the relevant physical condition we have to apply is to say that the entropy of the fluid elements must not decrease. There are many different formulations of entropy conditions for conservation laws, but in essence they can all be reduced to the statement that for a traveling discontinuity to be an "entropy solution" (*i.e.* a physically acceptable solution) the characteristic lines *must converge* at the discontinuity. If the characteristic lines instead diverge from an initial discontinuity, then the entropy solution is a rarefaction wave instead.[81]

In the previous discussion we have assumed that we have a single non-linear scalar conservation law, in which case the solution of the Riemann problem is either a shock wave or a rarefaction wave. There is in fact another possible solution known as a *contact discontinuity* that corresponds to the case of a linear conservation law for which the characteristic speed is constant and the initial discontinuity simply moves along characteristic lines.

When we deal with a system of conservation laws, as opposed to a simple scalar conservation law, an extra complication arises. In this case the Rankine–Hugoinot jump condition still has the form (7.7.9), but now both $[u]$ and $[F]$ are vectors so that we have in fact a set of jump conditions. However, the speed s remains a scalar, which means that not all possible jumps in the initial data are allowed as there might simply not exist a value of s that satisfies all the jump conditions at the same time. The simplest example of this corresponds to a linear system of conservation laws for which the flux vector is given by $F(u) = Mu$, with M a constant matrix. In this case the jump conditions reduce to

$$M\,[u] = s\,[u] \; . \tag{7.7.11}$$

This implies that $[u]$ must be an eigenvector of the characteristic matrix M, with s the corresponding eigenvalue. That is, only jumps corresponding to eigenvectors of M will result in a single propagating discontinuity, and they will move with the corresponding characteristic speed.

What happens if we take as initial data a jump $[u]$ that does not correspond to an eigenvector of M? After all, we are allowed to choose the initial data freely. In this case the initial discontinuity will "split" into a group of separate discontinuities traveling at different speeds. The original jump $[u]$ needs to be decomposed into a linear combination of eigenvectors of M, with each component traveling at its own characteristic speed. Since we are considering a linear system, all these discontinuities will correspond to simple contact discontinuities.

In the case of non-linear systems of equations like the Euler equations the situation is more complicated, but the general idea is the same. The general

[81]Rarefaction waves are often also called *rarefaction fans*, as characteristic lines "fan out" from the discontinuity.

solution of the Riemann problem involves separating an initial jump $[u]$ into a series of jumps, but now the different jumps might develop into shock waves that satisfy the entropy condition, rarefaction waves, or simple contact discontinuities. Contact discontinuities, in particular, can be expected to be present whenever we have a situation where a given eigenvalue $\lambda_i(u)$ is constant along the integral curves of the corresponding eigenvector $e_i(u)$ (*i.e* the curves that are tangent to the eigenvector), that is, if

$$\nabla_u \lambda_i \cdot e_i(u) = 0 \,, \tag{7.7.12}$$

with ∇_u the gradient of $\lambda_i(u)$ with respect to the u's (not with respect to the spatial coordinates). When this happens we say that the corresponding eigenfield w_i is *linearly degenerate.* If, on the other hand, the above expression is non-zero for some i, then the corresponding eigenfield w_i is called *genuinely non-linear.* This situation does in fact occur for the hydrodynamic equations, in which case we can see that all three eigenfields that propagate along the flow lines are linearly degenerate, while the sound waves are genuinely non-linear. For the linearly degenerate fields we can have at most contact discontinuities, for which the density ρ_0 and specific internal energy ϵ are discontinuous, while the pressure p and flow speed v^i remain continuous.

We will not describe here in any detail the theory behind the general solution of the Riemann problem for non-linear systems; the interested reader can see for example [187] and references therein. It is sufficient to say that the solution to the Riemann problem for the hydrodynamic equations in the Newtonian case is well known, see *e.g.* [105]. The corresponding solution in the relativistic case was recently found in the case of special relativity by Marti and Müller [199], and in the case of general relativity by Pons, Marti and Müller [229].[82] A particular case of the Riemann problem for the hydrodynamic equations, known as the *shock tube problem* and consisting of a fluid that is initially at rest with a discontinuity in both the density ρ_0 and the specific internal energy ϵ (or equivalently in the pressure p), is often used as a test of numerical hydrodynamic codes.

As a simple illustration, we will consider here a propagating shock in the case of the one-dimensional Newtonian hydrodynamic equations. The idea is to look for a solution corresponding to a single propagating shock moving to the right into a region in which the fluid is at rest. We will further assume that the fluid is an ideal gas with equation of state $p = (\gamma - 1)\epsilon\rho_0$. The Rankine–Hugoinot conditions in this case are

$$[\rho_0 v] = s\,[\rho_0] \,, \tag{7.7.13}$$

$$[Sv + p] = s\,[S] \,, \tag{7.7.14}$$

$$[(\mathcal{E} + p)\,v] = s\,[\mathcal{E}] \,, \tag{7.7.15}$$

[82] A Fortran code for computing the exact solution to the Riemann problem in the Newtonian case can be found in the recent book by Toro [292], and another one for the relativistic case can be found in the online version of the review paper by Marti [200].

with $S = \rho_0 v$, $\mathcal{E} = \rho_0(\epsilon + v^2/2)$, and where we are assuming that on the region to the right the fluid speed vanishes, *i.e.* $v_r = 0$. Notice that since we want to find a single shock wave, rather than specifying jumps in the main variables (ρ_0, ϵ, v) we will solve for the values of these variables on the left of the shock in terms of the values on the right and a given shock speed s.

We can solve the above system of equations either directly or by first going to a frame where the shock is at rest and then transforming back to the laboratory frame (the second route is simpler). The algebra is rather straightforward and not very illuminating, so we will not show it here. We finds that there are in fact two consistent solutions, the first of which corresponds to having a discontinuity of "size zero" propagating at an arbitrary speed. This solution is clearly not interesting as it corresponds to having no discontinuity at all. The non-trivial solution, on the other hand, takes the form:

$$\frac{\rho_l}{\rho_r} = \frac{M^2(\gamma+1)}{2 + M^2(\gamma-1)}, \tag{7.7.16}$$

$$\frac{\epsilon_l}{\epsilon_r} = \frac{\left[2\gamma M^2 - (\gamma-1)\right]\left[2 + M^2(\gamma-1)\right]}{M^2(\gamma+1)^2}, \tag{7.7.17}$$

$$v_l = \frac{2\left(M^2-1\right)c_r}{M(\gamma+1)}. \tag{7.7.18}$$

with $c_r = \sqrt{\gamma p_r/\rho_r} = \sqrt{\gamma(\gamma-1)\epsilon_r}$ the local speed of sound on the right of the shock, and where we have introduced the *Mach number* M defined as the ratio of the shock speed s over c_r

$$M := \frac{s}{c_r} = \frac{s}{\sqrt{\gamma(\gamma-1)\epsilon_r}}. \tag{7.7.19}$$

Using the above expressions, the jump in the pressure can also be found to be

$$\frac{p_l}{p_r} = \frac{2\gamma M^2 - (\gamma-1)}{\gamma+1}. \tag{7.7.20}$$

We can now invert the last relation to solve for the shock speed s in terms of the pressure ratio to find

$$s = c_r \left[\frac{\gamma+1}{2\gamma}\left(\frac{p_l}{p_r}\right) + \frac{\gamma-1}{2\gamma}\right]^{1/2}. \tag{7.7.21}$$

Now, for a shock to be moving to the right we must clearly have $p_l > p_r$, which from the above expression implies that $s > c_r$ $(M > 1)$, or in other words the shock must be supersonic as in enters the region to the right. As the shock moves, fluid elements that were initially at rest acquire a speed $v_l > 0$ (see Figure 7.2).

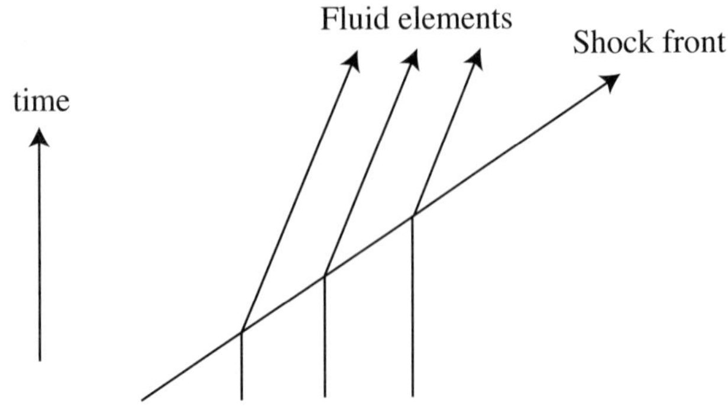

Fig. 7.2: Motion of a single shock wave. As fluid elements that are initially at rest cross the shock front they acquire a speed $v_l > 0$. The shock, however, still moves faster than the fluid elements so that $s > v_l$. The entropy of the fluid elements also increases as they pass through the shock.

Finally, from the discussion of Section 7.5 we can find that the change in entropy as a fluid element crosses the shock is given by

$$\frac{\sigma_l}{\sigma_r} = \ln \frac{p_l}{p_r} - \gamma \ln \frac{\rho_l}{\rho_r} . \tag{7.7.22}$$

Using the expressions above for p_l/p_r and ρ_l/ρ_r we can show that if $M > 1$ and $\gamma > 1$, then $\sigma_l > \sigma_r$ so that the entropy of the fluid elements increases as they move through the shock.

7.8 Imperfect fluids: viscosity and heat conduction

In the previous Sections we have always assumed that we have a perfect fluid. In some situations, however, it might be important to consider the effects of viscosity and heat conduction on the dynamics of the fluid. The study of such effects is known as *irreversible thermodynamics*, and in the Newtonian case leads to the well known Navier–Stokes equations for fluid motion which generalize the Euler equations to the case of non-perfect fluids. In this Section we will consider the relativistic generalization of irreversible thermodynamics. As the subject lies somewhat outside the main scope of this book, we will only discuss the basic physical ideas (for a more detailed discussion see for example the reviews by Maartens [194, 195]). It is important to keep in mind that the theory presented here is essentially phenomenological; a more fundamental treatment can only come from kinetic theory.

7.8.1 *Eckart's irreversible thermodynamics*

The generalization of the Newtonian theory of irreversible thermodynamics to the relativistic case was originally developed by Eckart [116].[83] Let us start by

[83] Landau and Lifshitz [184] give an equivalent formulation to that of Eckart that follows a somewhat different approach, particularly related to the interpretation of the 4-velocity u^μ. In

assuming that the state of a fluid element is always close to equilibrium so that we can still define a local temperature T. The state of the fluid will still be described by the rest mass energy density ρ_0, the specific internal energy ϵ, and the 4-velocity of the fluid elements u^μ, but now the pressure is assumed to include an extra contribution coming from the viscosity and known as the *bulk viscous pressure* Π,

$$p \to p + \Pi \; , \tag{7.8.1}$$

The non-viscous part of the pressure p is still assumed to be given by the same equation of state as before.

The stress-energy tensor of the fluid is generalized by considering the general decomposition of a symmetric tensor in terms of parallel and normal projections along the fluid's 4-velocity field u^μ. Doing this we find the following expression for the stress-energy tensor

$$T_{\mu\nu} = \rho_0 \left(1 + \epsilon\right) u_\mu u_\nu + \left(p + \Pi\right) h_{\mu\nu} + q_\mu u_\nu + q_\mu u_\nu + \pi_{\mu\nu} \; , \tag{7.8.2}$$

with $h_{\mu\nu} = g_{\mu\nu} + u_\mu u_\nu$ the projection operator into the fluid's rest frame, and where q^μ and $\pi_{\mu\nu}$ are a vector and a symmetric tensor such that

$$q_\mu u^\mu = 0 \; , \qquad \pi_{\mu\nu} u^\mu = 0 \; , \qquad \pi^\mu_\mu = 0 \; . \tag{7.8.3}$$

The vector q^μ can be interpreted as the energy flux in the fluid's frame, or in other words the heat flow. The tensor $\pi_{\mu\nu}$, on the other hand, represents the presence of possible anisotropic stresses (the pressure corresponds to the isotropic stresses). The quantities Π, q^μ and $\pi_{\alpha\beta}$ together are usually called the *thermodynamic fluxes*.

The dynamics of the fluid are still obtained from the conservation laws

$$\nabla_\mu \left(\rho_0 u^\mu\right) = 0 \; , \qquad \nabla_\mu T^{\mu\nu} = 0 \; . \tag{7.8.4}$$

Conservation of particles still leads to the same continuity equation as before, which for our purposes we will write in the form

$$\frac{d\rho_0}{d\tau} + \rho_0 \nabla_\mu u^\mu = 0 \; , \tag{7.8.5}$$

with $d/d\tau := u^\mu \nabla_\mu$. The conservation of energy-momentum, on the other hand, is now modified because of the presence of the extra dissipative terms. In particular, contracting the energy-momentum conservation with u^μ results in

$$\frac{d\rho}{d\tau} + \left(\rho + p + \Pi\right) \nabla_\mu u^\mu + \nabla_\mu q^\mu + q^\mu \, \frac{du_\mu}{d\tau} + \pi^{\mu\nu} \sigma_{\mu\nu} = 0 \; , \tag{7.8.6}$$

the Eckart approach used here u^μ corresponds to the flow of particles, while in the approach of Landau and Lifshitz it corresponds instead to the flow of energy. Of course, in the absence of heat conduction energy flows only with the particles, so that for a perfect fluid the two points of view are identical.

with $\rho = \rho_0(1+\epsilon)$, and where $\sigma_{\mu\nu}$ is the *shear tensor* associated with the fluid motion, *i.e* the symmetric tracefree part of the projection of $\nabla_\mu u_\nu$:

$$\sigma_{\mu\nu} := h^\alpha_\mu h^\beta_\nu \left(\nabla_{(\alpha} u_{\beta)} - \frac{1}{3}\, h_{\alpha\beta} \nabla_\lambda u^\lambda \right) . \tag{7.8.7}$$

It is clear that the energy-momentum conservation involves the thermodynamic fluxes Π, q^μ and $\pi_{\mu\nu}$. The system of equations will therefore not be closed unless we can relate these fluxes to quantities associated with the fluid state, also known as the *thermodynamic potentials.*

The standard approach due to Eckart starts by considering the entropy current S^μ and asking for its covariant divergence (*i.e.* the entropy production) to be non-negative, in accordance with the second law of thermodynamics:

$$\nabla_\mu S^\mu \geq 0 \, . \tag{7.8.8}$$

In the case of a perfect fluid the entropy current is simply given by $S^\mu = \sigma \rho_0 u^\mu$, with σ the specific entropy, and as we have seen the conservation equations directly imply $\nabla_\mu S^\mu = 0$. When there is heat flow q^μ, however, it contributes to the entropy flux with a term given q^μ/T ($dS = dQ/T$), with T the temperature of the fluid element, so that

$$S^\mu = \sigma \rho_0 u^\mu + \frac{q^\mu}{T} \, . \tag{7.8.9}$$

Calculating the divergence and using the conservation of particles we find that

$$T \nabla_\mu S^\mu = \rho_0 T \, \frac{d\sigma}{d\tau} - q^\mu \nabla_\mu \ln T + \nabla_\mu q^\mu \, . \tag{7.8.10}$$

Using now the conservation equations (7.8.5) and (7.8.6), together with the first law of thermodynamics in the form

$$T d\sigma = d\epsilon + p \, d \left(\frac{1}{\rho_0} \right) , \tag{7.8.11}$$

we can rewrite the entropy divergence as

$$T \nabla_\mu S^\mu = - \left[\Pi \, \nabla_\mu u^\mu + q^\mu \left(\frac{du_\mu}{d\tau} + \nabla_\mu \ln T \right) + \pi^{\mu\nu} \sigma_{\mu\nu} \right] . \tag{7.8.12}$$

It is now clear that we can guarantee that this divergence will always be positive if we postulate a linear relation between the thermodynamic fluxes and the thermodynamic potentials of the form

$$q_\mu = -\chi \left(T \, \frac{du_\mu}{d\tau} + h^\nu_\mu \nabla_\nu T \right) , \tag{7.8.13}$$

$$\Pi = -\zeta \nabla_\mu u^\mu \, , \tag{7.8.14}$$

$$\pi_{\mu\nu} = -2\eta \sigma_{\mu\nu} \, , \tag{7.8.15}$$

with χ, ζ and η positive parameters known respectively as the coefficients of *heat conduction*, *bulk viscosity* and *shear viscosity* (the projection operator applied to $\nabla_\mu T$ in the equation for q_μ is there to guarantee that $q_\mu u^\mu = 0$). The above expressions are the direct relativistic generalization of the corresponding Newtonian expressions (except for the correction term $T du_\mu/d\tau$ in the expression for q_μ). We can now substitute these laws for the thermodynamic fluxes into the expression for the stress-energy tensor and use the conservation laws to derive the relativistic version of the Navier–Stokes equations, but we will not go into such detail here.

The Eckart theory just presented has a serious drawback in the sense that it is not causal. This can be seen from the fact that if a thermodynamic potential is turned off, the corresponding flux vanishes instantaneously. To make this more concrete consider the problem of heat flow for a fluid at rest in flat spacetime. In that case we have $u^\mu = \delta^\mu_0$ and also $q^0 = 0$, so the conservation equation (7.8.6) reduces to

$$\partial_t \rho = -\sum_i \partial_i q_i \; , \tag{7.8.16}$$

For a fluid at rest we also have $\partial_t \rho = \rho_0 \partial_t \epsilon = \rho_0 c_V \partial_t T$, which implies

$$\rho_0 c_V \partial_t T = -\sum_i \partial_i q_i \; . \tag{7.8.17}$$

On the other hand, the heat flux is now given by $q_i = -\chi \partial_i T$, so that we find

$$\rho_0 c_V \partial_t T = \chi \nabla^2 T \; , \tag{7.8.18}$$

with ∇^2 the ordinary flat space Laplacian. This is nothing more than the standard heat equation. It is a parabolic equation which involves propagation of signals at an infinite speed, which is clearly in contradiction with the basic relativistic law that no physical signal can travel faster than the speed of light. We then see that Eckart's theory, though providing us with a covariant generalization of the Newtonian equations for irreversible thermodynamics, is nevertheless physically unsatisfactory.

7.8.2 *Causal irreversible thermodynamics*

Let us now consider how we can modify the Eckart theory in order to recover causality. The expression for the stress-energy tensor (7.8.2) is clearly correct since it is just the general decomposition of a symmetric tensor in terms of parallel and orthogonal projections to u^μ. The main problem with Eckart's theory is that the modification of the entropy flux is just too simple. We should also consider dissipative contributions to this flux, which according to kinetic theory should be second order in the thermodynamic fluxes. We then postulate the following form for the entropy flux

$$S^\mu = \sigma \rho_0 u^\mu + \frac{q^\mu}{T} - \frac{u^\mu}{2T} \left(\beta_0 \Pi^2 + \beta_1 q_\alpha q^\alpha + \beta_2 \pi_{\alpha\beta} \pi^{\alpha\beta} \right) \; , \tag{7.8.19}$$

with $\beta_i > 0$ new thermodynamic coefficients (one can in principle also include mixed terms, but we will not consider them here). The coefficients β_i need not be constant and might in fact depend on the fluid variables ρ and ϵ, but for simplicity we will take them to be constant in what follows.

We can now calculate the divergence of the entropy flux (7.8.19) in the same way as before to obtain

$$\begin{aligned} T\nabla_\mu S^\mu = & -\Pi \left[\nabla_\mu u^\mu + \beta_0 \frac{d\Pi}{d\tau} + \frac{\beta_0 T}{2} \nabla_\mu \left(\frac{u^\mu}{T} \right) \right] \\ & -q^\mu \left[\frac{du_\mu}{d\tau} + \nabla_\mu \ln T + \beta_1 \frac{dq_\mu}{d\tau} + \frac{\beta_1 T}{2} q_\mu \nabla_\nu \left(\frac{u^\nu}{T} \right) \right] \\ & -\pi^{\mu\nu} \left[\sigma_{\mu\nu} + \beta_2 \frac{d\pi_{\mu\nu}}{d\tau} + \frac{\beta_2 T}{2} \pi_{\mu\nu} \nabla_\lambda \left(\frac{u^\lambda}{T} \right) \right] . \end{aligned} \tag{7.8.20}$$

Again, if we want to guarantee that this divergence is non-negative, we postulate a linear relation between the terms inside square brackets and the thermodynamic fluxes:

$$\Pi = -\zeta \left[\nabla_\mu u^\mu + \beta_0 \frac{d\Pi}{d\tau} + \frac{\beta_0 T}{2} \nabla_\mu \left(\frac{u^\mu}{T} \right) \right] , \tag{7.8.21}$$

$$q_\mu = -\chi T \left[\frac{du_\mu}{d\tau} + h^\nu_\mu \nabla_\nu \ln T + \beta_1 h^\nu_\mu \frac{dq_\nu}{d\tau} + \frac{\beta_1 T}{2} q_\mu \nabla_\nu \left(\frac{u^\nu}{T} \right) \right] , \tag{7.8.22}$$

$$\pi_{\mu\nu} = -2\eta \left[\sigma_{\mu\nu} + \beta_2 h^\alpha_\mu h^\beta_\nu \frac{d\pi_{\alpha\beta}}{d\tau} + \frac{\beta_2 T}{2} \pi_{\mu\nu} \nabla_\lambda \left(\frac{u^\lambda}{T} \right) \right] , \tag{7.8.23}$$

with the coefficients χ, ζ and η the same as before, and where we have again introduced the projection operator h^μ_ν to guarantee that q^μ and $\pi_{\mu\nu}$ remain orthogonal to u^μ. The terms proportional to $\nabla_\nu (u^\nu/T)$ are usually omitted as it can be argued that they are small compared to the other terms. In that case we can rewrite the last expressions as

$$\tau_0 \frac{d\Pi}{d\tau} + \Pi = -\zeta \nabla_\mu u^\mu , \tag{7.8.24}$$

$$\tau_1 h^\nu_\mu \frac{dq_\nu}{d\tau} + q_\mu = -\chi \left(T \frac{du_\mu}{d\tau} + h^\nu_\mu \nabla_\nu T \right) , \tag{7.8.25}$$

$$\tau_2 h^\alpha_\mu h^\beta_\nu \frac{d\pi_{\alpha\beta}}{d\tau} + \pi_{\mu\nu} = -2\eta \sigma_{\mu\nu} , \tag{7.8.26}$$

where we have introduced the shorthand

$$\tau_0 = \zeta \beta_0 , \tag{7.8.27}$$

$$\tau_1 = \chi T \beta_1 , \tag{7.8.28}$$

$$\tau_2 = 2\eta \beta_2 . \tag{7.8.29}$$

The form of the relations (7.8.24)–(7.8.26) is known as the Maxwell–Cattaneo form of the transport equations (or rather, its relativistic generalization), and they are evolution equations instead of simple algebraic relations as before. The parameters τ_i play the role of relaxation times. We now see that if a thermodynamic potential is suddenly switched off, the corresponding flux will die away slowly in a time of order τ_i. The values of the τ_i, or correspondingly the β_i, can be roughly estimated as the mean collision time between particles. The theory just presented is known as the (truncated) *Israel–Stewart* theory of relativistic irreversible thermodynamics [167].

To see how this theory solves the causality problem, consider again heat flow in a static fluid in flat spacetime. As before, the conservation equation still implies

$$\rho_0 c_V \partial_t T = -\sum_i \partial_i q_i \; . \tag{7.8.30}$$

But the heat flux now satisfies the equation

$$\tau_1 \partial_t q_i = -\chi \partial_i T - q_i \; . \tag{7.8.31}$$

This implies that

$$\begin{aligned} \tau_1 \rho_0 c_V \partial_t^2 T &= -\tau_1 \sum_i \partial_i \partial_t q_i \\ &= \chi \nabla^2 T + \sum_i \partial_i q_i \; , \end{aligned} \tag{7.8.32}$$

and finally

$$\tau_1 \partial_t^2 T - \frac{\chi}{\rho_0 c_V} \nabla^2 T + \partial_t T = 0 \; . \tag{7.8.33}$$

Instead of the standard parabolic heat equation we have now obtained a damped wave equation. The new equation is clearly hyperbolic, so the speed of propagation is now finite.

8

GRAVITATIONAL WAVE EXTRACTION

8.1 Introduction

Gravitational waves, that is perturbations in the geometry of spacetime that propagate at the speed of light, are one of the most important predictions of general relativity. Though such gravitational radiation has not yet been detected directly, there is strong indirect evidence of its existence in the form of the now famous binary pulsar PSR 1913+16, whose change in orbital period over time matches to very good accuracy the value predicted by general relativity as a consequence of the emission of gravitational waves [164, 282]. Moreover, there is every reason to believe that the new generation of large interferometric observatories (LIGO, VIRGO, GEO 600, TAMA) will finally succeed in detecting gravitational radiation within the next few years.

Gravitational waves are one of the most important physical phenomena associated with the presence of strong and dynamic gravitational fields, and as such they are of great interest in numerical relativity. Gravitational radiation can carry energy and momentum away from an isolated system, and it encodes information about the physical properties of the system itself. Predicting the gravitational wave signal coming from astrophysical systems has been one of the main themes in numerical relativity over the years, because such predictions can be used as templates that can significantly improve the possibility of detection.

There are two main approaches to the extraction of gravitational wave information from a numerical simulation. For a number of years the traditional approach has been based on the theory of perturbations of a Schwarzschild spacetime developed originally by Regge and Wheeler [235], Zerilli [308, 309], and a number of other authors, and later recast as a gauge invariant framework by Moncrief [209]. In recent years, however, it has become increasingly common in numerical relativity to extract gravitational wave information in terms of the components of the Weyl curvature tensor with respect to a frame of null vectors, using what is known as the *Newman–Penrose formalism* [218]. In the following Sections, I will present a brief introduction to both these approaches, and will describe how we can calculate the energy and momentum radiated by gravitational waves in each case.

Finally, a word of warning. Unfortunately, though the main ideas and results presented here are well known, there are no standard conventions in either the definitions or the notation used for many of the concepts introduced here (in particular, sign conventions in the definitions of Weyl scalars, spacetime invariants, *etc.*, can be very different from author and to author). Here I will use of notation and definitions that are common, though not universal, and I will try to keep the conventions consistent throughout. However, the reader is advised not to mix expressions taken from different sources without first carefully comparing the conventions used.

8.2 Gauge invariant perturbations of Schwarzschild

In this Section we will briefly discuss some of the ideas behind the theory of perturbations of a Schwarszchild black hole, concentrating particularly on those aspects that are important for gravitational wave extraction. For a more detailed discussion the reader is directed to the classic book by Chandrasekar [96], or the recent reviews by Kokkotas and Schmidt [175], Martel and Poisson [197], and Nagar and Rezolla [211].

Let us start by assuming that we have a metric of the form

$$g_{\mu\nu} = g^{(0)}_{\mu\nu} + h_{\mu\nu} \; , \tag{8.2.1}$$

where $|h_{\mu\nu}| \ll 1$ is a small perturbation, and $g^{(0)}_{\mu\nu}$ is the Schwarzschild metric in standard coordinates

$$g^{(0)}_{\mu\nu}\, dx^\mu dx^\nu = -\left(1 - \frac{2M}{r}\right) dt^2 + \left(1 - \frac{2M}{r}\right)^{-1} dr^2 + r^2 d\Omega^2 \; . \tag{8.2.2}$$

Because of the spherical symmetry of the background metric, it is convenient to think of the full spacetime manifold $\mathcal{M}$ as the product of two sub-manifolds, namely a Lorentzian two-dimensional manifold M^2 associated with the coordinates (t, r), and the two-sphere of unit radius S^2 associated with (θ, φ). We can then write our full four-dimensional manifold as $\mathcal{M} = M^2 \times S^2$. Following [211], in what follows we will make this separation explicit by using lower case Latin indices $(a, b, \dots)$ to represent the coordinates in S^2, and upper case indices $(A, B, \dots)$ for the coordinates in M^2. We can then rewrite the metric as

$$ds^2 = g_{AB}\, dx^A dx^B + r^2 \Omega_{ab}\, dx^a dx^b \; , \tag{8.2.3}$$

with Ω_{ab} the metric of the unit two-sphere, $\Omega_{ab} = \mathrm{diag}(1, \sin^2\theta)$. We will also distinguish covariant derivatives in the full spacetime from covariant derivatives in the sub-manifolds in the following way: ∇_μ will represent covariant derivatives in $\mathcal{M}$, while D_A and D_a will denote covariant derivatives in M^2 and S^2 respectively.

8.2.1 *Multipole expansion*

At this point it is in fact convenient to expand the metric perturbations $h_{\mu\nu}$ in multipoles using spherical harmonics $Y^{l,m}(\theta, \varphi)$, and separate those multipoles according to their properties with respect to parity transformations (*i.e.* reflections through the origin) $(\theta, \varphi) \rightarrow (\pi - \theta, \pi + \varphi)$. We say that a given multipole is *even* or *polar* if it transforms as $(-1)^l$, and *odd* or *axial* if it transforms instead as $(-1)^{l+1}$. The names even and odd come from the fact that once we multiply with the corresponding spherical harmonic, even perturbations remain unchanged under a parity transformation, while odd perturbations change sign. The names polar and axial, on the other hand, refer to the behavior under a local change from a right handed to a left handed basis: Polar perturbations remain unchanged, while axial perturbations again change sign.

In order to decompose $h_{\mu\nu}$, we further need to introduce the scalar, vector and tensor spherical harmonics. These tensorial properties refer only to transformations of coordinates in the unit sphere, so that in this sense we can regard the components h_{AB} as three scalars, h_{Ab} as two vectors, and h_{ab} as one tensor. Scalar harmonics are the usual functions $Y^{l,m}(\theta,\varphi)$ (see Appendix D). Vector harmonics, on the other hand, come in two different types. The even vector harmonics are simply defined as the gradient of the scalar harmonics on the sphere

$$Y_a^{l,m} := D_a Y^{l,m} , \tag{8.2.4}$$

while the odd vector harmonics are

$$X_a^{l,m} := -\epsilon_a{}^b D_b Y^{l,m} = -\epsilon_{ac}\, \Omega^{cb} D_b Y^{l,m} , \tag{8.2.5}$$

where ϵ_{ab} is the Levi–Civita completely antisymmetric tensor on the two-sphere ($\epsilon_{\theta\varphi} = -\epsilon_{\varphi\theta} = \Omega^{1/2} = \sin\theta$).

Similarly, we can define tensor harmonics of even and odd type. Even tensor harmonics can be constructed in two ways, either by multiplying the scalar harmonics with the angular metric Ω_{ab}, or by taking a second covariant derivative of the $Y^{l,m}$:

$$\Omega_{ab} Y^{l,m} , \quad D_a D_b Y^{l,m} . \tag{8.2.6}$$

However, there is a subtle point here. It turns out that these functions do not form a linearly independent set, so instead of the $D_a D_b Y^{l,m}$, it is better to use the so-called *Zerilli–Mathews tensor harmonics* defined as

$$Z_{ab}^{l,m} := D_a D_b Y^{l,m} + \frac{1}{2}\, l\,(l+1)\, \Omega_{ab} Y^{l,m} . \tag{8.2.7}$$

Remembering now that the scalar spherical harmonics are eigenfunctions of the Laplace operator on the sphere, that is

$$D^a D_a Y^{l,m} = \left[\partial_\theta^2 + \frac{1}{\sin^2\theta}\, \partial_\varphi^2 + \cot\theta\, \partial_\theta \right] Y^{l,m} = -l\,(l+1)\, Y^{l,m} , \tag{8.2.8}$$

we can easily see that the Zerilli–Mathews tensor harmonics $Z_{ab}^{l,m}$ are in fact traceless. We can also show that the divergence of $Z_{ab}^{l,m}$ is given by

$$D^b Z_{ab}^{l,m} = -\frac{1}{2}(l-1)(l+2)\, Y_a^{l,m} , \tag{8.2.9}$$

where in order to derive this expression we need to use the fact that the Ricci tensor for the two-sphere is just equal to the angular metric $R_{ab} = \Omega_{ab}$.

Odd parity tensor harmonics, on the other hand, can be constructed in only one way, namely

$$X_{ab}^{l,m} = \frac{1}{2}\left(D_a X_b^{l,m} + D_b X_a^{l,m}\right) . \tag{8.2.10}$$

From the definition of the $X_a^{l,m}$ we can easily see that the odd tensor harmonics are again traceless. We also find for the divergence of $X_{ab}^{l,m}$

$$D^b X_{ab}^{l,m} = -\frac{1}{2}(l-1)(l+2)\, X_a^{l,m} . \tag{8.2.11}$$

At this point it is important to mention the fact that, since Y^{00} is a constant, both vector and tensor harmonics vanish for $l = 0$. On the other hand, for $l = 1$ the vector harmonics do not vanish, but the tensor harmonics can still be easily shown to vanish from the explicit expressions for Y^{1m}. This means that vector harmonics are only non-zero for $l \geq 1$, and tensor harmonics for $l \geq 2$. The scalar mode with $l = 0$ can be interpreted as a variation in the mass of the Schwarzschild spacetime, the odd vector mode with $l = 1$ as an infinitesimal contribution to the angular momentum (the "Kerr" mode [252]), while the scalar and even vector modes with $l = 1$ are just gauge [197]. Also, for any given value of l, the index m can only take integer values from $-l$ to l.

Having defined the vector and tensor harmonics, the perturbed metric is expanded in multipoles, and separated into its even sector given by

$$\left(h_{AB}^{l,m}\right)^{(e)} = H_{AB}^{l,m}\, Y^{l,m} , \tag{8.2.12}$$

$$\left(h_{Ab}^{l,m}\right)^{(e)} = H_A^{l,m}\, Y_b^{l,m} , \tag{8.2.13}$$

$$\left(h_{ab}^{l,m}\right)^{(e)} = r^2 \left(K^{l,m}\, \Omega_{ab} Y^{l,m} + G^{l,m} Z_{ab}^{l,m}\right) , \tag{8.2.14}$$

and its odd sector given by

$$\left(h_{AB}^{l,m}\right)^{(o)} = 0 , \tag{8.2.15}$$

$$\left(h_{Ab}^{l,m}\right)^{(o)} = h_A^{l,m}\, X_b^{l,m} , \tag{8.2.16}$$

$$\left(h_{ab}^{l,m}\right)^{(o)} = h^{l,m}\, X_{ab}^{l,m} , \tag{8.2.17}$$

where the coefficients $(H_{AB}^{l,m}, H_A^{l,m}, K^{l,m}, G^{l,m}, h_A^{l,m}, h^{l,m})$ are in general functions of r and t.

The vector and tensor harmonics introduced above are related to the spin-weighted spherical harmonics defined in Appendix D. In order to find this relation, consider the two unit complex vectors (these vectors will appear again in Section 8.5 when we discuss the Newman–Penrose formalism)

$$m_a := \frac{1}{\sqrt{2}} (1, i \sin\theta) , \qquad \bar{m}_a := \frac{1}{\sqrt{2}} (1, -i \sin\theta) , \tag{8.2.18}$$

where $\bar{z}$ denotes the complex conjugate of z. In terms of $(m_a, \bar{m}_a)$, we find that the vector harmonics $Y_a^{l,m}$ and $X_a^{l,m}$ can be written as

$$Y_a^{l,m} = \left(\frac{l(l+1)}{2}\right)^{1/2} \left[{}_{-1}Y^{l,m} \, m_a - {}_1Y^{l,m} \, \bar{m}_a\right] , \tag{8.2.19}$$

$$X_a^{l,m} = -i \left(\frac{l(l+1)}{2}\right)^{1/2} \left[{}_{-1}Y^{l,m} \, m_a + {}_1Y^{l,m} \, \bar{m}_a\right] . \tag{8.2.20}$$

Similarly, for the tensor harmonics we find

$$Z_{ab}^{l,m} = \frac{1}{2} \left(\frac{(l+2)!}{(l-2)!}\right)^{1/2} \left[{}_{-2}Y^{l,m} \, m_a m_b + {}_2Y^{l,m} \, \bar{m}_a \bar{m}_b\right] , \tag{8.2.21}$$

$$X_{ab}^{l,m} = -\frac{i}{2} \left(\frac{(l+2)!}{(l-2)!}\right)^{1/2} \left[{}_{-2}Y^{l,m} \, m_a m_b - {}_2Y^{l,m} \, \bar{m}_a \bar{m}_b\right] , \tag{8.2.22}$$

where in order to derive the last expressions we have again used the fact that the scalar harmonics are eigenfunctions of the Laplace operator on the sphere. Notice again that, since the spin-weighted harmonics are defined only for $l > |s|$, the above expressions are consistent with the fact that vector harmonics start with $l = 1$ and tensor harmonics with $l = 2$.

8.2.2 *Even parity perturbations*

As we have seen, even parity perturbations are characterized by the coefficients $(H_{AB}^{l,m}, H_A^{l,m}, K^{l,m}, G^{l,m})$. These coefficients are clearly coordinate dependent, and in particular change under infinitesimal coordinate transformations, *i.e.* gauge transformations, of the form

$$x^\mu \to x^\mu + \xi^\mu , \tag{8.2.23}$$

with $|\xi^\mu| << 1$. Such gauge transformations were already discussed in Chapter 1, and we can easily show that they result in a change in the metric perturbation of the form

$$h_{\mu\nu} \to h_{\mu\nu} - \nabla_\mu \xi_\nu - \nabla_\nu \xi_\mu . \tag{8.2.24}$$

The only difference between this expression and that of Chapter 1 is the fact that we now have a non-trivial background metric and, as a consequence, the gauge transformation involves covariant derivatives with respect to the background metric.

It turns out that we can in fact construct gauge invariant combinations of the coefficients $(H_{AB}^{l,m}, H_A^{l,m}, K^{l,m}, G^{l,m})$. In order to do this, let us consider an even-parity gauge transformation $\xi_\mu = (\xi_A, \xi_a)$, which can be expanded in terms of scalar and vector harmonics as

$$\xi_A = \sum_{l,m} E_A^{l,m} Y^{l,m} , \qquad \xi_a = \sum_{l,m} E^{l,m} Y_a^{l,m} , \tag{8.2.25}$$

We now need to use this expression to find the transformation of the coefficients corresponding to an even perturbation. We start from the expression for the mixed Christoffel symbols which can be easily shown to be

$$\Gamma^A_{Bc} = \Gamma^a_{BC} = 0 \,, \quad \Gamma^A_{bc} = -r r^A \Omega_{bc} \,, \quad \Gamma^a_{Bc} = \frac{r_B}{r}\, \delta^a_c \,, \tag{8.2.26}$$

where $r_A := D_A r$. Using this we find that

$$\nabla_A \xi_B = \sum_{l,m} \left(D_A E_B^{l,m} \right) Y^{l,m} , \tag{8.2.27}$$

$$\nabla_A \xi_b = \sum_{l,m} \left(D_A E^{l,m} - \frac{r_A}{r}\, E^{l,m} \right) Y_b^{l,m} , \tag{8.2.28}$$

$$\nabla_a \xi_B = \sum_{l,m} \left(E_B^{l,m} - \frac{r_B}{r}\, E^{l,m} \right) Y_a^{l,m} , \tag{8.2.29}$$

$$\nabla_a \xi_b = \sum_{l,m} \left[\left(D_a Y_b^{l,m} \right) E^{l,m} + r \Omega_{ab}\, r^C E_C^{l,m} Y^{l,m} \right] . \tag{8.2.30}$$

The gauge transformation of $(H_{AB}^{l,m}, H_A^{l,m}, K^{l,m}, G^{l,m})$ then takes the form

$$H_{AB}^{l,m} \rightarrow H_{AB}^{l,m} - D_A E_B^{l,m} - D_B E_A^{l,m} , \tag{8.2.31}$$

$$H_A^{l,m} \rightarrow H_A^{l,m} - E_A^{l,m} - D_A E^{l,m} + \frac{2 r_A}{r}\, E^{l,m} , \tag{8.2.32}$$

$$K^{l,m} \rightarrow K^{l,m} - \frac{2}{r}\, r^A E_A^{l,m} + \frac{l(l+1)}{r^2}\, E^{l,m} , \tag{8.2.33}$$

$$G^{l,m} \rightarrow G^{l,m} - \frac{2}{r^2}\, E^{l,m} . \tag{8.2.34}$$

We can now use these results to construct gauge invariant combinations of coefficients. Two such invariant combinations are [143]

$$\tilde{K}^{l,m} := K^{l,m} + \frac{1}{2}\, l(l+1)\, G^{l,m} - \frac{2}{r}\, r^A \varepsilon_A^{l,m} , \tag{8.2.35}$$

$$\tilde{H}_{AB}^{l,m} := H_{AB}^{l,m} - D_A \varepsilon_B^{l,m} - D_B \varepsilon_A^{l,m} , \tag{8.2.36}$$

with

$$\varepsilon_A^{l,m} := H_A^{l,m} - \frac{1}{2}\, r^2 D_A G^{l,m} . \tag{8.2.37}$$

Instead of looking for gauge invariant combinations we can choose to work on a specific gauge. For example, from the transformations above it is clear that we can choose $E^{l,m}$ and $E_A^{l,m}$ such that $G^{l,m} = H_A^{l,m} = 0$. This is known as the *Regge–Wheeler gauge*, and with that choice we clearly has $\tilde{K}^{l,m} = K^{l,m}$, $\tilde{H}_{AB}^{l,m} = H_{AB}^{l,m}$.

In terms of the gauge invariant perturbations $\tilde{K}^{l,m}$ and $\tilde{H}_{AB}^{l,m}$ we define the so-called *Zerilli–Moncrief master function* as

$$\Psi_{\text{even}}^{l,m} := \frac{2r}{l(l+1)} \left[\tilde{K}^{l,m} + \frac{2}{\Lambda} \left(r^A r^B \tilde{H}_{AB}^{l,m} - r r^A D_A \tilde{K}^{l,m} \right) \right] , \tag{8.2.38}$$

with $\Lambda := (l-1)(l+2) + 6M/r$. The importance of this definition comes from the fact that using the perturbed Einstein field equations we can show that $\Psi_{\text{even}}^{l,m}$ obeys a simple wave-like equation known as the *Zerilli equation*, or more generally the *even-parity master equation*:

$$\partial_t^2 \Psi_{\text{even}}^{l,m} - \partial_{r^*}^2 \Psi_{\text{even}}^{l,m} + V_{\text{even}}^{l,m} \, \Psi_{\text{even}}^{l,m} = S_{\text{even}}^{l,m} , \tag{8.2.39}$$

with r^* the tortoise radial coordinate defined as $r^* := 1 + 2M \ln{(r/2M - 1)}$, and where $V_{\text{even}}^{l,m}$ and $S_{\text{even}}^{l,m}$ are a potential and a source term that depend on r and l, but whose explicit expression is not important for the present discussion (but see *e.g.* [197]).

The Zerilli equation is specially useful in the frequency domain, for which we assume that $\Psi_{\text{even}}^{l,m}(t,r) = \exp(i\omega_n t) F(r)$, with ω_n some complex frequency. Substituting this into the Zerilli equation, and imposing boundary conditions such that we have purely outgoing waves at spatial infinity ($r^* \to \infty$) and purely ingoing waves at the black hole horizon ($r^* \to -\infty$), we obtain an eigenvalue problem for the complex frequencies ω_n. The different values of ω_n found in this way give us the allowed frequencies of oscillation of the perturbations (the real part of ω_n) and their corresponding decay rates (the imaginary part of ω_n). Such solutions are known as the *quasi-normal modes* of the black hole.

We will not go into a discussion of the quasi-normal modes of Schwarzschild, but we will mention a few important points. First, all the resulting modes are damped, with the higher frequency (shorter wavelength) modes damped the fastest. The so-called *fundamental mode* $l = 2$, which is often the most exited and decays the slowest, has a wavelength of $\text{Re}(\omega) \sim 16.8M$. When the black hole is excited (*e.g.* by throwing matter into it), it behaves much like a bell in the sense that it oscillates at its characteristic frequencies, losing energy in the form of gravitational waves, and eventually settling down to a stationary state. Such behavior is known as “ringing”, and we typically find that the late time gravitational wave signal coming from a system that has collapsed to a single black hole can always be matched to a superposition of the quasi-normal modes of the final black hole. The reader interested in studying these issues in more detail is directed to references [96, 175].

8.2.3 *Odd parity perturbations*

Just as we did for the even perturbations, we can also construct gauge invariant quantities for the case of odd perturbations. We then start by considering an odd gauge transformation $\xi_\mu = (\xi_A, \xi_a)$, which can now be expanded in terms of scalar and vector harmonics as

$$\xi_A = 0 \, , \qquad \xi_a = \sum_{l,m} E^{l,m} X_a^{l,m} \, . \tag{8.2.40}$$

Using this we now find that

$$\nabla_A \xi_B = 0 \, , \tag{8.2.41}$$

$$\nabla_A \xi_b = \sum_{l,m} \left(D_A E^{l,m} - \frac{r_A}{r} \, E^{l,m} \right) X_b^{l,m} \, , \tag{8.2.42}$$

$$\nabla_a \xi_B = - \sum_{l,m} \frac{r_B}{r} \, E^{l,m} X_a^{l,m} \, , \tag{8.2.43}$$

$$\nabla_a \xi_b = \sum_{l,m} \left(D_a X_b^{l,m} \right) E^{l,m} \, , \tag{8.2.44}$$

with r_A the same as before. The coefficients $(h_A^{l,m}, h^{l,m})$ for the odd perturbations then transform as

$$h_A^{l,m} \to h_A^{l,m} - D_A E^{l,m} + \frac{2 r_A}{r} \, E^{l,m} \, , \tag{8.2.45}$$

$$h^{l,m} \to h^{l,m} - 2 E^{l,m} \, . \tag{8.2.46}$$

Just as before, we can use these transformations to construct a gauge invariant quantity in the following way [143]

$$\tilde{h}_A^{l,m} := h_A^{l,m} - \frac{1}{2} \, D_A h^{l,m} + \frac{r_A}{r} \, h^{l,m} \, . \tag{8.2.47}$$

Notice that, from the transformations above, it is clear that we can always choose $E^{l,m}$ such that $h^{l,m} = 0$, in which case we will have $\tilde{h}_A^{l,m} = h_A^{l,m}$, corresponding to the Regge–Wheeler gauge.

In terms of $\tilde{h}_A^{l,m}$ we define the *Cunningham–Price–Moncrief master function:*

$$\begin{aligned} \Psi_{\rm odd}^{l,m} &:= \frac{2r}{(l-1)(l+2)} \, \epsilon^{AB} \left[D_A \tilde{h}_B^{l,m} - \frac{2 r_A}{r} \, \tilde{h}_B^{l,m} \right] \\ &= \frac{2r}{(l-1)(l+2)} \, \epsilon^{AB} \left[D_A h_B^{l,m} - \frac{2 r_A}{r} \, h_B^{l,m} \right] \, . \end{aligned} \tag{8.2.48}$$

The second equality shows that $\Psi_{\rm odd}^{l,m}$ in fact depends only on $h_A^{l,m}$ and not on $h^{l,m}$ (the contributions from $h^{l,m}$ cancel when contracted with the ϵ^{AB}), but it is nevertheless gauge invariant (we can also change $D_A h_B^{l,m}$ for $\partial_A h_B^{l,m}$ since the Christoffel symbols also drop out, but we leave the covariant derivative to show explicitly that $\Psi_{\rm odd}^{l,m}$ is covariant).

Again, using the perturbed Einstein field equations we can show that $\Psi_{\rm odd}^{l,m}$ obeys a wave-like equation of the form

$$\partial_t^2 \Psi_{\rm odd}^{l,m} - \partial_{r^*}^2 \Psi_{\rm odd}^{l,m} + V_{\rm odd}^{l,m} \, \Psi_{\rm odd}^{l,m} = S_{\rm odd}^{l,m} \,, \tag{8.2.49}$$

where again $V_{\rm odd}^{l,m}$ and $S_{\rm odd}^{l,m}$ are a potential and a source term whose explicit expression we will not write here. The above equation is known as the *Regge–Wheeler equation*, or more generally the *odd-parity master equation.*

8.2.4 *Gravitational radiation in the TT gauge*

We now need to relate the gauge invariant perturbations introduced in the last two sections with the gravitational waves amplitudes in the transverse-traceless (TT) gauge, h^+ and $h^\times$. As we will see in Section 8.9, this will allow us to compute the energy and momentum carried by the gravitational waves.

Consider first the even parity perturbations. If asymptotically we approach the TT gauge then we will find that h_{AB} and h_{Ab} decay much faster than h_{ab}. According to the multiple expansion we can then ignore the coefficients $H_{AB}^{l,m}$ and $H_A^{l,m}$, and concentrate only on $K^{l,m}$ and $G^{l,m}$. Consider now an orthonormal angular basis $(\hat{e}_\theta, \hat{e}_\varphi)$. From the discussion on gravitational waves of Chapter 1 we see that in this basis we will have

$$h^+ = \frac{1}{2}\left(h_{\hat{\theta}\hat{\theta}} - h_{\hat{\varphi}\hat{\varphi}}\right) = \frac{1}{2r^2}\left(h_{\theta\theta} - \frac{h_{\varphi\varphi}}{\sin^2\theta}\right) , \tag{8.2.50}$$

$$h^\times = h_{\hat{\theta}\hat{\varphi}} = \frac{h_{\theta\varphi}}{r^2\sin\theta} \, . \tag{8.2.51}$$

Using now the multiple expansion we find that this implies

$$(h^+)_{\rm even}^{l,m} = \frac{G^{l,m}}{2}\left(Z_{\theta\theta}^{l,m} - \frac{Z_{\varphi\varphi}^{l,m}}{\sin^2\theta}\right) , \tag{8.2.52}$$

$$(h^\times)_{\rm even}^{l,m} = G^{l,m}\left(\frac{Z_{\theta\varphi}^{l,m}}{\sin\theta}\right) . \tag{8.2.53}$$

On the other hand, from the traceless condition we have

$$\Omega^{ab}\left(K^{l,m}\Omega_{ab}Y^{l,m} + G^{l,m}Z_{ab}^{l,m}\right) = 0 \, . \tag{8.2.54}$$

But we have already seen that $Z_{ab}^{l,m}$ is itself traceless, so the last condition reduces to $K^{l,m} = 0$. The Zerilli–Moncrief function defined in (8.2.38) then simplifies to

$$\Psi_{\rm even}^{l,m} = r\, G^{l,m} \, . \tag{8.2.55}$$

We then see that the contribution from the even perturbations to the TT metric functions is given in terms of $\Psi^{l,m}_{\rm even}$ as

$$
\begin{aligned}
(h^+)^{l,m}_{\rm even} &= \frac{\Psi^{l,m}_{\rm even}}{2r} \left(Z^{l,m}_{\theta\theta} - \frac{Z^{l,m}_{\varphi\varphi}}{\sin^2\theta} \right) \\
&= \frac{\Psi^{l,m}_{\rm even}}{r} \left[\frac{\partial^2}{\partial\theta^2} + \frac{1}{2}\, l(l+1) \right] Y^{l,m} \,, \qquad (8.2.56)
\end{aligned}
$$

$$
\begin{aligned}
(h^\times)^{l,m}_{\rm even} &= \frac{\Psi^{l,m}_{\rm even}}{r} \left(\frac{Z^{l,m}_{\theta\varphi}}{\sin\theta} \right) \\
&= \frac{\Psi^{l,m}_{\rm even}}{r} \left(\frac{im}{\sin\theta} \right) \left[\frac{\partial}{\partial\theta} - \cot\theta \right] Y^{l,m} \,, \qquad (8.2.57)
\end{aligned}
$$

where we used the fact that $\partial_\varphi Y^{l,m} = im Y^{l,m}$.

Consider now the odd perturbations, for which we have

$$
(h^+)^{l,m}_{\rm odd} = \frac{h^{l,m}}{2r^2} \left(X^{l,m}_{\theta\theta} - \frac{X^{l,m}_{\varphi\varphi}}{\sin^2\theta} \right) \,, \qquad (8.2.58)
$$

$$
(h^\times)^{l,m}_{\rm odd} = \frac{h^{l,m}}{r^2} \left(\frac{X^{l,m}_{\theta\varphi}}{\sin\theta} \right) \,. \qquad (8.2.59)
$$

We now need to relate $h^{l,m}$ to $\Psi^{l,m}_{\rm odd}$. In this case, however, we can not just ignore $h^{l,m}_A$ in favor of $h^{l,m}$ since from the definition of $\Psi^{l,m}_{\rm odd}$, equation (8.2.48), we see that it in fact depends only on $h^{l,m}_A$ and not on $h^{l,m}$. However, in the TT gauge these quantities are related to each other. In order to see this, consider the transverse condition on $h_{\mu a}$:

$$
\nabla^\mu h_{\mu a} = 0 \,. \qquad (8.2.60)
$$

Using now the multipole expansion and the expression for the mixed Christoffel symbols we find that this implies

$$
X^{l,m}_a D^A \left(r^2 h^{l,m}_A \right) + h^{l,m} D^b X^{l,m}_{ab} = 0 \,. \qquad (8.2.61)
$$

Substituting now the divergence of $X^{l,m}_{ab}$ from equation (8.2.11), this becomes

$$
D^A \left(r^2 h^{l,m}_A \right) = \frac{1}{2}\,(l-1)(l+2)\, h^{l,m} \,. \qquad (8.2.62)
$$

Now, remember also that in the TT gauge we have the extra freedom of taking $h_{\mu\nu}$ to be purely spatial, so that $h_{\mu 0} = 0$. Also, asymptotically the metric g_{AB} is just the Minkowski metric. The previous expression then reduces to

$$\partial_r \left(r^2 h_r^{l,m} \right) = \frac{1}{2} \, (l-1)(l+2) \, h^{l,m} \, . \tag{8.2.63}$$

Moreover, we can also rewrite expression (8.2.48) for $\Psi^{l,m}_{\rm odd}$ as

$$\Psi^{l,m}_{\rm odd} = \frac{2r}{(l-1)(l+2)} \, \partial_t h_r^{l,m} \, . \tag{8.2.64}$$

Collecting results we find that

$$\partial_r \left(r \Psi^{l,m}_{\rm odd} \right) = \frac{2}{(l-1)(l+2)} \, \partial_t \partial_r \left(r^2 h_r^{l,m} \right) = \partial_t h^{l,m} \, . \tag{8.2.65}$$

For an outgoing wave we have $\partial_t h^{l,m} \sim -\partial_r h^{l,m}$, so that we can integrate the above expression to find

$$h^{l,m} \sim -r \, \Psi^{l,m}_{\rm odd} \, . \tag{8.2.66}$$

We can then rewrite the odd metric perturbations as

$$\begin{aligned} (h^+)^{l,m}_{\rm odd} &= - \frac{\Psi^{l,m}_{\rm odd}}{2r} \left(X^{l,m}_{\theta\theta} - \frac{X^{l,m}_{\varphi\varphi}}{\sin^2\theta} \right) \\ &= - \frac{\Psi^{l,m}_{\rm odd}}{r} \left(\frac{im}{\sin\theta} \right) \left[\frac{\partial}{\partial\theta} - \cot\theta \right] Y^{l,m} \, , \end{aligned} \tag{8.2.67}$$

$$\begin{aligned} (h^\times)^{l,m}_{\rm odd} &= - \frac{\Psi^{l,m}_{\rm odd}}{r} \left(\frac{X^{l,m}_{\theta\varphi}}{\sin\theta} \right) \\ &= \frac{\Psi^{l,m}_{\rm odd}}{r} \left[\frac{\partial^2}{\partial\theta^2} + \frac{1}{2} \, l(l+1) \right] Y^{l,m} \, . \end{aligned} \tag{8.2.68}$$

The full TT coefficients h^+ and $h^\times$ then take the final form

$$h^+ = \frac{1}{2r} \sum_{l,m} \left[\Psi^{l,m}_{\rm even} \left(Z^{l,m}_{\theta\theta} - \frac{Z^{l,m}_{\varphi\varphi}}{\sin^2\theta} \right) - \Psi^{l,m}_{\rm odd} \left(X^{l,m}_{\theta\theta} - \frac{X^{l,m}_{\varphi\varphi}}{\sin^2\theta} \right) \right] , \tag{8.2.69}$$

$$h^\times = \frac{1}{r} \sum_{l,m} \left[\Psi^{l,m}_{\rm even} \left(\frac{Z^{l,m}_{\theta\varphi}}{\sin\theta} \right) - \Psi^{l,m}_{\rm odd} \left(\frac{X^{l,m}_{\theta\varphi}}{\sin\theta} \right) \right] . \tag{8.2.70}$$

Notice that since the above expressions involve only tensor harmonics, the sum over l starts with $l = 2$, while the sum over m goes from $-l$ to l. In what follows we will always assume this to be the case, and will never write the summation limits explicitly.[84]

[84] For summations involving coefficients with indices $l' = l \pm 1$ and $m' = m \pm 1$, such as those that will appear when discussing the linear and angular momentum carried by gravitational waves in Section 8.9, we should only remember that the expansion coefficients vanish whenever $l' < 2$ and $|m'| > l$.

Notice that while the functions $(\Psi^{l,m}_{\rm even}, \Psi^{l,m}_{\rm even})$ are in general complex, the TT coefficients h^+ and $h^\times$ must be real. Using the properties of the spherical harmonics under complex conjugation we can easily show that this implies

$$\bar{\Psi}^{l,m}_{\rm even} = (-1)^m \, \Psi^{l,-m}_{\rm even} \,, \qquad \bar{\Psi}^{l,m}_{\rm odd} = (-1)^m \, \Psi^{l,-m}_{\rm odd} \,. \tag{8.2.71}$$

We can also rewrite the above expressions for h^+ and $h^\times$ in terms of spin-weighted spherical harmonics. For this it is in fact more convenient to consider the complex combination $H := h^+ - ih^\times$, for which we find

$$H = \frac{1}{2r} \sum_{l,m} \left(\frac{(l+2)!}{(l-2)!} \right)^{1/2} \left[\Psi^{l,m}_{\rm even} + i \Psi^{l,m}_{\rm odd} \right] \, {}_{-2}Y^{l,m} \,. \tag{8.2.72}$$

Before leaving this section it is important to mention one very common convention used in numerical relativity. We start by considering a different odd master function originally introduced by Moncrief [209]:

$$Q^{l,m}_{\rm M} := \frac{2r^A \tilde{h}^{l,m}_A}{r} = \frac{2r^A}{r} \left(h^{l,m}_A - \frac{1}{2} \, D_A h^{l,m} + \frac{r_A}{r} \, h^{l,m} \right) \,. \tag{8.2.73}$$

Notice that with this definition $Q^{l,m}_{\rm M}$ is clearly gauge invariant and a scalar. The Moncrief function just defined has traditionally been the most common choice for studying odd perturbations of Schwarzschild, and because of this, many existing numerical implementations use this function instead of $\Psi^{l,m}_{\rm odd}$. Using again (8.2.63), we can show that asymptotically and in the TT gauge $\tilde{Q}^{l,m}_{\rm M}$ reduces to

$$Q^{l,m}_{\rm M} \sim -\frac{2r}{(l-1)(l+1)} \, \partial^2_r h^{l,m}_r \sim -\frac{2r}{(l-1)(l+1)} \, \partial^2_t h^{l,m}_r \,, \tag{8.2.74}$$

so that

$$Q^{l,m}_{\rm M} \sim -\partial_t \Psi^{l,m}_{\rm odd} \,. \tag{8.2.75}$$

We finally introduce the following rescaling

$$Q^{l,m}_{\rm even} := \left(\frac{(l+2)!}{2(l-2)!} \right)^{1/2} \Psi^{l,m}_{\rm even} \,, \tag{8.2.76}$$

$$Q^{l,m}_{\rm odd} := \left(\frac{(l+2)!}{2(l-2)!} \right)^{1/2} Q^{l,m}_{\rm M} = -\left(\frac{(l+2)!}{2(l-2)!} \right)^{1/2} \partial_t \Psi^{l,m}_{\rm odd} \,. \tag{8.2.77}$$

Again, the fact that h^+ and $h^\times$ are real implies that

$$\bar{Q}^{l,m}_{\rm even} = (-1)^m \, Q^{l,-m}_{\rm even} \,, \qquad \bar{Q}^{l,m}_{\rm odd} = (-1)^m \, Q^{l,-m}_{\rm odd} \,. \tag{8.2.78}$$

In terms of $Q^{l,m}_{\rm even}$ and $Q^{l,m}_{\rm odd}$ we now find for the complex coefficient H:

$$H = \frac{1}{\sqrt{2}r} \sum_{l,m} \left[Q^{l,m}_{\rm even} - i \int_{-\infty}^{t} Q^{l,m}_{\rm odd} \, dt' \right] \, {}_{-2}Y^{l,m} \,. \tag{8.2.79}$$

8.3 The Weyl tensor

A second approach to the extraction of gravitational wave information is based on definition of the *Weyl tensor*. In a general n-dimensional space the Riemann tensor $R^\alpha{}_{\beta\mu\nu}$ has n^4 components, but all its symmetries reduce the number of independent components to only $n^2(n^2-1)/12$. On the other hand, the Ricci tensor $R_{\mu\nu} := R^\alpha{}_{\mu\alpha\nu}$, being symmetric in its two indices, has in general $n(n+1)/2$ independent components. Notice that, in two dimensions, both the Riemann and Ricci tensors have only one independent component so they are both proportional to the scalar curvature R, while in three dimensions they both have six independent components, and the Riemann is proportional to the Ricci.[85] However, for dimensions larger than three, as in the case of the four-dimensional physical spacetime, the Riemann tensor has more independent components than the Ricci tensor, which implies that it can be decomposed in terms of the Ricci and an additional object known as the Weyl tensor, and defined for an n-dimensional space as

$$C_{\alpha\beta\mu\nu} := R_{\alpha\beta\mu\nu} - \frac{2}{n-2}\left[g_{\alpha[\mu}R_{\nu]\beta} - g_{\beta[\mu}R_{\nu]\alpha}\right] + \frac{2}{(n-1)(n-2)}\, g_{\alpha[\mu}g_{\nu]\beta}R \,. \tag{8.3.1}$$

From the definition above it is clear that the Weyl tensor has the same symmetries as the Riemann tensor. Moreover, we can also easily see that it is traceless:

$$C^\alpha{}_{\mu\alpha\nu} = 0 \,. \tag{8.3.2}$$

In n dimensions, the Weyl tensor has $n(n+1)(n+2)(n-3)/12$ independent components (*i.e.* 10 independent components in four dimensions), and vanishes identically for $n \leq 3$. Notice that if we are in vacuum, and the Einstein field equations hold, then the Ricci tensor vanishes and the Riemann and Weyl tensors coincide.

Another important property of the Weyl tensor is related to its behavior under conformal transformations of the form

$$\tilde{g}_{ij} = \Omega\, g_{ij} \,. \tag{8.3.3}$$

It turns out that even though, under such a transformation, the two metrics g_{ij} and $\tilde{g}_{ij}$ have, in general, different Riemann curvature tensors, they nevertheless have the same Weyl tensor:

$$\tilde{C}^\alpha{}_{\beta\mu\nu} = C^\alpha{}_{\beta\mu\nu} \,. \tag{8.3.4}$$

In other words, the Weyl tensor is conformally invariant, and because of this it is often called the *conformal curvature tensor*.

[85] The formula $n(n+1)/2$ for the number of independent components of the Ricci tensor does not apply for dimensions less than three, since being constructed from the Riemann tensor, the Ricci tensor can not have more independent components. In the particular case of only one dimension the Riemann tensor in fact vanishes, *i.e.* a single line has no intrinsic curvature.

Let us now concentrate in the four-dimensional physical spacetime. We can write the Bianchi identities (1.10.2) in terms of the Weyl tensor as

$$\nabla_\alpha C^\alpha{}_{\beta\mu\nu} = \nabla_{[\mu} R_{\nu]\beta} + \frac{1}{6}\, g_{\beta[\mu} \nabla_{\nu]} R \,, \tag{8.3.5}$$

which through the Einstein field equations reduces to

$$\nabla_\alpha C^\alpha{}_{\beta\mu\nu} = 8\pi \left[\nabla_{[\mu} T_{\nu]\beta} + \frac{1}{3}\, g_{\beta[\mu} \nabla_{\nu]} T \right] . \tag{8.3.6}$$

We then see that, in vacuum, the Weyl tensor has zero divergence.

Given an arbitrary timelike unit vector n^μ, we define the *electric* $E_{\mu\nu}$ and *magnetic* $B_{\mu\nu}$ parts of the Weyl tensor as

$$E_{\mu\nu} := n^\alpha n^\beta \, C_{\alpha\mu\beta\nu} \,, \tag{8.3.7}$$

$$B_{\mu\nu} := n^\alpha n^\beta \, C^*_{\alpha\mu\beta\nu} \,, \tag{8.3.8}$$

where $C^*_{\alpha\mu\beta\nu}$ is the so-called *dual Weyl tensor*

$$C^*_{\alpha\beta\mu\nu} := \frac{1}{2}\, C_{\alpha\beta\lambda\sigma}\, \epsilon^{\lambda\sigma}{}_{\mu\nu} \,, \tag{8.3.9}$$

with $\epsilon_{\alpha\beta\mu\nu}$ the Levi–Civita completely antisymmetric tensor. The word "dual" refers in general to a transformation such that when applied twice it returns you to the original situation with at most a change of sign (or a factor $\pm i$). In this case we find that

$$C^{**}_{\alpha\beta\mu\nu} = -C_{\alpha\beta\mu\nu} \,. \tag{8.3.10}$$

To prove this we need to use the fact that the contraction of two Levi–Civita symbols in an n-dimensional space is given in general by

$$\epsilon^{\mu_1 \cdots \mu_p \alpha_1 \cdots \alpha_{n-p}} \epsilon_{\mu_1 \cdots \mu_p \beta_1 \cdots \beta_{n-p}} = (-1)^s p!\, (n-p)!\, \delta^{[\alpha_1}_{\beta_1} \cdots \delta^{\alpha_{n-p}]}_{\beta_{n-p}} \,, \tag{8.3.11}$$

where s is the number of negative eigenvalues of the metric (which in the case of a Lorentzian spacetime is 1).

The names of the electric and magnetic tensors come from the fact that when we rewrite the Bianchi identities in terms of E_{ij} and B_{ij}, the resulting equations have the same structure as Maxwell's equations.

The symmetries of Weyl imply that the electric and magnetic tensors are both symmetric, traceless, and spacelike in the sense that

$$n^\mu E_{\mu\nu} = 0 \,, \qquad n^\mu B_{\mu\nu} = 0 \,. \tag{8.3.12}$$

The symmetry of the magnetic part is not evident from its definition, but it is in fact not difficult to prove (see below). This means that both $E_{\mu\nu}$ and $B_{\mu\nu}$ have five independent components, giving a total of 10. Since this is precisely the number of independent components of the Weyl tensor, we can expect to be able to express $C_{\alpha\beta\mu\nu}$ in terms of the electric and magnetic tensors. This is indeed the case, and the Weyl tensor can in general be written as

$$C_{\alpha\beta\mu\nu} = 2\left[l_{\alpha[\mu}E_{\nu]\beta} - l_{\beta[\mu}E_{\nu]\alpha} - n_{[\mu}B_{\nu]\lambda}\epsilon^{\lambda}{}_{\alpha\beta} - n_{[\alpha}B_{\beta]\lambda}\epsilon^{\lambda}{}_{\mu\nu}\right] , \tag{8.3.13}$$

with

$$l_{\mu\nu} := g_{\mu\nu} + 2n_{\mu}n_{\nu} . \tag{8.3.14}$$

If we now take the vector n^{μ} to be the unit normal vector to the spacelike hypersurfaces in the 3+1 formalism, we can use the Gauss–Codazzi and Codazzi–Mainardi equations (2.4.1) and (2.4.2) to write the electric and magnetic tensors in 3+1 language as

$$E_{ij} = R_{ij} + KK_{ij} - K_{im}K^{m}{}_{j} - 4\pi\left[S_{ij} + \frac{\gamma_{ij}}{3}(4\rho - S)\right] , \tag{8.3.15}$$
$$B_{ij} = \epsilon_{i}{}^{mn}\left[D_m K_{nj} - 4\pi\gamma_{jm}j_n\right] , \tag{8.3.16}$$

with ρ, S_{ij} and j_i the energy density, stress tensor and momentum density measured by the Eulerian observers, and where ϵ_{ijk} is now the Levi–Civita tensor in three dimensions which is constructed from $\epsilon_{\alpha\beta\mu\nu}$ as [86]

$$\epsilon_{\beta\mu\nu} = n^{\alpha}\epsilon_{\alpha\beta\mu\nu} . \tag{8.3.17}$$

Notice that the Hamiltonian constraint (2.4.10) guarantees that the trace of E_{ij} vanishes, while the trace of B_{ij} can be seen to vanish trivially from the anti-symmetry of the Levi–Civita tensor. In contrast, the symmetry of E_{ij} is evident, while the symmetry of B_{ij} comes about through the momentum constraints. To see this consider the contraction of B_{ij} with the Levi–Civita tensor

$$\epsilon^{ija}B_{ij} = \epsilon^{ija}\epsilon_{imn}\left[D^m K^n_j - 4\pi\delta^m_j j^n\right] . \tag{8.3.18}$$

Now, from the general expression for the contraction of two Levi–Civita symbols given above we find

$$\epsilon^{iab}\epsilon_{imn} = \delta^a_m\delta^b_n - \delta^b_m\delta^a_n , \tag{8.3.19}$$

which implies

$$\begin{aligned} \epsilon^{ija}B_{ij} &= \left(\delta^j_m\delta^a_n - \delta^a_m\delta^j_n\right)\left[D^m K^n_j - 4\pi\delta^m_j j^n\right] \\ &= \left[D^m K^a_m - D^a K - 8\pi j^a\right] = 0 , \end{aligned} \tag{8.3.20}$$

where the last equality follows from the momentum constraints (2.4.11). This shows that B_{ij} is indeed a symmetric tensor.

[86] The definition of the three-dimensional ϵ_{ijk} is one of those few places where it actually makes a difference to define the spacetime coordinates as $(x^0, x^1, x^2, x^3) = (t, x, y, z)$ or as $(x^1, x^2, x^3, x^4) = (x, y, z, t)$. Notice that, even in Minkowski spacetime, if we take the first choice we find $\epsilon_{123} = n^{\mu}\epsilon_{\mu 123} = \epsilon_{0123} = 1$, while with the second choice we have instead $\epsilon_{123} = n^{\mu}\epsilon_{\mu 123} = \epsilon_{4123} = -1$. So that when we work with $(x^1, x^2, x^3, x^4) = (x, y, z, t)$ it is in fact better to define $\epsilon_{\alpha\beta\mu} := n^{\nu}\epsilon_{\alpha\beta\mu\nu}$.

8.4 The tetrad formalism

Up to this point we have assumed that the components of tensors are always expressed in terms of a coordinate basis $\{\vec{e}_\mu\}$. However, in many cases it is particularly useful to work instead with a basis that is independent of the coordinates. In particular, the use of such non-coordinate basis is at the heart of many important theoretical advances in relativity theory and, as we will see in the next Section, forms the foundation of the Newman–Penrose formalism.

At every point of spacetime, consider a set of four linearly independent vectors $\{\vec{e}_{(a)}\}$, where the Latin index in parenthesis identifies the different vectors (here we will modify our usual convention and allow Latin indices inside parenthesis to take values $(0, 1, 2, 3)$). Moreover, assume that these vectors are such that

$$\vec{e}_{(a)} \cdot \vec{e}_{(b)} = \eta_{(a)(b)} \,, \tag{8.4.1}$$

with $\eta_{(a)(b)}$ a constant matrix independent of the position in spacetime. In such a case the set of vectors $\{\vec{e}_{(a)}\}$ is called a *tetrad*. Notice that the $\eta_{(a)(b)}$ are nothing more than the components of the metric tensor in the tetrad basis. In the particular case when the tetrad is orthonormal, $\eta_{(a)(b)}$ reduces to the Minkowski tensor, but for the moment we will only assume that it is constant.

Since the vectors $\{\vec{e}_{(a)}\}$ are linearly independent, it is clear that the matrix $\eta_{(a)(b)}$ can be inverted. We denote its inverse by $\eta^{(a)(b)}$, so that we have

$$\eta^{(a)(c)} \eta_{(c)(b)} = \delta^{(a)}_{(b)} \,. \tag{8.4.2}$$

Let us now introduce a new set of vectors $\{\vec{e}^{\,(a)}\}$ defined as

$$\vec{e}^{\,(a)} = \eta^{(a)(b)} \vec{e}_{\,(a)}. \tag{8.4.3}$$

From this definition it is easy to show that these vectors are such that

$$\vec{e}^{\,(a)} \cdot \vec{e}_{(b)} = \delta^{(a)}_{(b)} \,. \tag{8.4.4}$$

Figure 8.1 shows a graphical representation of the relationship between the two sets of vectors.

The tetrad vectors $\{\vec{e}_{(a)}\}$ can be used as a basis, so that we can express any arbitrary vector $\vec{v}$ as

$$\vec{v} = v^{(a)} \vec{e}_{(a)} \,. \tag{8.4.5}$$

In order to solve for $v^{(a)}$, we multiply both sides of the above expression with $\{\vec{e}^{\,(b)}\}$ to obtain

$$\vec{e}^{\,(b)} \cdot \vec{v} = \vec{e}^{\,(b)} \cdot \left(v^{(a)} \vec{e}_{(a)} \right) = v^{(a)} \left(\vec{e}^{\,(b)} \cdot \vec{e}_{(a)} \right) \,. \tag{8.4.6}$$

Using now (8.4.4) we find

$$v^{(a)} = \vec{v} \cdot \vec{e}^{\,(a)} \,. \tag{8.4.7}$$

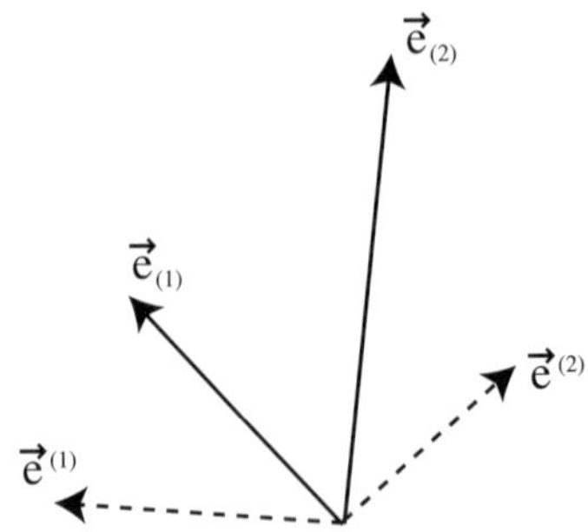

Fig. 8.1: Starting from the two vectors $(\vec{e}_{(1)}, \vec{e}_{(2)})$, we construct a new vectors $(\vec{e}^{\,(1)}, \vec{e}^{\,(2)})$ such that $\vec{e}_{(1)} \cdot \vec{e}^{\,(2)} = 0$, $\vec{e}_{(2)} \cdot \vec{e}^{\,(1)} = 0$, $\vec{e}_{(1)} \cdot \vec{e}^{\,(1)} = 1$, $\vec{e}_{(2)} \cdot \vec{e}^{\,(2)} = 1$.

The same can clearly be done with the vectors $\{\vec{e}^{\,(a)}\}$; we then have in general

$$\vec{v} = v^{(a)} \vec{e}_{(a)} = v_{(a)} \vec{e}^{\,(a)} , \tag{8.4.8}$$

with

$$v^{(a)} = \vec{v} \cdot \vec{e}^{\,(a)} , \qquad v_{(a)} = \vec{v} \cdot \vec{e}_{(a)} . \tag{8.4.9}$$

This implies, in particular, that the scalar product of any two vectors $\vec{v}$ and $\vec{u}$ takes the form

$$\vec{v} \cdot \vec{u} = v^\mu u_\mu = v^{(a)} u_{(a)} . \tag{8.4.10}$$

The scalar product can then be expressed either as the contraction of the co-variant and contra-variant spacetime components, or the contraction of the corresponding components in terms of the tetrad.

Using the above results we also find that

$$\vec{v} = v^{(a)} \vec{e}_{(a)} = (\vec{v} \cdot \vec{e}^{\,(a)}) \vec{e}_{(a)} = (v^\mu e^{(a)}{}_\mu)(e_{(a)}{}^\nu \vec{e}_\nu) , \tag{8.4.11}$$

where we have expressed the tetrad vector $\vec{e}_{(a)}$ in terms of the coordinate basis $\{\vec{e}_\mu\}$. The previous equation implies that

$$e^{(a)}{}_\mu \, e_{(a)}{}^\nu = \delta^\nu_\mu . \tag{8.4.12}$$

Together, equations (8.4.4) and (8.4.12) can be written as

$$e_{(a)}{}^\mu \, e^{(b)}{}_\mu = \delta^{(b)}_{(a)} , \tag{8.4.13}$$

$$e_{(a)}{}^\mu \, e^{(a)}{}_\nu = \delta^\mu_\nu . \tag{8.4.14}$$

A very important consequence of this is the following

$$g_{\mu\nu} = e_{(a)\mu} \, e^{(a)}{}_\nu = \eta^{(a)(b)} e_{(a)\mu} \, e_{(b)\mu} . \tag{8.4.15}$$

This expression allows us to recover the metric components in the coordinate frame in terms of the components of the tetrad vectors.

There are several advantages to the use of tetrads instead of a coordinate basis. In the first place, the components of the metric tensor are constant in a tetrad basis. But more important, since the tetrad is defined independently of the coordinates, the tetrad components of any geometric object behave as scalars with respect to coordinate changes.

We can also show that it is possible to go back and forth from the coordinate components of any tensor to its tetrad components by simply contracting the appropriate indices with the tetrad itself. For example, for a rank 2 tensor we have

$$T_{(a)(b)} = T_{\mu\nu} e_{(a)}{}^{\mu} e_{(b)}{}^{\nu} \,, \qquad T_{\mu\nu} = T_{(a)(b)} e^{(a)}{}_{\mu} e^{(b)}{}_{\nu} \,. \tag{8.4.16}$$

In general, the tetrad vectors can be used to define directional derivatives of a scalar function f as

$$f_{,(a)} := e_{(a)}{}^{\mu} \partial_\mu f \,. \tag{8.4.17}$$

Since the tetrad components of a vector are scalar functions, we can also find the directional derivatives of an arbitrary vector $\vec{A} = A^{(a)} \vec{e}_{(a)}$ in the following way

$$\begin{aligned} A_{(a),(b)} &:= e_{(b)}{}^{\mu} \partial_\mu A_{(a)} = e_{(b)}{}^{\mu} \partial_\mu \left(e_{(a)}{}^{\nu} A_\nu \right) \\ &= e_{(b)}{}^{\mu} \left(e_{(a)}{}^{\nu} \partial_\mu A_\nu + A_\nu \partial_\mu e_{(a)}{}^{\nu} \right) \,. \end{aligned} \tag{8.4.18}$$

In this expression, the partial derivatives can in fact be changed for covariant derivatives since the Christoffel symbols cancel out. We can then rewrite the directional derivative as

$$A_{(a),(b)} = e_{(a)}{}^{\nu} e_{(b)}{}^{\mu} \nabla_\mu A_\nu + e_{(b)}{}^{\mu} e_{(c)}{}^{\nu} A^{(c)} \nabla_\mu e_{(a)\nu} \,. \tag{8.4.19}$$

We now define the *Ricci rotation coefficients* as

$$\gamma_{(a)(b)(c)} := e_{(a)}{}^{\mu} e_{(c)}{}^{\nu} \nabla_\nu e_{(b)\mu} \,, \tag{8.4.20}$$

so the directional derivative takes the final form

$$A_{(a),(b)} = e_{(a)}{}^{\nu} e_{(b)}{}^{\mu} \nabla_\mu A_\nu + \gamma_{(c)(a)(b)} A^{(c)} \,. \tag{8.4.21}$$

From the definition of the Ricci rotation coefficients it is also easy to show that

$$\nabla_{(a)} \, \vec{e}_{(b)} \equiv e_{(a)}{}^{\mu} \nabla_\mu \vec{e}_{(b)} = \gamma^{(c)}{}_{(b)(a)} \vec{e}_{(c)} \,. \tag{8.4.22}$$

We then see that the $\gamma^{(c)}{}_{(b)(a)}$ are nothing more than the connection coefficients in the tetrad basis. Notice, however, that since the tetrad is not a coordinate basis, in general we will have that

$$\gamma^{(c)}{}_{(a)(b)} \neq \gamma^{(c)}{}_{(b)(a)} \,. \tag{8.4.23}$$

On the other hand, from the fact that $\vec{e}_{(a)} \cdot \vec{e}_{(b)} = \eta_{(a)(b)}$, with $\eta_{(a)(b)}$ constant, we can see that

$$\nabla_\mu \left(e_{(a)}{}^\nu e_{(b)\nu} \right) = 0 \quad \Rightarrow \quad e_{(a)}{}^\nu \nabla_\mu e_{(b)\nu} = -e_{(b)\nu} \nabla_\mu e_{(a)}{}^\nu \, , \tag{8.4.24}$$

which implies

$$\gamma_{(a)(b)(c)} = -\gamma_{(b)(a)(c)} \, . \tag{8.4.25}$$

The Ricci rotation coefficients are therefore antisymmetric in their first two indices. This means that in a general four-dimensional spacetime there are only 24 independent $\gamma_{(a)(b)(c)}$, which is in contrast with the 40 independent components of the Christoffel symbols $\Gamma^\alpha_{\mu\nu}$ in a coordinate basis. This is another advantage of the tetrad approach.

We now define the *intrinsic derivative* of a vector $\vec{A}$ as

$$A_{(a)|(b)} := e_{(a)}{}^\mu e_{(b)}{}^\nu \nabla_\nu A_\mu = A_{(a),(b)} - \gamma^{(c)}{}_{(a)(b)} A_{(c)} \, . \tag{8.4.26}$$

The intrinsic derivative is in fact nothing more than the covariant derivative expressed in the tetrad basis. Equation (8.4.26) can be generalized in a straightforward way to tensors of arbitrary rank.

The final ingredient is the expression for the Riemann curvature tensor in tetrad components. Starting from the Ricci identity for the tetrad vectors (equation (1.9.3)):

$$R^\alpha{}_{\beta\mu\nu} \, e_{(b)}{}^\beta = \nabla_\mu \nabla_\nu e_{(b)}{}^\alpha - \nabla_\nu \nabla_\mu e_{(b)}{}^\alpha \, , \tag{8.4.27}$$

and projecting onto the tetrad

$$R_{(a)(b)(m)(n)} = R_{\alpha\beta\mu\nu} e_{(a)}{}^\alpha e_{(b)}{}^\beta e_{(m)}{}^\mu e_{(n)}{}^\nu \, , \tag{8.4.28}$$

we find, after some algebra, that

$$\begin{aligned} R_{(a)(b)(m)(n)} = {} & \gamma_{(a)(b)(n),(m)} - \gamma_{(a)(b)(m),(n)} + \gamma_{(a)(b)(c)} \left(\gamma^{(c)}{}_{(m)(n)} - \gamma^{(c)}{}_{(n)(m)} \right) \\ & + \gamma_{(a)(c)(m)} \gamma^{(c)}{}_{(b)(n)} - \gamma_{(a)(c)(n)} \gamma^{(c)}{}_{(b)(m)} \, . \end{aligned} \tag{8.4.29}$$

Notice that the second term is different from zero only in the case when the rotation coefficients are not symmetric and vanishes for a coordinate basis, in which case the last expression reduces to the standard one (cf. equation (1.9.2)).

8.5 The Newman–Penrose formalism

8.5.1 *Null tetrads*

A variation on the tetrad formalism inspired by spinor techniques, and particularly well adapted to the study of gravitational radiation, was introduced by Newman and Penrose in 1962 [218]. The basic idea of the Newman–Penrose formalism is to introduce a tetrad of null vectors. In order to do this we first start from an orthonormal tetrad $\{\vec{e}_{(a)}\}$, so that the matrix $\eta_{(a)(b)}$ corresponds to the Minkowski matrix. In such a case, we can use equation (8.4.15) to rewrite the spacetime metric as

$$g_{\mu\nu} = -e_{(0)\mu} e_{(0)\nu} + e_{(1)\mu} e_{(1)\nu} + e_{(2)\mu} e_{(2)\nu} + e_{(3)\mu} e_{(3)\nu} \, . \tag{8.5.1}$$

Typically, we choose the vector $e^\mu_{(0)}$ as the unit normal to the spatial hypersurfaces $e^\mu_{(0)} = n^\mu$, $e^\mu_{(1)}$ as the unit radial vector in spherical coordinates

$e^\mu_{(1)} = e^\mu_r$, and $(e^\mu_{(2)}, e^\mu_{(3)})$ as unit vectors in the angular directions. Notice, however, that even in flat space the coordinate vectors e^μ_θ and e^μ_φ are not unit vectors, so they have to be normalized. Moreover, in a general spacetime e^μ_θ and e^μ_φ can not be expected to be orthogonal to e^μ_r or to each other, so a Gram–Schmidt orthogonalization procedure is required in order to construct $e^\mu_{(2)}$ and $e^\mu_{(3)}$.

Once we have an orthonormal basis, we can construct the two null vectors:

$$l^\mu := \frac{1}{\sqrt{2}} \left(e^\mu_{(0)} + e^\mu_{(1)} \right) , \quad k^\mu := \frac{1}{\sqrt{2}} \left(e^\mu_{(0)} - e^\mu_{(1)} \right) . \tag{8.5.2}$$

With the above choices for $e^\mu_{(0)}$ and $e^\mu_{(1)}$, l^μ is an outgoing null vector while k^μ is ingoing.[87] As long as we only consider real quantities we can not construct two more null vectors that are at the same time independent of l^μ and k^μ and orthogonal to them, but this is no longer the case if we allow for complex vectors, in which case we can define the vectors:

$$m^\mu := \frac{1}{\sqrt{2}} \left(e^\mu_{(2)} + i e^\mu_{(3)} \right) , \quad \bar{m}^\mu := \frac{1}{\sqrt{2}} \left(e^\mu_{(2)} - i e^\mu_{(3)} \right) . \tag{8.5.3}$$

The four complex vectors $(l^\mu, k^\mu, m^\mu, \bar{m}^\mu)$ form what is known as a *null tetrad*, and are such that

$$l_\mu l^\mu = k_\mu k^\mu = m_\mu m^\mu = \bar{m}_\mu \bar{m}^\mu = 0 , \tag{8.5.4}$$
$$l_\mu m^\mu = l_\mu \bar{m}^\mu = k_\mu m^\mu = k_\mu \bar{m}^\mu = 0 , \tag{8.5.5}$$
$$l_\mu k^\mu = -m_\mu \bar{m}^\mu = -1 . \tag{8.5.6}$$

In terms of the null tetrad, the matrix $\eta_{(a)(b)}$ becomes

$$\eta_{(a)(b)} = \eta^{(a)(b)} = \begin{pmatrix} 0 & -1 & 0 & 0 \\ -1 & 0 & 0 & 0 \\ 0 & 0 & 0 & +1 \\ 0 & 0 & +1 & 0 \end{pmatrix} , \tag{8.5.7}$$

and the spacetime metric $g_{\mu\nu}$ takes the form

$$g_{\mu\nu} = -l_\mu k_\nu - k_\mu l_\nu + m_\mu \bar{m}_\nu + \bar{m}_\mu m_\nu . \tag{8.5.8}$$

In the Newman–Penrose formalism the directional derivative operators along the tetrad components are given special names:

$$D := l^\mu \nabla_\mu , \quad \Delta := k^\mu \nabla_\mu , \quad \delta := m^\mu \nabla_\mu , \quad \bar{\delta} := \bar{m}^\mu \nabla_\mu . \tag{8.5.9}$$

The Ricci rotation coefficients are now called *spin coefficients*, and are also given individual names. We will separate them into two groups:

[87] It is standard notation to use n^μ instead of k^μ as the ingoing null vector. Here, however, we reserve n^μ for the timelike unit normal to the spatial hypersurfaces.

$$\kappa := \gamma_{(3)(1)(1)} = m^\mu D l_\mu \; , \tag{8.5.10}$$
$$\tau := \gamma_{(3)(1)(2)} = m^\mu \Delta l_\mu \; , \tag{8.5.11}$$
$$\sigma := \gamma_{(3)(1)(3)} = m^\mu \delta l_\mu \; , \tag{8.5.12}$$
$$\rho := \gamma_{(3)(1)(4)} = m^\mu \bar{\delta} l_\mu \; , \tag{8.5.13}$$
$$\tau' := \gamma_{(4)(2)(1)} = \bar{m}^\mu D k_\mu \equiv -\pi \; , \tag{8.5.14}$$
$$\kappa' := \gamma_{(4)(2)(2)} = \bar{m}^\mu \Delta k_\mu \equiv -\nu \; , \tag{8.5.15}$$
$$\rho' := \gamma_{(4)(2)(3)} = \bar{m}^\mu \delta k_\mu \equiv -\mu \; , \tag{8.5.16}$$
$$\sigma' := \gamma_{(4)(2)(4)} = \bar{m}^\mu \bar{\delta} k_\mu \equiv -\lambda \; , \tag{8.5.17}$$

and

$$\epsilon := \frac{1}{2}\left(\gamma_{(2)(1)(1)} - \gamma_{(4)(3)(1)}\right) = \frac{1}{2}\left(k^\mu D l_\mu - \bar{m}^\mu D m_\mu\right) \; , \tag{8.5.18}$$
$$\gamma := \frac{1}{2}\left(\gamma_{(2)(1)(2)} - \gamma_{(4)(3)(2)}\right) = \frac{1}{2}\left(k^\mu \Delta l_\mu - \bar{m}^\mu \Delta m_\mu\right) \; , \tag{8.5.19}$$
$$\beta := \frac{1}{2}\left(\gamma_{(2)(1)(3)} - \gamma_{(4)(3)(3)}\right) = \frac{1}{2}\left(k^\mu \delta l_\mu - \bar{m}^\mu \delta m_\mu\right) \; , \tag{8.5.20}$$
$$\alpha := \frac{1}{2}\left(\gamma_{(2)(1)(4)} - \gamma_{(4)(3)(4)}\right) = \frac{1}{2}\left(k^\mu \bar{\delta} l_\mu - \bar{m}^\mu \bar{\delta} m_\mu\right) \; . \tag{8.5.21}$$

We then have 12 complex spin coefficients, corresponding to 24 independent real quantities, as expected. The different names introduced above have historical origins (notice in particular that the four coefficients $\{\tau', \kappa', \rho', \sigma'\}$ have in fact two different names).

Several of the spin coefficients have clear geometrical interpretations. For example, since $l^\nu \nabla_\nu l_\mu$ is a vector we can expand it in terms of the tetrad as

$$D l_\mu = l^\nu \nabla_\nu l_\mu = a l_\mu + b k_\mu + c m_\mu + d \bar{m}_\mu \; . \tag{8.5.22}$$

The values of the coefficients can be obtained by taking contractions with all four tetrad vectors and using conditions (8.5.4)–(8.5.6). We then find that

$$l^\nu \nabla_\nu l_\mu = -\left(\epsilon + \bar{\epsilon}\right) l_\mu + \bar{\kappa} m_\mu + \kappa \bar{m}_\mu \; . \tag{8.5.23}$$

This implies that if $\kappa = 0$, then the flow lines of l^μ are geodesics. Moreover, if we also have $\epsilon + \bar{\epsilon} = 0$ then those geodesics are parameterized with an affine parameter.[88] If we now change l^μ for k^μ, we find in a similar way that the k^μ flow lines are geodesics if $\kappa' = 0$, and have an affine parameter if $\gamma + \bar{\gamma} = 0$.

Moreover, by expanding $\nabla_\mu l_\nu$ as a linear combination of products of tetrad vectors we can also show that if $\epsilon + \bar{\epsilon} = 0$, then

$$\nabla_\mu l^\mu = (\rho + \bar{\rho}) \; . \tag{8.5.24}$$

[88] If a geodesic is not parameterized with an affine parameter then under parallel transport the tangent vector is allowed to change as long as it keeps its direction (locally) fixed, so that we have $l^\mu \nabla_\nu l_\mu \propto l_\mu$.

This shows that the real part of ρ corresponds to the expansion of the l^μ flow lines. Similarly, we can show that the shear of the l^μ flow lines (*i.e.* the change in shape) is related to $\sigma\bar{\sigma}$, and the rotation is related to $(\rho - \bar{\rho})$.

8.5.2 *Tetrad transformations*

The definition of the null tetrad $(l^\mu, k^\mu, m^\mu, \bar{m}^\mu)$ is based on the choice of the original orthonormal tetrad $\{\vec{e}_{(a)}\}$. We can, of course, change this tetrad by an arbitrary rotation in space plus a Lorentz boost in a given direction while still keeping it orthonormal. We then have six degrees of freedom corresponding to possible transformations of the tetrad that will not change the formalism just developed. Such transformations are usually separated into three distinct classes:

- Null rotations of class I which leave the vector $\vec{l}$ unchanged:
$$l^\mu \to l^\mu \,, \quad m^\mu \to m^\mu + a l^\mu \,, \quad \bar{m}^\mu \to \bar{m}^\mu + \bar{a} l^\mu \,, \\ k^\mu \to k^\mu + \bar{a} m^\mu + a \bar{m}^\mu + a\bar{a} l^\mu \,. \tag{8.5.25}$$
- Null rotations of class II which leave the vector $\vec{k}$ unchanged:
$$k^\mu \to k^\mu \,, \quad m^\mu \to m^\mu + b k^\mu \,, \quad \bar{m}^\mu \to \bar{m}^\mu + \bar{b} k^\mu \\ l^\mu \to l^\mu + \bar{b} m^\mu + b \bar{m}^\mu + b\bar{b} k^\mu \,. \tag{8.5.26}$$
- Null rotations of class III which leave the directions of $\vec{l}$ and $\vec{k}$ and the product $l_\mu k^\mu$ unchanged:
$$l^\mu \to \lambda^{-1} l^\mu \,, \quad k^\mu \to \lambda k^\mu \,, \quad m^\mu \to e^{i\theta} m^\mu \,, \quad \bar{m}^\mu \to e^{-i\theta} \bar{m}^\mu \,. \tag{8.5.27}$$

Notice that the six degrees of freedom correspond to the two real parameters (λ, θ), plus the two complex parameters (a, b).

The class III rotation can be easily interpreted – it represents a Lorentz boost in the $(\vec{e}_0, \vec{e}_1)$ plane plus a rotation in the $(\vec{e}_2, \vec{e}_3)$ plane. The parameter θ corresponds to the angle of rotation in the $(\vec{e}_2, \vec{e}_3)$ plane (actually $-\theta$), while λ is given by
$$\lambda = \left[(1 - v)/(1 + v)\right]^{1/2} \,, \tag{8.5.28}$$
with v the boost speed along the $\vec{e}_1$ direction. Because of this the class III null rotations are also called *spin-boost transformations.*

The rotations of class I and II are harder to interpret directly since they correspond to complicated combinations of boosts and three-dimensional rotations that leave either $\vec{l}$ or $\vec{k}$ unchanged. The important thing to remember is that any combination of boosts and rotations in arbitrary directions can be expressed as a combination of these three types of null rotations. For example, a little algebra shows that a simple rotation in the $(\vec{e}_1, \vec{e}_2)$ plane by an angle ϕ corresponds to a class I transformation with $a = (\cos(\phi) - 1)/\sin\phi$, followed by a class II transformation with $b = \sin(\phi)/2$, followed by a class III transformation with $\lambda = (\cos(\phi) + 1)/2$ and $\theta = 0$.

The tetrad rotations just described play a very important role in the Petrov classification of spacetimes that will be introduced in Section 8.7.

8.6 The Weyl scalars

As we have seen, the Weyl tensor has, in general, 10 independent components. In the Newman–Penrose formalism those components can be conveniently represented by five complex scalar quantities known as the *Weyl scalars*, and defined as (the sign convention in these definitions is by no means universal and is frequently reversed)

$$\Psi_0 := C_{(1)(3)(1)(3)} = C_{\alpha\beta\mu\nu}\, l^\alpha m^\beta l^\mu m^\nu \;, \tag{8.6.1}$$

$$\Psi_1 := C_{(1)(2)(1)(3)} = C_{\alpha\beta\mu\nu}\, l^\alpha k^\beta l^\mu m^\nu \;, \tag{8.6.2}$$

$$\Psi_2 := C_{(1)(3)(4)(2)} = C_{\alpha\beta\mu\nu}\, l^\alpha m^\beta \bar{m}^\mu k^\nu \;, \tag{8.6.3}$$

$$\Psi_3 := C_{(1)(2)(4)(2)} = C_{\alpha\beta\mu\nu}\, l^\alpha k^\beta \bar{m}^\mu k^\nu \;, \tag{8.6.4}$$

$$\Psi_4 := C_{(2)(4)(2)(4)} = C_{\alpha\beta\mu\nu}\, k^\alpha \bar{m}^\beta k^\mu \bar{m}^\nu \;. \tag{8.6.5}$$

Notice that, as with all other Newman–Penrose quantities, the Ψ_a are scalars with respect to coordinate transformations, but they clearly depend on the choice of the null tetrad. These five complex scalars are enough to specify all 10 independent components of the Weyl tensor. The symmetries of the Weyl tensor imply that all other possible contractions of $C_{\alpha\beta\mu\nu}$ with combinations of the tetrad vectors either vanish or can be expressed as combinations of the Ψ_a.

Similarly, the 10 independent components of the Ricci tensor can also be described in terms of the *Ricci scalars*, which are separated into four real scalars (again, the sign convention is not universal)

$$\Phi_{00} := \frac{1}{2}\, R_{(1)(1)} \;, \tag{8.6.6}$$

$$\Phi_{11} := \frac{1}{4}\left(R_{(1)(2)} + R_{(3)(4)}\right) \;, \tag{8.6.7}$$

$$\Phi_{22} := \frac{1}{2}\, R_{(2)(2)} \;, \tag{8.6.8}$$

$$\Lambda := \frac{1}{24}\, R = \frac{1}{12}\left(R_{(3)(4)} - R_{(1)(2)}\right) \;, \tag{8.6.9}$$

and three complex scalars

$$\Phi_{01} := \frac{1}{2}\, R_{(1)(3)} = \frac{1}{2}\, \bar{R}_{(1)(4)} = \bar{\Phi}_{10} \;, \tag{8.6.10}$$

$$\Phi_{02} := \frac{1}{2}\, R_{(3)(3)} = \frac{1}{2}\, \bar{R}_{(4)(4)} = \bar{\Phi}_{20} \;, \tag{8.6.11}$$

$$\Phi_{12} := \frac{1}{2}\, R_{(2)(3)} = \frac{1}{2}\, \bar{R}_{(2)(4)} = \bar{\Phi}_{21} \;. \tag{8.6.12}$$

We have introduced these Ricci scalars here for completeness, but will not use them in what follows.

Using the definition of the electric and magnetic parts of the Weyl tensor, equations (8.3.7) and (8.3.8), we can rewrite the Weyl scalars as

$$\Psi_0 = Q_{ij}\, m^i m^j \,, \tag{8.6.13}$$

$$\Psi_1 = -\frac{1}{\sqrt{2}}\, Q_{ij}\, m^i e_r^j \,, \tag{8.6.14}$$

$$\Psi_2 = \frac{1}{2}\, Q_{ij}\, e_r^i e_r^j \,, \tag{8.6.15}$$

$$\Psi_3 = \frac{1}{\sqrt{2}}\, Q_{ij}\, \bar{m}^i e_r^j \,, \tag{8.6.16}$$

$$\Psi_4 = Q_{ij}\, \bar{m}^i \bar{m}^j \,, \tag{8.6.17}$$

with

$$Q_{ij} := E_{ij} - iB_{ij} \,, \tag{8.6.18}$$

and where $\vec{e}_r$ is the unit radial vector. These expressions can be easily obtained starting from equation (8.3.13) and using the fact that, for any arbitrary three-dimensional vector v^i, the following relations hold:

$$\epsilon_{ijk}\, v^i e_r^j m^k = -iv_k m^k \,, \qquad \epsilon_{ijk}\, v^i e_r^j \bar{m}^k = +iv_k \bar{m}^k \,, \tag{8.6.19}$$

plus the fact that

$$g^{\mu\nu} B_{\mu\nu} = 0 \quad \Rightarrow \quad B_{ij} m^i \bar{m}^j = -\frac{1}{2} B_{ij} e_r^i e_r^j \,. \tag{8.6.20}$$

We can also invert these relations to express Q_{ij} in terms of the Ψ_a:

$$\begin{aligned} Q_{ij} = {} & \Psi_0\, \bar{m}_i \bar{m}_j + \Psi_4\, m_i m_j + \Psi_2\, (2r_i r_j - m_i \bar{m}_j - \bar{m}_i m_j) \\ & - \sqrt{2}\, \Psi_1\, (r_i \bar{m}_j + \bar{m}_i r_j) + \sqrt{2}\, \Psi_3\, (r_i m_j + m_i r_j) \,. \end{aligned} \tag{8.6.21}$$

The expressions given above for the Ψ_a in terms of the electric and magnetic tensors provide us with a particularly simple way of calculating these scalars in the 3+1 approach: We start from the 3+1 expressions for E_{ij} and B_{ij}, equations (8.3.15) and (8.3.16), and then use these tensors to construct the Ψ_a.

8.7 The Petrov classification

The Weyl scalars play a very important role in the so-called *Petrov classification* of spacetimes [225]. Notice first that the Weyl tensor can be completely specified in terms of the five scalars Ψ_a. On the other hand, the Ψ_a clearly depend on the choice of tetrad. We can then ask if it is possible to make a transformation of the tetrad that will result in one or more of the Ψ_a becoming zero.

Without loss of generality let us assume that $\Psi_4 \neq 0$, which can always be done if the spacetime is not flat. It is now important to see how the different Ψ_a change under class I and II transformations (class III transformations are not necessary for the following discussion). For a class I transformation the different Weyl scalars can be shown to transform as

$$\Psi_0 \to \Psi_0 \; , \tag{8.7.1}$$
$$\Psi_1 \to \Psi_1 + \bar{a}\Psi_0 \; , \tag{8.7.2}$$
$$\Psi_2 \to \Psi_2 + 2\bar{a}\Psi_1 + \bar{a}^2\Psi_0 \; , \tag{8.7.3}$$
$$\Psi_3 \to \Psi_3 + 3\bar{a}\Psi_2 + 3\bar{a}^2\Psi_1 + \bar{a}^3\Psi_0 \; , \tag{8.7.4}$$
$$\Psi_4 \to \Psi_4 + 4\bar{a}\Psi_3 + 6\bar{a}^2\Psi_2 + 4\bar{a}^3\Psi_1 + \bar{a}^4\Psi_0 \; , \tag{8.7.5}$$

There is one very important point to notice about the above transformations. Simple inspection shows that the transformations of $(\Psi_3, \Psi_2, \Psi_1, \Psi_0)$ can be obtained by taking subsequent derivatives with respect to $\bar{a}$ of the transformation of Ψ_4, with appropriate rescalings.

Similarly, under a class II transformation we find that

$$\Psi_0 \to \Psi_0 + 4b\Psi_1 + 6b^2\Psi_2 + 4b^3\Psi_3 + b^4\Psi^4 \; , \tag{8.7.6}$$
$$\Psi_1 \to \Psi_1 + 3b\Psi_2 + 3b^2\Psi_3 + b^3\Psi_4 \; , \tag{8.7.7}$$
$$\Psi_2 \to \Psi_2 + 2b\Psi_3 + b^2\Psi_4 \; , \tag{8.7.8}$$
$$\Psi_3 \to \Psi_3 + b\Psi_4 \; , \tag{8.7.9}$$
$$\Psi_4 \to \Psi_4 \; . \tag{8.7.10}$$

Again, the transformations of $(\Psi_1, \Psi_2, \Psi_3, \Psi_4)$ can be found by taking derivatives of the transformation of Ψ_0 with respect to b and rescaling them appropriately.

Let us concentrate now on the class II transformations. It is clear that after such a transformation we can make Ψ_0 vanish as long as we choose the parameter b as one of the roots of the following quartic equation

$$\Psi_0 + 4b\Psi_1 + 6b^2\Psi_2 + 4b^3\Psi_3 + b^4\Psi^4 = 0 \; . \tag{8.7.11}$$

The above equation has in general four complex roots. The resulting directions associated with the new vector l^μ,

$$l^\mu \to l^\mu + \bar{b}m^\mu + b\bar{m} + b\bar{b}k^\mu \; , \tag{8.7.12}$$

are known as the *principal null directions* of the Weyl tensor. When some of the roots of equation (8.7.11) coincide the spacetime is said to be *algebraically special*. This leads to the Petrov classification that separates different spacetimes into six types according to the number of distinct root of (8.7.11):

<u>Petrov type I</u>: All four roots are distinct: b_1, b_2, b_3, b_4. In this case we can make a transformation of class II, with b equal to any of the distinct roots, that will result in $\Psi_0 = 0$. Furthermore, we can later make a transformation of class I to make Ψ_4 also vanish while keeping $\Psi_0 = 0$. This is because, for a class I

transformation, Ψ_0 remains unchanged while Ψ_4 transforms according to (8.7.5), so that Ψ_4 will vanish if we choose $\bar{a}$ as any of the roots of the following equation

$$\Psi_4 + 4\bar{a}\Psi_3 + 6\bar{a}^2\Psi_2 + 4\bar{a}^3\Psi_1 + \bar{a}^4\Psi_0 = 0 \, . \tag{8.7.13}$$

Notice that this equation will in fact be cubic since after the first class II transformation we already have $\Psi_0 = 0$. For a Petrov type I spacetime we can then always choose a tetrad such that only (Ψ_1, Ψ_2, Ψ_3) are different from zero.

Petrov type II: Two roots coincide: $b_1 = b_2$, b_3, b_4. In this case it is clear that the derivative of (8.7.11) with respect to b must also vanish for $b = b_1$ (since it is a double root). But looking at the transformation of Ψ_1 we see that this implies that Ψ_1 will also become zero. We can then make both Ψ_0 and Ψ_1 vanish just by taking $b = b_1$. And as before, we can now use a transformation of class I to make Ψ_4 also vanish (but notice that equation (8.7.13) will now be quadratic since $\Psi_0 = \Psi_1 = 0$). Now, a class I transformation leaves Ψ_0 unaltered but in general changes Ψ_1 according to (8.7.2), but the new Ψ_1 is a linear combination of the previous values of Ψ_0 and Ψ_1, and since both vanish then Ψ_1 will still vanish after the transformation. For a type II spacetime we can then always choose a tetrad such that only (Ψ_2, Ψ_3) are different from zero.

Petrov type III: Three roots coincide: $b_1 = b_2 = b_3$, b_4. Following the same argument as before we see that by choosing now $b = b_1$ both the first and second derivatives of (8.7.11) with respect to b must vanish, so that we obtain $\Psi_0 = \Psi_1 = \Psi_2 = 0$. We can now perform a class I transformation to make Ψ_4 also vanish while still keeping $\Psi_0 = \Psi_1 = \Psi_2 = 0$ (equation (8.7.13) is now linear and has only one root). For a type III spacetime we can therefore always choose a tetrad such that only Ψ_3 is different from zero.

Petrov type N: All four roots coincide: $b_1 = b_2 = b_3 = b_4$. Again, the same argument as before implies that, with a class II transformation, we can make $\Psi_0 = \Psi_1 = \Psi_2 = \Psi_3 = 0$. However, now it is not possible to make Ψ_4 also vanish since after the class II transformation all other Ψ's are zero and there is no more room left to transform Ψ_4 (equation (8.7.13) now has no solutions). For a type N spacetime we can then choose a tetrad such that only Ψ_4 is non-zero.

Petrov type D: Two pairs of roots coincide: $b_1 = b_2$, $b_3 = b_4$. This case is particularly interesting so I have have left it until last. Notice first that since we have two double roots, then necessarily the transformation (8.7.6) for Ψ_0 must have the form

$$\Psi_0 \rightarrow \Psi_4 (b - b_1)^2 (b - b_2)^2 \, . \tag{8.7.14}$$

This must be true regardless of the values of the other Ψ's, as otherwise we would not have two double roots. Now, since we can obtain the transformations of the other Ψ's as derivatives of this transformation with respect to b with adequate rescalings, we find that

$$\Psi_1 \rightarrow \frac{\Psi_4}{2}(b-b_1)(b-b_2)(2b-b_1-b_2) , \tag{8.7.15}$$

$$\Psi_2 \rightarrow \frac{\Psi_4}{6}\left[6b^2 - 6b(b_1+b_2) + b_1^2 + b_2^2 + 4b_1b_2\right] , \tag{8.7.16}$$

$$\Psi_3 \rightarrow \frac{\Psi_4}{2}(2b-b_1-b_2) , \tag{8.7.17}$$

$$\Psi_4 \rightarrow \Psi_4 . \tag{8.7.18}$$

We can now substitute these new values into a class I transformation to find that after both transformations Ψ_4 becomes

$$\Psi_4 \rightarrow \Psi_4 \left[\bar{a}(b-b_1)+1\right]^2 \left[\bar{a}(b-b_2)+1\right]^2 . \tag{8.7.19}$$

Choose now $b = b_1$, so that, after the first transformation, we have $\Psi_0 = \Psi_1 = 0$ (it is a double root). Then, after the second transformation we find

$$\Psi_4 \rightarrow \Psi_4 \left[\bar{a}(b_1-b_2)+1\right]^2 . \tag{8.7.20}$$

This shows that the quadratic equation needed to make Ψ_4 vanish has a double root itself, namely $\bar{a} = 1/(b_2 - b_1)$. By taking now this value for $\bar{a}$ we can therefore make both Ψ_4 and Ψ_3 vanish while still keeping $\Psi_0 = \Psi_1 = 0$. The final result is that for a type D spacetime we can always choose a tetrad such that only Ψ_2 remains non-zero.

Petrov type O: The Weyl tensor vanishes identically (the spacetime is conformally flat).

Examples of D type spacetimes include both the Schwarzschild and Kerr solutions, while a plane gravitational wave spacetime is of type N. Both Minkowski, and the Friedmann–Robertson–Walker cosmological spacetimes are of type O.

Finally, there is a very important property of the spacetime associated with an isolated source that must be mentioned here. It turns out that, under very general conditions, it is possible to show that the asymptotic behavior of the different Weyl scalars is of the form [218]

$$\Psi_n \sim \frac{1}{r^{5-n}} , \tag{8.7.21}$$

so that the Riemann tensor behaves as

$$R \sim \frac{N}{r} + \frac{III}{r^2} + \frac{II}{r^3} + \frac{I}{r^4} . \tag{8.7.22}$$

This is known as the "peeling theorem", and in particular implies that the far field of any source of gravitational waves behaves locally as a plane wave.

8.8 Invariants I and J

In four dimensions we can in general construct 14 independent invariant quantities from the Riemann curvature tensor, starting from the scalar curvature R and considering polynomial contractions of $R_{\alpha\beta\mu\nu}$. These invariants can be understood as corresponding to the 20 independent components of Riemann, minus the six degrees of freedom associated with a general Lorentz transformation (the three components of the boost speed plus the three Euler angles corresponding to a rotation in space). In vacuum, however, only four of these invariants are non-zero. These four vacuum invariants can be expressed as the real and imaginary parts of two complex quantities commonly known as the I and J scalars and defined as

$$I := \frac{1}{2}\,\mathcal{C}_{\alpha\beta\mu\nu}\mathcal{C}^{\alpha\beta\mu\nu}\,, \qquad J := \frac{1}{6}\,\mathcal{C}_{\alpha\beta\lambda\sigma}\mathcal{C}^{\lambda\sigma}{}_{\mu\nu}\mathcal{C}^{\alpha\beta\mu\nu}\,, \tag{8.8.1}$$

where $\mathcal{C}_{\alpha\beta\mu\nu} := (C_{\alpha\beta\mu\nu} - i\,C^*_{\alpha\beta\mu\nu})/4$ is known as the *self dual* Weyl tensor, since clearly $\mathcal{C}_{\alpha\beta\mu\nu} = i\mathcal{C}^*_{\alpha\beta\mu\nu}$. The above expressions for I and J can be rewritten in terms of the electric and magnetic parts of Weyl as

$$\begin{aligned} I &= \frac{1}{2}\,Q_{ab}Q^{ab} \\ &= \frac{1}{2}\left[\left(E_{ab}E^{ab} - B_{ab}B^{ab}\right) - 2iE_{ab}B^{ab}\right], \end{aligned} \tag{8.8.2}$$

$$\begin{aligned} J &= -\frac{1}{6}\,Q_{ij}Q^i_kQ^{jk} \\ &= -\frac{1}{6}\left[E_{ij}\left(E^i_kE^{jk} - 3B^i_kB^{jk}\right) + iB_{ij}\left(B^i_kB^{jk} - 3E^i_kE^{jk}\right)\right], \end{aligned} \tag{8.8.3}$$

with Q_{ij} given by equation (8.6.18) as before.

The normalization used above for I and J is not universal but is nevertheless quite common, and is particularly convenient when we rewrite I and J in terms of the Weyl scalars Ψ_a. Using (8.6.21) we find that

$$I = 3\Psi_2^2 - 4\Psi_1\Psi_3 + \Psi_0\Psi_4\,, \tag{8.8.4}$$

$$J = \Psi_0\Psi_2\Psi_4 + 2\Psi_1\Psi_2\Psi_3 - \Psi_0\Psi_3^2 - \Psi_1^2\Psi_4 - \Psi_2^3\,. \tag{8.8.5}$$

The expression for J can in fact be written in more compact form as the following determinant

$$J = \begin{pmatrix} \Psi_0 & \Psi_1 & \Psi_2 \\ \Psi_1 & \Psi_2 & \Psi_3 \\ \Psi_2 & \Psi_3 & \Psi_4 \end{pmatrix}\,. \tag{8.8.6}$$

The invariants I and J can be very useful in the characterization of numerical spacetimes. For example, they can be used to compare a numerical solution to an exact solution, or two numerical solutions corresponding to the same physical system but computed in different gauges. In particular, for Petrov type D spacetimes like Schwarzschild and Kerr it is possible to choose a null tetrad such that the only non-zero Weyl scalar is Ψ_2. In such a case the above expressions imply that

$$I = 3\Psi_2^2 \,, \quad J = -\Psi_2^3 \qquad \Rightarrow \qquad I^3 = 27J^2 \,. \tag{8.8.7}$$

Notice that even if the Ψ's depend on the choice of the null tetrad, the I and J invariants do not (they are true scalars), so that for a type D spacetime we will always have $I^3 = 27J^2$, regardless of the choice of tetrad. This result is in fact more general, as it is not difficult to see that for a spacetime of type II we also have $I^3 = 27J^2$, while for spacetimes of types III and N we have $I = J = 0$.

Following Baker and Campanelli [41], we define the *speciality index* S as

$$S := 27J^2/I^3 \,. \tag{8.8.8}$$

We then find that $S = 1$ for type II or D spacetimes (for type III or N spacetimes S can not be defined as $I = J = 0$). More generally, we can use the deviation of S from unity as a measure of how close we are to a type II or D spacetime. For isolated systems the regions where S is close to unity will correspond to the wave zone where we have small perturbations of a Kerr background. Baker and Campanelli show that for small perturbations of a type D spacetime, the index S differs from unity by terms that are quadratic on the perturbation parameter, which means that when S is significantly different from unity we can no longer trust quantities derived from a first order theory.

8.9 Energy and momentum of gravitational waves

When studying the evolution of isolated systems that emit gravitational radiation, a very important question is how much energy and momentum (including angular momentum) is carried away by the radiation. This is not a trivial problem since, as we have already mentioned in Chapter 1, in general relativity there is no notion of the energy density of the gravitational field itself. This can be easily understood if we remember that in the Newtonian theory the energy density of gravity is proportional to $\nabla_i \phi \nabla^i \phi$, with ϕ the gravitational potential. Since in the Newtonian limit ϕ plays the role of the metric, we would expect that in relativity the energy of gravity would be a combination of quadratic terms in first derivatives of the metric. But this will not work in general since metric derivatives are combinations of Christoffel symbols, and these vanish in locally flat coordinates.

There are, however, some cases in which we can define a meaningful notion of the energy associated with the gravitational field. For example, we can define the total energy and momentum for asymptotically flat spacetimes (see Appendix A). More important for our present discussion is the fact that it is also possible to define the flux of energy and momentum carried away by gravitational waves in the weak field approximation.

8.9.1 *The stress-energy tensor for gravitational waves*

In order to find an expression for the energy and momentum carried by gravitational waves, we must first remember that gravitational waves arise as first order perturbations to a background spacetime. On the other hand, from our discussion in Chapter 1 we know that the stress-energy tensor associated with a scalar or electromagnetic field is in fact second order in the field variables (cf. equations (1.12.13) and (1.12.17)). This indicates that if we want to assign a stress-energy tensor to the gravitational wave we must consider second order perturbations of the metric. We then start again by considering small perturbations to a background metric but keeping terms to second order:

$$g_{\mu\nu} = g^{(0)}_{\mu\nu} + \epsilon h^{(1)}_{\mu\nu} + \epsilon^2 h^{(2)}_{\mu\nu} \, , \tag{8.9.1}$$

where as before $g^{(0)}_{\mu\nu}$ is the background metric (which we will assume to be a solution of the vacuum Einstein equations), and where we have introduced explicitly the parameter $\epsilon \ll 1$ to keep track of the order of the different terms.

The Ricci tensor can also be expanded in terms of ϵ in the form

$$R_{\mu\nu} = \epsilon R^{(1)}_{\mu\nu} + \epsilon^2 R^{(2)}_{\mu\nu} \, , \tag{8.9.2}$$

where we have already taken $R^{(0)}_{\mu\nu} = 0$ since by construction the background metric is a solution of the vacuum equations. A straightforward calculation shows that the different pieces of the Ricci tensor have the form[89]

$$R^{(1)}_{\mu\nu} = F^{(1)}_{\mu\nu}(h^{(1)}) \, , \tag{8.9.3}$$

$$R^{(2)}_{\mu\nu} = F^{(1)}_{\mu\nu}(h^{(2)}) + F^{(2)}_{\mu\nu}(h^{(1)}) \, , \tag{8.9.4}$$

where $F^{(1)}_{\mu\nu}$ and $F^{(2)}_{\mu\nu}$ are defined as

$$F^{(1)}_{\mu\nu}(h) := \frac{1}{2}\left(-h_{|\mu\nu} - h_{\mu\nu|\alpha}{}^{\alpha} + h_{\alpha\mu|\nu}{}^{\alpha} + h_{\alpha\nu|\mu}{}^{\alpha}\right) \, , \tag{8.9.5}$$

$$\begin{aligned} F^{(2)}_{\mu\nu}(h) :=& \frac{1}{4} h_{\alpha\beta|\mu} h^{\alpha\beta}{}_{|\nu} + \frac{1}{2} h^{\alpha\beta}\left(h_{\alpha\beta|\mu\nu} + h_{\mu\nu|\alpha\beta} - 2h_{\alpha(\mu|\nu)\beta}\right) \\ &+ h_{\nu}{}^{\alpha|\beta} h_{\mu[\alpha|\beta]} + \frac{1}{2}\left(h^{\alpha\beta}{}_{|\beta} - \frac{1}{2}h^{|\alpha}\right)\left(h_{\mu\nu|\alpha} - 2h_{\alpha(\mu|\nu)}\right) \, , \end{aligned} \tag{8.9.6}$$

with h the trace of $h_{\mu\nu}$, $h := h^{\mu}_{\mu}$ (indices are raised and lowered with $g^{(0)}_{\mu\nu}$), and where the bar denotes covariant derivative with respect to $g^{(0)}_{\mu\nu}$.

Once we have expanded the Ricci tensor we must then consider the vacuum Einstein equations for each power of ϵ:

$$R^{(0)}_{\mu\nu} = 0 \, , \tag{8.9.7}$$

$$R^{(1)}_{\mu\nu} = F^{(1)}_{\mu\nu}(h^{(1)}) = 0 \, , \tag{8.9.8}$$

$$R^{(2)}_{\mu\nu} = F^{(1)}_{\mu\nu}(h^{(2)}) + F^{(2)}_{\mu\nu}(h^{(1)}) = 0 \, . \tag{8.9.9}$$

[89]This can be most easily derived by considering $g_{\mu\nu}$ and $g^{(0)}_{\mu\nu}$ as two different metrics on the same manifold, and using the difference between the connection coefficients associated with each metric (which is a tensor) to find the difference between the corresponding Ricci tensors, see [206].

The first of these equations is automatically satisfied, while the second describes the dynamics of the gravitational waves. The third equation, on the other hand, can be rewritten in terms of the Einstein tensor as

$$F^{(1)}_{\mu\nu}(h^{(2)}) - \frac{1}{2} g^{(0)}_{\mu\nu} F^{(1)}(h^{(2)}) = 8\pi \, t_{\mu\nu} \; , \tag{8.9.10}$$

with

$$t_{\mu\nu} := -\frac{1}{8\pi} \left[F^{(2)}_{\mu\nu}(h^{(1)}) - \frac{1}{2} g^{(0)}_{\mu\nu} F^{(2)}(h^{(1)}) \right] \; . \tag{8.9.11}$$

When it is written in this way we see that the contribution to the Einstein tensor coming from $h^{(2)}_{\mu\nu}$ has $t_{\mu\nu}$ as its source. This suggests that $t_{\mu\nu}$ can be interpreted as the stress-energy tensor of the gravitational waves described by $h^{(1)}_{\mu\nu}$. There are several reasons why this is in fact a good definition. First, $t_{\mu\nu}$ is clearly symmetric and quadratic in $h^{(1)}_{\mu\nu}$, as expected. But moreover, from the Bianchi identities to order ϵ^2 we can in fact show that

$$t^{\mu\nu}{}_{|\mu} = 0 \; , \tag{8.9.12}$$

so that $t_{\mu\nu}$ is conserved in the background metric.

The quantity $t_{\mu\nu}$ is in fact not gauge invariant; that is, it changes under a transformation of the metric perturbation $h^{(1)}_{\mu\nu}$ of the form

$$h^{(1)}_{\mu\nu} \;\rightarrow\; h^{(1)}_{\mu\nu} - \xi_{\mu|\nu} - \xi_{\nu|\mu} \; , \tag{8.9.13}$$

corresponding to the infinitesimal coordinate transformation $x^\mu \rightarrow x^\mu + \xi^\mu$ with $|\xi^\mu| \ll 1$. Because of this $t_{\mu\nu}$ is called the *stress-energy pseudo-tensor* of the gravitational waves, where "pseudo-tensor" refers to the fact that its definition requires some extra structure such as a special coordinate system. In a general case when we do not consider small perturbations, an analogous quantity to $t_{\mu\nu}$ can also be defined which is known as the *Landau–Lifshitz pseudo-tensor* [183], and in fact our expression for $t_{\mu\nu}$ can be found by calculating the general Landau–Lifshitz pseudo-tensor for small perturbations of a fixed background.

The fact that $t_{\mu\nu}$ is not gauge invariant does not imply that it is devoid of any meaning. It is indeed true that we can always choose coordinates such that $t_{\mu\nu}$ vanishes at a given point, but in the presence of non-trivial gravitational waves it is in general not possible to make $t_{\mu\nu}$ vanish over a finite region of spacetime. This implies that the energy and momentum carried by the gravitational waves can not be precisely localized and only has physical meaning in a finite region. This idea can be made more precise by introducing the so-called *short wavelength approximation* in which we consider the average of $t_{\mu\nu}$ over a region that covers several wavelengths but is at the same time small compared with the characteristic lengths associated with the background metric (notice that this kind of averaging over several wavelengths is not exclusive to gravitational waves, and in fact is done when we calculates the energy and momentum of any type of wave). Through a rather lengthy calculation we can show that the averaged stress-energy tensor for a gravitational wave is given by (from now on we will drop the super-index (1) from $h^{(1)}_{\mu\nu}$)

$$T_{\mu\nu} := \langle t_{\mu\nu} \rangle = \frac{1}{32\pi} \left\langle \bar{h}_{\alpha\beta|\mu} \bar{h}^{\alpha\beta}{}_{|\nu} - \frac{1}{2} \bar{h}_{|\mu} \bar{h}_{|\nu} - 2 \bar{h}^{\alpha\beta}{}_{|\beta} \bar{h}_{\alpha(\mu|\nu)} \right\rangle , \tag{8.9.14}$$

where $\langle\,\rangle$ denotes an average over several wavelengths, and where $\bar{h}_{\mu\nu}$ is the trace-reverse of $h_{\mu\nu}$ defined as

$$\bar{h}_{\mu\nu} := h_{\mu\nu} - \frac{1}{2} g^{(0)}_{\mu\nu} h \, . \tag{8.9.15}$$

The tensor $T_{\mu\nu}$ just defined is known as the *Isaacson stress-energy tensor* [166]. It is also not difficult to show that $T_{\mu\nu}$ is in fact gauge invariant. In the particular case where we consider the transverse-traceless (TT) gauge corresponding to $h = h^{\alpha\beta}{}_{|\beta} = 0$, the Isaacson stress-energy tensor reduces to

$$T_{\mu\nu} = \frac{1}{32\pi} \left\langle h_{\alpha\beta|\mu} \, h^{\alpha\beta}{}_{|\nu} \right\rangle \, . \tag{8.9.16}$$

8.9.2 *Radiated energy and momentum*

In the previous Section we derived the stress-energy tensor associated with gravitational waves. We are now in a position to use this stress-energy tensor to calculate the energy and momentum lost by an isolated system in the form of gravitational radiation. We will start by rewriting the Isaacson stress-energy tensor in the TT gauge (8.9.16) in a more convenient way. Assume first that we are far from the source, so that we can describe the gravitational waves as spherical waves propagating in a flat background. As was already discussed in Chapter 1, in the TT gauge we find that, in locally Cartesian coordinates, the metric perturbation $h_{\mu\nu}$ can be written as

$$h_{\mu\nu} = h^{+} A^{+}_{\mu\nu} + h^{\times} A^{\times}_{\mu\nu} \, , \tag{8.9.17}$$

where $h^{+,\times}$ are the amplitudes of the two independent wave polarizations, and $A^{+,\times}_{\mu\nu}$ are constant symmetric polarization tensors such that

$$A_{\mu\nu} l^{\nu} = 0 \, , \quad A^{\mu}{}_{\mu} = 0 \, , \quad A_{\mu\nu} u^{\nu} = 0 \, , \tag{8.9.18}$$

with l^{μ} a null vector (the wave vector), and u^{μ} an arbitrary unit timelike vector. Let us now consider the spatial orthonormal basis $(\hat{e}_r, \hat{e}_\theta, \hat{e}_\varphi)$ induced by the spherical coordinates (r, θ, φ) (notice that this is not the spherical coordinate basis). If we choose $u^{\mu} = (1, 0, 0, 0)$ (the unit timelike vector in the flat background), and take l^{μ} as an outgoing radial null vector (since for an isolated source we only expect outgoing waves), we find that the only non-zero components of the polarization tensors are

$$A^{+}_{\hat{\theta}\hat{\theta}} = -A^{+}_{\hat{\phi}\hat{\phi}} = 1 \,, \tag{8.9.19}$$

$$A^{\times}_{\hat{\theta}\hat{\phi}} = A^{\times}_{\hat{\phi}\hat{\theta}} = 1 \,, \tag{8.9.20}$$

where we have chosen the normalization $A^{+}_{\hat{\theta}\hat{\theta}} = A^{\times}_{\hat{\theta}\hat{\phi}} = 1$. This implies that

$$A^{+}_{\mu\nu} A^{+\mu\nu} = A^{\times}_{\mu\nu} A^{\times\mu\nu} = 2 \,, \qquad A^{+}_{\mu\nu} A^{\times\mu\nu} = 0 \,. \tag{8.9.21}$$

Using this we can rewrite the Isaacson stress-energy tensor (8.9.16) in locally Cartesian coordinates as

$$T_{\mu\nu} = \frac{1}{16\pi} \left\langle \partial_\mu h^+ \partial_\nu h^+ + \partial_\mu h^\times \partial_\nu h^\times \right\rangle \,, \tag{8.9.22}$$

or equivalently

$$T_{\mu\nu} = \frac{1}{16\pi} \, \mathrm{Re} \left\langle \partial_\mu H \partial_\nu \bar{H} \right\rangle \,, \tag{8.9.23}$$

with $H = h^+ - ih^\times$, and where $\mathrm{Re}(z)$ denotes the real part of z.

It turns out that the complex quantity H can in fact also be written in terms of the Weyl scalar Ψ_4. In order to see this, notice first that if we are in vacuum far from the source of the gravitational waves the Weyl and Riemann tensors coincide. Using now the expression for the Riemann tensor in the linearized approximation from Chapter 1, equation (1.14.4), we can easily show that, for plane waves in the TT gauge traveling along the r direction, the Weyl scalars Ψ_a become

$$\Psi_1 = \Psi_2 = \Psi_3 = 0 \,, \tag{8.9.24}$$

$$\Psi_0 = -\frac{1}{4} \left(\partial_t^2 h^+ + 2\partial_t \partial_r h^+ + \partial_r^2 h^+ \right) - \frac{i}{4} \left(\partial_t^2 h^\times + 2\partial_t \partial_r h^\times + \partial_r^2 h^\times \right) , \tag{8.9.25}$$

$$\Psi_4 = -\frac{1}{4} \left(\partial_t^2 h^+ - 2\partial_t \partial_r h^+ + \partial_r^2 h^+ \right) + \frac{i}{4} \left(\partial_t^2 h^\times - 2\partial_t \partial_r h^\times + \partial_r^2 h^\times \right) . \tag{8.9.26}$$

Now, for outgoing waves we have $h = h(r-t)$, so that $\partial_r h = -\partial_t h$. The Weyl scalars then reduce to

$$\Psi_0 = \Psi_1 = \Psi_2 = \Psi_3 = 0 \,, \tag{8.9.27}$$

$$\Psi_4 = -\ddot{h}^+ + i\ddot{h}^\times = -\ddot{H} \,. \tag{8.9.28}$$

This implies that for outgoing waves we can write

$$H = -\int_{-\infty}^{t} \int_{-\infty}^{t'} \Psi_4 \, dt'' dt' \,. \tag{8.9.29}$$

(For ingoing waves we would have instead that $\partial_r h = \partial_t h$, so that the only non-vanishing Weyl scalar would be $\Psi_0 = -\ddot{\bar{H}}$.)

It is important to notice that in the above derivation we have assumed that the null tetrad is that obtained from the standard spherical coordinates in flat space, so the expressions are only valid asymptotically. At a finite radius there is in fact no "standard" choice of tetrad, though there have been some recent proposals for such a choice (see *e.g.* [216, 91]). However, any choice of tetrad that reduces to the standard flat space tetrad asymptotically should give the same results in the limit of large r.

We can now use equation (8.9.23) to find the flux of energy and momentum carried away from an isolated system by the gravitational waves. Consider first the flux of energy along the direction i, which is given in general by T^{0i}. The energy flux along the radial direction for gravitational waves will then be given in locally Cartesian coordinates by

$$\frac{dE}{dtdA} = T^{0r} = \frac{1}{16\pi} \operatorname{Re} \left\langle \partial^0 H \partial^r \bar{H} \right\rangle = -\frac{1}{16\pi} \operatorname{Re} \left\langle \partial_t H \partial_r \bar{H} \right\rangle , \tag{8.9.30}$$

with dA the area element orthogonal to the radial direction. Since for outgoing waves we have $\partial_r h = -\partial_t h$, we find that

$$\frac{dE}{dtdA} = \frac{1}{16\pi} \left\langle \dot{H} \dot{\bar{H}} \right\rangle = \frac{1}{16\pi} \left\langle |\dot{H}|^2 \right\rangle . \tag{8.9.31}$$

The last expression is manifestly real, so there is no need to ask for the real part.

If we now want the total flux of energy leaving the system at a given time we need to integrate over the sphere to find (see, *e.g.* [90])

$$\frac{dE}{dt} = \lim_{r\to\infty} \frac{r^2}{16\pi} \oint |\dot{H}|^2 d\Omega = \lim_{r\to\infty} \frac{r^2}{16\pi} \oint \left| \int_{-\infty}^{t} \Psi_4 \, dt' \right|^2 d\Omega , \tag{8.9.32}$$

where we have taken $dA = r^2 d\Omega$, with $d\Omega$ the standard solid angle element, and where the limit of infinite radius has been introduced since the Isaacson stress-energy tensor is only valid in the weak field approximation. Notice also that we have dropped the averaging since the integral over the sphere is already performing an average over space, plus the expression above is usually integrated over time to find the total energy radiated, which again eliminates the need to take an average.

Consider next the flux of momentum which corresponds to the spatial components of the stress-energy tensor T_{ij}. The flux of momentum i along the radial direction will then be given by

$$\frac{dP_i}{dtdA} = T_{ir} = \frac{1}{16\pi} \operatorname{Re} \left\langle \partial_i H \partial_r \bar{H} \right\rangle . \tag{8.9.33}$$

Now, if we are sufficiently far away the radiation can be locally approximated as a plane wave, so that $\partial_i H \simeq (x_i/r)\, \partial_r H$ (in effect we are assuming that for large r the angular dependence can be neglected). Using again the fact that for outgoing waves $\partial_r h = -\partial_t h$ we find that, for large r,

$$\frac{dP_i}{dtdA} \simeq \frac{1}{16\pi} \, l_i \left\langle |\dot{H}|^2 \right\rangle , \tag{8.9.34}$$

where $\vec{l}$ is the unit radial vector in flat space

$$\vec{l} = (\sin\theta\cos\varphi, \sin\theta\sin\varphi, \cos\theta) \ . \tag{8.9.35}$$

Notice that the above expression implies that the magnitude of the momentum flux is equal to the energy flux, which is to be expected for locally plane waves.

The total flux of momentum leaving the system will again be given by an integral over the sphere as (again, see [90])

$$\frac{dP_i}{dt} = \lim_{r\to\infty} \frac{r^2}{16\pi} \oint l_i \, |\dot{H}|^2 d\Omega = \lim_{r\to\infty} \frac{r^2}{16\pi} \oint l_i \left| \int_{-\infty}^{t} \Psi_4 \, dt' \right|^2 d\Omega \ . \tag{8.9.36}$$

Finally, let us consider the flux of angular momentum. Locally, the flux of the i component of the angular momentum along the radial direction should correspond to $\epsilon_{ijk} x^j T^{kr}$, with ϵ_{ijk} the three-dimensional Levi–Civita antisymmetric tensor and T^{ij} the stress-energy tensor of the field under study (this corresponds to $\vec{r} \times \vec{p}$ in vector notation). However, in the case of gravitational waves such an expression is in fact wrong since the averaging procedure that is used to derive the Isaacson stress-energy tensor ignores terms that go as $1/r^3$, and it is precisely such terms the ones that contribute to the flux of angular momentum. A correct expression for the flux of angular momentum due to gravitational waves was first derived by DeWitt in 1971, and in the TT gauge has the form (see *e.g.* [288])

$$\frac{dJ^i}{dt} = \lim_{r\to\infty} \frac{r^2}{32\pi} \oint \epsilon^{ijk} \left(x_j \partial_k h_{ab} + 2\delta_{aj} h_{bk} \right) \partial_r h^{ab} \, d\Omega \ . \tag{8.9.37}$$

Remember that in the TT gauge all the time components of $h_{\alpha\beta}$ vanish, so the summations above are only over spatial indices.

The last expression can be rewritten in a more compact and easy to interpret form if we introduce the angular vectors $\vec{\xi}_i$ associated with rotations around the three-coordinate axis. These vectors are clearly Killing fields of the flat metric, and in Cartesian coordinates have components given by $\xi_i^k = \epsilon_i{}^{jk} x_j$, where ξ_i^k represents the k component of the vector $\vec{\xi}_i$. In terms of the vectors $\vec{\xi}_i$ the flux of angular momentum can now be written as

$$\frac{dJ_i}{dt} = - \lim_{r\to\infty} \frac{r^2}{32\pi} \oint \left(\pounds_{\xi_i} h_{ab} \right) \partial_t h^{ab} \, d\Omega \ , \tag{8.9.38}$$

where $\pounds_{\xi_i} h_{ab}$ is the Lie derivative of h_{ab} with respect to $\vec{\xi}_i$, and where we have again taken $\partial_r h = -\partial_t h$ for outgoing waves. The appearance of the Lie derivative is to be expected on physical grounds, since for rotational symmetry around a given axis ($\pounds_{\xi_i} h_{ab} = 0$) we should find that the corresponding angular momentum flux vanishes.

In order to write the angular momentum flux in terms of H or ψ_4, as we have done with the flux of energy and linear momentum, we must now be careful to take correctly into account the effect of the Lie derivative in (8.9.38). The easiest way to do this is to work in spherical coordinates (r, θ, φ), in which case the angular vectors have components

$$\vec{\xi}_x = (0, -\sin\varphi, -\cos\varphi \cot\theta) \;, \tag{8.9.39}$$

$$\vec{\xi}_y = (0, +\cos\varphi, -\sin\varphi \cot\theta) \;, \tag{8.9.40}$$

$$\vec{\xi}_z = (0, 0, 1) \;. \tag{8.9.41}$$

It is clear that the vector $\vec{\xi}_z$ corresponds to one of the vectors of the coordinate basis, which implies that Lie derivatives along it reduce to simple partial derivatives. However, this is no longer the case for the vectors $\vec{\xi}_x$ and $\vec{\xi}_y$. In order to calculate these Lie derivatives let us start by introducing the two complex angular vectors $\vec{\xi}_\pm := \vec{\xi}_x \pm i\vec{\xi}_y$. We then have

$$\vec{\xi}_\pm = e^{\pm i\varphi} \left(0, \pm i, -\cot\theta\right) \;. \tag{8.9.42}$$

We will furthermore introduce an orthonormal spherical basis $(\hat{e}_r, \hat{e}_\theta, \hat{e}_\varphi)$, and define the two unit complex vectors (notice the counterintuitive signs in this definition)

$$\hat{e}_\pm := \frac{1}{\sqrt{2}} \left(\hat{e}_\theta \mp i\hat{e}_\varphi\right) \;, \tag{8.9.43}$$

so that

$$\hat{e}_\pm = \frac{1}{r\sqrt{2}} \left(0, 1, \mp i \csc\theta\right) \;. \tag{8.9.44}$$

The vectors $\hat{e}_\pm$ clearly correspond to the vectors $\bar{m}$ and m of the Newman–Penrose formalism. We can then show after some algebra that the Lie derivative of $\hat{e}_\pm$ with respect to $\vec{\xi}_\pm$ is given by

$$\pounds_{\xi_\pm} e^a_\pm = \mp \left(i e^{\pm i\varphi} \csc\theta\right) e^a_\pm \;. \tag{8.9.45}$$

Let us now rewrite the metric perturbation h_{ab} in the TT gauge in terms of the orthonormal basis as:

$$\begin{aligned} h_{ab} &= h^+ \left[(\hat{e}_\theta)_a(\hat{e}_\theta)_b - (\hat{e}_\varphi)_a(\hat{e}_\varphi)_b\right] + h^\times \left[(\hat{e}_\theta)_a(\hat{e}_\varphi)_b + (\hat{e}_\varphi)_a(\hat{e}_\theta)_b\right] \\ &= \left(h^+ - ih^\times\right)(\hat{e}_-)_a(\hat{e}_-)_b + \left(h^+ + ih^\times\right)(\hat{e}_+)_a(\hat{e}_+)_b \\ &= H\,(\hat{e}_-)_a(\hat{e}_-)_b + \bar{H}\,(\hat{e}_+)_a(\hat{e}_+)_b \;. \end{aligned} \tag{8.9.46}$$

We are now in a position to calculate the Lie derivative of h_{ab} with respect to $\vec{\xi}_\pm$. We find

$$\pounds_{\xi_\pm} h_{ab} = (\hat{e}_-)_a(\hat{e}_-)_b \, \hat{J}_\pm H + (\hat{e}_+)_a(\hat{e}_+)_b \, \hat{J}_\pm \bar{H} \; , \tag{8.9.47}$$

where we have defined the operators

$$\hat{J}_\pm := \xi^a_\pm \partial_a - is \, e^{\pm i\varphi} \csc\theta = e^{\pm i\varphi} \left[\pm i\partial_\theta - \cot\theta \, \partial_\varphi - is \csc\theta \right] \; , \tag{8.9.48}$$

with s the spin weight of the function on which the operator is acting: $s = -2$ for H, and $s = +2$ for $\bar{H}$ (see Appendix D). The last result implies that

$$\left(\pounds_{\xi_\pm} h_{ab} \right) \partial_t h^{ab} = \hat{J}_\pm H \, \partial_t \bar{H} + \hat{J}_\pm \bar{H} \, \partial_t H \; , \tag{8.9.49}$$

from which we find

$$\left(\pounds_{\xi_x} h_{ab} \right) \partial_t h^{ab} = \hat{J}_x H \, \partial_t \bar{H} + \hat{J}_x \bar{H} \, \partial_t H = 2\mathrm{Re} \left\{ \hat{J}_x H \, \partial_t \bar{H} \right\} \; , \tag{8.9.50}$$

$$\left(\pounds_{\xi_y} h_{ab} \right) \partial_t h^{ab} = \hat{J}_y H \, \partial_t \bar{H} + \hat{J}_y \bar{H} \, \partial_t H = 2\mathrm{Re} \left\{ \hat{J}_y H \, \partial_t \bar{H} \right\} \; , \tag{8.9.51}$$

with $\hat{J}_x := (\hat{J}_+ + \hat{J}_-)/2$ and $\hat{J}_y := -i(\hat{J}_+ - \hat{J}_-)/2$.

Collecting results, the flux of angular momentum becomes

$$\frac{dJ_i}{dt} = - \lim_{r\to\infty} \frac{r^2}{16\pi} \, \mathrm{Re} \left\{ \oint \hat{J}_i H \, \partial_t \bar{H} \, d\Omega \right\} \; , \tag{8.9.52}$$

with the different angular momentum operators defined as

$$\hat{J}_x = \frac{1}{2} \left(\hat{J}_+ + \hat{J}_- \right) = - \sin\varphi \, \partial_\theta - \cos\varphi \left(\cot\theta \, \partial_\varphi - is \csc\theta \right) \; , \tag{8.9.53}$$

$$\hat{J}_y = \frac{i}{2} \left(\hat{J}_- - \hat{J}_+ \right) = + \cos\varphi \, \partial_\theta - \sin\varphi \left(\cot\theta \, \partial_\varphi - is \csc\theta \right) \; , \tag{8.9.54}$$

$$\hat{J}_z = \partial_\varphi \; . \tag{8.9.55}$$

Notice that, except for a factor of $-i\hbar$, these are just the quantum mechanical angular momentum operators with the correct spin weight (see Appendix D).

Finally, the flux of angular momentum can be calculated in terms of Ψ_4 as (see [90, 246])

$$\frac{dJ_i}{dt} = - \lim_{r\to\infty} \frac{r^2}{16\pi} \, \mathrm{Re} \Bigg\{ \oint \left(\int_{-\infty}^{t} \bar{\Psi}_4 \, dt' \right) \times \hat{J}_i \left(\int_{-\infty}^{t} \int_{-\infty}^{t'} \Psi_4 \, dt'' dt' \right) d\Omega \Bigg\} \; . \tag{8.9.56}$$

8.9.3 *Multipole decomposition*

In the previous Section we derived expressions for the flux of energy and momentum carried by gravitational waves. Though these expressions are quite general, it is convenient to rewrite them in terms of a multipole expansions of either the complex metric perturbation H, or the Weyl scalar Ψ_4. The multipole expansion of H was already calculated in terms of the gauge invariant perturbations $(\Psi^{l,m}_{\rm even}, \Psi^{l,m}_{\rm odd})$ and $(Q^{l,m}_{\rm even}, Q^{l,m}_{\rm odd})$ in Section 8.2, and is given by equations (8.2.72) and (8.2.79). On the other hand, the Weyl scalar Ψ_4 can be easily shown to have spin weight $s = -2$ (see Appendix D), so that we can expand it as

$$\Psi_4 = \sum_{l,m} A^{l,m} \left({}_{-2}Y^{l,m}(\theta,\phi) \right) , \tag{8.9.57}$$

with $A^{l,m}$ the expansion coefficients given by

$$A^{l,m} = \oint \Psi_4 \left({}_{-2}\bar{Y}^{l,m}(\theta,\phi) \right) \, d\Omega \, . \tag{8.9.58}$$

Notice that since we are expanding over the harmonics of spin weight $s = -2$, the sum over l also starts at $l = 2$. Comparing the multipole expansion for Ψ_4 with the expansion for the metric perturbation (8.2.72), and using the fact that asymptotically $\Psi_4 = -\ddot{H}$, we can relate the coefficients $A^{l,m}$ to $(\Psi^{l,m}_{\rm even}, \Psi^{l,m}_{\rm odd})$ and $(Q^{l,m}_{\rm even}, Q^{l,m}_{\rm odd})$ in the following way

$$\begin{aligned} A^{l,m} &= -\frac{1}{2r} \left(\frac{(l+2)!}{(l-2)!} \right)^{1/2} \left(\ddot{\Psi}^{l,m}_{\rm even} + i \ddot{\Psi}^{l,m}_{\rm odd} \right) \\ &= -\frac{1}{\sqrt{2}\, r} \left(\ddot{Q}^{l,m}_{\rm even} - i \dot{Q}^{l,m}_{\rm odd} \right) . \end{aligned} \tag{8.9.59}$$

Using this we can translate expressions in terms of the $A^{l,m}$ directly into expressions in terms of $(\Psi^{l,m}_{\rm even}, \Psi^{l,m}_{\rm odd})$ and/or $(Q^{l,m}_{\rm even}, Q^{l,m}_{\rm odd})$.

Consider first the radiated energy given by equation (8.9.32). If we substitute the multipole expansion for Ψ_4, and use the orthogonality of the ${}_sY^{l,m}$ (equation (D.28) of Appendix D), we immediately find that

$$\frac{dE}{dt} = \lim_{r\to\infty} \frac{r^2}{16\pi} \sum_{l,m} \left| \int_{-\infty}^{t} A^{l,m} \, dt' \right|^2 , \tag{8.9.60}$$

or equivalently in terms of the multipole expansion for H

$$\begin{aligned} \frac{dE}{dt} &= \lim_{r\to\infty} \frac{1}{64\pi} \sum_{l,m} \frac{(l+2)!}{(l-2)!} \left(\left| \dot{\Psi}^{l,m}_{\rm even} \right|^2 + \left| \dot{\Psi}^{l,m}_{\rm odd} \right|^2 \right) \\ &= \lim_{r\to\infty} \frac{1}{32\pi} \sum_{l,m} \left(\left| \dot{Q}^{l,m}_{\rm even} \right|^2 + \left| Q^{l,m}_{\rm odd} \right|^2 \right) , \end{aligned} \tag{8.9.61}$$

where, in order to derive these last expressions, we must use the fact that, as a consequence of (8.2.71), we have

$$\sum_m \left(\dot{\Psi}^{l,m}_{\rm even} \ddot{\Psi}^{l,m}_{\rm odd} - \ddot{\Psi}^{l,m}_{\rm even} \dot{\Psi}^{l,m}_{\rm odd} \right) = 0 \; . \tag{8.9.62}$$

The calculation for the linear momentum flux is somewhat more complicated. Substituting the multipole expansion of Ψ_4 in (8.9.36) we find

$$\begin{aligned} \frac{dP_i}{dt} &= \lim_{r\to\infty} \frac{r^2}{16\pi} \sum_{l,m} \sum_{l',m'} \oint l_i \left({}_{-2}Y^{l,m} \right) \left({}_{-2}\bar{Y}^{l',m'} \right) d\Omega \\ &\times \int_{-\infty}^{t} A^{l,m}\, dt' \int_{-\infty}^{t} \bar{A}^{l',m'}\, dt' \; . \end{aligned} \tag{8.9.63}$$

In order to calculate the integral over the sphere, notice first that the components of the radial unit vector l_i can be expressed in terms of scalar (*i.e.* spin zero) spherical harmonics as

$$l_x = \sin\theta \cos\varphi = \sqrt{\frac{2\pi}{3}} \left[Y^{1,-1} - Y^{1,1} \right] , \tag{8.9.64}$$

$$l_y = \sin\theta \sin\varphi = i \sqrt{\frac{2\pi}{3}} \left[Y^{1,-1} + Y^{1,1} \right] , \tag{8.9.65}$$

$$l_z = \cos\theta = 2 \sqrt{\frac{\pi}{3}}\, Y^{1,0} \; . \tag{8.9.66}$$

We then see that the flux of linear momentum involves integrals over three spin-weighted spherical harmonics. Such integrals can be calculated using equation (D.30) of Appendix D. They involve the Wigner 3-lm symbols with $l_3 = 1$, which are also explicitly given in Appendix D.

Instead of P_x and P_y it turns out to be easier to work with the complex quantity $P_+ := P_x + iP_y$. After a straightforward, but rather long, calculation we finally arrive at the following expressions for the multipole expansion of the flux of linear momentum in terms of the $A^{l,m}$ coefficients

$$\begin{aligned} \frac{dP_+}{dt} &= \lim_{r\to\infty} \frac{r^2}{8\pi} \sum_{l,m} \int_{-\infty}^{t} dt'\, A^{l,m} \\ &\times \int_{-\infty}^{t} dt' \left(a_{l,m} \bar{A}^{l,m+1} + b_{l,-m} \bar{A}^{l-1,m+1} - b_{l+1,m+1} \bar{A}^{l+1,m+1} \right) , \end{aligned} \tag{8.9.67}$$

$$\begin{aligned} \frac{dP_z}{dt} &= \lim_{r\to\infty} \frac{r^2}{16\pi} \sum_{l,m} \int_{-\infty}^{t} dt'\, A^{l,m} \\ &\times \int_{-\infty}^{t} dt' \left(c_{l,m} \bar{A}^{l,m} + d_{l,m} \bar{A}^{l-1,m} + d_{l+1,m} \bar{A}^{l+1,m} \right) , \end{aligned} \tag{8.9.68}$$

where we have defined the quantities

$$a_{l,m} := \frac{\sqrt{(l-m)(l+m+1)}}{l(l+1)} , \tag{8.9.69}$$

$$b_{l,m} := \frac{1}{2l} \sqrt{\frac{(l-2)(l+2)(l+m)(l+m-1)}{(2l-1)(2l+1)}} , \tag{8.9.70}$$

$$c_{l,m} := \frac{2m}{l(l+1)} , \tag{8.9.71}$$

$$d_{l,m} := \frac{1}{l} \sqrt{\frac{(l-2)(l+2)(l-m)(l+m)}{(2l-1)(2l+1)}} . \tag{8.9.72}$$

Using now (8.9.59) we can also rewrite these expressions in terms of the multipole expansion for H. The calculation is again quite long, and in order to simplify the expressions we must make use several times of (8.2.71). The final result is [90]

$$\begin{aligned}\frac{dP_+}{dt} &= - \lim_{r\to\infty} \frac{1}{16\pi} \sum_{l,m} \frac{(l+2)!}{(l-2)!} \Bigg[i a_{l,m} \dot{\Psi}^{l,m}_{\rm even} \dot{\bar{\Psi}}^{l,m+1}_{\rm odd} \\ &+ b_{l+1,m+1} \left(\frac{(l+3)}{(l-1)} \right)^{1/2} \left(\dot{\Psi}^{l,m}_{\rm even} \dot{\bar{\Psi}}^{l+1,m+1}_{\rm even} + \dot{\Psi}^{l,m}_{\rm odd} \dot{\bar{\Psi}}^{l+1,m+1}_{\rm odd} \right) \Bigg] \end{aligned} \tag{8.9.73}$$

$$\begin{aligned} &= - \lim_{r\to\infty} \frac{1}{8\pi} \sum_{l,m} \Bigg[-i a_{l,m} \dot{Q}^{l,m}_{\rm even} \bar{Q}^{l,m+1}_{\rm odd} \\ &+ b_{l+1,m+1} \left(\dot{Q}^{l,m}_{\rm even} \dot{\bar{Q}}^{l+1,m+1}_{\rm even} + Q^{l,m}_{\rm odd} \bar{Q}^{l+1,m+1}_{\rm odd} \right) \Bigg] , \end{aligned} \tag{8.9.74}$$

$$\begin{aligned}\frac{dP_z}{dt} &= \lim_{r\to\infty} \frac{1}{32\pi} \sum_{l,m} \frac{(l+2)!}{(l-2)!} \Bigg[-i c_{l,m} \dot{\Psi}^{l,m}_{\rm even} \dot{\bar{\Psi}}^{l,m}_{\rm odd} \\ &+ d_{l+1,m} \left(\frac{(l+3)}{(l-1)} \right)^{1/2} \left(\dot{\Psi}^{l,m}_{\rm even} \dot{\bar{\Psi}}^{l+1,m}_{\rm even} + \dot{\Psi}^{l,m}_{\rm odd} \dot{\bar{\Psi}}^{l+1,m}_{\rm odd} \right) \Bigg] \end{aligned} \tag{8.9.75}$$

$$\begin{aligned} &= \lim_{r\to\infty} \frac{1}{16\pi} \sum_{l,m} \Bigg[i c_{l,m} \dot{Q}^{l,m}_{\rm even} \bar{Q}^{l,m}_{\rm odd} \\ &+ d_{l+1,m} \left(\dot{Q}^{l,m}_{\rm even} \dot{\bar{Q}}^{l+1,m}_{\rm even} + Q^{l,m}_{\rm odd} \bar{Q}^{l+1,m}_{\rm odd} \right) \Bigg] , \end{aligned} \tag{8.9.76}$$

For the flux of angular momentum we now go back to equation (8.9.56). Expressing Ψ_4 in terms of its multipole expansion and integrating over the sphere we find

[90] These expressions can be easily shown to be equivalent to those derived by Pollney *et al.* in [228], and Sopuerta *et al.* in [274].

9

NUMERICAL METHODS

9.1 Introduction

Field theories play a fundamental role in modern physics. From Maxwell's classical electrodynamics, to quantum field theories, through the Schrödinger equation, hydrodynamics and general relativity, the notion of a field as a physical entity on its own right has had profound implications in our understanding of the Universe. Fields are continuous functions of space and time, and the mathematical description of their dynamics must be done in the context of partial differential equations.

The partial differential equations associated with physical theories are in general impossible to solve exactly except in very idealized cases. This difficulty can have different origins, from the presence of irregular boundaries, to the existence of non-linear terms in the equations themselves. In order to solve this type of equation in general dynamical situations it becomes inevitable to use numerical approximations.

There are many different ways in which we can solve partial differential equations numerically. The most popular methods are three: *Finite differencing* [178, 208, 243], *finite elements* [179, 207] and *spectral methods* [281]. In this Chapter I will describe the main ideas behind finite differencing methods, since this is the most commonly used approach in numerical relativity (though not the only one; in particular spectral methods have become increasingly popular in recent years [70, 171, 173]).

Finally, I should mention the fact that, for simplicity, in this Chapter, I will only discuss methods for the numerical solution of systems of evolution equations of essentially "hyperbolic" type, and will not discuss the solution of elliptic equations, such as those needed for obtaining initial data. The numerical solution of elliptic equations is a very important subject in its own right, and even a very simple introduction would demand a full Chapter on its own. The interested reader can find a discussion of some of the basic techniques for solving elliptic equations in [230], but we should point out that even if simple methods are quite easy to code they tend to be extremely slow in practice. Fast and efficient algorithms for solving elliptic equations (such as *e.g.* multi-grid) are usually much more complex.

9.2 Basic concepts of finite differencing

When we study a field in a continuous spacetime, we are faced with considering an infinite and non-countable number of unknown variables: the value of the

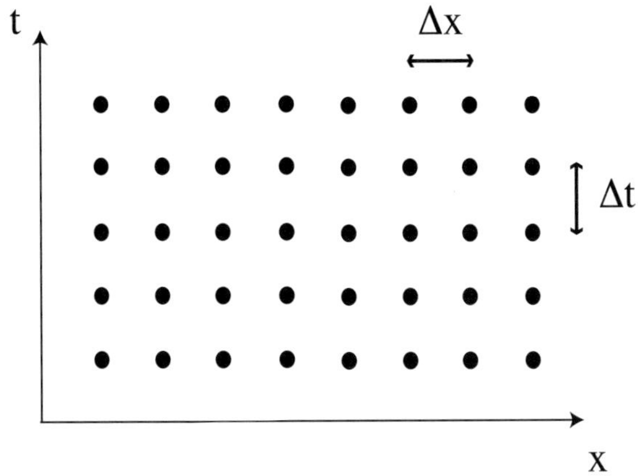

Fig. 9.1: Discretization of spacetime used in finite differencing.

field at every point of space and for all times. In order to find the value of the field using numerical approximations, the first thing that needs to be done is to reduce these unknowns to a finite number. There are several different ways of doing this. Spectral methods, for example, expand the solution as a finite linear combination of some appropriate basis functions. The variables to solve for are then the coefficients of such an expansion. A different approach is taken by finite differencing and finite elements. In both cases the number of variables is reduced by discretizing the domain of dependence of the functions, although using different strategies in each case.

The basic idea of finite differencing approximations is to substitute the continuous spacetime with a set of discrete points. This set of points is known as the computational *grid* or *mesh*. The distances in space between points on the grid do not necessarily have to be uniform, but in this Chapter we will assume for simplicity that they are. The time step between two consecutive levels is denoted by Δt, and the distance between two adjacent points by Δx. Figure 9.1 is a graphical representation of the *computational grid*. There is a huge literature on the subject of finite differencing; the interested reader can see, *e.g.* [178, 208, 243].

Once we have established the computational grid, the next step is to substitute our differential equations with a system of algebraic equations. This is done by approximating the differential operators by *finite differences* between the values of our functions at nearby points on the grid. In this way we obtain an algebraic equation at each grid point for each differential equation. These algebraic equations involve the values of the functions at the point under consideration and its nearest neighbors. The system of algebraic equations can then be solved in a simple way, but the price we have paid is that now we have a huge number of algebraic equations, so that a computer is needed in order to solve them all.

In order to make the ideas more concrete, let us write our differential equation in the general form

$$\mathcal{L}u = 0 \; , \tag{9.2.1}$$

where u denotes a set of functions of the spacetime coordinates (t, x^i), and $\mathcal{L}$ is some differential operator acting on u. Furthermore, let us denote by u_Δ the *discretized* approximation to u evaluated at the points of the computational grid, and by $\mathcal{L}_\Delta$ the finite difference version of the original differential operator. Here Δ can represent either Δx or Δt, which can in general be assumed to be proportional to each other. The finite difference version of our differential equation then takes the form

$$\mathcal{L}_\Delta u_\Delta = 0 \; . \tag{9.2.2}$$

The notation just introduced serves the purpose of indicating explicitly that a given finite difference approximation depends on the grid size parameter Δ. Indeed, since we are approximating a differential equation, the behavior of u_Δ in the limit when Δ vanishes is the crucial property of our approximation.

The *truncation error* of our finite difference approximation is defined as

$$\tau_\Delta := \mathcal{L}_\Delta u \; , \tag{9.2.3}$$

that is, the result of applying the finite difference operator $\mathcal{L}_\Delta$ to the solution of the original differential equation. Typically, the truncation error will not be zero, but it should approach zero as Δ becomes smaller. A related concept (often confused with the truncation error) is that of *solution error*, which is defined instead as the difference between the exact solution to the differential equation u and the exact solution to the finite difference equation u_Δ

$$\epsilon_\Delta := u - u_\Delta \; . \tag{9.2.4}$$

Of course, ϵ_Δ can only be evaluated at the points of our computational grid.

A minimum requirement for any given finite difference approximation is that, as the grid is refined (*i.e.* as Δt and Δx become smaller), the truncation errors should go to zero:

$$\lim_{\Delta \to 0} \tau_\Delta = 0 \; . \tag{9.2.5}$$

We then look for approximations that, in the continuous limit, approach the original differential equation and not a different one. When this happens locally we say that our approximation is *consistent*. In general, this property is quite easy to see from the structure of the finite difference approximation, and can often by checked by "eye". The important exceptions are situations where the coordinate system becomes singular, since proving consistency at a singular point might be non-trivial. For example, it is common for standard finite difference approximations to fail at the point $r = 0$ when using spherical coordinates.

Consistency is clearly fundamental for any finite difference approximation. When it fails, even if it is at just one point, we will not be able to recover the correct solution to the original differential equation. For a consistent finite difference approximation, in the continuum limit the truncation error typically approaches zero as a power of the discretization parameter Δ. We then say that a given approximation is of order n if [92]

$$\lim_{\Delta \to 0} \tau_\Delta \sim \Delta^n \; . \tag{9.2.6}$$

Consistency, however, is only a local property: A consistent finite difference approximation reduces *locally* to the differential equation in the continuum limit. In practice, we are really interested in a more global property. What we really look for is an approximation that improves *after a finite time* as the grid is refined. That is, the solution error ϵ_Δ at fixed time must go to zero in the continuum limit. This condition is known as *convergence*.

Convergence is clearly different from consistency, as consistent schemes can quite easily fail to converge. This can be understood if we remember that in the limit when Δt becomes zero, a finite time t can only be reached after an infinite number of time steps. This implies that even if the error in each time step is infinitesimal, its total integral could very well be finite. The numerical solution can even diverge and the solution error can in fact become infinite in the continuum limit! It is in general quite difficult to verify analytically if a given approximation scheme is convergent or not. Numerically, on the other hand, it is easy to check if the approximate solution is converging to something (that is, it is not diverging). The difficult part is to know if the numerical solution is converging to the exact solution and not to something else.

There is another very important property of finite differencing approximations. Independently of the behavior of the solution to the differential equation u, we must ask that the exact solution to the finite difference equations u_Δ should remain bounded after a given finite time t for any time step Δt. This requirement is known as *stability*, and implies that no component of the initial data should be amplified arbitrarily.

In order to give a more formal definition of stability, let us first introduce the following notation for the values of the finite difference approximation u_Δ on the computational grid:

$$(u_\Delta)^n_m := u_\Delta(t = n\Delta t, x = m\Delta x) \; . \tag{9.2.7}$$

Notice that, in general, x can represent several spatial dimensions, in which case the subindex m will in fact stand for several indices, one for each spatial dimension. Also, in order to simplify the notation, when we are considering only a fixed Δ we will simply write u^n_m.

[92] This is in contrast to spectral methods where the error approaches zero exponentially as the number of basis functions is increased. This is the main strength of spectral methods over finite differencing.

Let us now introduce the L^2 norm of our finite difference approximation as

$$||u_\Delta^n|| := \left[\Delta x \sum_m |(u_\Delta)_m^n|^2 \right]^{1/2} , \tag{9.2.8}$$

where the sum is over all points in the spatial grid. This norm is also commonly known as the *root mean square* (or simply *r.m.s.*) norm.

We will now say that the finite difference approximation is *stable* if for any $t > 0$ there exists a constant C_t such that

$$||u_\Delta^n|| \leq C_t ||u_\Delta^0|| , \tag{9.2.9}$$

for all $0 < n\Delta t < t$, in the limit when Δx and Δt go to zero. That is, in the continuum limit the norm of the finite difference solution up to a finite time t is bounded by the norm at $t = 0$ times a constant that is *independent* of Δx and Δt. Stability is a property of the system of finite difference equations, and is essentially the discrete version of the definition of well-posedness for a system of evolution equations (cf. equation (5.2.2)). An unstable finite difference approximation is simply useless in practice.

A fundamental result of the theory of finite difference approximations is the *Lax equivalence theorem* (for a proof see *e.g.* [243]):

Given an initial value problem that is mathematically well posed, and a finite difference approximation to it that is consistent, then stability is a necessary and sufficient condition for convergence.

This theorem is of great importance since it relates the final objective of any finite difference approximation, namely convergence to the exact solution, with a property that is usually much easier to prove: stability.

9.3 The one-dimensional wave equation

In order to make the concepts introduced in the previous Section more concrete, we will consider, as a simple example, the one-dimensional wave equation. This equation has a number of advantages. In the first place it can be solved exactly and the exact solution can then be used to compare with the numerical solution. Also, most fundamental equations in modern field theory can be seen as generalizations of one type of wave equation or another.

The one-dimensional wave equation (in flat space) has the following form:

$$\partial_t^2 \phi - c^2 \, \partial_x^2 \phi = 0 , \tag{9.3.1}$$

where ϕ is the wave function and c the wave speed.

9.3.1 *Explicit finite difference approximation*

Let us denote simply by (n, m) the point $(t = n\Delta t, x = m\Delta x)$. We can now approximate the differential operators that appear in equation (9.3.1) by using Taylor expansions of ϕ around the point (n, m). Consider, for example, the value of ϕ at the points $(n, m+1)$ and $(n, m-1)$:

$$\phi^n_{m+1} = \phi^n_m + (\partial_x \phi)\,\Delta x + \frac{1}{2}\left(\partial_x^2 \phi\right)(\Delta x)^2 + \frac{1}{6}\left(\partial_x^3 \phi\right)(\Delta x)^3 + \cdots\,, \tag{9.3.2}$$

$$\phi^n_{m-1} = \phi^n_m - (\partial_x \phi)\,\Delta x + \frac{1}{2}\left(\partial_x^2 \phi\right)(\Delta x)^2 - \frac{1}{6}\left(\partial_x^3 \phi\right)(\Delta x)^3 + \cdots\,, \tag{9.3.3}$$

where all derivatives are to be evaluated at the point (n, m). From these expressions it is easy to see that

$$\partial_x^2 \phi = \frac{\phi^n_{m+1} - 2\phi^n_m + \phi^n_{m-1}}{(\Delta x)^2} + \frac{(\Delta x)^2}{12}\left(\partial_x^4 \phi\right) + \cdots\,. \tag{9.3.4}$$

We can then approximate the second derivative as

$$\partial_x^2 \phi \simeq \frac{\phi^n_{m+1} - 2\phi^n_m + \phi^n_{m-1}}{(\Delta x)^2}\,. \tag{9.3.5}$$

From the above expressions we see that the truncation error is of order $(\Delta x)^2$, so this approximation is *second order accurate.*

The second derivative of ϕ with respect to t can be approximated in exactly the same way. In this way we find the following finite difference approximation for the wave equation:

$$\frac{\phi^{n+1}_m - 2\phi^n_m + \phi^{n-1}_m}{(c\Delta t)^2} - \frac{\phi^n_{m+1} - 2\phi^n_m + \phi^n_{m-1}}{(\Delta x)^2} = 0\,. \tag{9.3.6}$$

We can rewrite this equation in more compact form if we introduce the so-called *Courant parameter*, $\rho := \Delta t/\Delta x$.[93] Our approximation then takes the final form:

$$\left(\phi^{n+1}_m - 2\phi^n_m + \phi^{n-1}_m\right) - c^2\rho^2\left(\phi^n_{m+1} - 2\phi^n_m + \phi^n_{m-1}\right) = 0\,. \tag{9.3.7}$$

This equation has a very important property: It involves only one value of the wave function at the last time level; the value ϕ^{n+1}_m. We can then solve for this value in terms of values at previous time levels to obtain:

$$\phi^{n+1}_m = 2\phi^n_m - \phi^{n-1}_m + c^2\rho^2\left(\phi^n_{m+1} - 2\phi^n_m + \phi^n_{m-1}\right)\,. \tag{9.3.8}$$

Because of this property, the last approximation is known as an *explicit approximation.* If we know the values of the function ϕ at the time levels n and $n-1$,

[93] It is common to define the Courant parameter absorbing the wave speed into it as $\rho = c\Delta t/\Delta x$. This is very useful in the case of simple scalar equations, but becomes less so for systems of equations where we can have several different characteristic speeds.

we can use the last equation to calculate directly the values of ϕ at the new time level $n+1$. The process can then be iterated as many times as desired.

It is clear that all that is required in order to start the evolution is to know the values of the wave function at the first two time levels, and finding these two first levels is easy to do. As we are dealing with a second order equation, the initial data must include

$$f(x) := \phi(0,x) \; , \qquad g(x) := \partial_t \phi|_{t=0} \; . \tag{9.3.9}$$

The knowledge of $f(x)$ evidently gives us the first time level:

$$\phi^0_m = f(m\Delta x) \; . \tag{9.3.10}$$

For the second time level it is enough to approximate the first time derivative using finite differences. One possible approximation is given by

$$g(m\Delta x) = \frac{\phi^1_m - \phi^0_m}{\Delta t} \; , \tag{9.3.11}$$

from where we find

$$\phi^1_m = g(m\Delta x)\, \Delta t + \phi^0_m \; . \tag{9.3.12}$$

The previous expression, however, has one important drawback: From the Taylor expansion we can easily see that the truncation error for this expression is of order Δt, so the approximation is only first order. It is clear that we start our evolution with a first order error, the second order accuracy of the whole scheme will be lost. However, this problem is quite easy to fix. A second order approximation to the first time derivative is

$$g(m\Delta x) = \frac{\phi^1_m - \phi^{-1}_m}{2\Delta t} \; . \tag{9.3.13}$$

The problem now is that this expression involves the value of the function ϕ^{-1}_m, which is also unknown. But we already have one other equation that makes reference to ϕ^1_m and ϕ^{-1}_m; the approximation to the wave equation (9.3.7) evaluated at $n = 0$. We can then use these two equations to eliminate ϕ^{-1}_m and solve for ϕ^1_m. In this way we find the following second order approximation for the second time level

$$\phi^1_m = \phi^0_m + \frac{c^2 \rho^2}{2} \left(\phi^0_{m+1} - 2\phi^0_m + \phi^0_{m-1} \right) + \Delta t \, g(m\Delta x) \; . \tag{9.3.14}$$

Equations (9.3.10) and (9.3.14) give us all the information we require in order to start our evolution.

There is another important point that must be mentioned here. In order to reduce the total number of variables to a finite number, it is also necessary to consider a finite region of space with a finite number of points N, the so-called *computational domain*. It is therefore crucial to specify the boundary conditions that have to be applied at the edges of the computational domain. It is clear that the approximation to the wave equation (9.3.7) can not be used at the boundaries since it involves points outside the computational domain. There are many different ways to impose boundary conditions for the wave equation. One particularly simple choice is to assume that we have a periodic space, so we can simply choose as boundary conditions

$$\phi_0^n \equiv \phi_N^n \,. \tag{9.3.15}$$

This choice, apart from being extremely simple, is equivalent to using the interior approximation everywhere, so it allows us to concentrate on the properties of the interior scheme only. We will for the moment assume that our boundaries are indeed periodic, and will come back to the boundary issue later.

9.3.2 *Implicit approximation*

We have already introduced a simple finite difference approximation to the one-dimensional wave equation. The approximation that we found, however, is far from being unique. In principle, there are an infinite number of different ways to approximate the same differential equation using finite differences.

In order to simplify things, I will now introduce a more compact notation for the finite differences. Let us first define the spatial *forward* and *backward* difference operators

$$\Delta_x^+ \,\phi_m^n := \left(\phi_{m+1}^n - \phi_m^n\right) \,, \qquad \Delta_x^- \,\phi_m^n := \left(\phi_m^n - \phi_{m-1}^n\right) \,, \tag{9.3.16}$$

with analogous definitions for temporal difference operators. The *centered* difference operator is then defined as

$$\bar{\Delta}_x \,\phi_m^n := \frac{1}{2} \left(\Delta_x^+ + \Delta_x^-\right) \phi_m^n = \frac{1}{2} \left(\phi_{m+1}^n - \phi_{m-1}^n\right) \,. \tag{9.3.17}$$

We can now use these definitions to introduce the *second centered* difference operators

$$\Delta_x^2 \,\phi_m^n := \Delta_x^+ \Delta_x^- \,\phi_m^n = \phi_{m+1}^n - 2\phi_m^n + \phi_{m-1}^n \,. \tag{9.3.18}$$

(Do notice that with this notation $(\bar{\Delta}_x)^2 \neq \Delta_x^2$.)

Having defined these operators, we can now go back to the approximations we used for the differential operators that appear on the wave equation. Starting again from the Taylor series, it is possible to show that the second spatial derivative can be approximated more generally as

$$\partial_x^2 \phi \simeq \frac{1}{(\Delta x)^2} \, \Delta_x^2 \left[\frac{\theta}{2} \left(\phi_m^{n+1} + \phi_m^{n-1}\right) + (1-\theta)\, \phi_m^n \right] \,, \tag{9.3.19}$$

with θ some arbitrary parameter. The expression we had before, equation (9.3.5), can be recovered by taking $\theta = 0$. This new approximation corresponds to taking an average, with a certain weight, of the finite difference operators acting on different time levels. In the particular case when $\theta = 1$, the contribution from the middle time level in fact completely disappears.

If we now use the new approximation for the second spatial derivative, but keep the same approximation as before for the time derivative, we find the following finite difference approximation for the wave equation

$$\Delta_t^2\, \phi_m^n - c^2 \rho^2\, \Delta_x^2 \left[\frac{\theta}{2} \left(\phi_m^{n+1} + \phi_m^{n-1} \right) + (1-\theta)\, \phi_m^n \right] = 0 \; . \tag{9.3.20}$$

This is one possible generalization of (9.3.7), but not the only one. It is clear that we can play this game in many ways to obtain even more general approximations, all equally valid, and all second order (one can also find approximations that are fourth order accurate or even higher). The approximation given by (9.3.20) has a new and very important property: It involves not one, but three different values of ϕ at the last time level. This means that it is now not possible to solve for ϕ at the last time level explicitly in terms of its values in the two previous time levels. Because of this, the approximation (9.3.20) is known as an *implicit approximation.*

When we consider the equations for all the points in the grid, including the boundaries, it is possible to solve the full system by inverting a non-trivial matrix, which is of course a more time-consuming operation than the one needed for the explicit approximation.[94] However, in many cases implicit approximations turn out to have better properties than explicit ones, in particular related to the stability of the numerical scheme as we will see below.

9.4 Von Newmann stability analysis

A general method for studying the stability of systems of finite difference equations can be obtained directly from the definition of stability. We start by writing the finite difference equations as:

$$\mathbf{u}^{n+1} = \mathbf{B}\, \mathbf{u}^n \; , \tag{9.4.1}$$

where $\mathbf{u}^n$ is the solution vector at time level n, and $\mathbf{B}$ is an update matrix (in general sparse). It is important to notice that all finite difference approximation to a linear equation can be written in this way, even those that involve more than two time levels by simply introducing auxiliary variables. For example, for the three-level approximations to the wave equation introduced in the previous Sections we can define $u_m^n := \phi_m^n$ and $v_m^n := \phi_m^{n-1}$, and take the vector $\mathbf{u}^n$ to be given by $(u_1^n, v_1^n, ..., u_N^n, v_N^m)$ with N the total number of grid points.

If the matrix $\mathbf{B}$ has a complete set of eigenvectors, then the vector $\mathbf{u}^n$ can be written as a linear combination of them. The requirement for stability can then

[94]In the particular case of one spatial dimension, the resulting matrix is tridiagonal since each equation involves only a given point and its two nearest neighbors, and there are very efficient algorithms to invert such matrices. In more than one dimension, however, the matrix is no longer that simple, though it is still sparse (*i.e.* most of its entries are zero).

be reduced to asking for the matrix $\mathbf{B}$ not to amplify any of its eigenvectors, that is, we must ask for the magnitude of its largest eigenvalue of $\mathbf{B}$, known as its *spectral radius*, to be less than or equal to 1.

The stability analysis based on the idea just described is quite general, but it requires a knowledge of the entries of $\mathbf{B}$ over all space, including the boundary. There is, however, a very popular stability analysis method that, even though it can only be shown to give necessary conditions for stability, in many cases turns out to also give sufficient conditions. This method, originally introduced by von Newmann, is based on a Fourier decomposition.

To introduce von Newmann's method, we start by expanding the solution of (9.4.1) as a Fourier series

$$\mathbf{u}^n(\mathbf{x}) = \sum_{\mathbf{k}} \tilde{\mathbf{u}}^n(\mathbf{k})\, e^{i\,\mathbf{k}\cdot\mathbf{x}} \,, \tag{9.4.2}$$

where the sum is over all wave vectors $\mathbf{k}$ that can be represented on the grid.[95] If we now substitute this into the original equation (9.4.1) we find

$$\tilde{\mathbf{u}}^{n+1} = \tilde{\mathbf{B}}\, \tilde{\mathbf{u}}^n \,, \tag{9.4.3}$$

where now $\tilde{\mathbf{B}}$ is the Fourier transform of the original matrix $\mathbf{B}$, also known as the *amplification matrix*. The stability condition now corresponds to asking that no Fourier mode should be amplified, that is, for the spectral radius of $\tilde{\mathbf{B}}$ to be less than or equal to 1. This is von Newmann's stability condition.

It is important to stress the fact that in order to use this stability criteria we have assumed two things: 1) The boundary conditions are periodic, since otherwise we can not make a Fourier expansion, and 2) the entries of the matrix $\mathbf{B}$ are constant (i.e. independent of position), since otherwise it is not possible to decouple the different Fourier modes.

As an example of von Newmann's stability analysis we can now study the stability of the implicit approximation to the wave equation we derived previously, equation (9.3.20). Consider a Fourier mode of the form

$$\phi^n_m = \xi^n e^{imk\Delta x} \,. \tag{9.4.4}$$

If we substitute this back into the finite difference equation we find, after some algebra, a quadratic equation for ξ of the form

$$A\xi^2 + B\xi + C = 0 \,, \tag{9.4.5}$$

with coefficients given by

[95] The shortest wavelength that can be represented on the grid is clearly $2\Delta x$, also known as the *Nyquist limit*. This implies that the maximum value that any component of the wave vector can take is $\pi/\Delta x$.

$$A = c^2\rho^2\theta\left[\cos\left(k\Delta x\right) - 1\right] - 1 \, , \tag{9.4.6}$$
$$B = 2c^2\rho^2\left(1-\theta\right)\left[\cos\left(k\Delta x\right) - 1\right] + 2 \, , \tag{9.4.7}$$
$$C = c^2\rho^2\theta\left[\cos\left(k\Delta x\right) - 1\right] - 1 \, . \tag{9.4.8}$$

The two roots of this quadratic equation are, clearly

$$\xi_\pm = \frac{-B \pm \left(B^2 - 4AC\right)^{1/2}}{2A} \, , \tag{9.4.9}$$

and the general solution to the finite difference equation turns out to be

$$\phi^n_m = \sum_k \left[Z^+_k \left(\xi_+\left(k\right)\right)^n + Z^-_k \left(\xi_-\left(k\right)\right)^n\right] e^{imk\,\Delta x} \, , \tag{9.4.10}$$

where Z^+_k and Z^-_k are arbitrary constants.

On the other hand, from the fact that $A = C$ we can easily show that

$$|\xi_+\xi_-| = |C/A| = 1 \, . \tag{9.4.11}$$

This is a very important property, it implies that if the system is stable for all k, that is, if $|\xi_\pm(k)| \leq 1$ then the system will necessarily also be non-dissipative (the Fourier modes not only don't grow, they don't decay either). For the system to be stable we must then ask for

$$|\xi_+(k)| = |\xi_-(k)| = 1 \, . \tag{9.4.12}$$

It is easy to see that this will happen as long as

$$B^2 - 4AC \leq 0 \, . \tag{9.4.13}$$

Substituting now the values of the coefficients A, B and C into this expression we find the following stability condition:

$$c^2\rho^2\left(1-2\theta\right)\left[1 - \cos\left(k\Delta x\right)\right] - 2 \leq 0 \, . \tag{9.4.14}$$

As we want this to hold for all k, we must consider the case when the left hand side reaches its maximum value. If we take $\theta < 1/2$, this will happen for $k = \pi/\Delta x$, in which case the stability condition takes the simple form:

$$c^2\rho^2 \leq 1/\left(1-2\theta\right) \, . \tag{9.4.15}$$

For the explicit scheme we have $\theta = 0$, and the stability condition reduces to the well known *Courant–Friedrich–Lewy* (CFL) condition[96]

$$c\rho \leq 1 \quad \Rightarrow \quad c\Delta t \leq \Delta x \, . \tag{9.4.16}$$

The CFL condition has a clear geometric interpretation: The numerical domain of dependence must be larger than the physical domain of dependence, and

[96] For systems with N spatial dimensions the CFL stability condition is slightly modified and becomes instead $c\rho \leq 1/\sqrt{N}$.

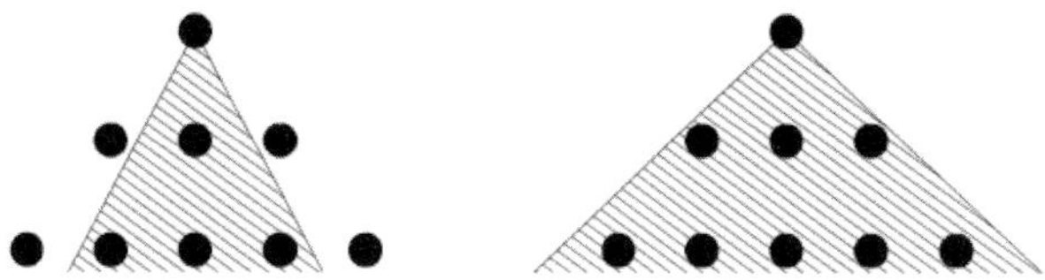

Fig. 9.2: CFL stability condition. For $c\Delta t \leq \Delta x$, the numerical domain of dependence is larger than the physical domain of dependence (shaded region), and the system is stable. For $c\Delta t > \Delta x$ we have the opposite situation, and the system is unstable.

not the other way around (see Figure 9.2). If this weren't the case, it would be impossible for the numerical solution to converge to the exact solution, since as the grid is refined there will always be relevant physical information that would remain outside the numerical domain of dependence. And, as we have seen, the Lax theorem implies that if there is no convergence then the system is unstable.

The argument we have just given clearly only applies to explicit schemes. This is because for an implicit scheme, the numerical domain of dependence is in fact the whole grid. In that case there is no simple geometric argument that can tell us what the stability condition should be.

In order to obtain the stability condition (9.4.15) we assumed that $\theta < 1/2$. If, on the other hand, we take $\theta \geq 1/2$ then we must go back to the general condition (9.4.14). However, in this case it is easy to see that the condition is always satisfied. This means that an implicit scheme with $\theta \geq 1/2$ is stable for all values of ρ, that is, it is *unconditionally stable.*

This takes us to one of the most important lessons of the theory of finite differencing: Simple schemes not always have the best stability properties.[97]

9.5 Dissipation and dispersion

The idea of decomposing the solution of our finite difference approximation in a Fourier series is not only useful to determine the stability of a scheme. It also gives us the opportunity to study other related properties of the solutions. Let us concentrate our attention on the case of one spatial dimension, and assume that we have a linear differential equation that admits solutions of the form

$$\phi(x,t) = e^{-i\omega t}\, e^{ikx} \,, \tag{9.5.1}$$

where for every real wave number k there exists a complex frequency ω given by

[97]This is even more so for systems of equations of parabolic type (such as the heat equation), for which explicit schemes are practically useless since the stability condition becomes $\Delta t \lesssim \Delta x^2$. The problem with this is that it means that if we reduce Δx by half, we must reduce Δt by a factor of four, so integrating to a desired finite time T quickly becomes prohibitive in terms of computer time.

$$\omega = \omega\,(k) \; . \tag{9.5.2}$$

The explicit form of this relation will of course be determined by the properties of our original differential equation. Clearly, the imaginary part of ω will give the rate of growth or decay of a given Fourier mode with time. A differential equation that admits solutions for which the imaginary part of ω is negative is called a *dissipative* equation.

It is also useful to define the concept of *dispersion*: A differential equation is said to be non-dispersive if the real part of ω is a linear function of k, and dispersive otherwise. For this reason, equation (9.5.2) is known as the *dispersion relation.*

The wave equation is one example of a differential equation that admits solutions of the form (9.5.1). In this case the dispersion relation is simply $\omega = \pm ck$, which implies that the wave equation is both non-dissipative and non-dispersive.

The dispersion relation contains within itself very important information about the behavior of the solution of a given differential equation. Consider, for example, a non-dissipative differential equation. Equation (9.5.1) implies that we will have solutions in the form of traveling sinusoidal waves. The speed of each wave will be given in terms of its wave number k and its frequency ω by the so-called *phase velocity* $v_p(k)$ defined as

$$v_p(k) := \frac{\omega}{k} \; . \tag{9.5.3}$$

In practice, however, we never deal with infinite plane waves but rather with localized wave packets. A wave packet that has a narrow Fourier transform, centered at a wavenumber k, doesn't travel with the phase velocity $v(k)$, but rather with a speed given by:

$$v_g(k) = \frac{d\omega}{dk} \; , \tag{9.5.4}$$

and known as the *group velocity.* Non-dispersive systems like the wave equation have the property that the phase velocity and group velocity are equal for all modes, $v_p(k) = v_g(k)$.

Even when we are dealing with a differential equation that is both non-dissipative and non-dispersive, it turns out that quite generally these properties are not preserved when we consider a finite difference approximation. These approximations are almost always dispersive, and are often also dissipative (notice that an instability can be interpreted as the result of negative dissipation).

The finite difference approximations to the simple one-dimensional wave equation discussed in the previous sections provide an excellent example of this type of phenomenon. Let us again consider the general implicit approximation to the 1D wave equation (equation (9.3.20)). Comparing equation (9.5.1) with the Fourier mode decomposition (9.4.4) used for the von Newmann stability, we see that the frequency ω is related to the parameter ξ through $\xi = e^{-i\omega\Delta t}$. Substituting the Fourier mode into equation (9.3.20) we now find

$$\cos(\omega\Delta t) = \frac{1 - c^2\rho^2(\theta - 1)[\cos(k\Delta x) - 1]}{1 - c^2\rho^2\theta[\cos(k\Delta x) - 1]} \,. \tag{9.5.5}$$

Introducing the adimensional quantities $K := k\Delta x$ and $\Omega := \omega\Delta t$, we can rewrite the last equation as

$$\Omega = \pm \arccos\left[\frac{1 - c^2\rho^2(\theta - 1)[\cos(K) - 1]}{1 - c^2\rho^2\theta[\cos(K) - 1]}\right] \,. \tag{9.5.6}$$

This expression is the numerical equivalent of the dispersion relation of the wave equation, and is called the *numerical dispersion relation*. Notice how it is far from being a simple linear function. In the limit when $K \ll 1$ it is not difficult to prove that this dispersion relation reduces to $\Omega = \pm c\rho K$, or in other words $\omega = \pm ck$, so we recover the correct dispersion relation for the wave equation.

The dispersion relation (9.5.6) can be used to study the dispersion and dissipation properties of our finite differencing approximation for the different values of the parameters ρ and θ. In the particular case of the explicit approximation ($\theta = 0$), the relation reduces to

$$\Omega = \pm \arccos\left(1 + c^2\rho^2[\cos(K) - 1]\right) \,. \tag{9.5.7}$$

If we now take also $c\rho = 1$ we find that $\Omega = \pm K$, that is, the finite difference approximation is neither dissipative nor dispersive and in fact has the same dispersion relation as the original differential equation. This remarkable fact can be understood if we calculate the truncation error associated with our finite difference approximation. It turns out that for $\theta = 0$ and $c\rho = 1$, the truncation error *vanishes exactly to all orders*, in other words the explicit finite difference approximation is *exact* for $c\rho = 1$ (this property of the explicit scheme goes away when we have more than one spatial dimension).

Figure 9.3 plots Ω as a function of $K = k\Delta x$ for the explicit scheme ($\theta = 0$) using three different values of the Courant parameter ρ (for simplicity we have taken $c = 1$). For $\rho = 1$ the dispersion relation is a straight line so that all modes propagate with the same speed. For the case $\rho = 0.8$ we see that the slope of the graph becomes smaller for larger values of K, so that the smaller wavelengths propagate more slowly than they should. This implies that a localized wave packet will disperse as it propagates, leaving a trail of smaller wavelengths behind. Finally, for the case $\rho = 1.2$ there are two things to notice. First, the slope of the graph gets larger for large values of K, so small wavelengths now propagate faster than they should. A wave packet would then disperse in the opposite way to before, with smaller wavelengths overtaking the main packet. Worse still, at $K \sim 2$ the slope in fact becomes infinite and the frequency Ω becomes purely imaginary (the imaginary part is also shown in the plot). Imaginary frequencies do not correspond to traveling waves any more but rather to

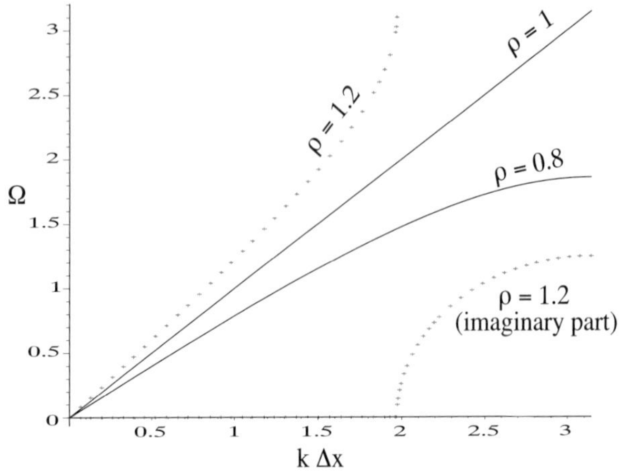

Fig. 9.3: Dispersion relation $\Omega(k\Delta x)$ for the explicit scheme ($\theta = 1$), and for three different values of the Courant parameter ρ (taking $c = 1$).

solutions that grow exponentially with time, *i.e.* the corresponding modes are unstable, which is to be expected since we are now violating the CFL condition.

9.6 Boundary conditions

In the previous sections I have discussed different finite difference approximations to the wave equation assuming that we are using periodic boundary conditions. However, it is clear that periodic boundary conditions generally do not correspond to what we expect from a realistic physical scenario. In the general case, we would expect to have one of the following situations:

1. The computational domain represents a finite region of space that is delimited by real physical boundaries where some kind of *exact* boundary conditions should be applied.
2. The computational domain represents a finite region of an infinite space. In this situation, the fact that we must necessarily have a finite computational domain implies that we must use some sort of artificial boundary condition. This condition, however, should allow the waves to leave the computational domain as cleanly as possible (there is no "real" boundary there). Numerical boundary conditions that have this property are generally known as *absorbing boundary conditions*, since their work is to absorb most of the energy of the waves that reach the boundaries, reflecting as little as possible back into the grid.[98]

[98]There is, in fact, a different approach to the problem of representing an infinite space using a finite computational domain. This approach consists of using a "compactification" of space, that is, introducing a new coordinate system that maps all of infinite space into a finite interval. In this way, the boundary of the finite differencing grid will correspond to physical

Absorbing boundary conditions come essentially in two different forms. The first consists of using a modified wave equation close to the boundaries of the grid, designed to attenuate the wave amplitude before the edge is reached. This can be easily done by introducing a dissipative term into the equation (see Section 9.9 below), and using a dissipation coefficient that grows gradually from zero to some finite value in a small strip close to the boundary. We can then fix the value of the wave function to zero at the very edge of the grid. Any small wave amplitude that still reaches the edge will be reflected and attenuated further still in the absorbing strip before reentering the central part of the grid. This type of absorbing boundary condition, generally known as *sponge filters*, can be very effective in the numerical study of wave phenomena in one and two dimensions. This approach suffers, however, from at least one important drawback: As the number of dimensions gets higher the number of points in the "small" strip close to the boundary increases rapidly. We might then find that in a three-dimensional problem almost as many grid points are needed in the absorbing region as in the interior, which is clearly undesirable since it represents a waste of computer resources. The second approach to the problem of designing absorbing boundary conditions is the use of the so-called *radiation boundary conditions*, also known as *outgoing wave boundary conditions*, and here we will concentrate on this type of boundary condition.

Consider a generic boundary condition of the form

$$\alpha u + \beta \, \partial_n u = \gamma \,, \tag{9.6.1}$$

where u is our unknown function (or functions), (α, β, γ) are given functions of space and time, and $\partial_n u$ denotes the derivative normal to the boundary under consideration. We further classify the different types of boundary condition in the following way:

1. $\beta = 0$. In this case the function u is specified on the boundary and we say that we have a boundary condition of *Dirichlet* type.
2. $\alpha = 0$. In this case the normal derivative of u is specified on the boundary and we have a boundary condition of *Newmann* type.
3. Both α and β are non-zero so that we have a differential relation for u at the boundary. We then say that we have a boundary condition of *mixed* type. Boundary conditions on finite domains are typically of Dirichlet or Newmann type, while radiation boundary conditions are of mixed type.

infinity, where the exact boundary conditions are often known (*e.g.* the wave function should be zero at spatial infinity if we have initial data of compact support). This approach works very well when we compactify on null hypersurfaces surfaces, or on hypersurfaces that are asymptotically null (as in the characteristic [301] and conformal [132, 165] approaches to numerical relativity). However, when we compactify on spatial hypersurfaces the loss in resolution as the waves approach the boundary of the grid produces a large numerical backscattering. Still, compactification on spatial hypersurfaces in combination with a sponge filter has been used in practice with good results (see *e.g.* [231]).

As an example of a radiation boundary condition, consider again the case of the one-dimensional wave equation. It is easy to see that the appropriate *exact* radiation boundary conditions are given by

$$\begin{aligned} &\partial_t \phi - c\,\partial_x \phi = 0\,, && \text{at the left boundary} \\ &\partial_t \phi + c\,\partial_x \phi = 0\,, && \text{at the right boundary} \end{aligned} \tag{9.6.2}$$

These are simple advection equations that represent waves traveling out from the computational domain on either side. For initial data of compact support they will allow the waves to leave the integration region very cleanly without any reflections. Of course, this is only true if we can apply these boundary conditions exactly. As soon as we substitute (9.6.2) for some finite difference approximation to it we generally find that outgoing waves are not completely eliminated and some amount of reflection is always present.

Without loss of generality, let us consider only the condition for the left boundary. There are many different ways of approximating this condition using finite differences. We can consider, for example, a forward difference in both time and space, resulting in the numerical boundary condition

$$\left(\phi_0^{n+1} - \phi_0^n\right) - c\rho\left(\phi_1^n - \phi_0^n\right) = 0\,, \tag{9.6.3}$$

where again $\rho = \Delta t/\Delta x$ is the Courant parameter, and where we have assumed that the numerical grid starts with the grid point $m = 0$. The approximation above is clearly explicit, but it is only first order accurate in both space and time (this approximation in fact corresponds to the upwind method for the advection equation that we will discuss in the next Section). A better approximation to the radiation boundary condition is obtained by using an implicit approximation for the spatial derivative, and at the same time introducing an averaged time difference to approximate the time derivative. Doing this we find

$$\left[\left(\phi_0^{n+1} - \phi_0^n\right) + \left(\phi_1^{n+1} - \phi_1^n\right)\right] - c\rho\left[\left(\phi_1^{n+1} - \phi_0^{n+1}\right) + \left(\phi_1^n - \phi_0^n\right)\right] = 0\,. \tag{9.6.4}$$

This approximation is now second order accurate in both space and time. The fact that it is implicit can seem at first glance to be a drawback, but one must remember that the boundary condition will be applied once we have already updated all the interior points. This means that we will already know the value of ϕ_1^{n+1} and we can simply solve the above equation for ϕ_0^{n+1}:

$$\phi_0^{n+1} = \frac{1}{1 + c\rho}\left[(c\rho - 1)\,\phi_1^{n+1} + (1 + c\rho)\,\phi_1^n + (1 - c\rho)\,\phi_0^n\right]\,. \tag{9.6.5}$$

In situations where we have more than one spatial dimension, radiation boundary conditions become more complicated. For example, for spherical waves in three dimensions we typically expect the solution to behave as $\phi = u(r - ct)/r$ for large r. This can be expressed in differential terms as

$$\partial_t \phi + c\,\partial_r \phi + c\,\phi/r = 0\,. \tag{9.6.6}$$

If we are using a Cartesian coordinate system, this equation will usually have to be applied at the boundaries of a cube, where the normal direction corresponds to one of the Cartesian directions x^i. In that case we can use the fact that for a spherically symmetric function, $\partial_i \phi = (x^i/r)\, \partial_r \phi$, and use as a boundary condition:

$$\frac{x^i}{r}\, \partial_t \phi + c\, \partial_i \phi - \frac{c x^i}{r^2} \phi = 0 \, . \tag{9.6.7}$$

This condition can now be applied at each of the Cartesian boundaries by taking x^i equal to (x, y, z). Of course, typically our function ϕ can not be expected to be spherically symmetric, but if the boundaries are sufficiently far away the angular derivatives will be much smaller than the radial derivative and can be safely ignored without introducing large reflections.

The specific boundary conditions described above only work well for simple systems like the wave equation. For more complex systems we would need to carefully consider what are the appropriate boundary conditions at the differential level, which will of course be determined by the type of problem being solved. For example, for hyperbolic systems of equations we should first decompose the solution into ingoing and outgoing modes at the boundary, and apply boundary conditions only to ingoing modes because applying them to outgoing modes would violate causality. Only after we have found adequate (and well-posed) boundary conditions at the differential level can we start constructing finite difference approximations to them.

9.7 Numerical methods for first order systems

In the previous Sections we have introduced the basic concepts behind finite differencing approximations, using as an example the one-dimensional wave equation. However, the systems of evolution equations that we are mainly interested in for numerical relativity are typically written as either mixed first order in time and second order in space systems like the ADM or BSSNOK equations, or fully first order systems like the Euler equations of hydrodynamics. Because of this we will review here some of the standard numerical methods used for this type of system. We will start by considering numerical methods for the one-dimensional advection equation

$$\partial_t u + v\, \partial_x u = 0 \, , \tag{9.7.1}$$

where u is the wave function and v some constant wave speed. For the finite differencing approximation to this equation we take the standard discretization of spacetime: $x_m = m \Delta x$, $t_n = n \Delta t$, $u^n_m = u(t_n, x_m)$, with n and m integers. The most obvious method for solving the advection equation numerically is to approximate $\partial_t u$ by a forward time difference, and $\partial_x u$ by a centered space difference:

$$\frac{u^{n+1}_m - u^n_m}{\Delta t} + v\, \frac{u^n_{m+1} - u^n_{m-1}}{2 \Delta x} = 0 \, . \tag{9.7.2}$$

Solving for u^{n+1}_m we then find

$$u_m^{n+1} = u_m^n - \frac{v\rho}{2} \left(u_{m+1}^n - u_{m-1}^n \right) , \tag{9.7.3}$$

where again ρ stands for the Courant parameter $\rho = \Delta t / \Delta x$. This approximation is known as the *forward Euler* method (also called FTCS for *forward time centered space*). This method has one very serious drawback: A von Newmann stability analysis shows that the method is *unconditionally unstable*, *i.e.* it is unstable for *any* choice of Δx and Δt, so it is useless in practice.

A stable method is obtained by using a backward time difference instead:

$$u_m^{n+1} = u_m^n - \frac{v\rho}{2} \left(u_{m+1}^{n+1} - u_{m-1}^{n+1} \right) . \tag{9.7.4}$$

This method, known as *backward Euler*, is stable but clearly implicit. Both Euler methods are also only first order accurate because of the off-centered time differences.

Remarkably, the stability problem with the standard Euler method can be fixed if instead of u_m^n we take the average $(u_{m+1}^n + u_{m-1}^n)/2$ in the first term of equation (9.7.3):

$$u_m^{n+1} = \frac{1}{2} \left(u_{m+1}^n + u_{m-1}^n \right) - \frac{v\rho}{2} \left(u_{m+1}^n - u_{m-1}^n \right) . \tag{9.7.5}$$

This method is known as *Lax–Friedrichs*, and although it is still only first order accurate, it is explicit and stable provided the CFL condition $v\rho \leq 1$ is satisfied. This is another example of how finite difference is sometimes an art as much as a science.

Other more sophisticated methods can be obtained from a Taylor series expansion of the form

$$\begin{aligned} u(t + \Delta t, x) &= u(t, x) + \Delta t \, \partial_t u + \frac{(\Delta t)^2}{2} \, \partial_t^2 u + \cdots \\ &= u(t, x) - v \Delta t \, \partial_x u + v^2 \frac{(\Delta t)^2}{2} \, \partial_x^2 u + \cdots , \end{aligned} \tag{9.7.6}$$

where in the second line we have used the original advection equation to exchange time derivatives for spatial derivatives. Approximating now the spatial derivatives using centered differences we obtain the so-called *Lax–Wendroff* finite difference approximation

$$u_m^{n+1} = u_m^n - \frac{v\rho}{2} \left(u_{m+1}^n - u_{m-1}^n \right) + \frac{v^2 \rho^2}{2} \left(u_{m+1}^n - 2u_m^n + u_{m-1}^n \right) . \tag{9.7.7}$$

Because of the Taylor expansion, the Lax–Wendroff method is second order accurate in both time and space. It is also clearly explicit and turns out to be stable if the CFL condition is satisfied.

The improved stability of Lax–Wendroff can in fact be intuitively understood. First, notice that Lax–Wendroff is basically the forward Euler method plus a correction term. This correction term is proportional to the second derivative of u, so we can see the method as a finite difference approximation to the equation

$$\partial_t u + v\,\partial_x u = \frac{v^2 \Delta t}{2}\,\partial_x^2 u\,. \tag{9.7.8}$$

But this is nothing more than the advection-diffusion equation. That is, Lax–Wendroff adds a diffusion (*i.e.* dissipative) term as a correction to forward Euler, improving its stability properties. The diffusion term vanishes in the continuum limit, so in the end we will still converge to the solution of the non-diffusive advection equation. A similar argument shows that the first order Lax–Friedrichs method also adds a diffusive correction term to forward Euler, only in this case we are in fact adding $\left[(\Delta x)^2/2\Delta t\right]\partial_x^2 u$.

There are still other quite standard finite difference approximations to the advection equation. For example, instead of centered spatial differences we can use one-sided spatial differences. However, when doing this we must be careful to use one-sided differences that respect the causal flow of information in order to obtain a stable system. This leads to the so-called *upwind* method that takes the following form

The name of the method comes from the fact that, in order to have stability, we must take finite differences along the direction of the flow. Upwind is again stable as long as the CFL condition is satisfied. It is also only first order accurate in both space and time. Nevertheless, it is at the heart of the modern shock capturing methods we will discuss in Section 9.10.

$$\begin{aligned} u_m^{n+1} &= u_m^n - v\rho\left(u_m^n - u_{m-1}^n\right) \qquad v \geq 0\,,\\ u_m^{n+1} &= u_m^n - v\rho\left(u_{m+1}^n - u_m^n\right) \qquad v \leq 0\,. \end{aligned} \tag{9.7.9}$$

Another quite common method is based on the idea of using centered differences in both space and time. This results in the following method:

$$u_m^{n+1} = u_m^{n-1} - v\rho\left(u_{m+1}^n - u_{m-1}^n\right)\,. \tag{9.7.10}$$

There are several things to notice about this method. First, in contrast to all other methods presented so far this is a three-level method. This method is known as *leapfrog*, as at each iteration we are using the centered time difference to "jump" over the middle time level. Leapfrog is second order accurate in both space and time, and stable if the CFL condition is satisfied. It can also be easily generalized to non-linear equations (which is not the case for Lax–Wendroff, for example). Still, it has two main drawbacks: First, since it is a three-level method, in order to start it we need to know the solution on the first two time levels. But for first order systems the initial data can only give us the very first time level and nothing more, so that leapfrog needs to be initialized using one of the other methods for at least one time step. Also, because of the way it is constructed it is not difficult to convince oneself that leapfrog in fact produces an evolution on two distinct, and decoupled, numerical grids, one of them taking even values of

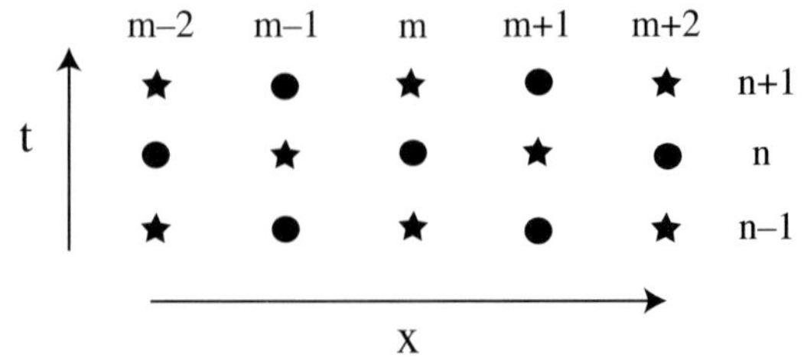

Fig. 9.4: Structure of the numerical grid for the leapfrog scheme. Notice how the points represented by the circles are decoupled from those represented by the stars, so in fact we have two distinct grids.

m for even values of n and odd values of m for odd values of n, and the other one doing the opposite (see Figure 9.4). This decoupling of the grid can lead to the development of numerical errors that have a checker board pattern in space and time typically known as *red-black errors.*

Finally, we will introduce one more implicit finite difference approximation to the advection equation. For this we simply take the average of the forward and backward Euler methods:

$$u_m^{n+1} = u_m^n - \frac{v\rho}{4} \left[\left(u_{m+1}^n - u_{m-1}^n \right) + \left(u_{m+1}^{n+1} - u_{m-1}^{n+1} \right) \right] . \tag{9.7.11}$$

This approximation is known as the *Crank–Nicholson* scheme. It turns out to be second order accurate in both time and space and stable for *all* values of ρ (*i.e* it is unconditionally stable). It also forms the basis for a method that has become very common in numerical relativity in the past few years and that we will discuss in the next Section.

Figure 9.5 shows a schematic representation of the different finite difference schemes introduced so far using the concept of *computational stencil* or *molecule*, *i.e.* a diagram that shows the relationship between the different grid points used in the approximation.

Up to this point we have considered only the simple scalar advection equation. However, the methods discussed above can be easily generalized to linear hyperbolic systems of the form

$$\partial_t u + A\, \partial_x u = 0 , \tag{9.7.12}$$

where now u denotes a set of unknowns $u = (u_1, u_2, ...)$, and A is some matrix with constant coefficients (the characteristic matrix). Most of the methods discussed above can be used directly by simply replacing the scalar function u with the vector of unknowns and the speed v with the characteristic matrix A. The CFL condition now takes the form

$$\max |\lambda_a| \; \Delta t \leq \Delta x , \tag{9.7.13}$$

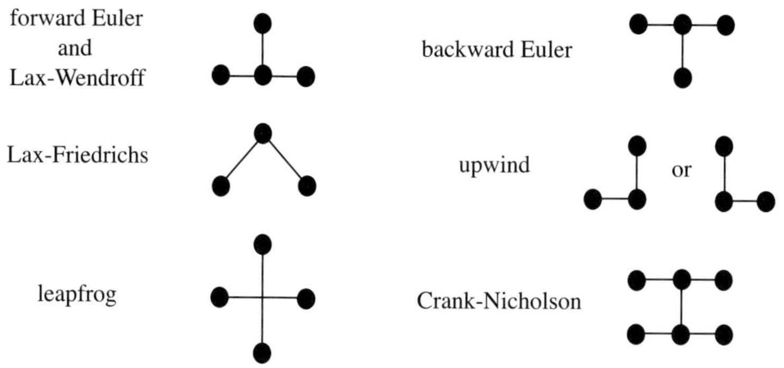

Fig. 9.5: Computational stencils for the different finite difference approximations to the advection equation.

where λ_a are the eigenvalues of A, so the time step is now restricted by the largest characteristic speed of the system. Notice that we are assuming that the system is hyperbolic, so that all eigenvalues are real.

We must take care, however, with one-sided methods such as upwind which would only be stable if all the eigenvalues of A had the same sign, and the method is used along the corresponding direction. In a more general case where A has eigenvalues of different signs we must first decompose the system into left-going and right-going fields, which can only be done if we have a strongly hyperbolic system (*i.e.* the matrix A has a complete set of eigenvectors, see Chapter 5). In such a case we separate the eigenvalues of A according to sign by defining

$$\lambda_a^+ = \max(\lambda_a, 0) \,, \qquad \lambda_a^- = \min(\lambda_a, 0) \,, \tag{9.7.14}$$

and constructing the matrices of positive and negative eigenvalues

$$\Lambda^+ = \text{diag}(\lambda_a^+, 0) \,, \qquad \Lambda^- = \text{diag}(\lambda_a^-, 0) \,. \tag{9.7.15}$$

A stable upwind method would then take the form

$$u_m^{n+1} = u_m^n - \rho \left[R\,\Lambda^+ R^{-1} \left(u_m^n - u_{m-1}^n \right) + R\,\Lambda^- R^{-1} \left(u_{m+1}^n - u_m^n \right) \right] \,, \tag{9.7.16}$$

with R the matrix of column eigenvectors of A. That is, each characteristic field is differentiated along the upwind direction associated with the sign of its corresponding characteristic speed.

The methods discussed here can in fact also be generalized to the case when the coefficients of the matrix A are functions of space and time (but now we must be careful with the Taylor expansion used to obtain the Lax–Wendroff scheme, as we will pick up terms coming from derivatives of A).

9.8 Method of lines

A different approach to solving systems of evolution equations is known as the *method of lines* (MOL). Here we start by first discretizing the spatial dimensions while leaving the time dimension continuous. For concreteness, let us assume that we have a single scalar partial differential equation of the form

$$\partial_t u = \mathcal{S}(u) \; , \tag{9.8.1}$$

with $\mathcal{S}$ some spatial differential operator. If we use standard finite differences for the spatial derivatives we can rewrite our original differential equation as a coupled system of ordinary differential equations of the form

$$\frac{d\mathbf{u}}{dt} = \mathbf{S}\,\mathbf{u} \; , \tag{9.8.2}$$

where now $\mathbf{u}$ is a vector constructed from the values of the function u in the different spatial grid points, and $\mathbf{S}$ is a matrix (typically sparse) that couples the different grid points. If the resulting system of ordinary differential equations is stable, we can use any standard numerical technique for solving them and we will obtain a stable evolution.

The method of lines has the advantage that it decouples the choice of spatial and time differencing, so that we can change the order of the spatial differencing or the time integration method independently of each other. It can also be generalized in a straightforward way to non-linear systems of equations, which is not always the case with some of the more standard methods (for example Lax–Wendroff). Moreover, it makes it quite easy to couple codes developed for different systems of equations, like for example the Euler equations for fluid dynamics and the 3+1 evolution equations for the spacetime. Because of this, codes based on the method of lines have become increasingly common in numerical relativity over the past decade.

As a simple example, consider again the advection equation (9.7.1), and assume that we discretize the spatial derivatives using centered differences. We then arrive at the following system of ordinary differential equations

$$\frac{du_m}{dt} = -\frac{v}{2\Delta x}\left(u_{m+1} - u_{m-1}\right) \; . \tag{9.8.3}$$

We can now solve these equations using, for example, second order Runge–Kutta which for a system of equations of the form

$$\frac{du}{dt} = S(u) \; , \tag{9.8.4}$$

takes the form

$$\begin{aligned} u^* &= u^n + \Delta t \; S(u^n)/2 \; , \\ u^{n+1} &= u^n + \Delta t \; S(u^*) \; . \end{aligned} \tag{9.8.5}$$

Second order Runge–Kutta can be summarized in the following way: Calculate the sources in the old time step and use them to advance the solution half a time step. Use now this intermediate step to recalculate the sources, go back and then advance the full time step.

Using Runge–Kutta we can now rewrite our method for solving the advection equation as

$$\begin{aligned} u^* &= u^n + \Delta t \, S(u^n)/2 \,, \\ u^{n+1} &= u^n + \Delta t \, S(u^*) \,. \end{aligned} \tag{9.8.6}$$

which after some algebra becomes

$$u_m^{n+1} = u_m^n - \frac{v\rho}{2}\left(u_{m+1}^n - u_{m-1}^n\right) + \frac{v^2\rho^2}{8}\left(u_{m+2}^n - 2u_m^n + u_{m-2}^n\right) , \tag{9.8.7}$$

where as before $\rho = \Delta t/\Delta x$. We can call this method MOL-RK2 for short. If we now compare the last expression with the Lax–Wendroff method discussed in the previous Section, equation (9.7.7), we see that it has essentially the same structure except for the fact that the correction term has now been approximated using twice the grid spacing. Unfortunately, a von Newmann stability analysis shows that the MOL-RK2 method just described is unstable for *any* value of ρ, so it is useless in practice.

A stable method is obtained if we use fourth order Runge–Kutta [180] instead, which corresponds to first recursively calculating the quantities

$$k_1 = S(u^n) \,, \tag{9.8.8}$$

$$k_2 = S(u^n + k_1\Delta t/2) \,, \tag{9.8.9}$$

$$k_3 = S(u^n + k_2\Delta t/2) \,, \tag{9.8.10}$$

$$k_4 = S(u^n + k_3\Delta t) \,, \tag{9.8.11}$$

and then taking

$$u^{n+1} = u^n + \frac{\Delta t}{6}\left(k_1 + 2k_2 + 2k_3 + k_4\right) \,. \tag{9.8.12}$$

Notice that fourth order Runge–Kutta requires four evaluations of the sources to advance one time step. We can call the method of lines obtained in this way MOL-RK4. Since this method gives us fourth order accuracy in time, it is natural to use it with fourth order spatial differencing as well.

Another very common choice for the time integration in the method of lines is the *iterative Crank–Nicholson* (ICN) scheme. The idea behind this method is to use an iterative scheme to approach the solution of the standard implicit Crank–Nicholson scheme described in the last Section (equation (9.7.11)). Viewed as a method of lines, the iteration can be described as follows

$$u^{*(1)} = u^n + \Delta t \, S(u^n) \,, \tag{9.8.13}$$

$$u^{*(l)} = u^n + \frac{\Delta t}{2}\left[S(u^n) + S(u^{*(l-1)})\right] , \quad l = 2, ..., N, \tag{9.8.14}$$

$$u^{n+1} = u^{*(N)} \,, \tag{9.8.15}$$

with $N \geq 2$. The method takes one initial forward Euler step, uses this to calculate an approximation to the source in the next time level, and then iterates using the average of the source in the old time level and the latest approximation to the new time level. It is clear that if the iterations converge in the limit $N \to \infty$, we will recover the implicit Crank–Nicholson scheme:

$$\frac{u^{n+1} - u^n}{\Delta t} = \frac{1}{2}\left[S(u^n) + S(u^{n+1})\right] . \tag{9.8.16}$$

Assuming that the source function $S(u)$ is a linear operator, the iterative method can be written formally as

$$u^{n+1} = u^n + \sum_{k=1}^{N} \frac{(\Delta t)^k}{2^{k-1}} S^k(u^n) . \tag{9.8.17}$$

Also, for linear source functions the iterations can be written in an entirely equivalent way as

$$\bar{u}^{*(l)} = u^n + \frac{\Delta t}{2} S(\bar{u}^{*(l-1)}) , \quad l = 1, ..., N-1, \tag{9.8.18}$$

$$u^{n+1} = u^n + \Delta t \, S(\bar{u}^{*(N-1)}) , \tag{9.8.19}$$

with $u^{*(l)}$ and $\bar{u}^{*(l)}$ related through $\bar{u}^{*(l)} = (u^n + u^{*(l)})$. The method now takes a series of half steps and a final full step. Viewed in this way we can see that the case $N = 2$ is in fact identical to second order Runge–Kutta. Notice that the two different versions of ICN are only equivalent for linear equations; for non-linear equations this equivalence is lost and since there is no *a priori* reason to prefer one version over the other it is a matter of personal choice which version is used in a given code.

For linear systems, it is also possible to show that the ICN method has the following important properties [14, 283]:

1. In order to obtain a stable scheme we must take *at least* three steps, that is $N \geq 3$. The case $N = 2$ is enough to achieve second order accuracy, but it is unstable (it is equivalent to second order Runge–Kutta). Stability in fact comes in pairs in terms of the number of steps: 1 and 2 steps are unstable, 3 and 4 are stable, 5 and 6 are unstable, and so on (see *e.g.* [283]).
2. The iterative scheme itself is only convergent if the standard CFL stability condition is satisfied, otherwise the iterations diverge [14].

These two results taken together imply that there is no reason (at least from the point of view of stability) to ever do more than three ICN steps. Three steps are already second order accurate, and provide us with a scheme that is stable as long as the CFL condition is satisfied.[99] Increasing the number of iterations will not improve the stability properties of the scheme any further. In particular,

[99]Some authors call the $N = 3$ method a "two-iteration" ICN scheme, since we do an initial Euler step and then iterate twice. This is identical to the "three-step" scheme discussed here.

we will never achieve the unconditional stability properties of the full implicit Crank–Nicholson scheme, since if we violate the CFL condition the iterations will diverge.

Three-step ICN became a workhorse method in numerical relativity for a number of years. Recently, however, the demand for higher accuracy has seen ICN replaced by fourth order Runge–Kutta with fourth order spatial differencing. Still, for applications that do not require very high accuracy, three-step ICN has proven to be a very robust method.

9.9 Artificial dissipation and viscosity

When we work with non-linear systems of equations like those of relativity or hydrodynamics, many of the stable numerical methods we have described in the previous Sections become slightly unstable because of the effects of both coefficients that depend on the dynamical variables, and lower order terms. Also, in the particular case of hydrodynamics, shock waves can develop that cause standard numerical algorithms to become very inaccurate owing to the appearance of spurious high frequency oscillations, known as *Gibbs phenomena*, near the shock front. In order to solve these problems we can add dissipative terms to the finite difference operators that act as "low pass filters" in the sense that they preferentially damp modes with wavelength similar to the grid spacing. Such high frequency modes are already unresolved and are therefore severely affected by truncation errors, and they are also frequently the source of the instabilities (though not always), so that we are often better off dissipating them away as otherwise they might become unstable and ruin the simulation. We have already seen an example of this when we discussed the stability properties of the Lax–Friedrichs and Lax–Wendroff schemes. Adding such dissipative terms to a numerical scheme goes under the name of *artificial dissipation* or *artificial viscosity*. Although these two terms are often used interchangeably, here I will use the name artificial dissipation for the more general idea and will keep the name artificial viscosity for a specific technique for dealing with shock waves in hydrodynamical simulations.

The standard way of adding artificial dissipation to a finite difference approximation is known as *Kreiss–Oliger dissipation* [178]. The basic idea is as follows: Assume that we have a finite difference scheme that can be written as

$$u_m^{n+1} = u_m^n + \Delta t \, S(u_m^n) \, , \tag{9.9.1}$$

with $S(u^n)$ some spatial finite difference operator. Let us now modify this scheme by adding a term of the form

$$u_m^{n+1} = u_m^n + \Delta t \, S(u_m^n) - \epsilon \, \frac{\Delta t}{\Delta x} \, (-1)^N \, \Delta_x^{2N} \left(u_m^n \right) \, , \tag{9.9.2}$$

with $\epsilon > 0$, $N \geq 1$ an integer, and where $\Delta_x^{2N} := (\Delta_x^+ \Delta_x^-)^N$ is the $2N$ centered difference operator. These Δ_x^{2N} operators appear in the finite difference

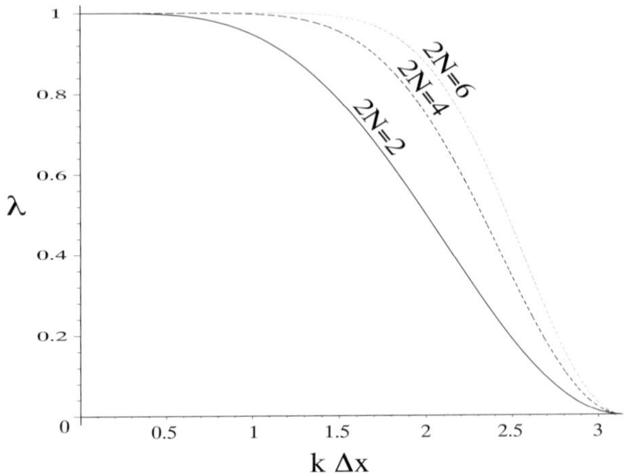

Fig. 9.6: Dissipation factor λ as a function of the wave number $k\Delta x$ for the case when $\epsilon\Delta t/\Delta x = 1/2^{2N}$ and $N = 1, 2, 3$. Notice how, as N increases, the dissipation factor approaches a step function at the Nyquist wavelength $k\Delta x = \pi$.

approximations to the high-order spatial derivatives $\partial_x^{2N} u$, and are very easy to construct by using the corresponding coefficients of the binomial expansion with alternated signs, for example:

$$\Delta_x^2 u_m^n = u_{m+1}^n - 2u_m^n + u_{m-1}^n \,, \tag{9.9.3}$$

$$\Delta_x^4 u_m^n = u_{m+2}^n - 4u_{m+1}^n + 6u_m^n - 4u_{m-1}^n + u_{m-2}^n \,, \tag{9.9.4}$$

$$\Delta_x^6 u_m^n = u_{m+3}^n - 6u_{m+2}^n + 15u_{m+1}^n - u_m^n + 15u_{m-1}^n - 6u_{m-2}^n + u_{m-3}^n \,. \tag{9.9.5}$$

The factor $(-1)^N$ in (9.9.2) guarantees that the extra term is dissipative if we take $\epsilon > 0$. Now, assume for the moment that $S(u_m^n) = 0$, then our finite difference scheme becomes

$$u_m^{n+1} = u_m^n - \epsilon \, \frac{\Delta t}{\Delta x} \, (-1)^N \, \Delta_x^{2N} \left(u_m^n\right) \,. \tag{9.9.6}$$

A von Newmann stability analysis shows that this scheme is stable as long as

$$0 \leq \epsilon \, \Delta t/\Delta x \leq 1/2^{2N-1} \,. \tag{9.9.7}$$

Moreover, if we choose ϵ such that $\epsilon \, \Delta t/\Delta x = 1/2^{2N}$, then the dissipation factor approaches a step function in Fourier space as N increases, so that it damps very strongly modes with wavelengths close to the grid spacing Δx, and leaves longer wavelengths essentially unaffected (see Figure 9.6).

Going back to the full finite difference scheme (9.9.2) we find that the stability depends on the explicit form of the operator S, but in general the extra term still introduces dissipation and has the effect of improving the stability of the original scheme for small positive values of ϵ (in practice it turns out that even quite small values of ϵ can result in important improvements in stability). The best strategy to improve stability would clearly be to add as much dissipation as possible without either making the scheme unstable (too large ϵ), or spoiling the accuracy of the scheme.

The crucial question then becomes: What is the effect of the dissipative term on the accuracy of the numerical scheme? In order to see this, let us rewrite (9.9.2) as

$$\frac{u_m^{n+1} - u_m^n}{\Delta t} = S(u_m^n) - \epsilon \, \frac{1}{\Delta x} \, (-1)^N \, \Delta_x^{2N} \, (u_m^n) \; . \tag{9.9.8}$$

In the limit of small Δt and Δx this can be replaced by

$$\partial_t u = S(u) - \epsilon \, (-1)^N \, (\Delta x)^{2N-1} \, \partial_x^{2N} u \; . \tag{9.9.9}$$

We then see that we have added a new term to the original differential equation that vanishes in the continuum limit as $(\Delta x)^{2N-1}$. In other words, if the original numerical scheme has an accuracy of order m, we would need to use a dissipative term such that $2N - 1 \geq m$ to be sure that we have not spoiled the accuracy of the scheme. We typically say that the "order" of the dissipative term is given by the number of derivatives in it, that is $2N$, so that if our original scheme was second order accurate we would need to add fourth order dissipation to it $(4 - 1 = 3 > 2)$, and if we had a fourth order scheme we must add sixth order dissipation $(6 - 1 = 5 > 4)$. Higher order dissipative terms clearly have problems when we get very close to the boundaries since we don't have enough points to apply the dissipative operators. This means that, close to the boundaries, we must either not use artificial dissipation, or we have to use one-sided dissipative operators.

A related, but more specialized, form of artificial dissipation is known as *artificial viscosity* and is a technique designed to deal with shock waves in hydrodynamical simulations. The main problem with the Euler equations is that the non-linearities give rise to the formation of shocks, *i.e.* propagating discontinuities that arise even from smooth initial data and that are associated with the fact that the characteristic lines cross. Of course, in a realistic physical situation we do not expect a shock wave to correspond to a true discontinuity, but rather to a sharp gradient in the hydrodynamical variables. Indeed, if we consider the presence of viscosity as is done in the Navier–Stokes equations, this is precisely what happens. However, in many physical situations, and in most astrophysical applications, the viscosity of the fluid is such that the characteristic width of a shock wave is much smaller than the size of the numerical grid, so the shock can not be resolved numerically. In such cases the numerical methods cause large spurious oscillations to appear close to the shock (Gibbs phenomena), and these oscillations can seriously affect the accuracy and stability of our finite difference approximation.[100] One way to fix this problem is to add by hand an artificial dissipation term that will smooth out the shock into a few grid zones.

There are several important considerations to be made when adding dissipative terms to the Euler equations. In the first place, if we just add a dissipation term with a constant coefficient we will be affecting the numerical solution everywhere. It would clearly be much better to use a dissipation term that will become larger in a zone where a shock is likely to develop, and smaller (or even zero) in other regions. Also, if we wish to mimic the physical situation in a realistic way, we need to make sure that the kinetic energy lost because of the presence of the dissipative term does not just vanish from the system, but is in fact converted into internal energy.

The original formulation of artificial viscosity is due to von Newmann and Richtmyer [220], and involves modifying the Euler equations by adding an extra viscous contribution Π to the pressure, such that

$$p \to p + \Pi \,. \tag{9.9.10}$$

This already guarantees that energy conservation will be maintained as the viscous term will transform kinetic energy into internal energy. In the original work of von Newmann and Richtmyer, the artificial viscosity term was taken to be of the form

$$\Pi_1 = c_1 \rho \left(\Delta x \, \partial_x v \right)^2 \,, \tag{9.9.11}$$

where $c_1 > 0$ is a constant coefficient, ρ is the density of the fluid at a given grid point, and v is the flow speed. This form of Π is motivated by considering the loss of kinetic energy when two fluid particles have a completely inelastic collision. A different type of artificial viscosity that vanishes less rapidly for small changes in v, and has the effect of reducing unphysical oscillations behind the shock, has the form

$$\Pi_2 = c_2 \rho \, c_s \Delta x \left| \partial_x v \right| \,, \tag{9.9.12}$$

with $c_2 > 0$ again a constant coefficient, and where c_s is the local speed of sound. A more generic form of artificial viscosity will then consist of adding together both types of terms:

$$\Pi = c_1 \rho \left(\Delta x \, \partial_x v \right)^2 + c_2 \rho \, c_s \Delta x \left| \partial_x v \right| \,. \tag{9.9.13}$$

The values of the constants c_1 and c_2 are adjusted for each simulation, with c_2 typically an order of magnitude smaller than c_1 (notice that the linear viscosity term introduces an truncation error of order Δx, while the error associated with the quadratic term has order $(\Delta x)^2$ instead). Also, in order to make sure that viscosity is only added in regions where shocks are likely to form, the coefficients

[100] The oscillations associated which such Gibbs phenomena are equivalent to those that arise in the Fourier expansion of a discontinuous function and have the same origin. As the resolution is increased their wavelength becomes smaller and they appear closer and closer to the shock front, but their amplitude remains essentially the same.

c_1 and c_2 are usually taken to be different from zero only for regions where the flow lines are converging, that is $\partial_x v < 0$, and are set to zero otherwise.

The simple form of artificial viscosity just described has been modified over the years, and considerably more sophisticated versions have been proposed, though the basic idea has remained (see for example the recent textbook by Wilson and Mathews [300] for the use of artificial viscosity in relativistic hydrodynamical simulations). Here, however, we will not go into the details of those more advanced artificial viscosity techniques and will refer the interested reader to any standard textbook on computational fluid dynamics.

9.10 High resolution schemes

As already mentioned in the last Section, standard finite difference methods have problems when dealing with discontinuous solutions such as those associated with hydrodynamical shock waves. Essentially, first order methods show severe dissipation close to a discontinuity, while second order methods introduce dispersion effects that produce large spurious oscillations (Gibbs phenomena). The rate of convergence can also be shown to drop near the discontinuity, to order $(\Delta t)^{1/2}$ for first order methods, and order $(\Delta t)^{2/3}$ for second order methods.

As we have seen, one way of dealing with discontinuous solutions is to add artificial viscosity to our finite difference methods. Indeed, such an approach is still quite common. The main disadvantage is that artificial viscosity tends to be somewhat *ad hoc* and must be tailored to the particular problem under study. Because of this we will consider here a more systematic approach to constructing numerical methods capable of dealing with discontinuous solutions. The discussion here will be very brief; for more details the reader can consult the books by LeVeque [187], and by Gustafsson, Kreiss and Oliger [157].

9.10.1 *Conservative methods*

As a simple example of an equation that gives rise to discontinuous solutions starting from smooth initial data let us consider again Burger's equation:

$$\partial_t u + u\, \partial_x u = 0 \,. \tag{9.10.1}$$

As was already discussed in Chapter 7, this equation is a simple non-linear generalization of the advection equation, where the wave speed is now given by the wave function u itself. Burger's equation is simple enough so that we can solve it exactly even for discontinuous initial data (the Riemann problem), and we find that for $u > 0$, and initial data such that u has a jump from some larger value at the left to a smaller value to the right, the discontinuity simply propagates at a constant speed that is the direct average of the values of u to the left and right (see Section 7.7 for a more detailed explanation).

Let us now assume that we approximate (9.10.1) using a simple generalization of the upwind method of the form

$$u_m^{n+1} = u_m^n - \frac{\Delta t}{\Delta x}\, u_m^n \left(u_m^n - u_{m-1}^n \right) \,. \tag{9.10.2}$$

This method is good for smooth solutions, and stable as long as $u > 0$ and $\Delta t/\Delta x \leq 1$. However, it *will not* converge to the correct solution of the Riemann problem for discontinuous initial data. What we find in practice is that even though the numerical solution looks qualitatively correct, with a propagating discontinuity smoothed out over a few grid points because of numerical dissipation, the discontinuity in fact moves at the wrong speed, and this does not improve with resolution.

Fortunately, there is a simple way to solve this problem. We must use a numerical method that is written in *conservation form*. That is, our numerical method must have the form

$$u_m^{n+1} = u_m^n - \frac{\Delta t}{\Delta x} \left[F_{m+1/2}^{n+1/2} - F_{m-1/2}^{n+1/2} \right] , \qquad (9.10.3)$$

with $F_{m+1/2}^{n+1/2}$ a function that depends on the values of u_m and u_{m+1} (plus maybe a few more nearest neighbors), but that has the same functional form as we move from one grid point to the next. The function F is known as the *numerical flux function*. Equation (9.10.3) can be understood as the numerical analog of the integral form of the conservation law if we think of u_m^n not as the value of $u(x,t)$ at a single grid point, but rather as the average over the grid cell.

We will usually obtain numerical methods in conservation form if we start from the conservative form of the differential equation. For Burger's equation we can start from

$$\partial_t u + \partial_x \left(u^2/2 \right) = 0 , \qquad (9.10.4)$$

which can easily be seen to be equivalent to (9.10.1). This leads to the following upwind method

$$u_m^{n+1} = u_m^n - \frac{\Delta t}{2\Delta x} \left[(u_m^n)^2 - (u_{m-1}^n)^2 \right] , \qquad (9.10.5)$$

where we have taken $F_{m+1/2}^{n+1/2} = (u_m^n)^2/2$. With this method we now find that discontinuities propagate at the correct speed.

This brings us to the Lax–Wendroff theorem (1960): *For hyperbolic systems of conservation laws, a numerical approximation written in conservation form, if it converges, will converge to a weak solution of the original system.*

9.10.2 *Godunov's method*

Godunov's method [146] is an approach to the finite difference approximation of hyperbolic conservation laws that is designed to work well on discontinuous solutions. It is based on the idea of updating the value of the function by solving Riemann problems at each cell interface. In this method, we define a piecewise constant function $\tilde{u}^n$ that has the value u_m^n over a given grid cell (with the cell interfaces located halfway between grid points), and we then solve Riemann problems at each cell interface (see Figure 9.7). This can be done without problems for times short enough so that the wave fronts from neighboring Riemann

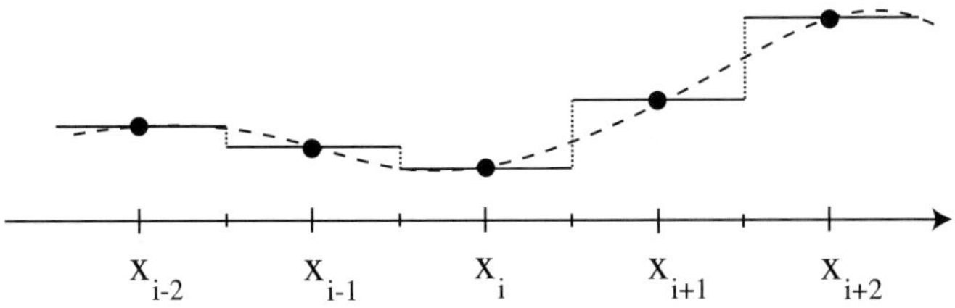

Fig. 9.7: Approximation of a function in piecewise constant form over the grid cells. The boundaries of the individual cells are halfway between the grid points. With Godunov's method we solve Riemann problems at each of the cell boundaries.

problems do not intersect each other (this is essentially the CFL condition). Finally, we use this solution to compute the numerical flux function in the following way

$$F^{n+1/2}_{m+1/2} = \frac{1}{\Delta t} \int_t^{t+\Delta t} f\left(\tilde{u}^n(x_{m+1/2}, t)\right) dt \,, \tag{9.10.6}$$

with $f(u)$ the exact flux function. Notice that this integral is in fact trivial since, by construction, $\tilde{u}^n$ is constant along the line $x = x_{m+1/2}$. If we denote this constant value by $u^*(u^n_m, u^n_{m+1})$ we will have

$$F^{n+1/2}_{m+1/2} = f\left(u^*(u^n_m, u^n_{m+1})\right) \,. \tag{9.10.7}$$

The problem now is to compute $u^*(u^n_m, u^n_{m+1})$: We know it is constant, but what is its value? In order to find u^* we must first determine the full wave structure of our system of equations and solve the Riemann problem.

For a system of linear conservation laws the Riemann problem is in fact very easy to solve and we find that Godunov's method reduces to the upwind method that takes into account the speed of propagation of each characteristic field. For a general non-linear *scalar* conservation law, on the other hand, Godunov's method reduces to

$$F_{m+1/2} = \begin{cases} \min f(u) \,, \; u_l \leq u \leq u_r & \text{if } u_l \leq u_r \,, \\ \max f(u) \,, \; u_l \geq u \geq u_r & \text{if } u_l \geq u_r \,. \end{cases} \tag{9.10.8}$$

For systems of non-linear conservation laws we must solve the full Riemann problem first. This can be very difficult in practice, so we often use approximate Riemann solvers.

Godunov's method, though quite elegant, does have one important disadvantage: Because of the piecewise constant form of the function $\tilde{u}^n$, the method is only first order accurate (as can also be seen from the fact that, for linear systems, it reduces to upwind).

9.10.3 *High resolution methods*

Finite difference approximations that are at least second order accurate on smooth solutions, and give well resolved, non-oscillatory solutions near discontinuities, are commonly known as *high resolution methods.* One natural (though not absolutely necessary) requirement for a high resolution method is that it should be *monotonicity preserving.* This means that if the initial data is monotonic, then the solution should retain this property for all time. In other words, a monotonicity preserving method would give rise to no spurious oscillations by construction.

However, the idea of constructing high resolution, monotonicity preserving, numerical methods is immediately faced with one very serious problem: Godunov has shown that a linear, monotonicity preserving, finite difference approximation is at most first order accurate. This implies that we must consider non-linear numerical methods if we want to have a high resolution monotonicity preserving method, even for linear systems of equations!

There are two related approaches to constructing non-linear monotonicity preserving methods that go under the names of *flux limiters* and *slope limiters.* Flux limiters use a high order flux F_H in smooth regions (*e.g.* Lax–Wendroff), and a low order flux F_L from a monotonic method (*e.g.* upwind) near a discontinuity. We can then write the full flux F in general as

$$F = F_H - (1 - \phi)\,(F_H - F_L) \ , \tag{9.10.9}$$

where the weight factor ϕ, known as the *limiter*, should be close to unity if the data is smooth, and should approach zero near a discontinuity. The question is then: How do we measure the smoothness of the data? The most natural possibility is to measure this smoothness by comparing the slope of the dynamical function u to the left and right of a given point by defining

$$\theta := \frac{u_m - u_{m-1}}{u_{m+1} - u_m} \ . \tag{9.10.10}$$

We can then take the limiter ϕ to be a function of θ.

As an example, let us go back to the advection equation (9.7.1) and consider the Lax–Wendroff method (9.7.7), which for our purposes we will rewrite as

$$u_m^{n+1} = u_m^n - v\rho\left(u_m^n - u_{m-1}^n\right) - \frac{v\rho}{2}\,(1 - v\rho)\left(u_{m+1}^n - 2u_m^n + u_{m-1}^n\right) , \tag{9.10.11}$$

with $\rho = \Delta t/\Delta x$ the Courant parameter. Here we have assumed that $v > 0$, which of course makes no difference for the stability of the Lax–Wendroff method, but does affect the way in which we have chosen to write it since we have clearly separated out an "upwind type" first term. We can now rewrite this last method in flux conservative form and simply read from it the form of the numerical flux. Doing this we find

$$F_{m+1/2}^{n+1/2} = v u_m^n + \frac{v}{2}\,(1 - v\rho)\left(u_{m+1}^n - u_m^n\right) \ . \tag{9.10.12}$$

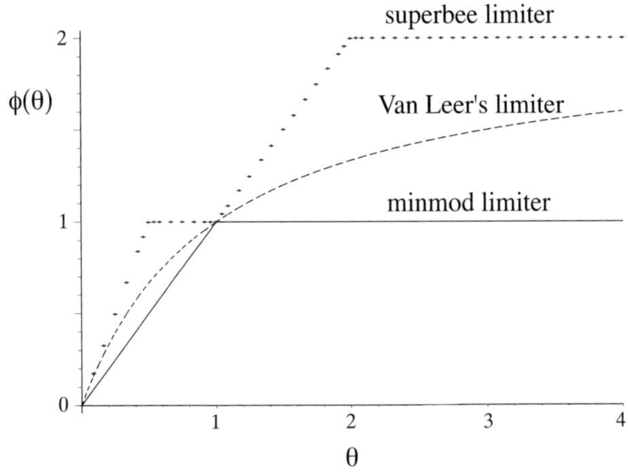

Fig. 9.8: Different limiter function $\phi(\theta)$: The solid line corresponds to Sweby's minmod limiter, the dashed line to Van Leer's limiter, and the dotted line to Roe's superbee limiter.

The idea now is to change the above flux to

$$F_{m+1/2}^{n+1/2} = v u_m^n + \frac{v}{2} (1 - v\rho) \left(u_{m+1}^n - u_m^n \right) \phi(\theta) \,, \tag{9.10.13}$$

that is, we have multiplied the second term with the limiter function ϕ. Notice that for $\phi = 1$ we recover the Lax–Wendroff flux, while for $\phi = 0$ the expression reduces to the upwind flux.

How do we choose the form of ϕ? There are in fact many possibilities that have been suggested in the literature. For example, Beam and Warming propose taking $\phi(\theta) = \theta$, though this is not really a flux limiter and only differs from Lax–Wendroff in the fact that the dissipative term is not centered but is upwinded instead. Other more interesting limiter functions are the following:

- Sweby's "minmod" limiter:

$$\phi(\theta) = \text{minmod}\,(1, \theta) \equiv \begin{cases} 0 \,, & \theta \leq 0 \,, \\ \theta \,, & 0 < \theta \leq 1 \,, \\ 1 \,, & \theta \geq 1 \,. \end{cases} \tag{9.10.14}$$

- Van Leer's limiter:

$$\phi(\theta) = \frac{\theta + |\theta|}{1 + |\theta|} \,. \tag{9.10.15}$$

- Roe's "superbee" limiter

$$\phi(\theta) = \max\left(0, \min(1, 2\theta), \min(\theta, 2)\right) \,. \tag{9.10.16}$$

Notice that all three limiters set $\phi(\theta) = 0$ for $\theta < 0$, which implies that at extrema the scheme will become only first order accurate. Also, all these limiters are constructed in such a way as to guarantee that the total variation of the numerical solution, defined as

$$TV(u) := \sum_m |u_m - u_{m-1}| \ , \tag{9.10.17}$$

never increases, *i.e.* they result in so-called *total-variation-diminishing* (TVD) schemes. By construction, TVD schemes can not give rise to spurious oscillations since these necessarily would increase the total variation of the function. Because of this, flux limiter methods can resolve discontinuities very sharply. The different limiters result in somewhat different behaviors at discontinuities, with the minmod limiter having more dissipation, the superbee limiter less, and van Leer's limiter somewhere in the middle.

Figure 9.8 shows the function $\phi(\theta)$ for the different limiters. It is in fact possible to show that $\phi(\theta)$ *must* lie in the region between the minmod and superbee limiters for the resulting scheme to be TVD. As can be seen from the figure, van Leer's limiter lies nicely in the middle of this region.

A related, but more geometric, approach to constructing high resolution methods is that of slope limiters. The idea here is to generalize Godunov's method by replacing the piecewise constant representation by something more accurate such as a piecewise linear function (or even a piecewise parabolic). For example, in a given cell we can approximate our function as

$$\tilde{u} = u_m + \sigma_m (x - x_m) \ , \qquad \text{for } x \in \left[x_{m-1/2}, x_{m+1/2}\right] \ , \tag{9.10.18}$$

with σ_m a slope to be determined in terms of the data. For a linear equation the obvious choice for this slope is

$$\sigma_m = \frac{u_{m+1} - u_m}{\Delta x} \ . \tag{9.10.19}$$

Now, in order to get a second order method we need to calculate the flux at the point $(x_m + \Delta x/2, t^n + \Delta t/2)$. For the advection equation we will have

$$\begin{aligned} F^{n+1/2}_{m+1/2} &= v\, u(x_m + \Delta x/2, t^n + \Delta t/2) \\ &= v u(x_m + (\Delta x - v\Delta t)/2, t^n) \\ &= v \left(u^n_m + \sigma^n_m \left(1 - v\rho\right) \Delta x/2\right) \\ &= v u^n_m + \frac{v}{2} \left(1 - v\rho\right) \left(u^n_{m+1} - u^n_m\right) \ . \end{aligned} \tag{9.10.20}$$

We then see that this choice of σ_m reduces to Lax–Wendroff, which as we know is bad for discontinuous solutions.

A much better choice is to "limit" the value of the slope σ_m. For example, we can choose the slope as

$$\begin{aligned}\sigma_m &= \left(\frac{1}{\Delta x}\right) \times \text{minmod}\left(u_{m+1} - u_m, u_m - u_{m-1}\right) \\ &= \left(\frac{u_{m+1} - u_m}{\Delta x}\right) \times \text{minmod}\left(1, \theta\right) \,, \end{aligned} \tag{9.10.21}$$

which for linear equations is equivalent to the minmod flux limiter. Similarly, we can use a slope of the form

$$\begin{aligned}\sigma_m &= \left(\frac{1}{\Delta x}\right) \left(\frac{2\left(u_{m+1} - u_m\right)\left(u_m - u_{m-1}\right)}{\left(u_{m+1} - u_m\right) + \left(u_m - u_{m-1}\right)}\right) \\ &= \left(\frac{u_{m+1} - u_m}{\Delta x}\right) \left(\frac{2\theta}{1 + \theta}\right) \,, \end{aligned} \tag{9.10.22}$$

for $\theta > 0$, and $\sigma_m = 0$ otherwise, which for linear equations reduces to van Leer's flux limiter.

There is also a family of limiter methods that relax the TVD condition in order to increase accuracy near extrema (where TVD methods become only first order accurate). These higher order methods are known as *essentially non-oscillatory* (ENO) methods, and instead of asking for the total variation to be non-increasing they allow it to increase as long as its growth is bounded by an exponential that is independent of the initial data. This means that such methods are *total variation stable*. However, we will not discuss ENO methods here.

As a final comment, it is important to stress the fact that in the previous discussion we have always assumed that the characteristic speed is positive. This has entered in the fact that, in order to limit the flux $F^{n+1/2}_{m+1/2}$ (or the slope of u in that region), we have considered the ratio $\theta = (u_m - u_{m-1})/(u_{m+1} - u_m)$. That is, we are considering the differences in the given cell and the cell that is directly to the left. In the case when the characteristic speed is negative we should have used instead $\theta = (u_{m+1} - u_m)/(u_{m+2} - u_{m+1})$. More generally, for a strongly hyperbolic system of equations we need to first diagonalize the system in order to find the eigenfields and eigenspeeds, and then use one or the other ratio for a given eigenfunction depending on the sign of its associated eigenspeed. Because of this "upwinding", limiter methods result in considerably more complex codes than standard methods.

9.11 Convergence testing

Before finishing this Chapter it is important to discuss what is perhaps the most important lesson related to any numerical calculation: A numerical calculation done at only one resolution (*i.e.* only one value of Δx and Δt) without studying how the solution behaves as the resolution is increased is *meaningless*. We must remember that numerical calculations are approximations to the underlying differential equations, and unless we have some idea of the size of the error in those approximations we have simply no way of knowing if the result of a given numerical calculation is already close to the correct solution, or if it is simply garbage.

In order to quantify the error we must carry out what is usually known as a *convergence test.* The key idea behind convergence testing is the observation by Richardson [242] that the solution of a stable finite difference approximation can be interpreted as a continuum function that can be expanded as a power series the discretization parameter Δ

$$u_\Delta(t,x) = u(t,x) + \Delta\, e_1(t,x) + \Delta^2 e_2(t,x) + \cdots \,, \tag{9.11.1}$$

where $u(t,x)$ is the solution of the original differential equation, and the $e_i(t,x)$ are error functions at different orders in Δ. For a first order accurate approximation we expect $e_1 \neq 0$, for a second order approximation we expect $e_1 = 0$ and $e_2 \neq 0$, *etc.*

Let us for a moment assume that we know the exact solution to the problem (as in the case of the one-dimensional wave equation). To test the convergence of our finite difference approximation we perform a calculation at two different resolutions Δ_1 and Δ_2, with $r \equiv \Delta_1/\Delta_2 > 1$, and calculate the solution error in each case

$$\epsilon_{\Delta_1} = u - u_{\Delta_1} \,, \qquad \epsilon_{\Delta_2} = u - u_{\Delta_2} \,. \tag{9.11.2}$$

Notice that in practice this error can only be calculated at the points corresponding to the numerical grid under consideration. We then find the r.m.s. norm of each solution error and calculate the ratio of both norms

$$c(t) := \frac{||\epsilon_{\Delta_1}||}{||\epsilon_{\Delta_2}||} \,. \tag{9.11.3}$$

This ratio is clearly a function of time only, and is known as the *convergence factor* (again, it can only be calculated at times when both solution errors are defined). If we have a finite difference approximation of order n we can now use the Richardson expansion to find that in the continuum limit the convergence factor will become

$$\lim_{\Delta\to 0} c(t) = \left(\frac{\Delta_1}{\Delta_2}\right)^n \equiv r^n \,. \tag{9.11.4}$$

The convergence test is typically done taking $r = 2$, so that we expect $c(t) \sim 2^n$. In practice we perform the calculation at several different resolutions and looks at the behavior of $c(t)$ as the resolution increases. If it is close (or getting closer) to the expected value we say that we are in the *convergence regime.*

Of course, the above procedure will only work if we know the exact solution. This is not true in most cases (after all, if it were true, we would not be using a numerical approximation), so the best thing that we can hope for is to prove that the numerical solution converges to some continuum function.[101] In order to do this we need to use at least three different resolutions $\Delta_1 > \Delta_2 > \Delta_3$. We then calculate the relative errors between different resolutions and define the convergence factor as

$$c(t) := \frac{||u_{\Delta_1} - u_{\Delta_2}||}{||u_{\Delta_2} - u_{\Delta_3}||} , \tag{9.11.5}$$

where the differences between solutions must be calculated at the points where the grids coincide (alternatively we can interpolate the coarse solution into the finer grid, but we must remember that interpolation also introduces errors). In the continuum limit we then expect the convergence factor to behave as

$$\lim_{\Delta \to 0} c(t) = \frac{\Delta_1^n - \Delta_2^n}{\Delta_2^n - \Delta_3^n} . \tag{9.11.6}$$

It is usual to take $\Delta_1/\Delta_2 = \Delta_2/\Delta_3 \equiv r$, in which case this again reduces to

$$\lim_{\Delta \to 0} c(t) = r^n . \tag{9.11.7}$$

Because we are taking norms of errors, the convergence test that we have just described is known as a *global* convergence test. It is also possible to perform so-called *local* convergence tests by simply plotting the relative errors $u_{\Delta_1} - u_{\Delta_2}$ and $u_{\Delta_2} - u_{\Delta_3}$ directly as functions of position in order to compare them by eye. Since for an order n scheme the Richardson expansion implies that both these errors will be proportional to $e_n(t, x)$, we expect that if the relative errors are rescaled appropriately they should lie approximately one on top of the other. For example, if we have $\Delta_1/\Delta_2 = \Delta_2/\Delta_3 = r$, then after we multiply $u_{\Delta_2} - u_{\Delta_3}$ with r^n it should coincide (roughly) with $u_{\Delta_1} - u_{\Delta_2}$.

Convergence testing not only allows us to verify that the errors are indeed becoming smaller at the correct rate as the grid is refined, but it also allows us to estimate the remaining solution error. Assume for example that we have an order n scheme and two numerical calculations with resolutions Δ_1 and Δ_2, then the Richardson expansion implies that

$$\begin{aligned} u_{\Delta_1} - u_{\Delta_2} &= e_n \left(\Delta_1^n - \Delta_2^n\right) + \mathcal{O}(\Delta^{n+1}) \\ &= e_n \Delta_2^n \left(r^n - 1\right) + \mathcal{O}(\Delta^{n+1}) \\ &\sim \epsilon_{\Delta_2} \left(r^n - 1\right) . \end{aligned} \tag{9.11.8}$$

We can then estimate the solution error on our highest resolution grid as

$$\epsilon_{\Delta_2} \sim \frac{1}{r^n - 1} \left(u_{\Delta_1} - u_{\Delta_2}\right) . \tag{9.11.9}$$

This error estimate is accurate to at least order $n+1$, and often to order $n+2$ since for centered schemes there are only even powers of n in the Richardson expansion.

[101] In some cases we in fact do know the exact solution, at least in part. For example, in the 3+1 formalism we know that the constraints must vanish, so we can check the convergence of the constraints to zero.

This allows us to put error bars in a numerical simulation. For example, if we have a second order accurate scheme and two numerical solutions at different resolutions with $r = \Delta_1/\Delta_2 = 2$, then the solution error on the highest resolution will be roughly one-third ($2^2 - 1 = 3$) of the difference between the two numerical solutions (assuming that we are already in the convergence regime).

As a final comment, the Richardson expansion also allows us to improve our numerical solution by simply using the results of two or more simulations with different resolutions to estimate the exact solution. For example, once we have estimated the solution error in our highest resolution we can see that

$$u = u_{\Delta_2} - \epsilon_{\Delta_2} + \mathcal{O}(\Delta^{n+1}) \,, \tag{9.11.10}$$

so that

$$u = u_{\Delta_2} - \frac{1}{r^n - 1}\left(u_{\Delta_1} - u_{\Delta_2}\right) + \mathcal{O}(\Delta^{n+1}) \sim \frac{r^n u_{\Delta_2} - u_{\Delta_1}}{r^n - 1} \,. \tag{9.11.11}$$

We then have an estimate of the final solution that is at least one order higher than our original approximation. This estimate is known as the *Richardson extrapolation* and can improve our solution significantly provided that we start with an error that is already small.[102]

I will finish by stressing again the importance of convergence tests in any numerical calculation: We should simply never believe a numerical calculation for which no convergence studies have been made.[103]

[102] As with any type of extrapolation, Richardson extrapolation is not very good if the initial error is very large, in which case it might even make things worse. So it should be used with care.

[103] Boyd goes so far as to define an "idiot" as someone who publishes a numerical calculation without checking it against an identical calculation with a different resolution [75].

10

EXAMPLES OF NUMERICAL SPACETIMES

10.1 Introduction

In this Chapter we will discuss the application of the techniques of numerical relativity to some simple spacetimes. We will consider in turn three different types of example, starting from 1+1 "toy" relativity, and moving on to the case of spherical and axial symmetry. We will discuss some of the special problems that arise in each case, such as the issue of gauge shocks in the 1+1 case, and the regularization of the equations in both spherical and axial symmetry, and will also give examples of numerical simulations of simple physical systems, such as the evolution of Schwarzschild, the collapse of a scalar field, and the evolution of a Brill wave spacetime.

10.2 Toy 1+1 relativity

As our first example, let us consider vacuum general relativity in one spatial dimension. It is well known that in such a case the gravitational field is trivial and there are no true dynamics. This is because in a two-dimensional manifold the Riemann curvature only has one independent component, namely the Ricci scalar, and this component vanishes in vacuum. However, we can still have non-trivial gauge dynamics that can be used as a simple example of the type of behavior we can expect in the higher dimensional case.

In particular, we will be interested in studying the gauge dynamics produced by a slicing condition of the Bonna–Masso family, equation (4.2.52)

$$\partial_t \alpha = -\alpha^2 f(\alpha)\, K \, , \tag{10.2.1}$$

where for simplicity we have taken the shift vector to be zero. Notice that since we only have one spatial dimension, in this case the trace of the extrinsic curvature is simply given by $K = K^x_x$.[104]

The ADM evolution equations in the 1+1 case, together with the Bona–Masso slicing condition, can be written in first order form as

$$\partial_t \alpha = -\alpha^2 f K \, , \tag{10.2.2}$$

$$\partial_t g = -2\alpha g K \, , \tag{10.2.3}$$

and

[104]This means, in particular, that maximal slicing is trivial since setting $K = K^x_x = 0$ in vacuum immediately implies $\alpha = 1$, so that we do not even have gauge dynamics.

$$\partial_t D_\alpha + \partial_x (\alpha f K) = 0 \,, \tag{10.2.4}$$

$$\partial_t D_g + \partial_x (2\alpha K) = 0 \,, \tag{10.2.5}$$

$$\partial_t K + \partial_x (\alpha D_\alpha / g) = \alpha \left(K^2 - D_\alpha D_g / 2g \right) \,, \tag{10.2.6}$$

where we have defined $g := g_{xx}$, $D_\alpha := \partial_x \ln \alpha$ and $D_g := \partial_x \ln g$.

Before going any further, it is important to notice that the evolution equation for K can in fact be rewritten as a conservation law in the following way

$$\partial_t \left(g^{1/2} K \right) + \partial_x \left(\alpha D_\alpha / g^{1/2} \right) = 0 \,. \tag{10.2.7}$$

If we now define the vector $\vec{v} := (D_\alpha, D_g, \tilde{K})$, with $\tilde{K} := g^{1/2} K$, then the evolution equations for the first order variables can be written as a fully conservative system of the form

$$\partial_t \vec{v} + \partial_x (\mathbf{M}\, \vec{v}) = 0 \,, \tag{10.2.8}$$

with the characteristic matrix $\mathbf{M}$ given by:

$$\mathbf{M} = \begin{pmatrix} 0 & 0 & \alpha f / g^{1/2} \\ 0 & 0 & 2\alpha / g^{1/2} \\ \alpha / g^{1/2} & 0 & 0 \end{pmatrix} . \tag{10.2.9}$$

Now, in Chapter 5 it was shown that the ADM evolution equations are not strongly hyperbolic in the 3+1 case, even when using the Bona–Masso slicing condition. However, this is no longer true in the case of 1+1 and we find that the matrix $\mathbf{M}$ defined above has eigenvalues

$$\lambda_0 = 0 \,, \qquad \lambda_\pm = \pm \alpha \left(f/g \right)^{1/2} \,, \tag{10.2.10}$$

with corresponding eigenvectors

$$\vec{e}_0 = (0, 1, 0) \,, \qquad \vec{e}_\pm = \left(f, 2, \pm f^{1/2} \right) . \tag{10.2.11}$$

Since the eigenvalues are clearly real for $f > 0$, and the two eigenvectors are linearly independent, the system (10.2.8) is in fact strongly hyperbolic. The eigenfunctions of the system are defined by (see Chapter 5)

$$\vec{\omega} = \mathbf{R}^{-1} \vec{v} \,, \tag{10.2.12}$$

with $\mathbf{R}$ the matrix of column eigenvectors. Using an adequate choice of normalization we find that

$$\omega_0 = D_\alpha / f - D_g / 2 \,, \qquad \omega_\pm = \tilde{K} \pm D_\alpha / f^{1/2} \,, \tag{10.2.13}$$

which can be easily inverted to give

$$\tilde{K} = \frac{(\omega_+ + \omega_-)}{2} \,, \tag{10.2.14}$$

$$D_\alpha = \frac{f^{1/2}\,(\omega_+ - \omega_-)}{2} \,, \tag{10.2.15}$$

$$D_g = \frac{(\omega_+ - \omega_-)}{f^{1/2}} - 2\omega_0 \,. \tag{10.2.16}$$

It is important to notice that with the eigenfunctions scaled as above, their evolution equations also turn out to be conservative and have the simple form

$$\partial_t \vec{\omega} + \partial_x \left(\mathbf{\Lambda}\, \vec{\omega} \right) = 0 \,, \tag{10.2.17}$$

with $\mathbf{\Lambda} := \text{diag}\,\{\lambda_i\}$. If, however, the eigenfunctions are rescaled in the way $\omega_i' = F_i(\alpha, g)\,\omega_i$, then the evolution equations for the ω_i' will in general no longer be conservative, and non-trivial sources will be present. The important point is that there is in fact one normalization in which the equations are conservative.

10.2.1 *Gauge shocks*

Although the 1+1 evolution equations derived above are strongly hyperbolic and therefore well-posed, it turns out that in the general case they can give rise to singular solutions (*i.e.* gauge pathologies). These singular solutions are very similar to the shock waves that develop in hydrodynamics, and because of this they are known as *gauge shocks* [2, 3, 4, 19].

There are at least two different ways in which we can see how gauge shocks develop. Let us first consider the evolution equations for the traveling eigenfunctions $\omega_\pm$:

$$\partial_t \omega_\pm + \partial_x \left(\lambda_\pm \omega_\pm \right) = 0 \,. \tag{10.2.18}$$

We can rewrite these equations as

$$\partial_t \omega_\pm + \lambda_\pm \partial_x \omega_\pm = -\omega_\pm \partial_x \lambda_\pm \,. \tag{10.2.19}$$

Using now the expressions for $\lambda_\pm$, and denoting $f' \equiv df/d\alpha$, we find

$$\begin{aligned} \partial_x \lambda_\pm &= \mp \frac{\alpha f^{1/2}}{2 g^{3/2}}\, \partial_x g \pm \frac{f^{1/2}}{g^{1/2}} \left(1 + \frac{\alpha f'}{2f} \right) \partial_x \alpha \\ &= \lambda_\pm \left[\left(f + \frac{\alpha f'}{2} \right) \frac{D_\alpha}{f} - \frac{D_g}{2} \right] \\ &= \lambda_\pm \left[\left(f - 1 + \frac{\alpha f'}{2} \right) \frac{\omega_+ - \omega_-}{2 f^{1/2}} + \omega_0 \right] , \end{aligned} \tag{10.2.20}$$

and finally

$$\begin{aligned} &\partial_t \omega_\pm + \lambda_\pm \partial_x \omega_\pm = \\ &\qquad \lambda_\pm \omega_\pm \left[\left(1 - f - \frac{\alpha f'}{2} \right) \frac{\omega_+ - \omega_-}{2 f^{1/2}} - \omega_0 \right] . \end{aligned} \tag{10.2.21}$$

Assume now that we are in a region such that $\omega_0 = \omega_- = 0$. It is clear that ω_0 will not be excited since it does not evolve, nor will ω_- be excited since from the equation above we see that all its sources vanish. The evolution equation for ω_+ then simplifies to

$$\partial_t \omega_+ + \lambda_+ \partial_x \omega_+ = \frac{\lambda_+}{2f^{1/2}} \left(1 - f - \frac{\alpha f'}{2} \right) \omega_+^2 \, . \tag{10.2.22}$$

But this equation implies that ω_+ will blow up along its characteristics unless the term in parenthesis vanishes:

$$1 - f - \frac{\alpha f'}{2} = 0 \, . \tag{10.2.23}$$

We will call this the *shock avoiding condition.* This condition can in fact be easily integrated to give

$$f(\alpha) = 1 + \kappa/\alpha^2 \, , \tag{10.2.24}$$

with κ an arbitrary constant. Notice that harmonic slicing corresponds to $f = 1$ which is of this form, but $f = \text{const.} \neq 1$ is not. Notice also that 1+log slicing, for which $f = 2/\alpha$, though not of the form (10.2.24), nevertheless satisfies condition (10.2.23) when $\alpha = 1$ (we will come back to the case of 1+log slicing below).

In some cases it is even possible to predict exactly when a blow up will occur. In order to see this we first need to define the rescaled eigenfunctions $\Omega_\pm := \alpha \omega_\pm / g^{1/2}$. For their evolution equations we now find

$$\begin{aligned} \partial_t \Omega_\pm + \lambda_\pm \partial_x \Omega_\pm = & \left(1 - f - \frac{\alpha f'}{2} \right) \frac{\Omega_\pm^2}{2} \\ & + \left(1 - f + \frac{\alpha f'}{2} \right) \frac{\Omega_\pm \Omega_\mp}{2} \, . \end{aligned} \tag{10.2.25}$$

Notice that with this new scaling, all contributions from ω_0 to the sources have disappeared. If we now assume that we have initial data such that $\Omega_- = 0$, then the evolution equation for Ω_+ simplifies to

$$\partial_t \Omega_+ + \lambda_+ \partial_x \Omega_+ = \left(1 - f - \frac{\alpha f'}{2} \right) \frac{\Omega_+^2}{2} \, , \tag{10.2.26}$$

which can be rewritten as

$$\frac{d\Omega_+}{dt} = \left(1 - f - \frac{\alpha f'}{2} \right) \frac{\Omega_+^2}{2} \, , \tag{10.2.27}$$

with d/dt the derivative along the characteristic. The last equation has a very important property: For constant f the coefficient of the quadratic source term is itself also constant. In that case the equation can be easily integrated to find (assuming $f \neq 1$)

$$\Omega_+ = \frac{\Omega_+^0}{1 - (1 - f)\, \Omega_+^0 \, t/2} \, , \tag{10.2.28}$$

where $\Omega_+^0 \equiv \Omega_+(t = 0)$. The solution will then blow up at a finite time given by $t^* = 2/[(1 - f)\, \Omega_+^0]$. Clearly, this time will be in the future if $(1 - f)\, \Omega_+^0 > 0$, otherwise it will be in the past. Since, in general, Ω_+^0 will not be constant in space, the first blow-up will occur at the time

$$t^* = 2/[(1-f)\,\max(\Omega_+^0(x))] \,. \tag{10.2.29}$$

This shows that blow-ups can easily develop when we use the Bona–Masso slicing condition with a function $f(\alpha)$ such that (10.2.23) does not hold. Of course, this does not imply that such blow-ups will always develop, since their appearance will clearly depend on the specific form of the initial data being used, but it does show that we should not be terribly surprised if they do.

There is another way in which we can see how these blow-ups develop that provides us with a more direct geometrical interpretation of the problem. Consider now the evolution of the eigenspeeds themselves along their corresponding characteristic lines. From (10.2.10) we find that

$$\partial_t \lambda_\pm = \pm \partial_t \left(\alpha f^{1/2} / g^{1/2} \right) , \tag{10.2.30}$$

and using the evolution equations for α and g this reduces to

$$\begin{aligned} \partial_t \lambda_\pm &= \frac{\alpha \lambda_\pm}{g^{1/2}} \left(1 - f - \frac{\alpha f'}{2} \right) \tilde{K} \\ &= \frac{\alpha \lambda_\pm}{g^{1/2}} \left(1 - f - \frac{\alpha f'}{2} \right) \frac{(\omega_+ + \omega_-)}{2} \,. \end{aligned} \tag{10.2.31}$$

In a similar way we find for the spatial derivative

$$\begin{aligned} \partial_x \lambda_\pm &= \lambda_\pm \left[\frac{D_\alpha}{f} \left(f + \frac{\alpha f'}{2} \right) - \frac{D_g}{2} \right] \\ &= \lambda_\pm \left[\left(1 - f - \frac{\alpha f'}{2} \right) \frac{(\omega_- - \omega_+)}{2 f^{1/2}} + \omega_0 \right] . \end{aligned} \tag{10.2.32}$$

The last two equations together imply that

$$\partial_t \lambda_\pm + \lambda_\pm \partial_x \lambda_\pm = \frac{\alpha \lambda_\pm}{g^{1/2}} \left[\left(1 - f - \frac{\alpha f'}{2} \right) \omega_\mp \pm f^{1/2} \omega_0 \right] . \tag{10.2.33}$$

If we now consider again a region where $\omega_- = \omega_0 = 0$, the last equation reduces to

$$\partial_t \lambda_+ + \lambda_+ \partial_x \lambda_+ = 0 \,. \tag{10.2.34}$$

But this is nothing more than Burgers' equation, the prototype for studying shock formation (see Chapter 7). This shows that the blow-ups studied above correspond to places where the characteristic lines cross. At those points the spatial derivative of λ_+ becomes infinite, and since this derivative is given in terms of eigenfields, some of the eigenfields must blow up. In other words, the blow-ups associated with the Bona–Masso slicing condition develop precisely as shocks do, and it is because of this that we refer to them as gauge shocks.

At this point we could worry about a possible inconsistency in our analysis. The fact that we obtain Burgers' equation independently of the value of the function $f(\alpha)$ would seem to indicate that gauge shocks should develop even if the shock avoiding condition (10.2.23) is satisfied. But this not the case since from the derivation above we can easily see that if (10.2.23) is satisfied, and we are in a region with $\omega_0 = 0$, then $\partial_t \lambda_+ = \partial_x \lambda_+ = 0$. So Burgers' equation, though formally still correct, becomes trivial.

A very important question is what happens to the geometry of the spacetime being evolved when a gauge shock develops. Since we are simply considering the evolution of a non-trivial foliation of Minkowski spacetime, it is clear that the background geometry remains perfectly regular. Therefore, the only thing that can become pathological are the spatial hypersurfaces that determine the foliation. As the numerical examples of the following Section will show, the formation of a gauge shock indicates that the spatial hypersurface has in fact become non-smooth (it has developed a kink).

10.2.2 *Approximate shock avoidance*

Before showing some numerical examples of the formation of gauge shocks, let us go back to the issue of what form of the function $f(\alpha)$ we should use in order to avoid gauge shocks. As already mentioned, the shock avoiding condition (10.2.23) can be trivially integrated and we find that in order to avoid gauge shocks it is necessary to choose $f(\alpha)$ of the form (10.2.24). In fact, in that expression we must take $\kappa > 0$, to be sure that f remains positive for small values of α, as otherwise the system of evolution equations would stop being hyperbolic.

Now, for $\kappa > 0$ and small α we find that $f \sim 1/\alpha^2$, so from our discussion on singularity avoidance in Chapter 4 we see that the slicing will be strongly singularity avoiding. However, we can also see that in this case we are in the regime for which the lapse can easily become negative. This means that the solution (10.2.24) has a serious drawback for any non-zero value of κ, since it can allow the lapse to become negative as it collapses to zero. This would seem to imply that the requirements of avoiding gauge shocks while at the same time guaranteeing that the lapse will not become negative are mutually incompatible, except in the particular case of harmonic slicing $f = 1$. On the other hand, harmonic slicing is only marginally singularity avoiding and is therefore not desirable when evolving black hole spacetimes (unless we are interested in studying the singularity itself).

In order to look for a useful slicing condition we will therefore relax our requirements and look for approximate solutions of the shock avoiding condition (10.2.23). We start by assuming that the lapse is very close to 1, that is

$$\alpha = 1 + \epsilon \,, \tag{10.2.35}$$

with $\epsilon \ll 1$. Notice that the limit above applies to situations that are close to flat space, but generally does not apply to strong field regions (like the region close to a black hole) where the lapse can be expected to be very different from 1. However, in such regions considerations about singularity avoidance are probably more important. Our aim is to find slicing conditions that can avoid singularities in strong field regions, and at the same time do not have a tendency to generate shocks in weak field regions.

We can now expand f in terms of ϵ as

$$f = a_0 + a_1 \epsilon + \mathcal{O}(\epsilon^2) = a_0 + a_1(\alpha - 1) + \mathcal{O}(\epsilon^2) \,, \tag{10.2.36}$$

and look for solutions to (10.2.23) to lowest order in ϵ. Substituting (10.2.36) into (10.2.23) we find

$$1 - a_0 - a_1/2 + \mathcal{O}(\epsilon) = 0 \,. \tag{10.2.37}$$

This means that if we want condition (10.2.23) to be satisfied to lowest order in ϵ we must ask for

$$a_1 = 2\,(1 - a_0) \,, \tag{10.2.38}$$

which implies

$$\begin{aligned} f &= a_0 + 2\,(1 - a_0)\,\epsilon + \mathcal{O}(\epsilon^2) \\ &= (3a_0 - 2) + 2\,(1 - a_0)\,\alpha + \mathcal{O}(\epsilon^2) \,. \end{aligned} \tag{10.2.39}$$

Here we must remember that (10.2.39) is just an expansion for small ϵ. Any form of the function $f(\alpha)$ that has the same expansion to first order in ϵ will also satisfy condition (10.2.23) to lowest order. For example, one family of such functions emerges if we ask for $f(\alpha)$ to have the form

$$f = \frac{p_0}{1 + q_1 \epsilon} \,. \tag{10.2.40}$$

It is not difficult to show that for this to have an expansion of the form (10.2.39) we must ask for

$$p_0 = a_0 \,, \qquad q_1 = 2\,(a_0 - 1)\,/a_0 \,, \tag{10.2.41}$$

which implies

$$f = \frac{a_0^2}{a_0 + 2\,(a_0 - 1)\,\epsilon} = \frac{a_0^2}{(2 - a_0) + 2\,(a_0 - 1)\,\alpha} \,. \tag{10.2.42}$$

Notice that if we take $a_0 = 1$ we recover harmonic slicing. But there is one other case that is of special interest: For $a_0 = 2$ the previous solution reduces to

$$f = 2/\alpha \,, \tag{10.2.43}$$

which corresponds to a member of the 1+log family. The crucial observation here is that, as already mentioned in Chapter 4, this specific member of the 1+log family is precisely the one that has been found empirically to be very robust in black hole simulations, and because of this it is considered as the *standard* 1+log slicing. The fact that it is the only member of the 1+log family that satisfies condition (10.2.23), even approximately, means that we should expect it to be particularly well behaved.

Notice, however, that standard 1+log slicing avoids gauge shocks only if the lapse function α is close to 1, so we should not be surprised to find that 1+log develops gauge shocks when the lapse has collapsed to values close to 0, as is typical in regions close to a black hole. Moreover, the analysis presented here applies only to the simple 1+1-dimensional case, and while gauge shocks will certainly still exist in the higher dimensional case (since after all 1+1 is just a special case of 3+1), we could easily expect other potential gauge pathologies to appear. Nevertheless, empirically it has been found that standard 1+log is (perhaps surprisingly) extremely robust in black hole simulations, and because of this it has become the *de facto* standard slicing condition when studying, for example, inspiral black hole collisions. Still, there is evidence that, at least in some cases, black hole simulations using a 1+log slicing condition do develop gauge problems [141], but whether such problems are gauge shocks in the sense discussed here (*i.e.* singularities associated with the crossing of characteristics) or some different type of gauge pathology still remains an open question.

10.2.3 *Numerical examples*

We will now show some examples of numerical simulations of a 1+1 vacuum spacetime to see how gauge shocks can appear very easily, and how the shock avoiding condition derived in the previous Section does indeed eliminate them.

We start by describing the initial data used in the simulations. Since the argument presented in the previous Section that allows us to predict the precise time of formation of a gauge shock depends on having only a mode propagating in a specific direction, we will be interested in constructing initial data corresponding to such a purely traveling mode.

Notice that the simplest way to construct initial data is to just take a trivial initial slice in Minkowski coordinates, and use a non-trivial initial lapse function to introduce a later distortion. However, since in this case the initial extrinsic curvature vanishes, we can see from (10.2.13) that, inevitably, modes traveling in both directions will be excited. We then conclude that in order to have waves propagating in only one direction we are forced to consider a non-trivial initial slice.

The initial data considered here has already been discussed in [2], and is constructed as follows: We start by considering an initial slice given in Minkowski coordinates (t_M, x_M) as

$$t_M = h(x_M) , \tag{10.2.44}$$

with h some profile function that decays rapidly (*e.g.* a Gaussian function). It is then not difficult to show that if we keep $x = x_M$ as our spatial coordinate, the spatial metric and extrinsic curvature turn out to be

$$g = 1 - h'^2 \quad \Rightarrow \quad D_g = -2h'h''/g \, , \tag{10.2.45}$$

$$K_{xx} = -h''/\sqrt{g} \quad \Rightarrow \quad \tilde{K} = -h''/g \, . \tag{10.2.46}$$

Assume now that we want to have waves traveling to the right (the opposite situation can be done in an analogous way). This means that we want $\omega_- = 0$, which in turn implies that

$$D_\alpha = \sqrt{f}\,\tilde{K} = -\sqrt{f}\,h''/g \, . \tag{10.2.47}$$

This results in the following differential equation for α:

$$\partial_x \alpha = -\alpha\sqrt{f}\,h''/\left(1 - h'^2\right) \, . \tag{10.2.48}$$

In the particular case when f is a constant the above equation can be easily integrated to find

$$\alpha = \left(\frac{1 - h'}{1 + h'}\right)^{\sqrt{f}/2} \, . \tag{10.2.49}$$

From this we can reconstruct the rescaled eigenfunction Ω_+, and use equation (10.2.29) to predict the time of shock formation given a specific form of $h(x)$. In all the simulations shown here, the profile function h has been chosen to be a simple Gaussian of the form

$$h(x) = e^{-x^2}/10 \, . \tag{10.2.50}$$

But do notice that even if we keep $h(x)$ always the same, changing the value of f will change the initial lapse (10.2.49).

The evolution code used for the simulations is based on a method of lines integration with three-step iterative Crank–Nicholson. For the spatial differentiation the code uses a slope limiter method of van Leer's type (see Chapter 9). No artificial dissipation is introduced. Since the slope limiter method requires information about the direction of propagation, the code evolves the eigenfields $(\omega_0, \omega_\pm)$ directly, and the variables $(D_\alpha, D_g, \tilde{K})$ are later reconstructed from these.

Consider first an evolution with $f = 0.5$. Figure 10.1 shows the numerical evolution of the eigenfield ω_+ and the lapse function α, using a resolution of $\Delta x = 0.003125$ and a Courant factor of $\Delta t/\Delta x = 0.5$. For the initial data used here, according to (10.2.29) a blow-up is expected at time $T^* \simeq 9.98$. The plot shows both the initial data (dotted lines) and the numerical solution at time $t = 10$ (solid lines), just after the expected blow-up. We clearly see how the eigenfield ω_+ has developed a large spike at the center of the pulse, while the lapse has developed a corresponding sharp gradient (the numerical solution for ω_+ does not in fact become infinite because the numerical method has some inherent dissipation). Notice also that after $t = 10$ the pulse has propagated a distance of ~ 7, in accordance with the fact that the characteristic speed in this case is $\lambda = \alpha\sqrt{f/g} \sim \sqrt{f} \sim 7.07$.

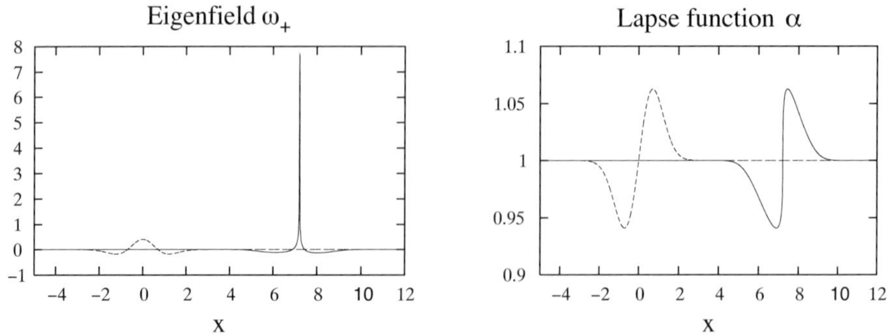

Fig. 10.1: Evolution of the eigenfield ω_+ and the lapse function α for the case where $f = 0.5$. The dotted lines show the initial data, and the solid lines the solution at $t = 10$, just after the expected blow-up.

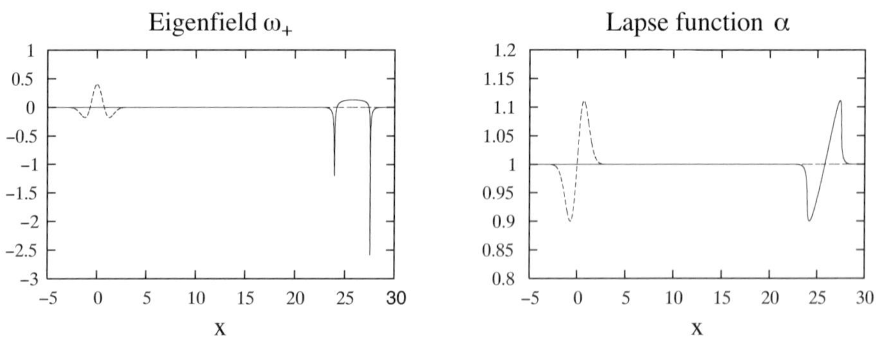

Fig. 10.2: Evolution of the eigenfield ω_+ and the lapse function α for the case where $f = 1.5$. The dotted lines show the initial data, and the solid lines the solution at $t = 21$, just after the expected blow-up.

Next, consider the case where $f = 1.5$. For the same profile function $h(x)$ as before, equation (10.2.29) now predicts a blow-up at time $T^* = 20.84$. Figure 10.2 shows the numerical evolution of ω_+ and α in this case, using the same resolution and Courant parameter as before. The plot again shows both the initial data (dotted lines), and the numerical solution at time $t = 21$ (solid lines), just after the expected blow-up. Notice that ω_+ has now developed two large spikes at the front and back of the pulse, with corresponding sharp gradients developing in the lapse function α. The sign of the spikes in ω_+ has also been reversed. Notice also that in this case the pulse has traveled a distance of ~ 25 after $t = 21$, corresponding to a characteristic speed of $\lambda \sim \sqrt{f} \sim 1.22$.

We then see that for both cases with $f = \text{const.} \neq 1$, gauge shocks develop as expected. For completeness, Figure 10.3 shows a simulation of the same initial data, but now using harmonic slicing corresponding to $f = 1$. The plot shows

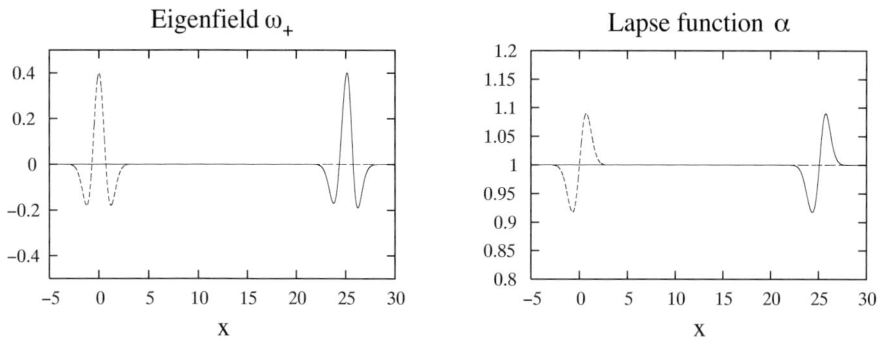

Fig. 10.3: Evolution of the eigenfield ω_+ and the lapse function α for the case of harmonic slicing, corresponding to $f = 1.0$. The dotted lines show the initial data, and the solid lines the solution at $t = 25$.

the solution after $t = 25$. One can clearly see that the pulse now propagates with a speed ~ 1, maintaining its initial profile so that no shock forms.

A more interesting example corresponds to the case where we choose the function $f(\alpha)$ as a member of the shock avoiding family (10.2.24): $f = 1 + \kappa/\alpha^2$. However, since f is now not constant, if we want to have a purely right-propagating pulse we can not take a lapse of the form (10.2.49) anymore. Fortunately, it turns out that in this case we can still integrate equation (10.2.48) exactly. One finds

$$\alpha = \frac{1}{2}\left[C\sqrt{\frac{1-h'}{1+h'}} - \frac{\kappa}{C}\sqrt{\frac{1+h'}{1-h'}}\right] , \tag{10.2.51}$$

with C an integration constant. If we now ask for the lapse to become 1 far away, then this constant must take the value $C = 1 + \sqrt{1+\kappa}$.

Figure 10.4 shows a simulation for a shock avoiding slicing with $\kappa = 0.5$. Since $\alpha \sim 1$, in this case we expect the pulse to propagate at a speed of $\lambda \sim \sqrt{f} = \sqrt{1+\kappa} \sim 1.22$, *i.e.* essentially the same speed as in the case $f = 1.5$. From the figure we see that the pulse propagates maintaining its initial profile, showing that this form of f does indeed avoid the formation of gauge shocks.

At this point we could worry about the fact that even if the simulations show sharp gradients and large spikes, this does not in itself prove that a true singularity has developed at the predicted time. In fact, since the numerical method has some inherent dissipation, we would not expect to see the true blow-up numerically (moreover, the slope limiter method is designed precisely so that such large gradients can be handled without problems). However, there is at least one way to show numerically that a true blow-up is developing at a specific time: Since after the blow-up the differential equation itself makes no sense anymore, we should not expect the numerical solution to converge after this time.

In order to see this we can consider, for example, the convergence of the constraint $C_\alpha := D_\alpha - \partial_x \ln \alpha$. Numerically this constraint will not vanish, but

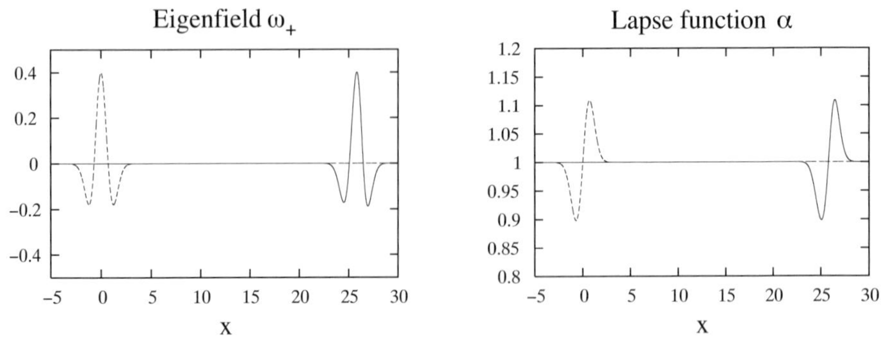

Fig. 10.4: Evolution of the eigenfield ω_+ and the lapse function α for the case of a shock avoiding slicing of the form (10.2.24) with $\kappa = 0.5$. The dotted lines show the initial data, and the solid lines the solution at $t = 21$.

it should converge to zero as the resolution is increased. Define now the convergence factor as the ratio of the r.m.s norm of C_α for a run at a given resolution and another run at twice the resolution. Before the blow-up this convergence factor should approach 4 (since the evolution code is second order accurate), but after the blow-up it should drop below 1, indicating loss of convergence. As an example of this, let us consider again the simulation with $f = 0.5$ discussed above. Figure 10.5 shows a plot of the convergence factors as a function of time using five different resolutions: $\Delta x = 0.05, 0.025, 0.0125, 0.00625, 0.003125$. The figure shows that as the resolution is increased the convergence factors approach the expected value of 4 for $t < 10$ (remember that in this case a shock is expected at $T^* \simeq 9.98$), but after that time they drop below 1, as expected. Moreover, as the resolution is increased the convergence factor resembles more and more a step function, corresponding to second order convergence before the blow-up, and no convergence afterwards. The behavior of the convergence factor is very similar for the case with $f = 1.5$, with the loss of convergence now centered around $t \sim 21$, as expected.

Before leaving this Section, there is one final point that should be addressed. As we have already mentioned, since we are evolving a foliation of Minkowski spacetime it is clear that the background geometry remains perfectly regular, so when a gauge shock develops the only thing that can become pathological are the spatial hypersurfaces that define the foliation. Figure 10.6 shows a comparison of the initial hypersurface and the final hypersurface at $t = 10$, as seen in Minkowski spacetime, using again the data from the same numerical simulation with $f = 0.5$ discussed above (for an easier comparison the final slice has been moved back in time so that it lies on top of the initial slice at the boundaries). The hypersurfaces are reconstructed by explicitly keeping track of the position of the normal observers in the original Minkowski coordinates during the evolution. Notice how the initial slice is very smooth (it has a Gaussian profile), while the

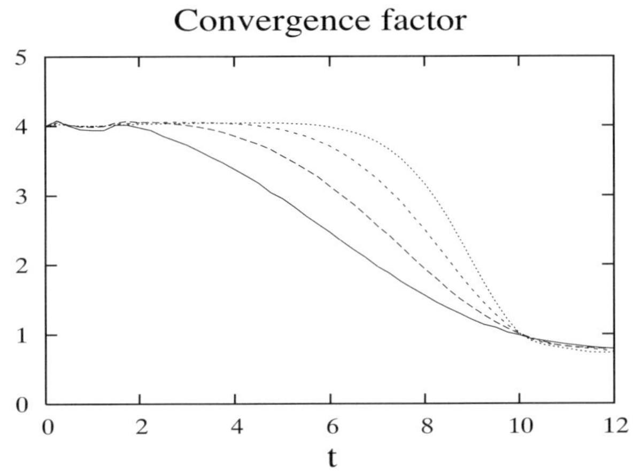

Fig. 10.5: Convergence of the constraint C_α for the simulation with $f = 0.5$ done at the five different resolutions: $\Delta x = 0.05, 0.025, 0.0125, 0.00625, 0.003125$. As the resolution increases, the convergence factors approach the expected value of 4 for $t < 10$, but after that time they drop sharply indicating loss of convergence.

final slice has developed a sharp kink. This shows that the formation of a gauge shock indicates that the hypersurface, though still spacelike everywhere, is no longer smooth (its derivative is now discontinuous).

10.3 Spherical symmetry

As our second example we will consider the case of spherically symmetric spacetimes – a case that, though still relatively simple, nevertheless has many important physical applications. This case is interesting for several reasons. In the first place, even though it already represents a full three-dimensional spacetime (as opposed to the 1+1 case studied in the previous sections), the dynamical variables depend only on the radial coordinate, so the numerical codes are still only one-dimensional. Because of this it is standard in numerical relativity to refer codes for spherically symmetric spacetimes as "one-dimensional" (1D) codes. On the other hand, many interesting astrophysical systems, in particular stars and black holes, are spherically symmetric to a very good approximation, so that spherical symmetry already allows us to study interesting physical problems such as the stability of stars, gravitational collapse, horizon dynamics, *etc.* Perhaps the main drawback of spherical symmetry is the fact that in such a case there are no gravitational waves, so we are missing one key element of general relativity. Still, it was precisely in spherical symmetry that the first major contribution of numerical relativity to the understanding of general relativity was made: the discovery of critical phenomena in gravitational collapse by Choptuik [98].

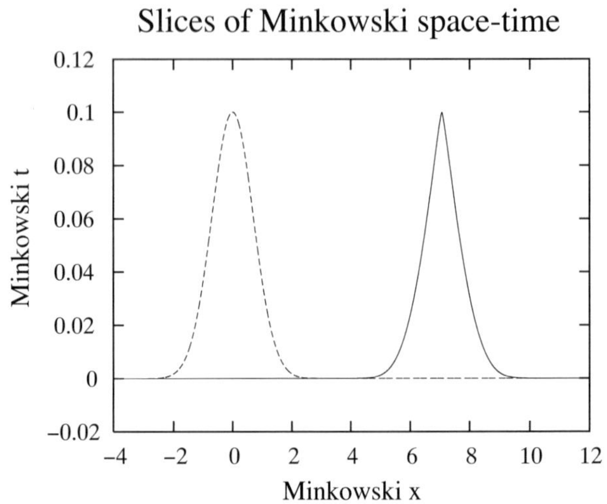

Fig. 10.6: Initial hypersurface (dotted line) and final hypersurface at $t = 10$ (solid line), as seen in Minkowski coordinates, for the simulation with $f = 0.5$ discussed in the text. The final slice has been moved back in time so that it lies on top of the initial slice at the boundaries. The initial slice is smooth, while the final slice has a sharp kink.

10.3.1 *Regularization*

Let us start by writing the general form of the spatial metric in spherical symmetry as

$$dl^2 = A(r,t)dr^2 + r^2 B(r,t)\, d\Omega^2 \, , \tag{10.3.1}$$

with A and B positive metric functions, and $d\Omega^2 = d\theta^2 + \sin^2\theta d\varphi^2$ the solid angle element. Notice that we have already factored out the r^2 dependency from the angular metric functions. This has the advantage of making explicit the dependency on r of these geometric quantities, and making the regularization of the resulting equations easier.

As we will deal with the Einstein equations in first order form, we will introduce the auxiliary quantities:

$$D_A := \partial_r \ln A \, , \qquad D_B := \partial_r \ln B \, . \tag{10.3.2}$$

In order to simplify the equations, we will also work with the mixed components of the extrinsic curvature, $K_A := K^r_r$, $K_B := K^\theta_\theta = K^\varphi_\varphi$.

Before writing down the 3+1 evolution equations in spherical symmetry, we must first remember that the spherical coordinates introduced above are in fact singular at the origin, and this coordinate singularity can be the source of serious numerical problems caused by the lack of regularity of the geometric variables there. The problem arises because of the presence of terms in the evolution equations that go as $1/r$ near the origin. The regularity of the metric guarantees the exact cancellations of such terms at the origin, thus ensuring well behaved solutions. This exact cancellation, however, though certainly true for analytical solutions, usually fails to hold for numerical solutions. One then finds that the $1/r$ terms do not cancel and the numerical solution becomes ill-behaved near $r = 0$: It not only fails to converge at the origin, but it can also easily turn out to be violently unstable after just a few time steps.

The usual way to deal with the regularity problem is to use the so-called *areal* (or *radial*) gauge, where the radial coordinate r is chosen in such a way that the proper area of spheres of constant r is always $4\pi r^2$ (so that $B = 1$ during the whole evolution). If, moreover, we also choose a vanishing shift vector we end up in the so-called *polar-areal* gauge [48, 97], for which the lapse is forced to satisfy a certain ordinary differential equation in r. The name *polar* comes from the fact that, for this gauge choice, there is only one non-zero component of the extrinsic curvature tensor, namely K_{rr} [48]. In the polar-areal gauge the problem of achieving the exact cancellation of the $1/r$ terms is reduced to imposing the boundary condition $A = 1$ at $r = 0$, which can be easily done if we solve for A directly from the Hamiltonian constraint (which in this case reduces to an ordinary differential equation in r), and ignore its evolution equation. If we do this in vacuum, we end up inevitably with Minkowski spacetime in the usual coordinates (we can also recover Schwarzschild by working in isotropic coordinates and factoring out the conformal factor analytically).

The main drawback of the standard approach is that the gauge choice has been completely exhausted. In particular, the polar-areal gauge can not penetrate apparent horizons, since inside an apparent horizon it is impossible to keep the areas of spheres fixed without a non-trivial shift vector.[105] One would then like to have a way of dealing with the regularity issue that allows more generic gauge choices to be made. Because of this I will discuss here a more general regularization procedure originally introduced in [18, 29].

There are in fact two different types of regularity conditions that the variables $\{A, B, D_A, D_B, K_A, K_B\}$ must satisfy at $r = 0$. The first type of conditions are simply those imposed by the requirement that the different variables should be well defined at the origin, and imply the following behavior for small r:

$$\begin{aligned} A &\sim A^0 + \mathcal{O}(r^2)\,, & B &\sim B^0 + \mathcal{O}(r^2)\,, \\ D_A &\sim \mathcal{O}(r)\,, & D_B &\sim \mathcal{O}(r)\,, \\ K_A &\sim K_A^0 + \mathcal{O}(r^2)\,, & K_B &\sim K_B^0 + \mathcal{O}(r^2)\,, \end{aligned}$$

with $\{A^0, B^0, K_A^0, K_B^0\}$ in general functions of time. These regularity conditions are in fact quite easy to implement numerically. For example, we can use a finite differencing grid that staggers the origin, and then obtain boundary data on a fictitious point at $r = -\Delta r/2$ by demanding for $\{A, B, K_A, K_B\}$ to be even functions at $r = 0$, and for $\{D_A, D_B\}$ to be odd.

[105] The polar-areal gauge has in fact been used successfully in many situations, particularly in the study of critical collapse to a black hole, where the presence of the black hole is identified by the familiar "collapse of the lapse" even if no apparent horizon can be found [98].

The second type of regularity conditions is considerably more troublesome. In order to see the problem, we will first write the ADM evolution equations in first order form for the case of spherical symmetry, which for vanishing shift take the form

$$\partial_t A = -2\alpha A K_A \,, \tag{10.3.3}$$

$$\partial_t B = -2\alpha B K_B \,, \tag{10.3.4}$$

$$\partial_t D_A = -2\alpha [K_A D_\alpha + \partial_r K_A] \,, \tag{10.3.5}$$

$$\partial_t D_B = -2\alpha [K_B D_\alpha + \partial_r K_B] \,, \tag{10.3.6}$$

$$\begin{aligned} \partial_t K_A = & -\frac{\alpha}{A} \left[\partial_r (D_\alpha + D_B) + D_\alpha^2 - \frac{D_\alpha D_A}{2} + \frac{D_B^2}{2} - \frac{D_A D_B}{2} \right. \\ & \left. - A K_A (K_A + 2K_B) - \frac{1}{r}(D_A - 2D_B) \right] + 4\pi\alpha M_A \,, \end{aligned} \tag{10.3.7}$$

$$\begin{aligned} \partial_t K_B = & -\frac{\alpha}{2A} \left[\partial_r D_B + D_\alpha D_B + D_B^2 - \frac{D_A D_B}{2} - \frac{1}{r}(D_A - 2D_\alpha - 4D_B) \right. \\ & \left. - \frac{2\,(A - B)}{r^2 B} \right] + \alpha K_B (K_A + 2K_B) + 4\pi\alpha M_B \,, \end{aligned} \tag{10.3.8}$$

where, as before, we have introduced $D_\alpha := \partial_r \ln \alpha$, and where M_A and M_B are matter terms given by

$$M_A = 2S_B - S_A - \rho \,, \qquad M_B = S_A - \rho \,, \tag{10.3.9}$$

with ρ the energy density of the matter, S_{ij} the stress tensor, and where we have defined $S_A := S^r_r$, $S_B := S^\theta_\theta$. On the other hand, the Hamiltonian and momentum constraints take the form

$$\begin{aligned} H = & -\partial_r D_B + \frac{1}{r^2 B}\,(A - B) + A K_B\,(2K_A + K_B) \\ & + \frac{1}{r}\,(D_A - 3D_B) + \frac{D_A D_B}{2} - \frac{3D_B^2}{4} - 8\pi A \rho = 0 \,, \end{aligned} \tag{10.3.10}$$

$$M = -\partial_r K_B + (K_A - K_B) \left[\frac{1}{r} + \frac{D_B}{2} \right] - 4\pi j_A = 0 \,, \tag{10.3.11}$$

where $j_A := j_r$ is just the momentum density of matter in the radial direction.

Since $\{D_\alpha, D_A, D_B\}$ go as r near the origin, terms of the type D/r are in fact regular and represent no problem. However, both in the Hamiltonian constraint, and in the evolution equation for K_B there is a term of the form $(A - B)/r^2$, while in the momentum constraint there is a term of the form $(K_A - K_B)/r$, and, given the behavior of these variables near the origin, these terms would seem to blow up. The reason why this does not in fact happen is that, near the origin, we must also ask for the extra regularity conditions

$$A - B \sim \mathcal{O}(r^2) \,, \qquad K_A - K_B \sim \mathcal{O}(r^2) \,, \tag{10.3.12}$$

or in other words $A^0 = B^0$, $K_A^0 = K_B^0$. These conditions arise as a consequence of the fact that space must remain locally flat at $r = 0$. The local flatness condition implies that near $r = 0$ it must be possible to write the metric as

$$dl^2{}_{R\sim 0} = dR^2 + R^2 d\Omega^2 \; , \tag{10.3.13}$$

with R a radial coordinate that measures proper distance from the origin. A local transformation of coordinates from R to r takes the metric into the form

$$dl^2{}_{r\sim 0} = \left(\frac{dR}{dr} \right)^2_{r=0} \left(dr^2 + r^2 d\Omega^2 \right) \; , \tag{10.3.14}$$

which implies that $A^0 = B^0$ and, since this must hold for all time, also that $K_A^0 = K_B^0$.

Implementing numerically both the symmetry regularity conditions and the local flatness regularity conditions at the same time is not entirely trivial. The reason for this is that at $r = 0$ we now have three boundary conditions for just two variables, *i.e.* both the derivatives of A and B must vanish, plus A and B must be equal to each other (and similarly for K_A and K_B). The boundary conditions for the exact equations are also over-determined, but in that case the consistency of the equations implies that if they are satisfied initially they remain satisfied for all time. In the numerical case, however, this is not true owing to truncation errors, and very rapidly one of the three boundary conditions fails to hold. Notice also that, from the above equations, we can easily see why the polar-areal gauge has no serious regularity problem. In that gauge we have $B = 1$ by construction. If we now impose the boundary condition $A(r = 0) = 1$, and solve for $A(r)$ by integrating the Hamiltonian constraint (ignoring the evolution equations), then the $(A - B)/r^2$ term causes no trouble.

There are several different ways in which we can solve the regularity problem. When we discuss axial symmetry in the following sections we will introduce a very general way to deal with the regularity issue, but here we will follow a somewhat simpler approach. We start by introducing an auxiliary variable defined as [18, 29]

$$\lambda := \frac{1}{r} \left(1 - \frac{A}{B} \right) . \tag{10.3.15}$$

Local flatness then implies that the variable λ has the following behavior near the origin

$$\lambda \sim \mathcal{O}(r) \, . \tag{10.3.16}$$

Just as before, this condition on λ can also be easily imposed numerically by using a grid that staggers the origin, and asking for λ to be odd across $r = 0$. In this way λ serves the dual purpose of absorbing the ill-behaved terms in our equations and solving the problem of the over-determined boundary conditions:

The extra boundary condition is imposed on λ itself. In terms of λ, the evolution equation for K_B takes the form

$$\begin{aligned}\partial_t K_B = & -\frac{\alpha}{2A}\left[\partial_r D_B + D_\alpha D_B + D_B^2 - \frac{D_A D_B}{2} - \frac{1}{r}(D_A - 2D_\alpha - 4D_B)\right. \\ & \left. + \frac{2\lambda}{r}\right] + \alpha K_B(K_A + 2K_B) + 4\pi\alpha M_B \,,\end{aligned} \tag{10.3.17}$$

while the Hamiltonian constraint becomes

$$\begin{aligned}H = & -\partial_r D_B - \frac{\lambda}{r} + AK_B\left(2K_A + K_B\right) \\ & + \frac{1}{r}\left(D_A - 3D_B\right) + \frac{D_A D_B}{2} - \frac{3D_B^2}{4} - 8\pi A\rho \,.\end{aligned} \tag{10.3.18}$$

Now, if λ is considered as a new independent degree of freedom we still need to add an evolution equation for it. Such an evolution equation can be obtained directly from the definition of λ:

$$\partial_t \lambda = \frac{2\alpha A}{B}\left(\frac{K_A - K_B}{r}\right) \,. \tag{10.3.19}$$

Unfortunately, this last equation presents us with a new problem since it clearly has the dangerous term $(K_A - K_B)/r$, but this term can be removed with the help of the momentum constraint (10.3.11) to find

$$\partial_t \lambda = \frac{2\alpha A}{B}\left[\partial_r K_B - \frac{D_B}{2}(K_A - K_B) + 4\pi j_A\right] \,, \tag{10.3.20}$$

which is now regular at the origin.

The regularized ADM evolution equations are then given by (10.3.5)–(10.3.7), with (10.3.8) replaced by (10.3.17), plus the evolution equation for λ given by (10.3.20).

10.3.2 *Hyperbolicity*

Having found a regular version of the ADM evolution equations, the next step is to construct a strongly hyperbolic evolution system. As we have already discussed in Chapter 5, the 3+1 ADM evolution equations are generally only weakly hyperbolic and are therefore not well-posed. However, in our discussion on 1+1 we mentioned the fact that, in that simple case, the ADM equations are in fact strongly hyperbolic. It turns out that in the case of spherical symmetry something similar happens, and the ADM equations are already strongly hyperbolic in most cases, though not in all, and the one exception is so important that the system must still be modified.

It is clear that for the hyperbolicity analysis we still need to say something about the lapse function, and as usual we will choose the Bona–Masso slicing condition

$$\partial_t \alpha = -\alpha^2 f(\alpha) K = -\alpha^2 f(\alpha)(K_A + 2K_B) , \qquad (10.3.21)$$

which implies

$$\partial_t D_\alpha = -\partial_r \left[\alpha f(\alpha)(K_A + 2K_B) \right] . \qquad (10.3.22)$$

Consider then the evolution system for the variables $(D_\alpha, D_A, D_B, K_A, K_B)$ given by equations (10.3.22) and (10.3.5)–(10.3.8). Notice that for the hyperbolicity analysis of this system the regularity issue is not relevant.

One finds that in general the system as it stands is strongly hyperbolic, with the following characteristic structure:

- There is one eigenfield with eigenspeed $\lambda = 0$ given by

$$w_0 = D_\alpha / f - (D_A + 2D_B)/2 . \qquad (10.3.23)$$

- There are two eigenfields with eigenspeeds $\lambda = \pm\alpha/\sqrt{A}$ (*i.e.* the coordinate speed of light) given by

$$w_\pm^l = A^{1/2} K_B \mp D_B/2 . \qquad (10.3.24)$$

- Finally, there are two eigenfields with eigenspeeds $\lambda = \pm\alpha\sqrt{f/A}$ (gauge speeds) given by

$$w_\pm^f = A^{1/2} \left(K_A + 2\,\frac{f+1}{f-1}\,K_B \right) \mp \left(\frac{D_\alpha}{f^{1/2}} + 2\,\frac{D_B}{f-1} \right) . \qquad (10.3.25)$$

However, as can be clearly seen from the above expressions, there is one particular case where hyperbolicity fails. If we choose harmonic slicing corresponding to $f = 1$, then the eigenfields $w_\pm^f$ become ill-defined. In fact, by multiplying $w_\pm^f$ with $f - 1$, we can see that for $f = 1$ the eigenfields $w_\pm^f$ and $w_\pm^l$ become proportional to each other, so that we do not have a complete set anymore and the system is only weakly hyperbolic. Since harmonic slicing is such an important gauge condition, having a system that fails to be strongly hyperbolic in that case is clearly unacceptable.

There are many different ways in which we can modify the evolution equations to obtain a strongly hyperbolic system for all $f > 0$. For example, one possibility would be to use the BSSNOK system adapted to the special case of spherical symmetry (this would probably be a very good choice, particularly when dealing with black hole spacetimes). However, for simplicity we will consider here a much simpler alternative. We start by making a change of variables, so that instead of using D_A and K_A as fundamental variables, we take $\widetilde{D} = D_A - 2D_B$ and $K = K_A + 2K_B$. We then rewrite the evolution equations in terms of the new variables, and use the Hamiltonian and momentum constraints to eliminate the terms proportional to $\partial_r D_B$ and $\partial_r K_B$ from the evolution equations of K and $\widetilde{D}$ respectively. Doing this we find

$$\partial_t \widetilde{D} = -2\alpha \left[\partial_r K + D_\alpha \left(K - 4K_B \right) \right. \\ \left. - 4\left(K - 3K_B \right) \left(\frac{1}{r} + \frac{D_B}{2} \right) + 16\pi j_A \right] , \tag{10.3.26}$$

$$\partial_t K = -\frac{\alpha}{A} \left(\partial_r D_\alpha + D_\alpha^2 + \frac{2D_\alpha}{r} - \frac{D_\alpha \widetilde{D}}{2} \right) \\ + \alpha \left(K^2 - 4KK_B + 6K_B^2 \right) + 4\pi\alpha \left(\rho + S_A + 2S_B \right) . \tag{10.3.27}$$

In terms of the original variables this is equivalent to adding $-2(\alpha/A)H$ to the evolution equation for K_A, and $8\alpha M$ to the evolution equation for D_A. In this way we obtain a new system of evolution equations that is strongly hyperbolic for all $f > 0$, and has the following characteristic structure:

- One eigenfield with eigenspeed $\lambda = 0$ given by

$$w_0 = D_\alpha / f - \widetilde{D}/2 . \tag{10.3.28}$$

- Two eigenfields with eigenspeeds $\lambda = \pm\alpha/\sqrt{A}$ given by

$$w_\pm^l = A^{1/2} K_B \mp D_B/2 . \tag{10.3.29}$$

- Two eigenfields with eigenspeeds $\lambda = \pm\alpha\sqrt{f/A}$ given by

$$w_\pm^f = A^{1/2} K \mp D_\alpha / f^{1/2} . \tag{10.3.30}$$

There is still one important issue that should be addressed here. In the discussion of the hyperbolicity of the original ADM evolution equations we mentioned the fact that the regularity problem was irrelevant for this analysis. However, when modifying the system by adding multiples of the constraints we need to be more careful. Adding multiples of the Hamiltonian constraint in fact represents no problem, as the introduction of the variable λ already regularized this constraint, as seen in (10.3.18). The momentum constraint, however, is not regularized as it still includes the term $(K_A - K_B)/r$. One could try to play the same game as before and introduce yet another auxiliary variable of the form $\Lambda = (K_A - K_B)/r$ to absorb this term (this is essentially what we will do below when considering the case of axial symmetry). However, it turns out that in this case this is not really necessary.

Let us then consider some arbitrary first order formulation of the Einstein evolution equations in spherical symmetry that has the generic form

$$\partial_t u_i = q_i(u, v) , \tag{10.3.31}$$

$$\partial_t v_i = M_i^j(u)\, \partial_r v_j + p_i(u, v) , \tag{10.3.32}$$

where $u = (\alpha, A, B, \lambda)$ and $v = (D_\alpha, D_A, D_B, K_A, K_B)$. The source terms q and p are assumed not to depend on derivatives of any of the fields. The formulation might be hyperbolic or not, depending on the characteristic structure of the matrix M. We will assume that we have arrived at such a formulation by adding multiples of the Hamiltonian and momentum constraints to the evolution equations for the v's. This means that we can expect that the source terms p_i will in general contain terms proportional to $(K_A - K_B)/r$. We will then rewrite the evolution equations for the v_i as

$$\partial_t v_i = M_i^j(u)\, \partial_r v_j + p_i'(u,v) + \frac{f_i(u)}{r}\,(K_A - K_B)\,. \tag{10.3.33}$$

Here we are assuming that the coefficient $f_i(u)$ of the $(K_A - K_B)/r$ terms depends on the u's, but not on the v's, which will typically be the case. Using now equation (10.3.19) we find

$$\partial_t v_i = M_i^j(u)\, \partial_r v_j + p_i'(u,v) + \frac{f_i(u)B}{2\alpha A}\,\partial_t \lambda\,, \tag{10.3.34}$$

which implies

$$\partial_t \left(v_i - \frac{f_i(u)B}{2\alpha A}\,\lambda \right) = M_i^j(u)\, \partial_r v_j + p_i'(u,v) - \lambda\, \partial_t \left(\frac{f_i(u)B}{2\alpha A} \right)\,. \tag{10.3.35}$$

If we now define

$$v_i' := v_i - \frac{f_i(u)B}{2\alpha A}\,\lambda\,, \tag{10.3.36}$$

we can transform the last equation into

$$\partial_t v_i' = M_i^j(u)\, \partial_r v_j + p_i'(u,v) - \lambda F_i\,, \tag{10.3.37}$$

with $F_i^t := \partial_t\left(f_i(u)B/2\alpha A\right)$. Notice that F_i^t so defined will involve no spatial derivatives of u's or v's. The final step is to substitute the spatial derivative of v_j for that of v_j' to find

$$\begin{aligned} \partial_t v_i' &= M_i^j(u)\, \partial_r v_j' + p_i'(u,v) - \lambda F_i^t + M_i^j(u)\, \partial_r \left(\frac{f_j(u)B}{2\alpha A}\,\lambda \right) \\ &= M_i^j(u)\, \partial_r v_j' + p_i'(u,v) + \lambda \left(F_i^r - F_i^t \right) + \left(\frac{M_i^j(u) f_j(u) B}{2\alpha A} \right) \partial_r \lambda\,, \end{aligned} \tag{10.3.38}$$

with $F_i^r = M_i^j(u)\, \partial_r [f_j(u)B/2\alpha A]$. Using now the fact that

$$\partial_r \lambda = -\frac{1}{r} \left[\lambda + \frac{A}{B}\,(D_A - D_B) \right]\,, \tag{10.3.39}$$

we finally find

$$\begin{aligned} \partial_t v_i' = {} & M_i^j(u)\, \partial_r v_j' + p_i'(u,v) + \lambda \left(F_i^r - F_i^t \right) \\ & - \frac{M_i^j(u) f_j(u) B}{2\alpha A r} \left[\lambda + \frac{A}{B}\,(D_A - D_B) \right]\,. \end{aligned} \tag{10.3.40}$$

This last system is now regular, and has precisely the same characteristic structure as the original system. What we have done is transform the original evolution equations for the v_i variables into evolution equations for the new v_i' variables, for which the principal part is the same and the source terms are regular. Notice that typically only some of the $f_i(u)$ will be different from zero, so we do not need to transform all variables.

In the particular case of the strongly hyperbolic system introduced above we have only used the momentum constraint to modify the evolution equation for $\widetilde{D}$, so that this variable must be replaced with

$$\widetilde{U} := \widetilde{D} - 4B\lambda/A \,. \tag{10.3.41}$$

The evolution equation for $\widetilde{U}$ then becomes

$$\begin{aligned} \partial_t \widetilde{U} = & -2\alpha \left[\partial_r K + D_\alpha \left(K - 4K_B \right) \right. \\ & \left. - \, 2 \left(K - 3K_B \right) \left(D_B - 2\lambda B/A \right) + 16\pi j_A \right] \,. \end{aligned} \tag{10.3.42}$$

10.3.3 *Evolving Schwarzschild*

As our first example of numerical simulations of a spherically symmetric spacetime we will consider evolutions of a single Schwarzschild black hole. One could think that, since the Schwarzschild spacetime is static, there would simply be no evolution in such a numerical simulation. This is not true for two reasons: In the first place, the Schwarzschild spacetime is only static outside the black hole horizon; inside the horizon the solution is dynamic and the singularity is reached in a finite proper time. Second, as we have already seen in the previous Section, even in static spacetimes it is interesting to study the artificial evolution induced by a non-trivial gauge choice. Studying the case of Schwarzschild allows us to acquire valuable experience that can later be used in truly dynamic situations, such as gravitational collapse or the collision of compact objects, where we expect to find black holes during the evolution even if none were present initially.

As always, we must start with the choice of initial data. It is clear that simply taking the Schwarzschild metric at $t = 0$ in standard coordinates is not a good choice since this metric is singular at the horizon. A much better choice is to use isotropic coordinates. In such coordinates the spatial metric has the form (see equation (6.2.3))

$$dl^2 = \psi^4 \left(dr^2 + r^2 d\Omega^2 \right) \,, \tag{10.3.43}$$

with the conformal factor given by $\psi = 1 + M/2r$, and where r is related to the standard "areal" Schwarzschild radius through $r_{\text{Schwar}} = r\psi^2 = r\left(1 + M/2r\right)^2$.

As we have already mentioned in Chapter 6, there are several different ways in which we could choose to deal with the singularity at $r = 0$ (remember that this is a coordinate singularity associated with the compactification of infinity on the other side of the Einstein–Rosen bridge, and not the physical singularity). For example, we could place a boundary at the throat, which at $t = 0$ coincides with the horizon and is located at $r = M/2$, and use an isometry boundary condition. Alternatively, we could excise the black hole interior. For simplicity, however, we will use here the static puncture evolution technique (see Section 6.3). We then start by extracting analytically the singular conformal factor and defining new dynamical variables as:

$$\tilde{A} := A/\psi^4 \,, \qquad \tilde{B} := B/\psi^4 \,, \tag{10.3.44}$$

$$\tilde{D}_A := D_A - 4\,\partial_r \ln\psi \,, \qquad \tilde{D}_B := D_B - 4\,\partial_r \ln\psi \,, \tag{10.3.45}$$

(notice that the variables K_A and K_B are not rescaled). We now rewrite the ADM equations, or any equivalent strongly hyperbolic reformulation, in terms of the new variables. In terms of our rescaled variables, the initial data for Schwarzschild is simply:

$$\tilde{A} = \tilde{B} = 1 \,, \qquad \tilde{D}_A = \tilde{D}_B = 0 \,. \tag{10.3.46}$$

As the Schwarzschild metric is static, and the shift vector is zero in isotropic coordinates, the initial extrinsic curvature is just

$$K_A = K_B = 0 \,. \tag{10.3.47}$$

We still have to choose the gauge condition. For the shift vector we simply choose to keep it equal to zero, while for the lapse we will use either 1+log slicing, which corresponds to equation (10.3.21) with $f = 2/\alpha$, or maximal slicing (see equation (4.2.8)), which in this case reduces to:

$$\frac{1}{\tilde{A}\psi^4}\left[\partial_r^2\alpha + \left(\frac{2}{r} + \tilde{D}_B - \frac{\tilde{D}_A}{2} + 2\partial_r \ln\psi\right)\partial_r\alpha\right] = K_A^2 + 2K_B^2 \,. \tag{10.3.48}$$

This equation must be integrated numerically every time step. As it is a second order differential equation in r, it requires two boundary conditions. The boundary condition at infinity (or in practice, at the edge of the computational domain) is just:

$$\partial_r\alpha|_{r\to\infty} = \frac{1-\alpha}{r} \,. \tag{10.3.49}$$

This is a Robin type boundary condition that demands that, as $r \to \infty$, the lapse behaves as $\alpha = 1 + \mathcal{O}(r^{-1})$. That is, the lapse approaches 1 (its value in flat space) as $1/r$. The other boundary condition has to be given in the interior, and there are three interesting possibilities. The first possibility is to ask for $\alpha(r=0) = -1$. In this case it is in fact possible to show that there is an exact solution to the maximal slicing equation given by:

$$\alpha = \frac{1 - M/2r}{1 + M/2r} \,. \tag{10.3.50}$$

This lapse function is antisymmetric with respect to the throat of the wormhole at $r = M/2$, and is precisely the lapse that gives us a static solution, that is, it is the standard Schwarzschild lapse written in terms of the isotropic radius. This choice is not very interesting for our purposes, since besides taking us to a completely static situation, it does not penetrate inside the black hole horizon. Another possibility is to ask for:

$$\partial_r \alpha|_{r=0} = 0 . \tag{10.3.51}$$

This no longer results in the Schwarzschild lapse and gives us a dynamical situation (even if the dynamics are just a result of the gauge choice). Also, this case does penetrate the black hole horizon. This is in fact the choice we will use in the numerical simulations shown below.

A third possibility is to ask for the lapse to be symmetric at the throat of the wormhole at $r = M/2$. This choice also results in dynamical evolution and penetrates the horizon. In the case of a puncture evolution the throat is not a natural boundary of the computational domain, and this makes the symmetric lapse hard to use. However, if instead of a puncture evolution we choose to locate a boundary at the throat and impose isometry conditions, then the symmetric lapse becomes a natural choice. Notice that, as was already discussed in Section 4.2.3, the maximal slicing equation can in fact be solved analytically in the case of Schwarzschild, so we know what to expect from a numerical simulation.

With the ingredients we have introduced so far we are now ready to do a numerical simulation. The simulations shown here use a code based on the regularized strongly hyperbolic system introduced in the last Section, with simple second order centered differences in space and a three-step iterative Crank–Nicholson time integrator. All simulations have been performed using a grid with 1000 points in the radial direction, and an interval of $\Delta r = 0.01$. The time step used was $\Delta t = \Delta r/2$, and a total of 4000 time steps have been calculated in order to reach a final time of $t = 10$. The initial data corresponds to a black hole of mass $M = 1$, so that in these units the simulation reaches a time equal to $t = 10M$.

Let us first consider the case of maximal slicing. Figure 10.7 shows the evolution of the lapse every $t = 1M$ (this figure is in fact identical to Figure 6.3 of Chapter 6). From the figure we can see how the lapse evolves rapidly toward zero in the region close to the puncture ($r = 0$). This is the well known "collapse of the lapse", and happens because the maximal slicing condition does not allow the volume elements to change, and close to the singularity the only way to achieve this is to freeze the time evolution. In the central regions the lapse in fact approach zero exponentially.

Figure 10.8 shows the evolution of the radial metric $\tilde{A}$ for the same simulation (again, this is identical to Figure 6.5 of Chapter 6). We see how the radial metric is growing in the region close to the black hole horizon. This phenomenon is also well known, and is called slice stretching. What happens here is a combination of two effects. In the first place, as the lapse has collapsed in the central regions, time freezes there while it keeps moving forward outside, so the result is that the hypersurfaces stretch. Also, even if the lapse remained equal to 1 during the whole evolution, we should expect a growth in the radial metric, owing to

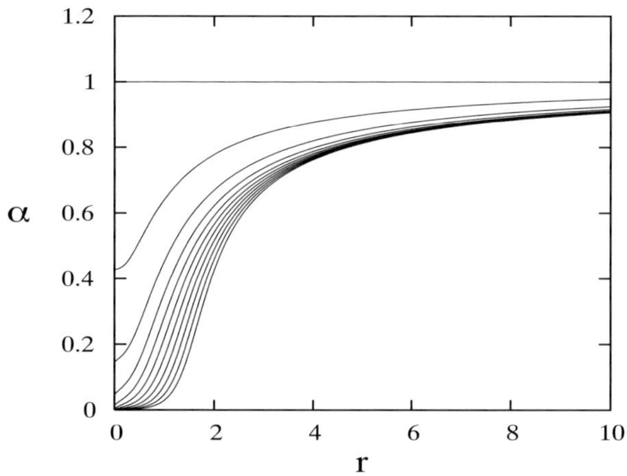

Fig. 10.7: Evolution of the lapse function α for Schwarzschild using maximal slicing. The value of α is shown every $t = 1M$.

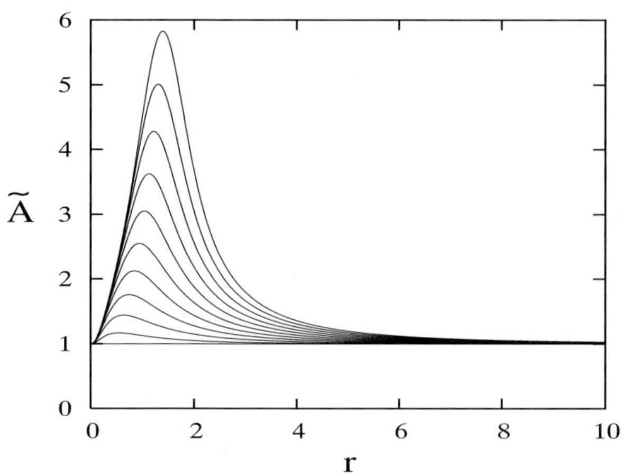

Fig. 10.8: Evolution of the conformal metric variable $\tilde{A}$ for Schwarzschild using maximal slicing. The value of the metric is shown every $t = 1M$.

the fact that the normal observers at different distances from the black hole fall with different accelerations, so the distance between them increases (remember that since the shift is zero in this simulation, our coordinates are tied to these observers).

From the figures it is clear that, with the gauge conditions that we have chosen, the Schwarzschild spacetime does not appear static. This can be seen even more dramatically if we study the position of the horizon during the evolution. Figure 10.9 shows the radius of the black hole horizon r_h as a function of time.

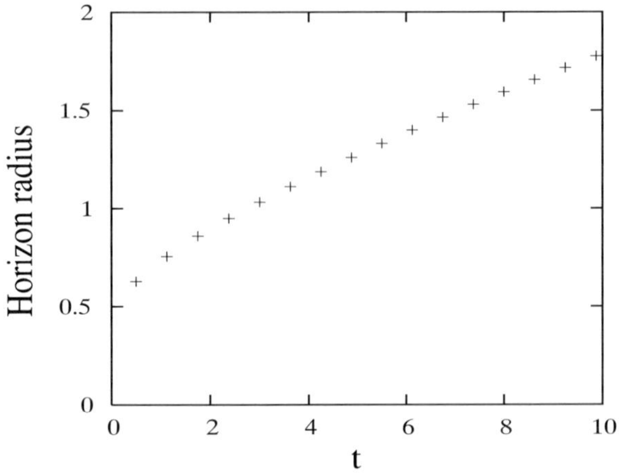

Fig. 10.9: Horizon radius as a function of time. At $t = 0$ the horizon is located at $r_h = 0.5$, and by $t = 10$ it has moved out to $r_h = 1.8$.

At $t = 0$ the horizon is located at $r_h = 0.5$, as corresponds to the isotropic coordinates (for $M = 1$), but during the evolution the horizon moves outward, and by $t = 10$ it is already at $r_h \sim 1.8$. This apparent growth is not real; it is simply a coordinate effect caused by the fact that our radial coordinate r is tied to the normal observers, and these observers are falling into the black hole. To convince ourselves that the growth is only apparent, Figure 10.10 shows the horizon area $a_h = 4\pi r_h^2 \psi^4 \tilde{B}$ as a function of time during the simulation. We see that the area remains constant at $a_h \simeq 50.2$ for the duration of the run, a value that corresponds to the area of the horizon of a Schwarzschild black hole of unit mass, namely $a_h = 16\pi \sim 50.2$.

Finally, let us consider an evolution using 1+log slicing instead of maximal slicing. Since 1+log corresponds to an evolution equation for α we need to specify the initial value of the lapse, and the obvious choice is to take $\alpha = 1$. Figure 10.11 shows the evolution of the lapse function for a numerical simulation of the Schwarzschild spacetime starting from puncture initial data and using 1+log slicing (again, this is just Figure 6.4 of Chapter 6). Notice that something very interesting has happened in this evolution: The lapse has failed to collapse at the puncture. The reason for this is that the speed of propagation for the 1+log slicing is $v = (2\alpha/A)^{1/2} = (2\alpha/\tilde{A}\psi^4)^{1/2}$, which vanishes at the puncture so that no information can reach that point. The 1+log slicing condition then keeps the lapse equal to one at both asymptotic ends. Of course, this behavior can be improved if we use a pre-collapsed lapse that is already zero at the puncture initially, such as *e.g.* $\alpha = 1/\psi^2$. The use of such a pre-collapsed lapse is now typical of black hole simulations that use the moving puncture method (as opposed to the static puncture approach used here), as otherwise the large gradients that develop in the lapse can cause the simulations to fail quite rapidly.

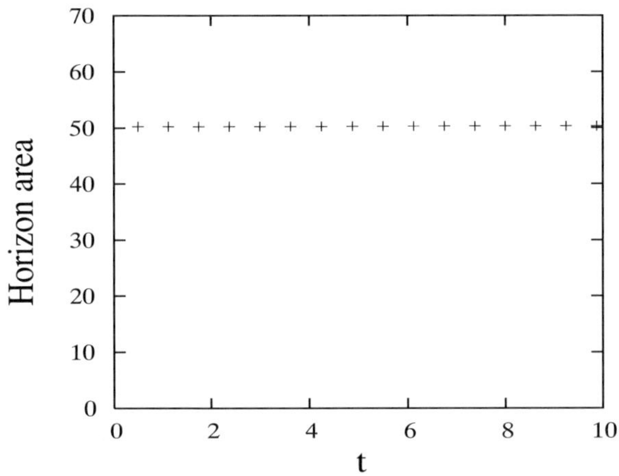

Fig. 10.10: The area of the horizon remains constant during the evolution, with a value close to $a_h = 16\pi \sim 50.2$.

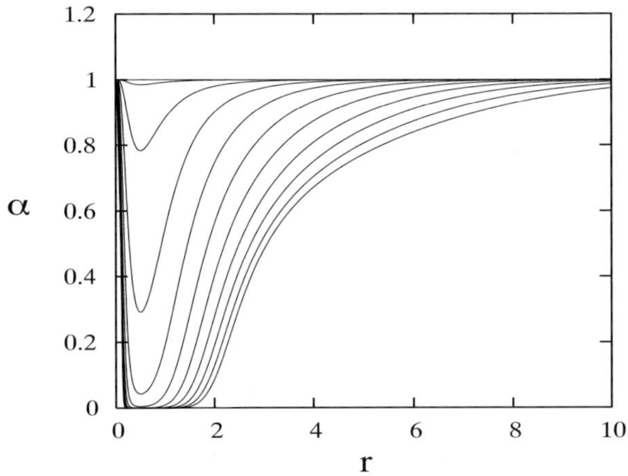

Fig. 10.11: Evolution of the lapse function α for Schwarzschild using 1+log slicing. The value of α is shown every $t = 1M$.

10.3.4 *Scalar field collapse*

For our second example of a simulation in spherically symmetry we will consider a non-vacuum case where the matter corresponds to a simple massless scalar field. When we consider matter, there are two extra ingredients that must be added to the picture. First, we need to know how to construct the energy density ρ, momentum density j_a, and stress tensor S_{ab} that appear in the ADM equations, plus we also need to know the evolution equations for the matter field itself.

Let us first construct the ADM matter quantities (ρ, j_a, S_{ab}). Even though the simulations shown below will be restricted to the case of a massless scalar field, for completeness we will consider the general expressions for a scalar field with an arbitrary self-interaction potential $V(\phi)$. We start from the stress-energy tensor for the scalar field which is given by

$$T_{\mu\nu} = \nabla_\mu \phi \, \nabla_\nu \phi - \frac{g_{\mu\nu}}{2} \left(\nabla_\alpha \phi \nabla^\alpha \phi + 2V \right) , \qquad (10.3.52)$$

with $g_{\mu\nu}$ the spacetime metric. Notice that for a scalar field whose self-interaction includes only a mass term we will have $V(\phi) = m^2 \phi^2 / 2$, while for the massless case we have $V(\phi) = 0$. Using the spherical metric (10.3.1) we find that

$$\rho = n^\mu n^\nu T_{\mu\nu} = \frac{1}{2A} \left(\frac{\Pi^2}{B^2} + \Psi^2 \right) + V , \qquad (10.3.53)$$

$$j_A = -n^\mu T_{\mu r} = - \frac{\Pi \Psi}{A^{1/2} B} , \qquad (10.3.54)$$

$$S_A = \gamma^{rr} T_{rr} = \frac{1}{2A} \left(\frac{\Pi^2}{B^2} + \Psi^2 \right) - V , \qquad (10.3.55)$$

$$S_B = \gamma^{\theta\theta} T_{\theta\theta} = \frac{1}{2A} \left(\frac{\Pi^2}{B^2} - \Psi^2 \right) - V , \qquad (10.3.56)$$

where we have defined

$$\Pi := \left(A^{1/2} B / \alpha \right) \partial_t \phi , \qquad \Psi := \partial_r \phi . \qquad (10.3.57)$$

Having found the form of the matter terms that appear in the 3+1 evolution equations, let us go back to the issue of the evolution equation for the scalar field itself. Starting from the conservation law $\nabla_\nu T^{\mu\nu} = 0$, we can easily show that the scalar field must evolve through the Klein–Gordon equation:

$$\Box \phi = \partial_\phi V \quad \Rightarrow \quad \partial_\mu \left((-g)^{1/2} \, \partial^\mu \phi \right) = (-g)^{1/2} \partial_\phi V , \qquad (10.3.58)$$

with g the determinant of $g_{\mu\nu}$, which is given in terms of the lapse and the determinant of the spatial metric as $g = -\alpha^2 \gamma$. In the particular case of spherical symmetry the Klein–Gordon equation reduces to

$$\partial_t \Pi = \frac{1}{r^2} \, \partial_r \left(\frac{\alpha B r^2}{A^{1/2}} \, \Psi \right) - \alpha A^{1/2} B \, \partial_\phi V . \qquad (10.3.59)$$

Notice that this only gives us an evolution equation for Π. The evolution equation for Ψ, on the other hand, can be obtained from the fact that the partial derivatives of ϕ commute. The final system of equations is then

$$\partial_t \phi = \frac{\alpha}{A^{1/2} B} \, \Pi , \qquad (10.3.60)$$

$$\partial_t \Psi = \partial_r \left(\frac{\alpha}{A^{1/2} B} \, \Pi \right) , \qquad (10.3.61)$$

$$\partial_t \Pi = \frac{1}{r^2} \, \partial_r \left(\frac{\alpha B r^2}{A^{1/2}} \, \Psi \right) - \alpha A^{1/2} B \, \partial_\phi V . \qquad (10.3.62)$$

This system of evolution equations couldn't look simpler. However, its approximation in terms of finite differences presents us with a beautiful example of how the naive use of simple centered spatial differences can sometimes fail to be consistent at places where the coordinate system becomes singular, such as the point $r = 0$ in our spherical coordinates. The problem comes from the evolution equation for Π. Consider a general term of the form

$$T = \frac{1}{r^2}\, \partial_r \left(r^2 f(r) \right) \,, \tag{10.3.63}$$

with $f(r)$ some arbitrary smooth function of r that for small r behaves as $f = ar$. This implies that close to the origin we will have $T = 3a$. Now consider a centered finite difference approximation to T, and assume for simplicity that the grid staggers the origin so that $r_0 = -\Delta r/2$, $r_1 = \Delta r/2$, $r_2 = 3\Delta r/2$, *etc.* (similar results are obtained if the grid is not staggered, but we then have to do something special at the point $r = 0$). At the point $r_1 = \Delta r/2$ we will have

$$T_1 = \frac{1}{r_1^2} \left[\frac{r_2 f_2 - r_0 f_0}{2\Delta r} \right] = \frac{a}{r_1^2} \left[r_2^3 - r_0^3 \right] = 7a \,. \tag{10.3.64}$$

But this is clearly very different from the expected value of $T = 3a$. This means that, close to the origin, the finite difference approximation has a serious problem, and even though we can expect this problem to be confined to a very small region around $r = 0$, it can still make the entire numerical scheme go unstable.

What has gone wrong? The problem can be easily understood if we notice that, since our finite difference approximation is second order, the truncation error should have the form $\tau_\Delta \sim (\Delta r)^2\, \partial_r^2 T$. But close to the origin we have, to leading order, $\partial_r^2 T \sim 4f(r)/r^3 = 4a/r^2$, so the truncation error becomes $\tau_\Delta \sim a(\Delta r/r)^2$. Now, close to the origin we also have $r \sim \Delta r$, which implies that the truncation error remains finite regardless of how small Δr becomes, so the finite difference approximation is inconsistent!

Fortunately, there is a well known trick that can fix this problem [127]. Notice first that, quite generally,

$$\frac{1}{r^2}\, \partial_r = 3\, \partial_{r^3} \,. \tag{10.3.65}$$

Let us then define $T' = 3\, \partial_{r^3} (r^2 f(r))$, and consider the centered finite difference approximation to this new expression. At the point $r_1 = \Delta r/2$ we will have

$$T_1' = 3 \left(\frac{r_2^2 f_2 - r_0^2 f_0}{r_2^3 - r_0^3} \right) = 3a \,. \tag{10.3.66}$$

We now find the correct value at the origin. Since this trick is consistent for all r, we can in fact use the above approximation everywhere, even for large r. Similar problems can also arise when finite differencing the 3+1 evolution equations of general relativity, so we must always be careful about the consistency of finite difference approximations at places where the coordinate systems become singular.

We will now consider a couple of numerical simulations for the case of a massless scalar field. The code used for these simulations uses centered second order differences in space for both the evolution of the geometry and the evolution of the scalar field, with the exception of the trick described above for the evolution equation of Π. As before, the time integration is done with a three-step iterative Crank–Nicholson scheme. As for the gauge conditions, the simulations shown below are done with maximal slicing and zero shift.

The code also uses fourth order Kreiss–Oliger dissipation on all variables (equation (9.9.2) with $N = 2$), with a dissipation coefficient of $\epsilon = 0.05$. Adding some dissipation is in fact necessary, to obtain stable simulations, as otherwise terms of the form D/r, with $D = (D_\alpha, D_A, D_B)$, though formally regular, have nevertheless a tendency to introduce small instabilities close to the origin.

For the scalar field we take an initial configuration of the form

$$\phi = ar^2 e^{-(r-5)^2} , \tag{10.3.67}$$

with a an amplitude parameter. We also take $\Pi(t = 0) = 0$, so that the initial pulse is expected to separate into two smaller pulses traveling in opposite directions.

In order to be consistent with Einstein's equations we now need to solve the Hamiltonian and momentum constraints (10.3.10) and (10.3.11) for this scalar field configuration. Notice first that since $\Pi = 0$ we also have $j_A = 0$, so that the momentum constraint (10.3.11) can be trivially solved by taking $K_A = K_B = 0$. In order to solve the Hamiltonian constraint (10.3.10) we first choose $B = 1$, in which case this constraint reduces to

$$\partial_r A = A\left(\frac{1}{r}(1 - A) + 8\pi r A\rho\right) . \tag{10.3.68}$$

This equation is solved numerically for A using the fact that, for our initial data, $\rho = \Psi^2/2A$. For our simulations we solve the above equation with a second order Runge–Kutta method. The integration is done starting from the origin and taking as boundary condition $A(r = 0) = 1$.

We will consider first a simulation with a scalar field amplitude of $a = 0.001$. This amplitude is in fact already very large, but not quite large enough to cause the scalar field to collapse to form a black hole. Figure 10.12 shows three snapshots of the evolution of the scalar field ϕ in this case: The solid line shows the initial data, the dashed line the solution after $t = 5$, and the dotted line the solution after $t = 20$. From the figure we see that the initial pulse first separates into two smaller pulses traveling in opposite directions, as expected. By $t = 5$, the outward moving pulse has moved to $r \sim 10$, while the inward moving pulse has reached the origin. By $t = 20$, the outward moving pulse has moved to

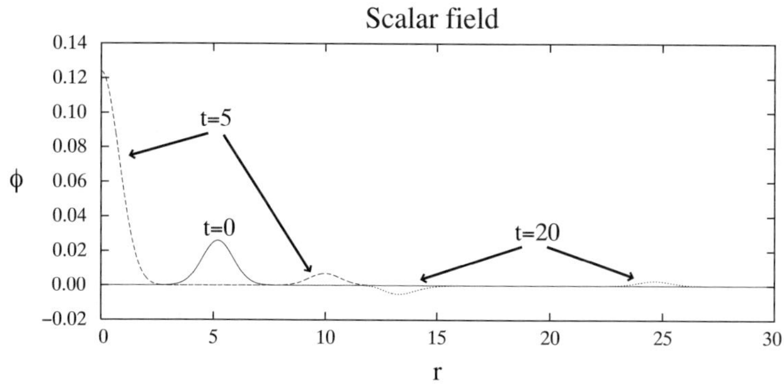

Fig. 10.12: Evolution of the scalar field ϕ for an initial configuration with amplitude $a = 0.001$.

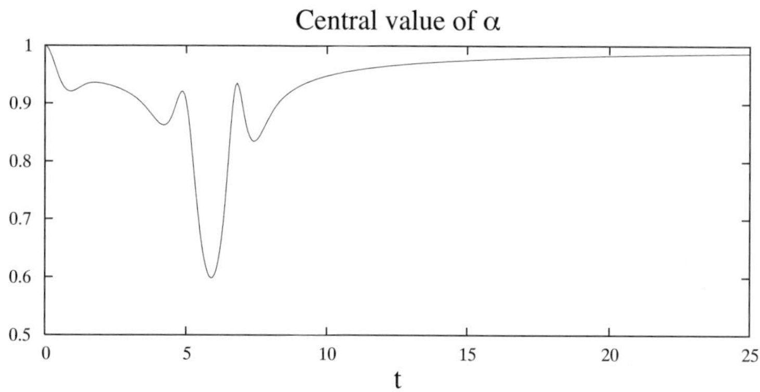

Fig. 10.13: Central value of the lapse as a function of time for an initial scalar field configuration with amplitude $a = 0.001$.

$r \sim 25$, while the pulse that was originally moving inward has imploded through the origin, changing sign in the process, and is now also moving outward having reached $r \sim 14$. The evolution then proceeds with both pulses moving out and leaving flat space behind.

As we can see, the evolution of ϕ behaves much as we would expect for an evolution in flat space, so we might be tempted to think that spacetime is almost flat during the whole evolution. This is in fact not so, and there is a period when spacetime has a large curvature as the inward moving pulse reaches the origin. In order to see this, Figure 10.13 shows the value of the lapse at the origin as a function of time. Notice that, just after the inward moving pulse reaches the origin ($t \sim 5$), the central value of the lapse drops to ~ 0.6, indicating that significant curvature has developed. However, since the scalar field density is not large enough to collapse to a black hole the central value of the lapse eventually

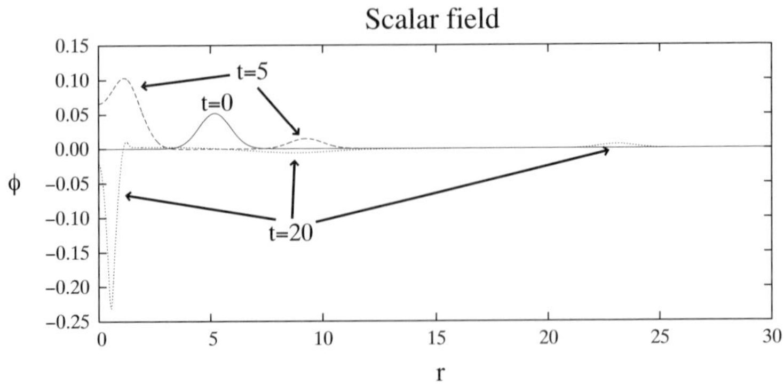

Fig. 10.14: Evolution of the scalar field ϕ for an initial configuration with amplitude $a = 0.002$.

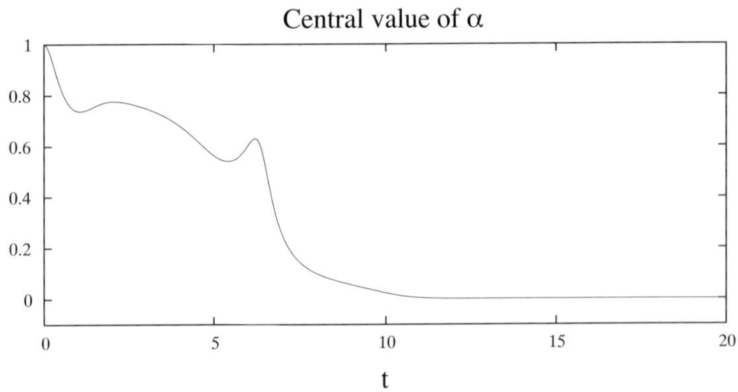

Fig. 10.15: Central value of the lapse as a function of time for an initial scalar field configuration with amplitude $a = 0.002$.

returns to 1 after a few oscillations.

Next we consider a simulation with an initial scalar field configuration with twice the amplitude as before, that is $a = 0.002$. Figure 10.14 again shows the three snap-shots of the evolution of the scalar field for times $t = 0$, $t = 5$, and $t = 20$. As before, the initial pulse separates into two smaller pulses, with the outward moving pulse behaving much as it did in the previous simulation. The inward moving pulse, however, now behaves quite differently. It initially does move inward and by $t = 5$ has almost reached the origin (though it is moving more slowly than before). It later implodes through the origin and changes sign, but now most of the pulse remains frozen close to the origin (owing to the collapse of the lapse as we will see below), and only a very small portion manages to escape after a considerable delay.

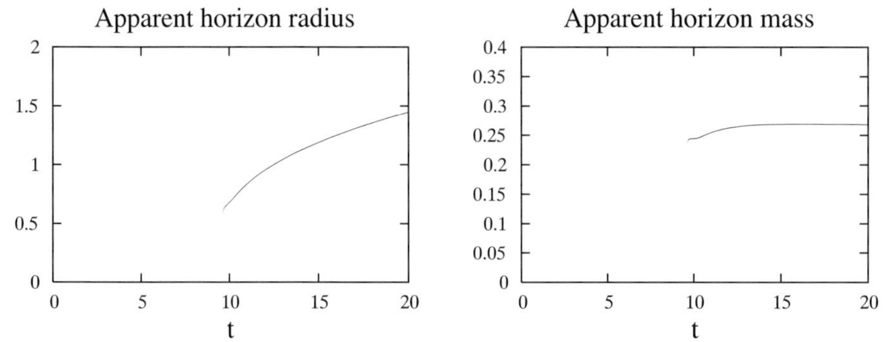

Fig. 10.16: Coordinate radius and mass of the apparent horizon found for the simulation with an initial scalar field amplitude of $a = 0.002$.

What has happened in this case is that the inward going pulse has in fact collapsed to a black hole. This can again be seen more clearly by looking at the evolution of the central value of the lapse function which is shown in Figure 10.15. Notice how, after a couple of bounces, the central lapse collapses to zero, indicating the presence of a black hole. Since the lapse is now zero in the central regions, the evolution stops there, freezing the scalar field configuration.

Of course, the collapse of the lapse, though a strong indicator, is not in itself proof of the presence of a black hole. In order to be sure, we should look for an apparent horizon, and indeed such an apparent horizon is found during this simulation. Figure 10.16 shows the coordinate radius r_{AH} and mass M_{AH} of the apparent horizon during the evolution (the horizon mass is defined as $M_{AH} = \sqrt{A_{AH}/16\pi}$, with A_{AH} the horizon area). Notice how the horizon first appears at $t \sim 9.5$ with a coordinate radius of $r_{AH} \sim 0.6$, and after that the horizon grows in coordinate space until at $t = 20$ it has a radius of $r_{AH} \sim 1.5$. The horizon mass, on the other hand, starts at $M_{AH} \sim 0.24$ and rapidly stabilizes at a value of $M_{AH} \sim 0.27$. Notice that for this data set the initial ADM mass can be easily calculated and turns out to be $M_{ADM} = 0.54$. The fact that the final black hole has half the initial ADM mass is to be expected, because the piece of the initial pulse that started moving outward has managed to escape.

The ADM mass can be calculated by using, for example, the "Schwarzschild-like mass" defined in equation (A.26) of Appendix A, and this gives the exact ADM mass for a Schwarzschild spacetime at any radius. Since our initial scalar field configuration is restricted to a small region, it is clear that outside this region the spacetime will be Schwarzschild and this approach will give us the correct ADM mass.

A second method for calculating the ADM mass of our spacetime can be found by first defining a function $m(r)$ that is related to the radial metric $A(r)$ through

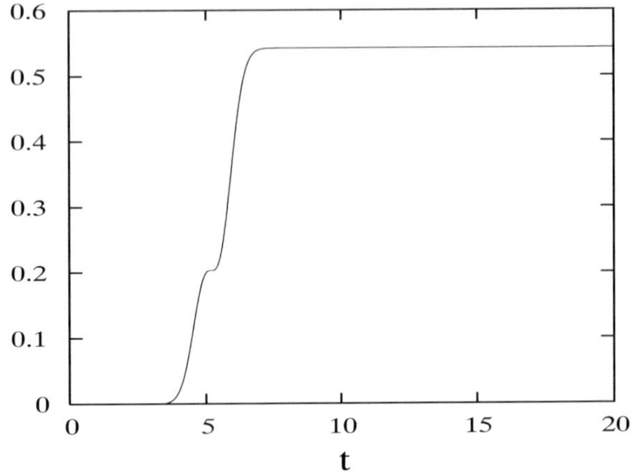

Fig. 10.17: Mass function $m(r)$ at $t = 0$ for the scalar field configuration with amplitude $a = 0.002$.

$$A(r) = \frac{1}{1 - 2m(r)/r} \, , \tag{10.3.69}$$

and then rewriting the Hamiltonian constraint at $t = 0$, equation (10.3.68), in terms of $m(r)$. Doing this we find that the Hamiltonian constraint simplifies to:

$$\partial_r m = 4\pi \rho r^2 \, . \tag{10.3.70}$$

Notice that, since at $t = 0$ we are using the areal radius ($B = 1$), in the vacuum region we must recover the Schwarzschild metric, which implies that $m(r)$ must in fact be constant and equal to the total ADM mass. We can then find the ADM mass by simply integrating the above equation up to a point where the scalar field is negligible:

$$m(r) = \int_0^r 4\pi \rho r^2 dr \, , \qquad M_{ADM} = \lim_{r \to \infty} m(r) \, . \tag{10.3.71}$$

It is quite remarkable that the above integral is precisely the expression for the Newtonian mass contained inside a sphere of radius r. This is a general result in spherical symmetry as long as $B = 1$ and $K_A = K_B = 0$. In that case we also finds that $m(r)$ is precisely the Schwarzschild-like mass (A.26), so that both mass measures coincide for all r. Figure 10.17 shows a plot of the function $m(r)$ at $t = 0$. From the figure we can clearly see that $m(r)$ increases monotonically in the region where the scalar field is non-zero, and settles down to a constant value of ~ 0.54.

Simulations similar to those presented here, but focusing on the threshold of black hole formation (*i.e.* the smallest value of the amplitude a for which a black hole is formed), led Choptuik to the discovery of critical phenomena in gravitational collapse [98]. We will not discuss these important results here, but the interested reader can look at *e.g.* [153] and references therein.

10.4 Axial symmetry

As our last example of numerical spacetimes we will consider the case of axial symmetry. Axisymmetric spacetimes have been studied in numerical relativity since very early on. The pioneering work of Hahn and Lindquist in 1964 [158], as well as the first successful calculations of head-on black hole collisions of Smarr and Eppley in the 1970s [123, 270, 271], were done assuming axial symmetry, and as early as 1983 there was already a beautiful review article on axisymmetric numerical relativity by Bardeen and Piran [48]. However, regularity issues at both the origin and the axis of symmetry, no doubt worsened by the fact that researchers at the time were still not aware of the well-posedness problems of the standard ADM formulation, made axisymmetric simulations particularly difficult to keep well behaved. As a consequence of this, axisymmetric codes were practically abandoned as soon as computers became powerful enough to be able to handle full three-dimensional simulations in the early 1990s. Nevertheless, because of the introduction of well-posed strongly hyperbolic formulations and a better understanding of the regularity problem, it is now possible to go back to considering axisymmetric codes. Such codes have the advantage that, being only two-dimensional, they demand considerably less computer resource than full three-dimensional codes, and also they allow one to isolate the simulations from possible purely three-dimensional effects (such as *e.g.* non-axisymmetric instabilities).

10.4.1 *Evolution equations and regularization*

There are at least two different approaches to deriving evolution equations in axisymmetry and study their regularity conditions. The simplest approach, and the one we will consider here, is to take the standard 3+1 decomposition and restrict it to the case of axial symmetry. However, for completeness it is important to mention a different approach that is more geometric in nature and is based on a decomposition of spacetime that first projects out the axial symmetry, and only later does an ADM-like split of the resulting three-dimensional spacetime. This approach, known as the $(2+1)+1$ decomposition, was originally suggested by Geroch [144] and used by Nakamura and collaborators [213], and is frequently used to this day (see *e.g.* [99]). However, as here we are just interested in some simple aspects of axisymmetric simulations we will not discuss it further.

The first step in considering axisymmetric spacetimes is to decide on the choice of the spatial coordinates to be used. Two different possibilities seem natural, namely spherical coordinates (r, θ, φ) and cylindrical coordinates (ρ, z, φ), which are related to Cartesian coordinates (x, y, z) through

$$r = \sqrt{x^2 + y^2 + z^2}\,, \tag{10.4.1}$$

$$\rho = \sqrt{x^2 + y^2} \tag{10.4.2}$$

$$\theta = \arctan\left(z/\rho\right)\,, \tag{10.4.3}$$

$$\varphi = \arctan\left(x/y\right)\,. \tag{10.4.4}$$

Spherical coordinates have the advantage of being well adapted to the asymptotic boundary conditions, as well as to the propagation of gravitational waves, but they make the regularity conditions more complicated since we have to worry about regularity both at the origin $r = 0$, and at the axis of symmetry $\theta = 0$. Because of this, here we will consider only cylindrical coordinates, for which the only regularity problems are associated with the axis of symmetry $\rho = 0$.

The next step is to consider the form of the spatial metric. In the case of spherical symmetry it was in fact possible to choose the spatial metric as diagonal, with the two angular metric components proportional to each other so that only two metric components were truly independent. One might then think that such a simplification of the metric is also possible in axisymmetry. Unfortunately, a little thought can convince us that this is not so in the general case, and that all six spatial metric components must be considered independently. For example, asymptotically the metric component $\gamma_{z\varphi}$ corresponds to the $h_\times$ polarization of gravitational waves, so it will in general not vanish. Also, there is no reason to assume that the coordinate lines associated with z and ρ should be orthogonal, so that $\gamma_{\rho z}$ will in general not vanish. Finally, if there is angular momentum in the spacetime there will be dragging of inertial frames so that even if initially we have $\gamma_{\rho\varphi} = 0$, this will not necessarily remain so during evolution as the ρ coordinate lines can be dragged differentially around the axis of symmetry. In summary, all three off-diagonal components of the metric will in general be non-zero. Nevertheless, it is possible to *impose* the condition that $\gamma_{\rho\varphi} = 0$ during the entire evolution by choosing appropriately the shift vector component β^φ (see *e.g.* [48]), and this is often done in practice.

Here, however, we will follow a different route and simplify the metric by considering only a restricted class of axisymmetric spacetimes, namely those that have zero angular momentum and no odd-parity gravitational waves (*i.e.* $h_\times = 0$), so that we can always take $\gamma_{z\varphi} = \gamma_{\rho\varphi} = 0$. In such a case the spatial metric can be simplified to

$$\begin{aligned} dl^2 &= \gamma_{\rho\rho} d\rho^2 + \gamma_{zz} dz^2 + 2\gamma_{\rho z} d\rho dz + \gamma_{\varphi\varphi} d\varphi^2 \\ &\equiv A\, d\rho^2 + B\, dz^2 + 2\rho C\, d\rho dz + \rho^2 T\, d\varphi^2 \,, \end{aligned} \tag{10.4.5}$$

with the quantities (A, B, C, T) functions of (ρ, z) only, and where we have explicitly extracted some factors of ρ in order to make the regularization procedure somewhat simpler. We can also assume that $\beta^\varphi = 0$, so that there are only two non-zero shift components. Notice that with the above restrictions we can still study a wide variety of interesting physical systems, such as the collapse of non-rotating stars, and even non-trivial strong gravitational wave spacetimes with even-parity such as the Brill waves that we will discuss in the following Section.

Let us now consider the regularity of the metric functions at the axis of symmetry. Just as in spherical symmetry, there are two different types of regularity conditions. In the first place, axial symmetry implies that the metric should remain unchanged under the transformation $\rho \to -\rho$, which implies that (A, B, C, T) must all be even (and smooth) functions of ρ. As before, this can be easily implemented numerically by staggering the origin, introducing a fictitious point $\rho = -\Delta\rho/2$, and setting the values of the metric variables (A, B, C, T) at this point by just copying them from the point $\rho = +\Delta\rho/2$.

The second type of regularity conditions are again associated with the fact that space must be locally flat at $\rho = 0$. In order to see where these conditions come from, let us start from the metric components in Cartesian coordinates. In this case axisymmetry implies that the metric must be invariant under reflections about the x and y axes and under exchange of x for y, while local flatness implies that it must also be smooth on the axis. These two requirements together imply that for fixed z we must have

$$\gamma_{xx} \sim k_\rho + \mathcal{O}(x^2 + y^2) \sim k_\rho + \mathcal{O}(\rho^2) \,, \tag{10.4.6}$$
$$\gamma_{yy} \sim k_\rho + \mathcal{O}(x^2 + y^2) \sim k_\rho + \mathcal{O}(\rho^2) \,, \tag{10.4.7}$$
$$\gamma_{zz} \sim k_z + \mathcal{O}(x^2 + y^2) \sim k_z + \mathcal{O}(\rho^2) \,, \tag{10.4.8}$$
$$\gamma_{xy} \sim \mathcal{O}(xy) \sim \mathcal{O}(\rho^2) \,, \tag{10.4.9}$$
$$\gamma_{xz} \sim \mathcal{O}(x) \sim \mathcal{O}(\rho) \,, \tag{10.4.10}$$
$$\gamma_{yz} \sim \mathcal{O}(y) \sim \mathcal{O}(\rho) \,, \tag{10.4.11}$$

where k_ρ and k_z are constants. Let us now consider a transformation to cylindrical coordinates (ρ, z, φ): $x = \rho \cos\varphi$, $y = \rho \sin\varphi$, $z = z$. Under such a transformation we find

$$\gamma_{\rho\rho} = \gamma_{xx} \cos^2\varphi + \gamma_{yy} \sin^2\varphi + 2\gamma_{xy} \sin\varphi \cos\varphi \,, \tag{10.4.12}$$
$$\gamma_{zz} = \gamma_{zz} \,, \tag{10.4.13}$$
$$\gamma_{\varphi\varphi} = \rho^2 \left(\gamma_{xx} \sin^2\varphi + \gamma_{yy} \cos^2\varphi - 2\gamma_{xy} \sin\varphi \cos\varphi \right) \,, \tag{10.4.14}$$
$$\gamma_{\rho z} = \gamma_{xz} \cos\varphi + \gamma_{yz} \sin\varphi \,, \tag{10.4.15}$$
$$\gamma_{\rho\varphi} = \rho \left(\gamma_{yy} - \gamma_{xx} \right) \sin\varphi \cos\varphi + \rho\, \gamma_{xy} \left(\cos^2\varphi - \sin^2\varphi \right) \,, \tag{10.4.16}$$
$$\gamma_{z\varphi} = \rho \left(-\gamma_{xz} \sin\varphi + \gamma_{yz} \cos\varphi \right) \,. \tag{10.4.17}$$

From the behavior of the different Cartesian metric components near the axis we then see that

$$\gamma_{\rho\rho} \sim k_\rho + \mathcal{O}(\rho^2) \,, \tag{10.4.18}$$
$$\gamma_{zz} \sim k_z + \mathcal{O}(\rho^2) \,, \tag{10.4.19}$$
$$\gamma_{\varphi\varphi} \sim \rho^2 \left(k_\rho + \mathcal{O}(\rho^2) \right) \,, \tag{10.4.20}$$
$$\gamma_{\rho z} \sim \mathcal{O}(\rho) \,, \tag{10.4.21}$$
$$\gamma_{\rho\varphi} \sim \mathcal{O}(\rho^3) \,, \tag{10.4.22}$$
$$\gamma_{z\varphi} \sim \mathcal{O}(\rho^2) \,. \tag{10.4.23}$$

Going back to the definition of the functions (A, B, C, T), we see that, close to the axis, A and T are such that $A - T \sim \mathcal{O}(\rho^2)$, so that they can be written in general as

$$A := H + \rho^2 J \; , \qquad T := H - \rho^2 J \; , \tag{10.4.24}$$

where H and J are regular functions that are even in ρ. The expressions above can be easily inverted to find

$$H := \frac{A+T}{2} \; , \qquad J := \frac{A-T}{2\rho^2} \; . \tag{10.4.25}$$

Since the above relation between A and T must hold for all time, the components of the extrinsic curvature must have a similar behavior, so that we can write (again assuming no angular momentum and no odd-parity gravitational waves)

$$K_{ij} = \begin{pmatrix} K_A & \rho K_C & 0 \\ \rho K_C & K_B & 0 \\ 0 & 0 & \rho^2 K_T \end{pmatrix} \; , \tag{10.4.26}$$

with $K_A = K_H + \rho^2 K_J, K_T = K_H - \rho^2 K_J$, and where (K_H, K_J, K_B, K_C) are even functions of ρ. The regularity conditions above have been derived rather informally; a more formal proof based on solving the Killing equation for axial symmetry can be found in *e.g.* [244].

A regular system of equations in axial symmetry can now be found by simply evolving directly the variables (H, J, K_H, K_J), instead of (A, T, K_A, K_T). The evolution equations for K_H and K_J can be obtained from those of K_A and K_T (*i.e.* directly from the ADM equations or any hyperbolic reformulation of them). The resulting equations are very long and we will not write them here, but they are trivial to obtain.

There is still, however, one important point that should be mentioned. Even if the evolution equations for K_H and K_J are regular, this regularity is not necessarily apparent and care must be taken with some terms that are irregular on their own but become regular when added together. Such terms typically have the form

$$\frac{1}{\rho^2} \left(\partial_\rho D - \frac{D}{\rho} \right) \; , \tag{10.4.27}$$

with D the derivative of either the lapse, or some combination of the metric coefficients (A, B, C, T), with respect to either ρ or z. Since such derivatives behave as $\mathcal{O}(\rho)$ near the axis, the term above seems to be ill-behaved. However, it is easy to see that

$$\frac{1}{\rho^2} \left(\partial_\rho D - \frac{D}{\rho} \right) = \frac{1}{\rho} \, \partial_\rho \left(\frac{D}{\rho} \right) \; . \tag{10.4.28}$$

The term on the right hand side is now clearly regular since $D/\rho \sim 1 + \mathcal{O}(\rho^2)$. The specific form of such terms depends on the details of the formulation being used (*i.e.* ADM, BSSNOK, NOR, *etc.*), and the equations must be carefully inspected before writing a numerical code (see *e.g.* [245]).

10.4.2 *Brill waves*

As an example of the evolution of an axially symmetric spacetime we will consider a non-trivial system consisting of strong non-linear gravitational waves in vacuum, commonly known as *Brill waves* [78, 124, 125]. The construction of such a spacetime starts by considering an axisymmetric initial slice with a metric of the form

$$ds^2 = \Psi^4 \left[e^{2q} \left(d\rho^2 + dz^2 \right) + \rho^2 d\varphi^2 \right] , \tag{10.4.29}$$

with (ρ, z, φ) cylindrical coordinates, and where both q and Ψ are functions of (t, ρ, z) only. The function q is quasi-arbitrary, and must only satisfy the following boundary conditions

$$q\,|_{\rho=0} = 0 , \tag{10.4.30}$$

$$\partial_\rho^n q\,|_{\rho=0} = 0 \qquad \text{for odd } n , \tag{10.4.31}$$

$$q\,|_{r\to\infty} = O\left(r^{-2}\right) , \tag{10.4.32}$$

where $r = \sqrt{\rho^2 + z^2}$. In order to find the conformal factor Ψ, we first impose the condition of time symmetry, which implies that the momentum constraints are identically satisfied. The Hamiltonian constraint, on the other hand, takes the form

$$D^2_{\text{flat}} \Psi + \frac{1}{4} \left(\partial_\rho^2 q + \partial_z^2 q \right) \Psi = 0 , \tag{10.4.33}$$

with D^2_{flat} the flat space Laplacian. Since this is an elliptic equation we must also say something about the boundary conditions. At infinity we must clearly have $\Psi = 1$; however since our computational domain is finite we must ask for boundary condition at a finite distance. Notice that we expect that far away the solution should approach a Schwarzschild spacetime, which implies that Ψ must behave asymptotically as $\Psi \sim 1 + M/2r$. This provides us with our boundary condition for large r.

Once a function q has been chosen, all we need to do is solve the above elliptic equation numerically for Ψ. Different forms of the function q have been used by different authors [124, 125, 163, 267]. Here we will consider the one introduced by Holz and collaborators in [163], which has the form

$$q = a\rho^2 e^{-(\rho^2 + z^2)} , \tag{10.4.34}$$

with a a constant that determines the initial amplitude of the wave (for small a the waves should disperse to infinity, while for large a they should collapse to form a black hole).

Before considering any evolutions, let us first discuss the solution of the initial data, which involves solving the elliptic equation (10.4.33) numerically. In Chapter 9 we did not discuss the solution of elliptic equations, but here we can present a quick and dirty method that is very easy to code (it is also extremely inefficient, so the reader should not use it for any serious applications). The

a	$M_{\rm ADM}$
1	0.0338 ± 0.0006
2	0.126 ± 0.001
3	0.270 ± 0.002
4	0.459 ± 0.003
5	0.698 ± 0.004
6	0.991 ± 0.005
10	2.91 ± 0.01
12	4.67 ± 0.02

Table 10.1 *ADM mass for Brill wave initial data with different amplitudes a.*

basic idea is to transform equation (10.4.33) into a wave-like equation in some fictitious time τ of the form

$$\partial_\tau^2 \Psi = D_{\rm flat}^2 \Psi + \frac{1}{4} \left(\partial_\rho^2 q + \partial_z^2 q \right) \Psi \; , \tag{10.4.35}$$

and then evolve this equation until a stationary state is reached, taking as initial data $\Psi = 1$ and using an outgoing wave boundary condition of the form

$$\Psi = 1 + f(r - t)/r \; . \tag{10.4.36}$$

Notice that the once we reach a stationary state we will have found a solution of our original elliptic equation. This solution will in fact also have the correct boundary condition since, for a stationary situation, we must have that $f(r - t) = k$, with k a constant that will correspond to half the ADM mass of our spacetime. Of course, since "errors" (*i.e.* the non-stationary part of the solution) must leave the computational domain through the boundaries with little dissipation, the algorithm just presented is not particularly fast, and in fact becomes slower when we push the boundaries further out. Standard iterative methods use a similar idea but instead of a wave equation they transform the problem into a parabolic equation so that errors dissipate away (these methods are also rather slow unless we use more complex algorithms such as multi-grid, but we will not discuss this issue here).

Table 10.1 shows the ADM masses obtained for Brill wave initial data with different amplitudes a. These masses have been calculated using the fact that asymptotically we should have $M = 2r(\Psi - 1)$. The calculation of the mass is done at the edge of the computational domain, both along the axis of symmetry $\rho = 0$ and along the equator $z = 0$, and the difference between these two values is used as an estimation of the error in the mass estimate. It is interesting to note that at the resolutions used here (see below), we find that this error estimate is in fact larger than the truncation error (obtained by convergence studies). The masses shown here are in agreement with previously published values, see *e.g.* [8, 163]. These results were obtained using a resolution of $\Delta\rho = \Delta z = 0.05$, with the boundaries located at $\rho, z = \pm 20$.

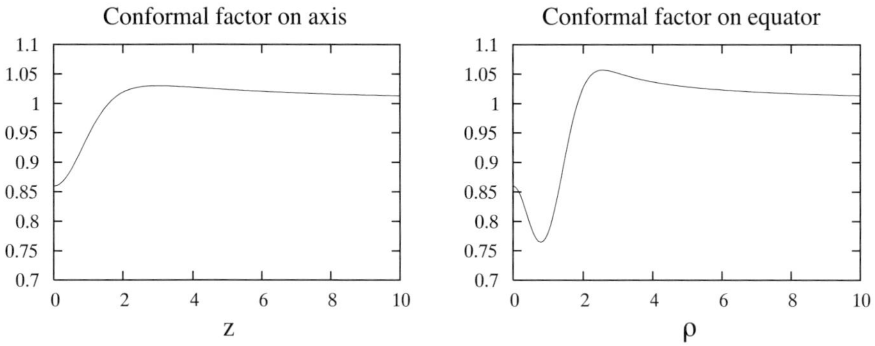

Fig. 10.18: Initial profile of the conformal factor Ψ for a Brill wave with amplitude $a = 3$, along both the axis $\rho = 0$ and the equator $z = 0$.

Previous studies have shown that Brill waves of this type have an apparent horizon already present in the initial data for $a \gtrsim 12$ [8, 267]. Incidentally, Brill waves also provide a very nice example of a regular vacuum spacetime that nevertheless has a non-zero (and positive) ADM mass. These configurations are also extremely compact, notice that for $a = 6$ the mass is $M \sim 1$ while the "radius", which can be estimated as the place where the function q drops to less than 10% of its maximum value, is of order $R \sim 2.2$. For an amplitude of $a = 12$ we find that the mass is $M \sim 4.7$ while the radius remains the same, implying that $M/R \sim 2$, so we shouldn't be surprised to find that an apparent horizon is already present in this case.

Here we will only consider one example of a dynamical simulation for the case of a Brill wave with amplitude $a = 3$. The initial data for Ψ can be seen in Figure 10.18, which shows plots along both the axis $\rho = 0$ and the equator $z = 0$. We will evolve this initial data using a 1+log slicing condition and vanishing shift. The code used here uses second order finite differencing, and is based on the Nagy–Ortiz–Reula (NOR) formulation discussed in Chapter 5, but adapted to the case of curvilinear coordinates (see *e.g.* [245]).

Figure 10.19 shows a plot of the value of the lapse at the origin as a function of time for three different resolutions $\Delta\rho = \Delta z = (0.1, 0.05, 0.025)$, using a Courant parameter of $\Delta t/\Delta\rho = 0.2$, and with the boundaries located at $\rho, z = \pm 10$. The first thing to notice is that at the lowest resolution of $\Delta\rho = 0.1$ the central value of the lapse is very different from the other two cases, indicating that this resolution is still too low to give us a good idea of the correct solution (this simulation in fact crashed at $t \sim 4.7$). At the two higher resolutions, however, the central value of the lapse already behaves quite similarly, with an initial drop to a value of just below 0.5, and a later rise back toward one. Figure 10.20, on the other hand, shows the root mean square of the Hamiltonian constraint as a function of time for the same three resolutions. The Figure clearly shows that the code is converging to second order, as expected.

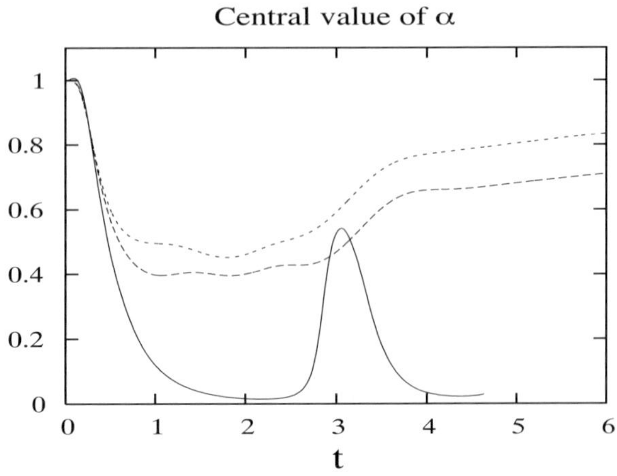

Fig. 10.19: Time evolution of the central value of the lapse α for a Brill wave with amplitude $a = 3$, using three different resolutions. The solid line corresponds to $\Delta\rho = 0.1$, the dashed line to $\Delta\rho = 0.05$, and the dotted line to $\Delta\rho = 0.025$.

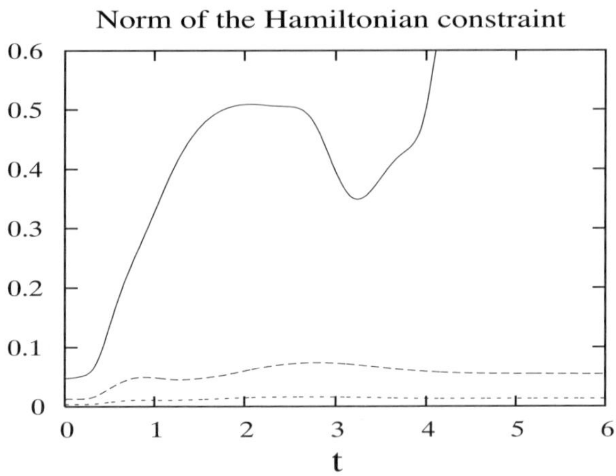

Fig. 10.20: Time evolution of the root mean square of the Hamiltonian constraint for a Brill wave with amplitude $a = 3$ using three different resolutions. The solid line corresponds to $\Delta\rho = 0.1$, the dashed line to $\Delta\rho = 0.05$, and the dotted line to $\Delta\rho = 0.025$.

From the fact that the lapse drops to below 0.5 at the early stages of the evolution we can conclude that a Brill wave with an amplitude of $a = 3$ is already very strong. However, since the lapse later returns to 1, we see that this wave is still not quite strong enough to collapse to a black hole. In fact, previous numerical studies have determined that the threshold for black hole

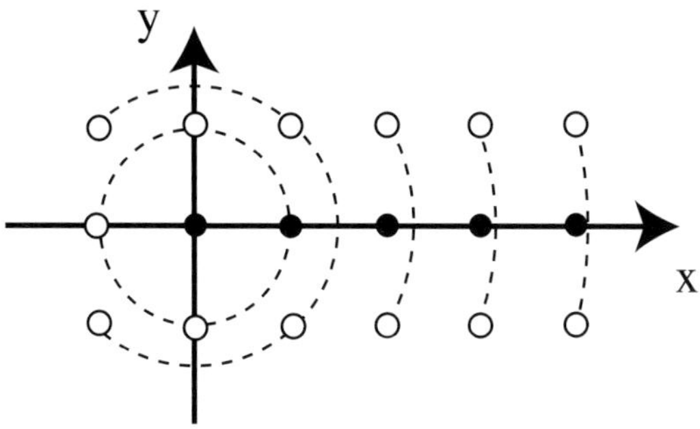

Fig. 10.21: The Cartoon approach: The points on the $y = 0$ plane are evolved using standard three-dimensional finite differencing. Those on the adjacent planes are then obtained by rotating around the z axis and interpolating.

formation corresponds to an amplitude of $a \sim 5$ [5]. Simulations with such large amplitudes, however, are extremely hard to do since the dynamics are so strong that numerical instabilities become very difficult to control, and because of this we will not consider them here.[106]

10.4.3 *The "Cartoon" approach*

Before finishing this Chapter it is important to mention a different approach to the evolution of axisymmetric spacetimes. This approach in commonly known as the "Cartoon" technique [9], and consists of evolving an axisymmetric spacetime using a three-dimensional code with a computational domain that is only a few grid points thick (typically 3 or 5) in the y direction, *i.e.* we consider only a thin slab consisting of the $y = 0$ plane plus one or two more planes on either side.[107] The field variables in the central $y = 0$ plane can then be updated using standard Cartesian finite differencing, while those in the extra $y \neq 0$ planes can be computed from the data in the central plane by making tensor rotations around the z axis, and then interpolating to the desired grid point (see Figure 10.21). In other words, the boundary conditions on the y direction are obtained by enforcing axial symmetry on all tensor quantities.

In order to apply the Cartoon approach it is necessary to consider the effect on tensor quantities of a rotation around the z axis. Assume that we are using cylindrical coordinates (ρ, z, φ). A rotation around the z axis can then be seen as the following change of coordinates

$$\rho' = \rho\,, \qquad \varphi' = \varphi + \varphi_0\,, \qquad z' = z\,, \tag{10.4.37}$$

[106] In fact, the code described above already requires a very large Kreiss–Oliger dissipation parameter in order to handle the $a = 3$ case.

[107] The original idea, as well as the name, were first suggested by S. Brandt. The name makes reference to the fact the we use a Cartesian code to do a two-dimensional simulation, that is Cartesian-2D or simply "Cartoon".

or equivalently in Cartesian coordinates

$$x' = x \cos\varphi_0 - y \sin\varphi_0 \,, \qquad y' = x \sin\varphi_0 + y \cos\varphi_0 \,, \qquad z' = z \,. \tag{10.4.38}$$

To see the effect of this change of coordinates on tensors we will consider as a particular example the case of a rank 2 tensor T_{ij} (for tensors of different rank the procedure is entirely analogous). Let us start from the Jacobian for the transformation of the Cartesian coordinates:

$$\Lambda(\varphi_0)^i_j = \frac{\partial x^{i'}}{\partial x^j} = \begin{pmatrix} \cos\varphi_0 & -\sin\varphi_0 & 0 \\ \sin\varphi_0 & \cos\varphi_0 & 0 \\ 0 & 0 & 1 \end{pmatrix} . \tag{10.4.39}$$

Under this change of coordinates the tensor T_{ij} will transform as

$$T_{ij}(x', y', z') = \Lambda^a_i \Lambda^b_j T_{ab}(x, y, z) \,. \tag{10.4.40}$$

Notice that this is just the standard transformation law for tensors. The crucial observation here is that, because of the axisymmetry assumption, we can in fact use the above equation to *define* the value of of T_{ij} outside the $y = 0$ plane:

$$T_{ij}(\rho \cos\varphi, \rho \sin\varphi, z) := \Lambda^a_i \Lambda^b_j T_{ab}(\rho, 0, z) \,. \tag{10.4.41}$$

Typically, we will know the position (x, y) of the grid points where we need data in the planes $y \neq 0$, so that we need to find the corresponding points in the $y = 0$ plane before applying the above expression. In general, the desired point in the central plane will not coincide with any grid point, so one-dimensional interpolation along the x axis must be used in order to obtain the values of the different tensor components at that point. Notice that if we want to preserve the order of accuracy of our finite difference method the interpolation should be at least one order higher, in other words for a second order code we should use cubic interpolation. Notice also that, at the boundaries, extrapolation, as opposed to interpolation, would seem to be required (see Figure 10.21). Extrapolation, on the other hand, is undesirable since it tends to introduce instabilities. However, it turns out that extrapolation is not really necessary, since we can first apply the physical boundary on the plane $y = 0$, rotate to obtain all *interior* points for the planes $y \neq 0$, and finally apply again the physical boundary condition on those planes.

The extensive use of interpolation in the Cartoon approach might also seem to be a problem, since we could expect it to introduce large numerical errors, or even cause the code to become unstable. The complexity of the equations coupled with the interpolation make a detailed stability analysis very difficult in practice, but empirical evidence shows that this approach is quite robust.

The main advantage of the Cartoon approach is clearly the fact that it completely bypasses the issue of the coordinate singularities on the axis and origin, making the regularity problem irrelevant. On the other hand, it has the disadvantage that it requires us to write a full three-dimensional code, which is considerably more time-consuming than writing an axisymmetric code. However, if we already have a three-dimensional code available, adapting it to do axisymmetric simulations using the Cartoon method should be quite straightforward. The main idea can, of course, also be used for spherical symmetry by considering a thin pencil of grid points centered around the x axis that is only a few grid points thick along the y and z directions, though spherical symmetry is usually sufficiently simple on its own so that using the Cartoon method is probably not necessary.

APPENDIX A

TOTAL MASS AND MOMENTUM IN GENERAL RELATIVITY

The concept of energy is of fundamental importance in most physical theories. Indeed, even in relativity the stress-energy tensor of matter $T^{\mu\nu}$ plays a key role both in the conservation laws, from which the dynamics can often by fully determined, and as a source of the gravitational field. In general relativity, the stress-energy tensor satisfies a conservation law of the form $\nabla_\nu T^{\mu\nu} = 0$, but in contrast to special relativity this local conservation law does not lead to global conservation of energy integrated over a finite volume. The reason is clear, as $T^{\mu\nu}$ represents only the energy of matter and does not take into account the contribution from the gravitational field. However, it turns out that, in relativity, we can in general not define the energy density of the gravitational field itself.

On the other hand, in general relativity we can in fact define the *total energy* of an isolated system in a meaningful way. We will consider here two different approaches to defining both the total energy and the momentum of an asymptotically flat spacetime. The presentation here will be brief and not rigorous; for a more rigorous discussion see *e.g.* [31, 222, 306].

The first approach is motivated by considering weak gravitational fields for which $g_{\mu\nu} = \eta_{\mu\nu} + h_{\mu\nu}$, with $|h_{\mu\nu}| \ll 1$. In such a case, it is clear that the total mass (energy) and momentum can be defined as the integral of the energy density ρ and momentum density j^i of matter, respectively

$$M = \int \rho \, dV \,, \qquad P_i = \int j_i \, dV \,. \tag{A.1}$$

Using now the Hamiltonian and momentum constraints of the 3+1 formalism, equations (2.4.10) and (2.4.11), plus the fact that, in the weak field limit, the extrinsic curvature K_{ij} is itself a small quantity, we can rewrite this as (in all following expressions sum over repeated indices is understood)

$$M = \frac{1}{16\pi} \int R \, dV \,, \qquad P_i = \frac{1}{8\pi} \int \partial_j \left(K_{ij} - \delta_{ij} K \right) dV \,, \tag{A.2}$$

where R is the Ricci scalar of the spatial metric, which in the linearized theory is given by

$$R = \partial_j \left(\partial_i h_{ij} - \partial_j h \right) \,, \tag{A.3}$$

with h the trace of h_{ij}. Notice that both the mass and momentum are now written as volume integrals of a divergence, so using Gauss' theorem we can rewrite them as

$$M = \frac{1}{16\pi} \oint_S \left(\partial_i h_{ij} - \partial_j h \right) dS^j \, , \quad P_i = \frac{1}{8\pi} \oint_S \left(K_{ij} - \delta_{ij} K \right) dS^j \, , \qquad \text{(A.4)}$$

where the integrals are calculated over surfaces outside the matter sources, and where $\gamma_{ij} = \delta_{ij} + h_{ij}$ is the spatial metric and $dS^i = s^i dA$, with s^i the unit outward-pointing normal vector to the surface and dA the area element. The expressions above are called the *ADM mass* and *ADM momentum* of the spacetime (ADM stands for Arnowitt, Deser, and Misner [31]). Notice that the ADM mass depends only on the behavior of the spatial metric, while the ADM momentum depends on the extrinsic curvature instead. We should stress the fact that for these expressions to hold we must be working with quasi-Minkowski, *i.e.* Cartesian type, coordinates.

In order to derive the ADM integrals above, we started from the weak field theory and used Gauss' theorem to transform the volume integrals into surface integrals. In the case of strong gravitational fields, the total mass and momentum can not be expected to be given directly by the volume integrals of the energy and momentum densities, precisely because these integrals fail to take into account the effect of the gravitational field. However, we in fact define the total mass and momentum of an isolated system through its gravitational effects on faraway test masses, and far away the weak field limit does hold, so the surface integrals above will still give the correct values we physically associate with the total mass and momentum, but they must be evaluated at infinity to guarantee that we are in the weak field regime. We then define the ADM mass and momentum in the general case as

$$M_{ADM} = \frac{1}{16\pi} \lim_{r \to \infty} \oint_S \left(\delta^{ij} \partial_i h_{jk} - \partial_k h \right) dS^k \, , \qquad \text{(A.5)}$$

$$P^i_{ADM} = \frac{1}{8\pi} \lim_{r \to \infty} \oint_S \left(K^i_j - \delta^i_j K \right) dS^j \, , \qquad \text{(A.6)}$$

Using the same idea we can also construct a measure of the total angular momentum J^i of the system starting from the density of angular momentum given by $\epsilon_{ijk} x^j j^k$, with x^i Cartesian coordinates (this is just $\vec{r} \times \vec{j}$ in standard three-dimensional vector notation). The ADM angular momentum then becomes

$$J^i_{ADM} = \frac{1}{8\pi} \lim_{r \to \infty} \oint_S \epsilon^{ijk} x_j \left(K_{kl} - \delta_{kl} K \right) dS^l \, . \qquad \text{(A.7)}$$

Notice that, in fact, in the limit $r \to \infty$, the term proportional to K can be dropped out since at spatial infinity the Cartesian vector $\vec{x}$ is collinear with the area element $d\vec{S}$ so that $\epsilon^{ijl} x_j dS_l = 0$. The integral can then be rewritten as

$$J^i_{ADM} = \frac{1}{8\pi} \lim_{r \to \infty} \oint_S \epsilon^{ijk} x_j K_{kl} \, dS^l \, . \qquad \text{(A.8)}$$

In particular, for the projection along the z axis we find

$$J^z_{ADM} = \frac{1}{8\pi} \lim_{r\to\infty} \oint_S \epsilon^{zjk} x_j K_{kl} \, dS^l$$
$$= \frac{1}{8\pi} \lim_{r\to\infty} \oint_S \varphi^k K_{kl} \, dS^l \,, \tag{A.9}$$

with $\vec{\varphi}$ the coordinate basis vector associated with the azimuthal angle φ.

There is a very important property of the ADM mass and momentum defined above. Since these quantities are defined at spatial infinity i^0, and since any gravitational waves taking energy and momentum away from the system will instead reach null infinity $\mathscr{J}^+$, the ADM mass and momentum will remain constant during the evolution of the system.

The expression for the ADM mass given in (A.5) has the disadvantage that it is not covariant and must be calculated in Cartesian-type coordinates. This can in fact be easily remedied. In an arbitrary curvilinear coordinate system define the tensor $h_{ij} := \gamma_{ij} - \gamma^0_{ij}$, with γ^0_{ij} the metric of flat space expressed in the same coordinates. It is easy to convince oneself that h_{ij} does indeed transform like a tensor. We can now rewrite the ADM mass as

$$M_{ADM} = \frac{1}{16\pi} \lim_{r\to\infty} \oint_S \left(D^0_j h^j_k - D^0_k h \right) dS^k \,, \tag{A.10}$$

where D^0_i is the covariant derivative with respect to the flat metric γ^0_{ij} in the corresponding curvilinear coordinates, and where indices are raised and lowered also with γ^0_{ij}. Notice that the tensor h_{ij} is not unique because we can still make infinitesimal coordinate transformations that will change h_{ij} without changing $D^0_j h^j_k - D^0_k h$ (gauge transformations in the context of linearized theory).

From the last result we can also obtain a particularly useful expression for the ADM mass:

$$M_{ADM} = \frac{1}{8\pi} \lim_{r\to\infty} \oint_S \left(k - k^0\right) dA \,, \tag{A.11}$$

where dA is the area element of the surface S, k is the trace of the extrinsic curvature of S, and k^0 is the trace of the extrinsic curvature of S when embedded in flat space.[108] It is important to mention that the surface S, when embedded in flat space, must have the same intrinsic curvature as it did when embedded in curved space. The derivation of this expression is not difficult and can be obtained by considering an adapted coordinate system, with one of the coordinates measuring proper distance along the normal direction and the other two coordinates transported orthogonally off S; see for example [227] (in 3+1 terms, we would have a "lapse" equal to 1 and zero "shift", but in this case the orthogonal direction is spacelike and the surface is bi-dimensional).

[108] This expression for the ADM mass is due to Katz, Lynden-Bell, and Israel [169] (see also [227], but notice that our definition of extrinsic curvature has a sign opposite to that reference).

In the case when the spatial metric is conformally flat, *i.e.* $\gamma_{ij} = \psi^4 \delta_{ij}$, the expression for the ADM mass simplifies considerably and reduces to [222]

$$M_{ADM} = -\frac{1}{2\pi} \lim_{r\to\infty} \oint_S \partial_j \psi \, dS^j \,, \tag{A.12}$$

where we have assumed that far away ψ becomes unity.

There is another common approach to defining global energy and momentum that works for spacetimes that have a Killing field ξ^μ. I will give here a rough derivation of the corresponding expressions; a more formal derivation can be found in *e.g.* [108]. Notice first that for any antisymmetric tensor $A^{\mu\nu}$ we have

$$|g|^{1/2} \nabla_\nu A^{\mu\nu} = \partial_\nu \left(|g|^{1/2} A^{\mu\nu} \right) \,, \tag{A.13}$$

$$|g|^{1/2} \nabla_\mu \nabla_\nu A^{\mu\nu} = \partial_\mu \partial_\nu \left(|g|^{1/2} A^{\mu\nu} \right) = 0 \,. \tag{A.14}$$

Integrating the second expression above and using Gauss' theorem we find

$$\begin{aligned} 0 &= \int_\Omega \nabla_\mu \nabla_\nu A^{\mu\nu} d\Omega = \int_\Omega |g|^{1/2} \nabla_\mu \nabla_\nu A^{\mu\nu} d^4x \\ &= \int_\Omega \partial_\mu \partial_\nu \left(|g|^{1/2} A^{\mu\nu} \right) d^4x = \oint_{\partial\Omega} \partial_\nu \left(|g|^{1/2} A^{\mu\nu} \right) \hat{n}_\mu \, d^3x \,, \end{aligned} \tag{A.15}$$

where Ω is some four-dimensional region of spacetime, $\partial\Omega$ is its boundary and $\hat{n}^\mu$ is the unit outward-pointing normal vector to the boundary, with the norm defined using a flat metric. Assume now that Ω is the region bounded by two spacelike hypersurfaces Σ_1 and Σ_2 (with Σ_1 to the future of Σ_2), and a timelike cylindrical world-tube σ. We then find

$$\begin{aligned} \int_{\Sigma_1} \partial_\nu \left(|g|^{1/2} A^{\mu\nu} \right) \hat{n}_\mu \, d^3x - \int_{\Sigma_2} \partial_\nu \left(|g|^{1/2} A^{\mu\nu} \right) \hat{n}_\mu \, d^3x \\ + \int_\sigma \partial_\nu \left(|g|^{1/2} A^{\mu\nu} \right) \hat{n}_\mu \, d^3x = 0 \,, \end{aligned} \tag{A.16}$$

where in both the integrals over Σ_1 and Σ_2 the normal vector $\hat{n}^\mu$ is taken to be future pointing.

Take now $A^{\mu\nu} = \nabla^\mu \xi^\nu$, which will be antisymmetric if ξ^μ is a Killing vector. For the integrals over the timelike cylinder we then have

$$\begin{aligned} \int_\sigma \partial_\nu \left(|g|^{1/2} \nabla^\mu \xi^\nu \right) \hat{n}_\mu \, d^3x &= \int_\sigma |g|^{1/2} \left(\nabla^\nu \nabla^\mu \xi_\nu \right) \hat{n}_\mu \, d^3x \\ &= \int_\sigma |g|^{1/2} R^{\mu\nu} \xi_\nu \hat{n}_\mu \, d^3x \\ &= 8\pi \int_\sigma |g|^{1/2} \left(T^{\mu\nu} - \frac{1}{2} g^{\mu\nu} T \right) \xi_\nu \hat{n}_\mu \, d^3x \,, \end{aligned} \tag{A.17}$$

where $R_{\mu\nu}$ is the Ricci tensor and $T_{\mu\nu}$ the stress-energy tensor of matter. In the second line above we used the Ricci identity to relate the commutator of covariant derivatives to the Riemann tensor (1.9.3) plus the fact that ξ^μ is a Killing field, and in the third line the Einstein field equations. If we now assume that we have an isolated source and the cylindrical world-tube is outside it then the above integral clearly vanishes. Equation (A.16) then implies that

$$\int_{\Sigma_1} \partial_\nu \left(|g|^{1/2} \nabla^\mu \xi^\nu \right) \hat{n}_\mu \, d^3x = \int_{\Sigma_2} \partial_\nu \left(|g|^{1/2} \nabla^\mu \xi^\nu \right) \hat{n}_\mu \, d^3x \,, \qquad \text{(A.18)}$$

This means that the spatial integral is in fact independent of the hypersurface Σ chosen, or in other words, the spatial integral is a conserved quantity.

To proceed further we choose coordinates such that $\hat{n}^\mu = \delta^\mu_0$. Using the fact that $\nabla^\mu \xi^\nu$ is antisymmetric, and applying again Gauss' theorem, the spatial integral becomes

$$\begin{aligned} \int_\Sigma \partial_i \left(|g|^{1/2} \nabla_0 \xi^i \right) d^3x &= \oint_{\partial\Sigma} |g|^{1/2} \nabla_0 \xi^i \, \hat{s}_i \, d^2x \\ &= \oint_{\partial\Sigma} |g|^{1/2} \nabla^\mu \xi^\nu \, \hat{s}_\nu \hat{n}_\mu \, d^2x \,, \end{aligned} \qquad \text{(A.19)}$$

where $\partial\Sigma$ is now the two-dimensional boundary of Σ and $\hat{s^\mu}$ is the spatial unit outward-pointing normal vector to $\partial\Sigma$. In the last expression we can notice that $|g|^{1/2} \hat{n}_\mu \hat{s}_\nu \, d^2x = n_\mu s_\nu \, dA$, where now n^μ and s^μ are unit vectors with respect to the full curved metric $g_{\mu\nu}$, and dA is the proper area element of $\partial\Sigma$. We can then write the integral as

$$I_K \left(\vec{\xi} \right) := -\frac{1}{4\pi} \oint_{\partial\Sigma} s_\mu n_\nu \nabla^\mu \xi^\nu \, dA \,. \qquad \text{(A.20)}$$

This is known as the *Komar integral* [176] and as we have seen is a conserved quantity (the quantity $A_{\mu\nu} = \nabla_\mu \xi_\nu$ is also frequently called the *Komar potential*).[109] The integral can in fact be calculated over any surface $\partial\Sigma$ outside the matter sources. The normalization factor $-1/4\pi$ has been chosen to ensure that, for the case of a timelike Killing field that has unit magnitude at infinity, the Komar integral will coincide with the ADM mass of the spacetime. If, on the other hand, ξ^μ is an axial vector associated with an angular coordinate ϕ (*i.e.* $\xi^\mu = \delta^\mu_\phi$), then the Komar integral turns out to be $-2J$, with J the total angular momentum (we can check that this is so by considering a Kerr black hole). The fact that, for a static spacetime, the Komar integral and the ADM mass coincide can be used to derive a general relativistic version of the virial theorem, but we will not consider this issue here (the interested reader can see *e.g.* [69]).

[109] The Komar integral is usually written in differential form notation as

$$I_K = -\frac{1}{4\pi} \oint \nabla^\mu \xi^\nu dS_{\mu\nu} = -\frac{1}{8\pi} \oint \epsilon_{\mu\nu\alpha\beta} \nabla^\mu \xi^\nu \, dx^\alpha \wedge dx^\beta \,.$$

In fact, in the last expression, the area element $dx^\alpha \wedge dx^\beta$ is frequently not even written.

Incidentally, we can use the same derivation that allowed us to show that the integral over the timelike cylinder σ vanishes to rewrite the Komar integral as a volume integral of the stress-energy of matter as

$$I_K\left(\vec{\xi}\right) := 2\int_\Sigma \left(T^{\mu\nu} - \frac{1}{2}g^{\mu\nu}T\right) n_\mu \xi_\nu \, dV \; , \tag{A.21}$$

with dV the proper volume element of the spatial hypersurface.

The Komar integral can in fact also be used to define mass at null infinity $\mathscr{J}^+$ in the case of non-stationary asymptotically flat spacetimes. This is known as the *Bondi mass*, and is defined as

$$M_B := -\frac{1}{4\pi} \lim_{S\to\mathcal{J}} \oint_S s_\mu n_\nu \nabla^\mu \xi^\nu \, dA \; , \tag{A.22}$$

where now ξ^μ is assumed to be the generator of an asymptotic time translation symmetry, and where the surface of integration is now taken to approach a cross section $\mathcal{J}$ of $\mathscr{J}^+$ (when ξ^μ is not an exact Killing field we must ask for an extra normalization condition, see *e.g.* [295]). While the ADM mass measures the total energy available in a spacetime, the Bondi mass represents the remaining energy at a retarded time. This means that, in particular, the Bondi mass can be used to calculate the change in the total energy of an isolated system that goes through a phase in which it radiates gravitational waves.

Before finishing this Appendix it is important to mention that the ADM expressions for mass and momentum given above, though correct, in practice converge very slowly as r goes to infinity, so that if we evaluate them at a finite radius in a numerical simulation the errors might be quite large. We can of course evaluate them at several radii and then extrapolate to the asymptotic value. For the mass, however, there are other approaches that work well in practice. One such approach is based on the use of the so-called *Hawking mass*, which is a quasi-local measure of energy defined for any given closed surface S as (see *e.g.* [276])

$$M_{\rm H} := \sqrt{\frac{A}{16\pi}} \left(1 + \frac{1}{16\pi}\oint_S H_{\rm in} H_{\rm out} dS\right) \; , \tag{A.23}$$

where A is the area of the surface, and $H_{\rm in}$ and $H_{\rm out}$ are the expansions of the ingoing and outgoing null geodesics which are given in terms of 3+1 quantities as (see equation (6.7.8))

$$H_{\rm in} = K_{mn}s^m s^n - K - D_m s^m \; , \tag{A.24}$$

$$H_{\rm out} = K_{mn}s^m s^n - K + D_m s^m \; , \tag{A.25}$$

with $\vec{s}$ being the unit outward-pointing normal vector to S, and D_i the standard three-dimensional covariant derivative. The Hawking mass is defined by thinking that the presence of a mass must cause light rays to converge, and it turns out that the only gauge invariant measure of this on the surface is given by $H_{\rm in}H_{\rm out}$.[110] We would then expect the mass to be given by an expression of the form $M = A + B \oint H_{\rm in}H_{\rm out} dS$, with the constants A and B fixed by looking at special cases.

For spheres in Minkowski spacetime the Hawking mass vanishes identically since $H_{\rm in} = -H_{\rm out} = 2/r$. For spheres in Schwarzschild (in standard coordinates) we have $H_{\rm in} = -H_{\rm out} = 2(1 - 2M/r)^{1/2}/r$ and $A = 4\pi r^2$, so that $M_{\rm H} = M$. That is, the Hawking mass gives us the correct mass at all radii, which makes it a more useful measure of energy than the ADM mass. The Hawking mass in general is not always positive definite, and is not always monotonic either, but for sufficiently "round" surfaces (and particularly in spherical symmetry) both these properties can be shown to be satisfied if the dominant energy condition holds.

Another less formal but very useful approach to obtaining a local expression for the mass of the spacetime is based on the fact that many astrophysically relevant spacetimes will not only be asymptotically flat, but they will also be asymptotically spherically symmetric. In such a case we know that the spacetime will approach Schwarzschild far away. For Schwarzschild we can easily prove the following exact relation between the mass M and the area A of spheres

$$M = \left(\frac{A}{16\pi} \right)^{1/2} \left[1 - \frac{(dA/dr)^2}{16\pi g_{rr} A} \right] , \tag{A.26}$$

where r is some arbitrary radial coordinate. In numerical simulations of spacetimes that are asymptotically spherically symmetric we can calculate the area A of coordinate spheres at finite radii, the rate of change of the area with radius dA/dr, and the average radial metric over the sphere $\bar{g}_{rr}$, and then use the above expression to estimate the mass. In practice we often find that as r is increased this "Schwarzschild-like" mass converges very rapidly to the correct ADM mass, much more rapidly than the ADM integral itself. This works even for spacetimes with non-zero angular momentum like Kerr, as the angular momentum terms in the metric decay faster than the mass term.

[110]More specifically, the product of the Newman–Penrose spin coefficients ρ and ρ' is gauge invariant under a spin-boost transformation of the null tetrad, and in general it is given by $\rho\rho' = H_{\rm in}H_{\rm out}/8$.

APPENDIX B

SPACETIME CHRISTOFFEL SYMBOLS IN 3+1 LANGUAGE

In the derivation of 3+1 equations we often need to write the 4-metric of spacetime and its associated Christoffel symbols in 3+1 language. The 4-metric in terms of 3+1 quantities has the form

$$g_{00} = -\left(\alpha^2 - \gamma_{ij}\,\beta^i\beta^j\right) , \tag{B.1}$$

$$g_{0i} = \gamma_{ij}\,\beta^j = \beta_i , \tag{B.2}$$

$$g_{ij} = \gamma_{ij} , \tag{B.3}$$

and the corresponding inverse metric is

$$g^{00} = -1/\alpha^2 , \tag{B.4}$$

$$g^{0i} = \beta^i/\alpha^2 , \tag{B.5}$$

$$g^{ij} = \gamma^{ij} - \beta^i\beta^j/\alpha^2 . \tag{B.6}$$

From this we can obtain the following expressions for the 4-Christoffel symbols in terms of 3+1 quantities

$$\Gamma^0_{00} = \left(\partial_t\,\alpha + \beta^m\partial_m\,\alpha - \beta^m\beta^n K_{mn}\right)/\alpha , \tag{B.7}$$

$$\Gamma^0_{0i} = \left(\partial_i\alpha - \beta^m K_{im}\right)/\alpha , \tag{B.8}$$

$$\Gamma^0_{ij} = -K_{ij}/\alpha , \tag{B.9}$$

$$\begin{aligned}\Gamma^l_{00} &= \alpha\partial^l\alpha - 2\alpha\beta^m K^l_m - \beta^l\left(\partial_t\alpha + \beta^m\partial_m\alpha - \beta^m\beta^n K_{mn}\right)/\alpha \\ &\quad + \partial_t\beta^l + \beta^m\, D_m\beta^l ,\end{aligned} \tag{B.10}$$

$$\Gamma^l_{m0} = -\beta^l\left(\partial_m\alpha - \beta^n K_{mn}\right)/\alpha - \alpha K^l_m + D_m\beta^l , \tag{B.11}$$

$$\Gamma^l_{ij} = {}^{(3)}\Gamma^l_{ij} + \beta^l K_{ij}/\alpha , \tag{B.12}$$

with D_i the covariant derivative associated with the spatial metric γ_{ij}, and ${}^{(3)}\Gamma^l_{ij}$ the corresponding three-dimensional Christoffel symbols. The 3-covariant derivatives of the shift that appear in these expressions come from partial derivatives of the metric along the time direction contained in the $\Gamma^\alpha_{\mu\nu}$.

The contracted Christoffel symbols $\Gamma^\alpha := g^{\mu\nu}\Gamma^\alpha_{\mu\nu}$ then become

$$\Gamma^0 = -\frac{1}{\alpha^3}\left(\partial_t\alpha - \beta^m\partial_m\alpha + \alpha^2 K\right) , \tag{B.13}$$

$$\begin{aligned}\Gamma^i &= {}^{(3)}\Gamma^i + \frac{\beta^i}{\alpha^3}\left(\partial_t\alpha - \beta^m\partial_m\alpha + \alpha^2 K\right) \\ &\quad - \frac{1}{\alpha^2}\left(\partial_t\beta^i - \beta^m\partial_m\beta^i + \alpha\partial^i\alpha\right) .\end{aligned} \tag{B.14}$$

APPENDIX C

BSSNOK WITH NATURAL CONFORMAL RESCALING

The BSSNOK formulation of the 3+1 evolution equations described in Section 2.8 is based on a conformal rescaling of the metric and extrinsic curvature. However, in its standard version this formulation uses a rescaling of the extrinsic curvature that is in fact not the most natural rescaling for this quantity. Here I will present a version of the BSSNOK equations that uses the natural rescaling for a tracefree symmetric tensor.

We start by summarizing the standard BSSNOK equations. We perfom a conformal transformation of the spatial metric and extrinsic curvature of the form

$$\tilde{\gamma}_{ij} := e^{-4\phi}\,\gamma_{ij} \,, \tag{C.1}$$

$$\tilde{A}_{ij} := e^{-4\phi}\left(K_{ij} - \frac{1}{3}\gamma_{ij}K\right) \,, \tag{C.2}$$

where the conformal factor is taken to be $\phi = \frac{1}{12}\ln\gamma$, so that the conformal metric $\tilde{\gamma}_{ij}$ has unit determinant. In the previous expression K is the trace of K_{ij}, which implies that the tensor $\tilde{A}_{ij}$ is traceless. We also introduces the auxiliary variables

$$\tilde{\Gamma}^i = \tilde{\gamma}^{jk}\tilde{\Gamma}^i_{jk} = -\partial_j\tilde{\gamma}^{ij} \,, \tag{C.3}$$

where the second equality follows from the fact that $\tilde{\gamma}_{ij}$ has unit determinant.

The full system of evolution equations then becomes

$$\frac{d}{dt}\,\tilde{\gamma}_{ij} = -2\alpha\tilde{A}_{ij} \,, \tag{C.4}$$

$$\frac{d}{dt}\,\phi = -\frac{1}{6}\alpha K \,, \tag{C.5}$$

$$\frac{d}{dt}\,K = -D_iD^i\alpha + \alpha\left(\tilde{A}_{ij}\tilde{A}^{ij} + \frac{1}{3}K^2\right) + 4\pi\alpha\left(\rho + S\right) \,, \tag{C.6}$$

$$\begin{aligned}\frac{d}{dt}\,\tilde{A}_{ij} &= e^{-4\phi}\left\{-D_iD_j\alpha + \alpha R_{ij} + 4\pi\alpha\left[\gamma_{ij}\left(S-\rho\right) - 2S_{ij}\right]\right\}^{\mathrm{TF}} \\ &\quad + \alpha\left(K\tilde{A}_{ij} - 2\tilde{A}_{ik}\tilde{A}^k{}_j\right) \,,\end{aligned} \tag{C.7}$$

$$\begin{aligned}\frac{d}{dt}\,\tilde{\Gamma}^i &= \tilde{\gamma}^{jk}\partial_j\partial_k\beta^i + \frac{1}{3}\,\tilde{\gamma}^{ij}\partial_j\partial_k\beta^k - 2\tilde{A}^{ij}\partial_j\alpha \\ &\quad + 2\alpha\left(\tilde{\Gamma}^i_{jk}\tilde{A}^{jk} + 6\tilde{A}^{ij}\partial_j\phi - \frac{2}{3}\tilde{\gamma}^{ij}\partial_jK - 8\pi e^{4\phi}j^i\right) \,,\end{aligned} \tag{C.8}$$

where TF denotes the tracefree part of the expression inside the brackets, and where we have used the Hamiltonian constraint to eliminate the Ricci scalar from the evolution equation for K, and the momentum constraints to eliminate the divergence of $\tilde{A}^{ij}$ from the evolution equation for $\tilde{\Gamma}^i$. In the previous expressions we have $d/dt := \partial_t - \pounds_{\vec{\beta}}$, with $\pounds_{\vec{\beta}}$ the Lie derivative with respect to the shift that must be calculated for tensor densities: ψ, a scalar density of weight $1/6$, and $\tilde{\gamma}_{ij}$ and $\tilde{A}_{ij}$, tensor densities with weight $-2/3$. Also, even though the $\tilde{\Gamma}^i$ is strictly speaking not a vector, its Lie derivative is understood as that corresponding to a vector density of weight $2/3$. Finally, the Ricci tensor is separated into two contributions in the following way:

$$R_{ij} = \tilde{R}_{ij} + R^{\phi}_{ij} \,, \tag{C.9}$$

where $\tilde{R}_{ij}$ is the Ricci tensor associated with the conformal metric $\tilde{\gamma}_{ij}$:

$$\begin{aligned} \tilde{R}_{ij} = & -\frac{1}{2}\tilde{\gamma}^{lm}\partial_l\partial_m\tilde{\gamma}_{ij} + \tilde{\gamma}_{k(i}\partial_{j)}\tilde{\Gamma}^k + \tilde{\Gamma}^k\tilde{\Gamma}_{(ij)k} \\ & + \tilde{\gamma}^{lm}\left(2\tilde{\Gamma}^k_{l(i}\tilde{\Gamma}_{j)km} + \tilde{\Gamma}^k_{im}\tilde{\Gamma}_{klj}\right) \,, \end{aligned} \tag{C.10}$$

and where R^{ϕ}_{ij} is given by ϕ:

$$R^{\phi}_{ij} = -2\tilde{D}_i\tilde{D}_j\phi - 2\tilde{\gamma}_{ij}\tilde{D}^k\tilde{D}_k\phi + 4\tilde{D}_i\phi\,\tilde{D}_j\phi - 4\tilde{\gamma}_{ij}\tilde{D}^k\phi\,\tilde{D}_k\phi \,, \tag{C.11}$$

with $\tilde{D}_i$ the covariant derivative associated with the conformal metric.

The Hamiltonian and momentum constraints also take the form

$$R = \tilde{A}_{ij}\tilde{A}^{ij} - \frac{2}{3}K^2 + 16\pi\rho \,, \tag{C.12}$$

$$\partial_j\tilde{A}^{ij} = -\tilde{\Gamma}^i_{jk}\tilde{A}^{jk} - 6\tilde{A}^{ij}\partial_j\phi + \frac{2}{3}\tilde{\gamma}^{ij}\partial_j K + 8\pi e^{4\phi}j^i \,. \tag{C.13}$$

Now, in Chapter 3 it was shown that the natural conformal transformation for the tracefree extrinsic curvature is in fact $\bar{A}^{ij} = \psi^{10}A^{ij}$ ($\bar{A}_{ij} = \psi^2 A_{ij}$), with $\phi = \ln\psi$. We will then consider the following conformal transformation

$$\bar{\gamma}_{ij} := e^{-4\phi}\,\gamma_{ij} \,, \tag{C.14}$$

$$\bar{A}_{ij} := e^{+2\phi}\left(K_{ij} - \frac{1}{3}\gamma_{ij}K\right) \,. \tag{C.15}$$

We will furthermore introduce the densitized lapse

$$\bar{\alpha} = \alpha\gamma^{-1/2} = e^{-6\phi}\alpha \,. \tag{C.16}$$

With this new conformal scaling, the evolution equations become instead (do notice that in some places there is $\bar{\alpha}$ and in others just α)

$$\frac{d}{dt}\,\bar{\gamma}_{ij} = -2\bar{\alpha}\bar{A}_{ij} \; , \tag{C.17}$$

$$\frac{d}{dt}\,\phi = -\frac{e^{6\phi}}{6}\,\bar{\alpha}K \; , \tag{C.18}$$

$$\frac{d}{dt}\,K = -D_i D^i \alpha + \bar{\alpha}\left[e^{-6\phi}\bar{A}_{ij}\bar{A}^{ij} + \frac{1}{3}e^{6\phi}K^2 + 4\pi e^{6\phi}\left(\rho + S\right)\right] \; , \tag{C.19}$$

$$\frac{d}{dt}\,\bar{A}_{ij} = e^{2\phi}\left\{-D_i D_j \alpha + \alpha R_{ij} + 4\pi\alpha\left[\gamma_{ij}\left(S - \rho\right) - 2S_{ij}\right]\right\}^{\mathrm{TF}} - 2\bar{\alpha}\bar{A}_{ik}\bar{A}^k{}_j \; , \tag{C.20}$$

$$\frac{d}{dt}\,\bar{\Gamma}^i = \tilde{\gamma}^{jk}\partial_j\partial_k\beta^i + \frac{1}{3}\,\tilde{\gamma}^{ij}\partial_j\partial_k\beta^k - 2\bar{A}^{ij}\partial_j\bar{\alpha} + 2\bar{\alpha}\left(\tilde{\Gamma}^i_{jk}\tilde{A}^{jk} - \frac{2}{3}e^{6\phi}\bar{\gamma}^{ij}\partial_j K - 8\pi e^{10\phi}j^i\right) \; , \tag{C.21}$$

and the constraints become

$$R = e^{-12\phi}\bar{A}_{ij}\bar{A}^{ij} - \frac{2}{3}\,K^2 + 16\pi\rho \; , \tag{C.22}$$

$$\partial_j\bar{A}^{ij} = -\bar{\Gamma}^i_{jk}\bar{A}^{jk} + \frac{2}{3}e^{6\phi}\bar{\gamma}^{ij}\partial_j K + 8\pi e^{10\phi}j^i \; . \tag{C.23}$$

Notice that with this new rescaling, a term involving derivatives of the conformal factor ϕ has disappeared from both the evolution equation for $\bar{\Gamma}^i$ and the momentum constraints.

APPENDIX D

SPIN-WEIGHTED SPHERICAL HARMONICS

Consider the Laplace operator written in spherical coordinates as

$$\nabla^2 f = \frac{1}{r^2}\,\partial_r^2\,\left(r^2 f\right) + \frac{1}{r^2}\,L^2 f\;, \tag{D.1}$$

where L^2 is the angular operator

$$L^2 f := \frac{1}{\sin\theta}\,\partial_\theta\left(\sin\theta\,\partial_\theta f\right) + \frac{1}{\sin^2\theta}\,\partial_\varphi^2 f\;. \tag{D.2}$$

By separation of variables, we can show that f will be a solution of the Laplace equation $\nabla^2 f = 0$ if it can be written as

$$f(r,\theta,\varphi) = r^l g_1(\theta,\varphi) + \frac{1}{r^{l+1}}\,g_2(\theta,\varphi)\;, \tag{D.3}$$

where $g_{1,2}(\theta,\varphi)$ are eigenfunctions of the L^2 operator such that

$$L^2 g = -l\,(l+1)\,g\;. \tag{D.4}$$

The solutions of the last equation are known as *spherical harmonics*. They are usually denoted by $Y^{l,m}(\theta,\varphi)$ and have the form

$$Y^{l,m}(\theta,\varphi) = \left[\frac{(2l+1)}{4\pi}\,\frac{(l-m)!}{(l+m)!}\right]^{1/2} P^{l,m}\left(\cos\theta\right)e^{im\varphi}\;, \tag{D.5}$$

where l and m are integers such that $|m| \leq l$, and $P^{l,m}(z)$ are the associated Legendre polynomials. The $Y^{l,m}(\theta,\varphi)$ are orthogonal to each other when integrated over a sphere, and the normalization chosen above is such that

$$\oint Y^{l,m}(\theta,\varphi)\,\bar{Y}^{l',m'}(\theta,\varphi)\,d\Omega = \delta_{ll'}\delta_{mm'}\;, \tag{D.6}$$

where $d\Omega = \sin\theta d\theta d\varphi$ is the standard area element of the sphere. Using the fact that the associated Legendre functions are such that

$$P^{l,-m}(z) = (-1)^m\frac{(l-m)!}{(l+m)!}\,P^{l,m}(z)\;, \tag{D.7}$$

we can show that the complex conjugate of the $Y^{l,m}$ is given by

$$\bar{Y}^{l,m}(\theta,\varphi) = (-1)^m\,Y^{l,-m}(\theta,\varphi)\;. \tag{D.8}$$

When we work with non-scalar functions defined on the sphere we introduce the so-called *spin-weighted spherical harmonics* as generalizations of the ordinary spherical harmonics. Spin-weighted spherical harmonics were first introduced by Newman and Penrose [219] for the study of gravitational radiation, but they can also be used to study solutions of the Maxwell equations, the Dirac equation, or in fact dynamical equations for fields of arbitrary spin.

Consider a complex function f on the sphere that might correspond to some combination of components of a tensorial (or spinorial) object in the orthonormal basis $(\hat{e}_r, \hat{e}_\theta, \hat{e}_\varphi)$ induced by the spherical coordinates (r, θ, φ). We will say that f has spin weight s if, under a rotation of the angular basis $(\hat{e}_\theta, \hat{e}_\varphi)$ by an angle ψ, it transforms as $f \rightarrow e^{-is\psi} f$. A trivial example is a scalar function whose spin weight is clearly zero. A more interesting example corresponds to a three-dimensional vector $\vec{v}$ with components $(v^{\hat{r}}, v^{\hat{\theta}}, v^{\hat{\phi}})$. Notice that these components are different from those in the coordinate basis (which is not orthonormal), and are related to them through $(v^{\hat{r}}, v^{\hat{\theta}}, v^{\hat{\phi}}) = (v^r, rv^\theta, r\sin\theta\, v^\phi)$. Define now two unit complex vectors as

$$\hat{e}_\pm := (\hat{e}_\theta \mp i\hat{e}_\varphi) / \sqrt{2} \, . \tag{D.9}$$

The vector $\vec{v}$ can then be written as

$$\vec{v} = v^0 \hat{e}_r + v^+ \hat{e}_+ + v^- \hat{e}_- \, , \tag{D.10}$$

where

$$v^0 := v^{\hat{r}} \, , \qquad v^+ := \left(v^{\hat{\theta}} + i v^{\hat{\phi}} \right) / \sqrt{2} \, , \qquad v^- := \left(v^{\hat{\theta}} - i v^{\hat{\phi}} \right) / \sqrt{2} \, . \tag{D.11}$$

By considering a rotation of the vectors $(\hat{e}_\theta, \hat{e}_\varphi)$ by an angle ψ it is now easy to see that v^0 has spin weight zero, while the spin weight of $v^\pm$ is ± 1.

The spin-weighted spherical harmonics, denoted by ${}_sY^{l,m}(\theta, \varphi)$, form a basis for the space of functions with definite spin weight s. They can be introduced in a number of different ways. We can start by defining the operators

$$\begin{aligned} \eth f &:= -\sin^s\theta \left(\partial_\theta + \frac{i}{\sin\theta}\, \partial_\varphi \right) \left(f \sin^{-s}\theta \right) \\ &= -\left(\partial_\theta + \frac{i}{\sin\theta}\, \partial_\varphi - s\cot\theta \right) f \, , \end{aligned} \tag{D.12}$$

$$\begin{aligned} \bar{\eth} f &:= -\sin^{-s}\theta \left(\partial_\theta - \frac{i}{\sin\theta}\, \partial_\varphi \right) \left(f \sin^{s}\theta \right) \\ &= -\left(\partial_\theta - \frac{i}{\sin\theta}\, \partial_\varphi + s\cot\theta \right) f \, , \end{aligned} \tag{D.13}$$

where s is the spin weight of f. The spin-weighted spherical harmonics are then defined for $|m| \leq l$ and $l \geq |s|$ in terms of the standard spherical harmonics as

$${}_sY^{l,m} := \left[\frac{(l-s)!}{(l+s)!} \right]^{1/2} \eth^s \left(Y^{l,m} \right) \, , \qquad +l \geq s \geq 0 \, , \tag{D.14}$$

$${}_sY^{l,m} := (-1)^s \left[\frac{(l+s)!}{(l-s)!} \right]^{1/2} \bar{\eth}^{-s} \left(Y^{l,m} \right) \, , \qquad -l \leq s \leq 0 \, . \tag{D.15}$$

In particular we have ${}_0Y^{l,m} = Y^{l,m}$. The above definition implies that

$$\eth \left({}_sY^{l,m}\right) = +\left[(l-s)(l+s+1)\right]^{1/2} {}_{s+1}Y^{l,m} \,, \tag{D.16}$$
$$\bar{\eth} \left({}_sY^{l,m}\right) = -\left[(l+s)(l-s+1)\right]^{1/2} {}_{s-1}Y^{l,m} \,. \tag{D.17}$$

Because of this, $\eth$ and $\bar{\eth}$ are known as the *spin raising* and *spin lowering* operators. We also find that

$$\bar{\eth}\eth \left({}_sY^{l,m}\right) = -\left[l(l+1) - s(s+1)\right] {}_sY^{l,m} \,, \tag{D.18}$$
$$\eth\bar{\eth} \left({}_sY^{l,m}\right) = -\left[l(l+1) - s(s-1)\right] {}_sY^{l,m} \,, \tag{D.19}$$

so the ${}_sY^{l,m}$ are eigenfunctions of the operators $\bar{\eth}\eth$ and $\eth\bar{\eth}$, which are generalizations of L^2. For a function with zero spin weight we in fact find that $L^2 f = \bar{\eth}\eth f = \eth\bar{\eth} f$.

From the above definitions it is possible to show that the complex conjugate of the ${}_sY^{l,m}$ is given by

$${}_s\bar{Y}^{l,m}(\theta,\varphi) = (-1)^{s+m} {}_{-s}Y^{l,-m}(\theta,\varphi) \,, \tag{D.20}$$

which is just the generalization of (D.8) to the case of non-zero spin.

We can also find generalizations of the standard angular momentum operators for the case of non-zero spin weight by looking for operators $\hat{J}_z$ and $\hat{J}_\pm$ such that (here we are ignoring the factor $-i\hbar$ that normally appears in quantum mechanics) [112]

$$\hat{J}_z \, {}_sY^{l,m} = im \, {}_sY^{l,m}, \tag{D.21}$$
$$\hat{J}_\pm \, {}_sY^{l,m} = i\left[(l \mp m)(l+1 \pm m)\right]^{1/2} {}_sY^{l,m\pm 1}. \tag{D.22}$$

We then find that such operators must have the form

$$\hat{J}_z = \partial_\varphi \,, \tag{D.23}$$
$$\hat{J}_\pm = e^{\pm i\varphi}\left[\pm i\partial_\theta - \cot\theta \, \partial_\varphi - is\csc\theta\right] \,. \tag{D.24}$$

The operators for the x and y components of the angular momentum are then simply obtained from $\hat{J}_\pm = \hat{J}_x \pm i\hat{J}_y$, so that we find:

$$\hat{J}_x = \left(\hat{J}_+ + \hat{J}_-\right)/2 \,, \quad \hat{J}_y = -i\left(\hat{J}_+ - \hat{J}_-\right)/2 \,, \tag{D.25}$$

The ${}_sY^{l,m}$ can also be constructed in terms of the so-called *Wigner rotation matrices* d^l_{ms}, which are defined in quantum mechanics as the following matrix elements of the operator for rotations around the y axis:

$$d^l_{ms}(\theta) := \left\langle l,m \left| e^{-i\hat{J}_y\theta} \right| l,s \right\rangle \,. \tag{D.26}$$

We then find that the ${}_sY^{l,m}$ have the form

$$sY^{l,m}(\theta,\phi) = (-1)^m \left(\frac{2l+1}{4\pi}\right)^{1/2} e^{im\phi}\, d^l_{-ms}(\theta) \; . \tag{D.27}$$

Closed expressions for the rotation matrices d^l_{ms}, as well as their principal properties, are well known but we will not go into the details here (the interested reader can look at standard textbooks on quantum mechanics, *e.g.* [202, 294]).

There are several very important properties of the spin-weighted spherical harmonics that can be obtained directly from the properties of the rotation matrices d^l_{ms}. In the first place, just as the ordinary spherical harmonics, the different ${}_sY^{l,m}$ are orthonormal,

$$\oint {}_sY^{l,m}(\theta,\varphi)\; {}_{s'}\bar{Y}^{l',m'}(\theta,\varphi)\, d\Omega = \delta_{ss'}\delta_{ll'}\delta_{mm'} \; . \tag{D.28}$$

Also, for a given value of s, the ${}_sY^{l,m}$ form a complete set. This property can be expressed in the form

$$\sum_{l,m} {}_sY^{l,m}(\theta,\varphi)\; {}_s\bar{Y}^{l,m}(\theta',\varphi') = \delta(\varphi-\varphi')\,\delta(\cos\theta - \cos\theta') \; . \tag{D.29}$$

The integral of three spin-weighted spherical harmonics is also frequently needed (for example in the calculation of the momentum flux of gravitational waves) and can be expressed in general as

$$\oint {}_{s_1}Y^{l_1,m_1}(\theta,\varphi)\; {}_{s_2}Y^{l_2,m_2}(\theta,\varphi)\; {}_{s_3}Y^{l_3,m_3}(\theta,\varphi)\, d\Omega =$$
$$\left[\frac{(2l_1+1)\,(2l_2+1)\,(2l_3+1)}{4\pi}\right]^{1/2} \begin{pmatrix} l_1 & l_2 & l_3 \\ -s_1 & -s_2 & -s_3 \end{pmatrix} \begin{pmatrix} l_1 & l_2 & l_3 \\ m_1 & m_2 & m_3 \end{pmatrix} . \tag{D.30}$$

In the above expression we have used the Wigner 3-lm symbols, which are related to the standard Clebsch–Gordan coefficients $\langle l_1, m_1, l_2, m_2 | j_3, m_3\rangle$ through

$$\begin{pmatrix} l_1 & l_2 & l_3 \\ m_1 & m_2 & m_3 \end{pmatrix} = \frac{(-1)^{l_1-l_2-m_3}}{\sqrt{2l_3+1}} \langle l_1, m_1, l_2, m_2 | l_3, -m_3\rangle \; , \tag{D.31}$$

or equivalently

$$\langle l_1, m_1, l_2, m_2 | l_3, m_3\rangle = (-1)^{l_1-l_2+m_3} \sqrt{2l_3+1} \begin{pmatrix} l_1 & l_2 & l_3 \\ m_1 & m_2 & -m_3 \end{pmatrix} . \tag{D.32}$$

The Clebsch–Gordan coefficients arise from the addition of angular momentum in quantum mechanics, and correspond to the coefficients of the expansion of an eigenstate $|L, M\rangle$ with total angular momentum L and projection M, in terms of a basis formed by the product of the individual eigenstates $|l_1, m_1\rangle\, |l_2, m_2\rangle$. These coefficients are always real, and are only different from zero if

$$-l_i < m_i < l_i\ , \qquad |l_1 - l_2| < l_3 < l_1 + l_2\ . \tag{D.33}$$

The Clebsch–Gordan coefficients have some important symmetries, though these symmetries are easier to express in terms of the 3-lm symbols. In particular, the 3-lm symbols are invariant under an even permutation of columns, and pick up a factor $(-1)^{l_1+l_2+l_3}$ under an odd permutation. Also,

$$\begin{pmatrix} l_1 & l_2 & l_3 \\ -m_1 & -m_2 & -m_3 \end{pmatrix} = (-1)^{l_1+l_2+l_3} \begin{pmatrix} l_1 & l_2 & l_3 \\ m_1 & m_2 & m_3 \end{pmatrix} . \tag{D.34}$$

A closed expression for the Clebsch–Gordan coefficients was first found by Wigner (see *e.g.* [299]). This expression is somewhat simpler when written in terms of the 3-lm coefficients, and has the form

$$\begin{aligned} \begin{pmatrix} l_1 & l_2 & l_3 \\ m_1 & m_2 & m_3 \end{pmatrix} &= (-1)^{l_1-m_1}\, \delta_{m_1+m_2,-m_3} \\ &\times \left[\frac{(l_1+l_2-l_3)!\,(l_1+l_3-l_2)!\,(l_2+l_3-l_1)!\,(l_3+m_3)!\,(l_3-m_3)!}{(l_1+l_2+l_3+1)!\,(l_1+m_1)!\,(l_1-m_1)!\,(l_2+m_2)!\,(l_2-m_2)!} \right]^{1/2} \\ &\times \sum_{k\geq 0} \frac{(-1)^k}{k!} \left[\frac{(l_2+l_3+m_1-k)!\,(l_1-m_1+k)!}{(l_3-l_1+l_2-k)!\,(l_3-m_3-k)!\,(l_1-l_2+m_3+k)!} \right] . \end{aligned} \tag{D.35}$$

In the above expression the sum runs over all values of k for which the arguments inside the factorials are non-negative. Also, if the particular combination of $\{l_i, m_i\}$ is such that the arguments of the factorials outside of the sum are negative, then the corresponding coefficient vanishes. A more symmetric (though longer) expression that is equivalent to (D.35) was later derived by Racah [234], but we will not write it here.

In the general case, (D.35) is rather complicated, but this is not a serious problem as we can find tables of the most common coefficients in the literature, and even web-based "Clebsch–Gordan calculators". Moreover, in some special cases the coefficients simplify considerably. For example, in the case where $m_1 = l_1$, $m_2 = l_2$, and $l_3 = m_3 = l_1 + l_2$ we find

$$\begin{pmatrix} l_1 & l_2 & l_1+l_2 \\ l_1 & l_2 & l_1+l_2 \end{pmatrix} = \frac{1}{\sqrt{2(l_1+l_2)+1}} \quad \Rightarrow \quad \langle l_1, l_1, l_2, l_2 | l_1+l_2, l_1+l_2 \rangle = 1\ . \tag{D.36}$$

Another particularly interesting case corresponds to taking $l_3 = m_3 = 0$ (*i.e.* zero total angular momentum in quantum mechanics). In that case we find

$$\begin{pmatrix} l_1 & l_2 & 0 \\ m_1 & m_2 & 0 \end{pmatrix} = \langle l_1, m_1, l_2, m_2 | 0, 0 \rangle = \frac{(-1)^{l_1-m_1}}{\sqrt{2l_1+1}}\, \delta_{l_1,l_2}\, \delta_{m_1,-m_2}\ . \tag{D.37}$$

Taking this result, together with (D.20) and the fact that ${}_0Y^{00} = 1/\sqrt{4\pi}$, we can easily recover the orthonormality condition (D.28) from the integral of three ${}_sY^{l,m}$, (D.30).

The cases with $l_3 = 1$ are also interesting as they appear in the expression for the momentum radiated by gravitational waves. We find

$$\begin{aligned}\begin{pmatrix} l_1 & l_2 & 1 \\ m_1 & m_2 & 0 \end{pmatrix} &= (-1)^{l_1-m_1}\,\delta_{m_1+m_2,0} \\ &\times \left[\delta_{l_1,l_2}\left(\frac{2m_1}{\sqrt{(2l_1+2)(2l_1+1)(2l_1)}}\right)\right. \\ &+ \delta_{l_1,l_2+1}\left(\frac{(l_1+m_1)(l_1-m_1)}{l_1(2l_1+1)(2l_1-1)}\right)^{1/2} \\ &\left.- \delta_{l_1+1,l_2}\left(\frac{(l_2-m_2)(l_2+m_2)}{l_2(2l_2+1)(2l_2-1)}\right)^{1/2}\right], \end{aligned} \tag{D.38}$$

$$\begin{aligned}\begin{pmatrix} l_1 & l_2 & 1 \\ m_1 & m_2 & \pm 1 \end{pmatrix} &= (-1)^{l_1-m_1}\,\delta_{m_1+m_2,\mp 1} \\ &\times \left[\pm\,\delta_{l_1,l_2}\left(\frac{(l_1\mp m_1)(l_1\mp m_2)}{l_1(2l_1+2)(2l_1+1)}\right)^{1/2}\right. \\ &+ \delta_{l_1,l_2+1}\left(\frac{(l_1\mp m_1)(l_1\pm m_2)}{2l_1(2l_1+1)(2l_1-1)}\right)^{1/2} \\ &\left.+ \delta_{l_1+1,l_2}\left(\frac{(l_2\mp m_2)(l_2\pm m_1)}{2l_2(2l_2+1)(2l_2-1)}\right)^{1/2}\right]. \end{aligned} \tag{D.39}$$

REFERENCES

[1] Abrahams, A., Anderson, A., Choquet-Bruhat, Y., and York, J. Einstein and Yang-Mills theories in hyperbolic form without gauge-fixing. *Phys. Rev. Lett.*, 75:3377–3381, 1995.

[2] Alcubierre, M. The appearance of coordinate shocks in hyperbolic formulations of general relativity. *Phys. Rev. D*, 55:5981–5991, 1997.

[3] Alcubierre, M. Hyperbolic slicings of spacetime: singularity avoidance and gauge shocks. *Class. Quantum Grav.*, 20(4):607–624, 2003.

[4] Alcubierre, M. Are gauge shocks really shocks? *Class. Quantum Grav.*, 22:4071–4082, 2005.

[5] Alcubierre, M., Allen, G., Brügmann, B., Lanfermann, G., Seidel, E., Suen, W.-M., and Tobias, M. Gravitational collapse of gravitational waves in 3D numerical relativity. *Phys. Rev. D*, 61:041501 (R), 2000.

[6] Alcubierre, M., Allen, G., Brügmann, B., Seidel, E., and Suen, W.-M. Towards an understanding of the stability properties of the 3+1 evolution equations in general relativity. *Phys. Rev. D*, 62:124011, 2000.

[7] Alcubierre, M., Benger, W., Brügmann, B., Lanfermann, G., Nerger, L., Seidel, E., and Takahashi, R. 3D Grazing Collision of Two Black Holes. *Phys. Rev. Lett.*, 87:271103, 2001.

[8] Alcubierre, M., Brandt, S. R., Brügmann, B., Gundlach, C., Massó, J., Seidel, E., and Walker, P. Test-beds and applications for apparent horizon finders in numerical relativity. *Class. Quantum Grav.*, 17:2159–2190, 2000.

[9] Alcubierre, M., Brandt, S. R., Brügmann, B., Holz, D., Seidel, E., Takahashi, R., and Thornburg, J. Symmetry without symmetry: Numerical simulation of axisymmetric systems using Cartesian grids. *Int. J. Mod. Phys. D*, 10(3):273–289, 2001.

[10] Alcubierre, M. and Brügmann, B. Simple excision of a black hole in 3+1 numerical relativity. *Phys. Rev. D*, 63:104006, 2001.

[11] Alcubierre, M., Brügmann, B., Diener, P., Guzmán, F. S., Hawke, I., Hawley, S., Herrmann, F., Koppitz, M., Pollney, D., Seidel, E., and Thornburg, J. Dynamical evolution of quasi-circular binary black hole data. *Phys. Rev. D*, 72(4):044004, 2005.

[12] Alcubierre, M., Brügmann, B., Diener, P., Guzmán, F. S., Hawke, I., Hawley, S., Herrmann, F., Koppitz, M., Pollney, D., Seidel, E., and Thornburg, J. Dynamical evolution of quasi-circular binary black hole data. *Phys. Rev. D*, 72:044004, 5 August 2005.

[13] Alcubierre, M., Brügmann, B., Diener, P., Koppitz, M., Pollney, D., Seidel, E., and Takahashi, R. Gauge conditions for long-term numerical black hole evolutions without excision. *Phys. Rev. D*, 67:084023, 2003.

[14] Alcubierre, M., Brügmann, B., Dramlitsch, T., Font, J. A., Papadopoulos, P., Seidel, E., Stergioulas, N., and Takahashi, R. Towards a stable numerical evolution of strongly gravitating systems in general relativity: The conformal treatments. *Phys. Rev. D*, 62:044034, 2000.

[15] Alcubierre, M., Brügmann, B., Pollney, D., Seidel, E., and Takahashi, R. Black hole excision for dynamic black holes. *Phys. Rev. D*, 64:061501(R), 2001.

[16] Alcubierre, M., Corichi, A., González, J. A., Núñez, D., Reimann, B., and Salgado, M. Generalized harmonic spatial coordinates and hyperbolic shift conditions. *Phys. Rev. D*, 72:124018, 2005.

[17] Alcubierre, M., Corichi, A., González, J. A., Núñez, D., and Salgado, M. A hyperbolic slicing condition adapted to killing fields and densitized lapses. *Class. Quantum Grav.*, 20(18):3951–3968, 21 September 2003.

[18] Alcubierre, M. and González, J. A. Regularization of spherically symmetric evolution codes in numerical relativity. *Comp. Phys. Comm.*, 167:76–84, 2005. gr-qc/0401113.

[19] Alcubierre, M. and Massó, J. Pathologies of hyperbolic gauges in general relativity and other field theories. *Phys. Rev. D*, 57(8):R4511–R4515, 1998.

[20] Alcubierre, M. and Schutz, B. Time–symmetric ADI and causal reconnection: Stable numerical techniques for hyperbolic systems on moving grids. *J. Comput. Phys.*, 112:44, 1994.

[21] Anderson, A. and York, J. W. Fixing Einstein's equations. *Phy. Rev. Lett.*, 82:4384–4387, 1999.

[22] Anderson, A. and York, J. W. Hamiltonian time evolution for general relativity. *Phys. Rev. Lett.*, 81:1154–1157, 1998.

[23] Anninos, P., Bernstein, D., Brandt, S., Libson, J., Massó, J., Seidel, E., Smarr, L., Suen, W.-M., and Walker, P. Dynamics of apparent and event horizons. *Phys. Rev. Lett.*, 74(5):630–633, 30 January 1995.

[24] Anninos, P., Brandt, S. R., and Walker, P. New coordinate systems for axisymmetric black hole collisions. *Phys. Rev. D*, 57:6158–6167, 1998.

[25] Anninos, P., Camarda, K., Massó, J., Seidel, E., Suen, W.-M., and Towns, J. Three-dimensional numerical relativity: The evolution of black holes. *Phys. Rev. D*, 52(4):2059–2082, 1995.

[26] Anninos, P., Daues, G., Massó, J., Seidel, E., and Suen, W.-M. Horizon boundary conditions for black hole spacetimes. *Phys. Rev. D*, 51(10):5562–5578, 1995.

[27] Anninos, P., Hobill, D., Seidel, E., Smarr, L., and Suen, W.-M. The collision of two black holes. *Phys. Rev. Lett.*, 71(18):2851–2854, 1993.

[28] Anninos, P., Hobill, D., Seidel, E., Smarr, L., and Suen, W.-M. The head-on collision of two equal mass black holes. *Phys. Rev. D*, 52:2044–2058, 1995.

[29] Arbona, A. and Bona, C. Dealing with the center and boundary problems in 1d numerical relativity. *Comput. Phys. Commun.*, 118:229–235, 1999. gr-qc/9805084.

[30] Arbona, A., Bona, C., Massó, J., and Stela, J. Robust evolution system for numerical relativity. *Phys. Rev. D*, 60:104014, 1999. gr-qc/9902053.

[31] Arnowitt, R., Deser, S., and Misner, C. W. The dynamics of general relativity. In Witten, L., editor, *Gravitation: An introduction to current research*, pages 227–265. John Wiley, New York, 1962.

[32] Ashtekar, A., Beetle, C., Dreyer, O., Fairhurst, S., Krishnan, B., Lewandowski, J., and Wisniewski, J. Generic isolated horizons and their applications. *Phys. Rev. Lett.*, 85:3564–3567, 2000.

[33] Ashtekar, A., Beetle, C., and Fairhurst, S. Isolated horizons: A generalization of black hole mechanics. *Class. Quantum Grav.*, 16:L1–L7, 1999.

[34] Ashtekar, A., Beetle, C., and Fairhurst, S. Mechanics of isolated horizons. *Class. Quantum Grav.*, 17:253–298, 2000.

[35] Ashtekar, A., Fairhurst, S., and Krishnan, B. Isolated horizons: Hamiltonian evolution and the first law. *Phys. Rev. D*, 62:104025, 2000.

[36] Ashtekar, A. and Galloway, G. Some uniqueness results for dynamical horizons. *Advances in Theoretical and Mathematical Physics*, 9(1):1–30, 2005.

[37] Ashtekar, A. and Krishnan, B. Dynamical Horizons: Energy, angular momentum, fluxes, and balance laws. *Phys. Rev. Lett.*, 89:261101, 2002.

[38] Ashtekar, A. and Krishnan, B. Dynamical horizons and their properties. *Phys. Rev. D*, 68:104030, 2003.

[39] Ashtekar, A. and Krishnan, B. Isolated and dynamical horizons and their applications. *Living Rev. Relativity*, 7:10, 2004.

[40] Babiuc, M. C., Szilágyi, B., and J.Winicour. Harmonic initial-boundary evolution in general relativity. *Phys. Rev. D*, 73:064017, 2006.

[41] Baker, J. and Campanelli, M. Making use of geometrical invariants in black hole collisions. *Phys. Rev. D*, 62:127501, 2000.

[42] Baker, J., Brügmann, B., Campanelli, M., Lousto, C. O., and Takahashi, R. Plunge waveforms from inspiralling binary black holes. *Phys. Rev. Lett.*, 87:121103, 2001.

[43] Baker, J., Campanelli, M., Lousto, C. O., and Takahashi, R. Modeling gravitational radiation from coalescing binary black holes. *Phys. Rev. D*, 65:124012, 2002.

[44] Baker, J. G., Centrella, J., Choi, D.-I., Koppitz, M., and van Meter, J. Binary black hole merger dynamics and waveforms. *Phys. Rev. D*, 73:104002, 2006.

[45] Baker, J. G., Centrella, J., Choi, D.-I., Koppitz, M., and van Meter, J. Gravitational wave extraction from an inspiraling configuration of merging black holes. *Phys. Rev. Lett.*, 96:111102, 2006.

[46] Balakrishna, J., Daues, G., Seidel, E., Suen, W.-M., Tobias, M., and Wang, E. Coordinate conditions in three-dimensional numerical relativity. *Class. Quantum Grav.*, 13:L135–L142, 1996.

[47] Barcelo, C. and Visser, M. Twilight for the energy conditions? *Int. J. Mod. Phys. D*, 11:1553–1560, 2002.

[48] Bardeen, J. M. and Piran, T. General relativistic axisymmetric rotating systems: Coordinates and equations. *Phys. Reports*, 196:205–250, 1983.

[49] Baumgarte, T. W. Innermost stable circular orbit of binary black holes. *Phys. Rev. D*, 62:024018, 2000.

[50] Baumgarte, T. W. and Shapiro, S. L. On the numerical integration of Einstein's field equations. *Phys. Rev. D*, 59:024007, 1999.

[51] Beig, R. and Murchadha, N. Ó. Late time behavior of the maximal slicing of the Schwarzschild black hole. *Phys. Rev. D*, 57(8):4728–4737, 1998. gr-qc/9706046.

[52] Beig, R. The maximal slicing of a Schwarzschild black hole. *Ann. Phys.*, 11(5):507–510, 2000.

[53] Belinski, V., Khalatnikov, I., and Lifshitz, E. Oscillatory approach to a singular point in the relativistic cosmology. *Adv. Phys.*, 19:525–573, 1970.

[54] Bernstein, D. *A Numerical Study of the Black Hole Plus Brill Wave Spacetime.* PhD thesis, University of Illinois Urbana-Champaign, 1993.

[55] Bernstein, D., Hobill, D., and Smarr, L. Black hole spacetimes: Testing numerical relativity. In Evans, C., Finn, L., and Hobill, D., editors, *Frontiers in Numerical Relativity*, pages 57–73. Cambridge University Press, Cambridge, England, 1989.

[56] Bishop, N. T. The closed trapped region and the apparent horizon of two Schwarzschild black holes. *Gen. Rel. Grav.*, 14(9):717–723, 1982.

[57] Bishop, N. T. Numerical relativity: Combining the Cauchy and characteristic initial value problem. *Class. Quantum Grav.*, 10:333–341, 1993.

[58] Bishop, N. T., Isaacson, R., Maharaj, M., and Winicour, J. Black hole data via a Kerr-Schild approach. *Phys. Rev. D*, 57:6113–6118, 1998.

[59] Bona, C., Ledvinka, T., Palenzuela, C., and Zacek, M. General-covariant evolution formalism for numerical relativity. *Phys. Rev. D*, 67:104005, 2003.

[60] Bona, C., Ledvinka, T., Palenzuela, C., and Zacek, M. A symmetry-breaking mechanism for the Z4 general-covariant evolution system. *Phys. Rev. D*, 69:064036, 2004.

[61] Bona, C., Ledvinka, T., Palenzuela-Luque, C., and Zacek, M. Constraint-preserving boundary conditions in the Z4 numerical relativity formalism. *Class. Quantum Grav.*, 22:2615–2634, 2005.

[62] Bona, C. and Massó, J. Einstein's evolution equations as a system of balance laws. *Phys. Rev. D*, 40(4):1022–1026, 1989.

[63] Bona, C. and Massó, J. Hyperbolic evolution system for numerical relativity. *Phys. Rev. Lett.*, 68:1097, 1992.

[64] Bona, C. and Massó, J. Numerical relativity: evolving space-time. *International Journal of Modern Physics C: Physics and Computers*, 4:883, 1993.

[65] Bona, C., Massó, J., Seidel, E., and Stela, J. New Formalism for Numerical Relativity. *Phys. Rev. Lett.*, 75:600–603, 1995.

[66] Bona, C., Massó, J., Seidel, E., and Stela, J. First order hyperbolic formalism for numerical relativity. *Phys. Rev. D*, 56:3405–3415, 1997.

[67] Bona, C. and Palenzuela, C. Dynamical shift conditions for the Z4 and BSSN hyperbolic formalisms. *Phys. Rev. D*, 69:104003, 2004.

[68] Bona, C. and Palenzuela-Luque, C. *Elements of Numerical Relativity.* Springer-Verlag, Berlin, 2005.

[69] Bonazzola, S. and Gourgoulhon, E. A formulation of the virial theorem in general relativity. *Class. Quantum Grav.*, 11:1775–1784, 1994.

[70] Bonazzola, S. and Marck, J.-A. Pseudo-spectral methods applied to gravitational collapse. In Evans, C., Finn, L., and Hobill, D., editors, *Frontiers in Numerical Relativity*, pages 239–253. Cambridge University Press, Cambridge, England, 1989.

[71] Bondi, H., van der Burg, M. G. J., and Metzner, A. W. K. Gravitational waves in general relativity VII. Waves from axi-symmetric isolated systems. *Proc. R. Soc. London*, A269:21–52, 1962.

[72] Booth, I. and Fairhurst, S. Horizon energy and angular momentum from a hamiltonian perspective. *Class. Quantum Grav.*, 22:4515–4550, 2005.

[73] Bowen, J. M. General form for the longitudinal momentum of a spherically symmetric source. *Gen. Rel. Grav.*, 11(3):227–231, 1979.

[74] Bowen, J. M. and York, J. W. Time-asymmetric initial data for black holes and black hole collisions. *Phys. Rev. D*, 21(8):2047–2056, 1980.

[75] Boyd, J. P. *Chebyshev and Fourier Spectral Methods (Second Edition, Revised)*. Dover Publications, New York, 2001.

[76] Brandt, S. and Brügmann, B. A simple construction of initial data for multiple black holes. *Phys. Rev. Lett.*, 78(19):3606–3609, 1997.

[77] Brandt, S. and Seidel, E. The evolution of distorted rotating black holes III: Initial data. *Phys. Rev. D*, 54(2):1403–1416, 1996.

[78] Brill, D. S. On the positive definite mass of the Bondi-Weber-Wheeler time-symmetric gravitational waves. *Ann. Phys. (N. Y.)*, 7:466–483, 1959.

[79] Brill, D. S. and Lindquist, R. W. Interaction energy in geometrostatics. *Phys. Rev.*, 131(1):471–476, 1963.

[80] Brügmann, Bernd, Tichy, Wolfgang, Jansen, and Nina. Numerical simulation of orbiting black holes. *Phys. Rev. Lett.*, 92:211101, 2004.

[81] Brügmann, B. Adaptive mesh and geodesically sliced Schwarzschild spacetime in 3+1 dimensions. *Phys. Rev. D*, 54(12):7361–7372, 1996.

[82] Brügmann, B. Binary black hole mergers in 3D numerical relativity. *Int. J. Mod. Phys. D*, 8:85, 1999.

[83] Brügmann, B., González, J. A., Hannam, M., Husa, S., Sperhake, U., and Tichy, W. Calibration of moving puncture simulations. 2006. gr-qc/0610128.

[84] Bruhat, Y. Theoreme d'existence pour certains systemes d'equations aux derivees partielles non lineaires. *Acta Mathematica*, 88:141–225, 1952.

[85] Čadež, A. *Colliding Black Holes*. PhD thesis, University of North Carolina at Chapel Hill, Chapel Hill, North Carolina, 1971.

[86] Čadež, A. Apparent horizons in the two-black-hole problem. *Ann. Phys.*, 83:449–457, 1974.

[87] Calabrese, G., Lehner, L., and Tiglio, M. Constraint-preserving boundary conditions in numerical relativity. *Phys. Rev. D*, 65:104031, 2002.

[88] Calabrese, G., Pullin, J., Reula, O., Sarbach, O., and Tiglio, M. Well posed constraint-preserving boundary conditions for the linearized Einstein equations. *Communications in Mathematical Physics*, 240:377–395, 2003.

[89] Calabrese, G. and Sarbach, O. Detecting ill posed boundary conditions in general relativity. *J. Math. Phys*, 44:3888–3889, 2003.

[90] Campanelli, M. and Lousto, C. O. Second order gauge invariant gravitational perturbations of a Kerr black hole. *Phys. Rev. D*, 59:124022, 1999.

[91] Campanelli, M., Kelly, B. J., and Lousto, C. O. The Lazarus project II: Space-like extraction with the quasi-Kinnersley tetrad. *Phys. Rev. D*, 73:064005, 2006.

[92] Campanelli, M., Lousto, C. O., and Zlochower, Y. The last orbit of binary black holes. *Phys. Rev. D*, 73:061501(R), 2006.

[93] Campanelli, M., Lousto, C. O., Marronetti, P., and Zlochower, Y. Accurate evolutions of orbiting black-hole binaries without excision. *Phys. Rev. Lett.*, 96:111101, 2006.

[94] Campanelli, M., Lousto, C. O., Zlochower, Y., Krishnan, B., and Merritt, D. Spin flips and precession in black-hole-binary mergers. *Phys. Rev.*, D75:064030, 2007.

[95] Caudill, M., Cook, G. B., Grigsby, J. D., and Pfeiffer, H. P. Circular orbits and spin in black-hole initial data. *Phys. Rev. D*, 74:064011, 2006.

[96] Chandrasekhar, S. *The Mathematical Theory of Black Holes.* Oxford University Press, Oxford, England, 1983.

[97] Choptuik, M. W. Consistency of finite-difference solutions to Einstein's equations. *Phys. Rev. D*, 44:3124–3135, 1991.

[98] Choptuik, M. W. Universality and scaling in gravitational collapse of massless scalar field. *Phys. Rev. Lett.*, 70:9, 1993.

[99] Choptuik, M. W., Hirschmann, E. W., Liebling, S. L., and Pretorius, F. An axisymmetric gravitational collapse code. *Class. Quantum Grav.*, 20:1857–1878, 2003.

[100] Choquet-Bruhat, Y. and York, J. Geometrical well posed systems for the Einstein equations. *C. R. Acad. Sc. Paris*, 321:1089, 1995.

[101] Cook, G. B. Three-dimensional initial data for the collision of two black holes II: Quasi-circular orbits for equal-mass black holes. *Phys. Rev. D*, 50(8):5025–5032, 1994.

[102] Cook, G. B. Initial data for numerical relativity. *Living Rev. Relativity*, 3:5, 2000.

[103] Cook, G. B. Corotating and irrotational binary black holes in quasi-circular orbits. *Phys. Rev. D*, 65:084003, 2002.

[104] Cook, G. B., Huq, M. F., Klasky, S. A., Scheel, M. A., Abrahams, A. M., Anderson, A., Anninos, P., Baumgarte, T. W., Bishop, N., Brandt, S. R., Browne, J. C., Camarda, K., Choptuik, M. W., Correll, R. R., Evans, C. R., Finn, L. S., Fox, G. C., Gómez, R., Haupt, T., Kidder, L. E., Laguna, P., Landry, W., Lehner, L., Lenaghan, J., Marsa, R. L., Masso, J., Matzner, R. A., Mitra, S., Papadopoulos, P., Parashar, M., Rezzolla, L., Rupright, M. E., Saied, F., Saylor, P. E., Seidel, E., Shapiro, S. L., Shoemaker, D., Smarr, L., Suen, W. M., Szilágyi, B., Teukolsky, S. A., van Putten, M. H. P. M., Walker, P., Winicour, J., and York Jr, J. W. Boosted three-

dimensional black-hole evolutions with singularity excision. *Phys. Rev. Lett.*, 80:2512–2516, 1998.

[105] Courant, R. and Friedrichs, K. O. *Supersonic flows and shock waves.* Springer, Berlin, 1976.

[106] Dain, S., Lousto, C. O., and Takahashi, R. New conformally flat initial data for spinning black holes. *Phys. Rev. D*, 65:104038, 2002.

[107] Daues, G. E. *Numerical Studies of Black Hole Spacetimes.* PhD thesis, Washington University, St. Louis, Missouri, 1996.

[108] de Felice, F. and Clarke, C. J. S. *Relativity on curved manifolds.* Cambridge University Press, 1990.

[109] Diener, P. A new general purpose event horizon finder for 3D numerical spacetimes. *Class. Quantum Grav.*, 20(22):4901–4917, 2003.

[110] Diener, P., Herrmann, F., Pollney, D., Schnetter, E., Seidel, E., Takahashi, R., Thornburg, J., and Ventrella, J. Accurate evolution of orbiting binary black holes. *Phys. Rev. Lett.*, 96:121101, 2006.

[111] Dirac, P. A. M. Fixation of coordinates in the hamiltonian theory of gravitation. *Phys. Rev.*, 114:924, 1959.

[112] Dray, T. The relationship between monopole harmonics and spin-weighted spherical harmonics. *J. Math. Phys.*, 26:1030–1033, 1985.

[113] Dreyer, O., Krishnan, B., Shoemaker, D., and Schnetter, E. Introduction to Isolated Horizons in Numerical Relativity. *Phys. Rev. D*, 67:024018, 2003.

[114] Duez, M. D., Marronetti, P., Shapiro, S. L., and Baumgarte, T. W. Hydrodynamic simulations in 3+1 general relativity. *Phys. Rev. D*, 67:024004, 2003.

[115] Eardley, D. M. and Smarr, L. Time functions in numerical relativity: Marginally bound dust collapse. *Phys. Rev. D*, 19:2239, 1979.

[116] Eckart, C. The thermodynamics of irreversible processes III: Relativistic theory of the simple fluid. *Phys. Rev.*, 58:919, 1940.

[117] Eddington, A. S. A comparison of Whitehead's and Einstein's formulas. *Nature*, 113:192, 1924.

[118] Einstein, A. Ist die trägheit eines körpers von seinem energieinhalt abhängig? *Ann. Phys.*, 18:639–641, 1905.

[119] Einstein, A. Zur elektrodynamik bewegter körper. *Ann. Phys.*, 17:891–921, 1905.

[120] Einstein, A. Die feldgleichungen der gravitation. *Preuss. Akad. Wiss. Berlin, Sitzungsber.*, pages 844–847, 1915.

[121] Einstein, A. Zur algemeinen relativitätstheorie. *Preuss. Akad. Wiss. Berlin, Sitzungsber.*, pages 778–786, 1915.

[122] Emden, R. *Gaskugeln, Anwendungen der mechanischen Wärmetheorie.* Teubner, Leipzig, 1907.

[123] Eppley, K. R. *The numerical evolution of the collision of two black holes.* PhD thesis, Princeton University, Princeton, New Jersey, 1975.

[124] Eppley, K. R. Evolution of time-symmetric gravitational waves: Initial data and apparent horizons. *Phys. Rev. D*, 16(6):1609–1614, 1977.

[125] Eppley, K. R. Pure gravitational waves. In Smarr, L., editor, *Sources of gravitational radiation*, page 275. Cambridge University Press, Cambridge, England, 1979.

[126] Estabrook, F., Wahlquist, H., Christensen, S., DeWitt, B., Smarr, L., and Tsiang, E. Maximally slicing a black hole. *Phys. Rev. D*, 7(10):2814–2817, 1973.

[127] Evans, C. An approach for calculating axisymmetric gravitational collapse. In Centrella, J., editor, *Dynamical Spacetimes and Numerical Relativity*, pages 3–39. Cambridge University Press, Cambridge, England, 1986.

[128] Finkelstein, D. Past-future asymmetry of the gravitational field of a point particle. *Phys. Rev.*, 110:965–967, 1958.

[129] Font, J. A. Numerical hydrodynamics in general relativity. *Living Rev. Relativity*, 6:4, 2003.

[130] Font, J., Ibáñez, J., Martí, J., and Marquina, A. Multidimensional relativistic hydrodynamics: Characteristic fields and modern high-resolution shock-capturing schemes. *Astron. Astrophys.*, 282:304, 1994.

[131] Friedrich, H. Hyperbolic reductions for Einstein's equations. *Class. Quantum Grav.*, 13:1451–1469, 1996.

[132] Friedrich, H. Conformal Einstein evolution. *Lect. Notes Phys.*, 604:1–50, 2002.

[133] Friedrich, H. and Nagy, G. The initial boundary value problem for Einstein's vacuum field equations. *Commun. Math. Phys.*, 201:619–655, 1999.

[134] Friedrichs, K. O. On the laws of relativistic electromagneto-fluid dynamics. *Commun. Pure Appl. Math.*, 27:749–808, 1974.

[135] Frittelli, S. and Reula, O. First-order symmetric-hyperbolic Einstein equations with arbitrary fixed gauge. *Phys. Rev. Lett.*, 76:4667–4670, 1996.

[136] Frittelli, S. Note on the propagation of the constraints in standard 3+1 general relativity. *Phys. Rev. D*, 55:5992–5996, 1997.

[137] Frittelli, S. and Gomez, R. Einstein boundary conditions for the Einstein equations in the conformal-traceless decomposition. *Phys. Rev. D*, 70:064008, 2004.

[138] Frittelli, S. and Gomez, R. Einstein boundary conditions in relation to constraint propagation for the initial-boundary value problem of the Einstein equations. *Phys. Rev. D*, 69:124020, 2004.

[139] Garat, A. and Price, R. H. Nonexistence of conformally flat slices of the Kerr spacetime. *Phys. Rev. D*, 61:124011, 2000.

[140] Garfinkle, D. Harmonic coordinate method for simulating generic singularities. *Phys. Rev. D*, 65:044029, 2002.

[141] Garfinkle, D., Gundlach, C., and Hilditch, D. Comments on Bona-Masso type slicing conditions in long-term black hole evolutions. 2007. arXiv:0707.0726.

[142] Gentle, A., Holz, D., Kheyfets, A., Laguna, P., Miller, W., and Shoemaker, D. Constant crunch coordinates for black hole simulations. *Phys. Rev. D*, 63:064024, 2001.

[143] Gerlach, U. and Sengupta, U. Gauge-invariant perturbations on most general spherically symmetric space-times. *Phys. Rev. D.*, 19:2268–2272, 1979.

[144] Geroch, R. A method for generating solutions of Einstein's equations. *J. Math. Phys.*, 12:918, 1971.

[145] Gleiser, R. J., Nicasio, C. O., Price, R. H., and Pullin, J. Evolving the Bowen-York initial data for spinning black holes. *Phys. Rev. D*, 57:3401–3407, 1998.

[146] Godunov, S. K. A difference method for numerical calculations of discontinuous solutions of the equations of hydrpdynamics. *Mat. Sb.*, 47:271, 1959. in Russian.

[147] Gómez, R., Lehner, L., Marsa, R., Winicour, J., Abrahams, A. M., Anderson, A., Anninos, P., Baumgarte, T. W., Bishop, N. T., Brandt, S. R., Browne, J. C., Camarda, K., Choptuik, M. W., Cook, G. B., Correll, R., Evans, C. R., Finn, L. S., Fox, G. C., Haupt, T., Huq, M. F., Kidder, L. E., Klasky, S. A., Laguna, P., Landry, W., Lenaghan, J., Masso, J., Matzner, R. A., Mitra, S., Papadopoulos, P., Parashar, M., Rezzolla, L., Rupright, M. E., Saied, F., Saylor, P. E., Scheel, M. A., Seidel, E., Shapiro, S. L., Shoemaker, D., Smarr, L., Szilágyi, B., Teukolsky, S. A., van Putten, M. H. P. M., Walker, P., and York Jr, J. W. Stable characteristic evolution of generic three-dimensional single-black-hole spacetimes. *Phys. Rev. Lett.*, 80:3915–3918, 1998.

[148] Gourgoulhon, E. A generalized Damour-Navier-Stokes equation applied to trapping horizons. *Phys. Rev.*, D72:104007, 2005.

[149] Gourgoulhon, E. and Jaramillo, J. L. A 3+1 perspective on null hypersurfaces and isolated horizons. *Physics Reports*, 423(4–5):159–294, February 2006.

[150] Grandclément, P., Gourgoulhon, E., and Bonazzola, S. Binary black holes in circular orbits. II. Numerical methods and first results. *Phys. Rev. D*, 65:044021, 2002.

[151] Gundlach, C. and Walker, P. Causal differencing of flux-conservative equations applied to black hole spacetimes. *Class. Quantum Grav.*, 16:991–1010, 1999.

[152] Gundlach, C. Pseudo-spectral apparent horizon finders: An efficient new algorithm. *Phys. Rev. D*, 57:863–875, 1998.

[153] Gundlach, C. Critical phenomena in gravitational collapse. *Living Rev. Relativity*, 2:4, 1999.

[154] Gundlach, C. and Martin-Garcia, J. M. Symmetric hyperbolicity and consistent boundary conditions for second-order Einstein equations. *Phys. Rev. D*, 70:044032, 2004.

[155] Gundlach, C. and Martin-Garcia, J. M. Hyperbolicity of second-order in space systems of evolution equations. *Class. Quantum Grav.*, 23:S387–S404, 2006.

[156] Gundlach, C. and Martin-Garcia, J. M. Well-posedness of formulations of the Einstein equations with dynamical lapse and shift conditions. *Phys. Rev. D*, 74:024016, 2006.

[157] Gustafsson, B., Kreiss, H.-O., and Oliger, J. *Time dependent problems and difference methods*. Wiley, New York, 1995.

[158] Hahn, S. G. and Lindquist, R. W. The two body problem in geometrodynamics. *Ann. Phys.*, 29:304–331, 1964.

[159] Hannam, M., Husa, S., Pollney, D., Brugmann, B., and O'Murchadha, N. Geometry and regularity of moving punctures. *Phys. Rev. Lett.*, 99:241102, 2007.

[160] Hawking, S. W. The event horizon. In DeWitt, C. and DeWitt, B. S., editors, *Black Holes*, pages 1–55. Gordon and Breach, New York, 1973.

[161] Hawking, S. W. and Ellis, G. F. R. *The large scale structure of spacetime.* Cambridge University Press, Cambridge, England, 1973.

[162] Hayward, S. A. General laws of black hole dynamics. *Phys. Rev. D*, 49(12):6467–6474, 15 June 1994.

[163] Holz, D., Miller, W., Wakano, M., and Wheeler, J. Coalescence of primal gravity waves to make cosmological mass without matter. In Hu, B. L. and Jacobson, T. A., editors, *Directions in General Relativity: Proceedings of the 1993 International Symposium, Maryland; Papers in honor of Dieter Brill*, page 339, Cambridge, England, 1993. Cambridge University Press.

[164] Hulse, R. and Taylor, J. Discovery of a pulsar in a binary system. *Astrophys. J.*, 195:L51–L53, 1975.

[165] Husa, S. Numerical relativity with the conformal field equations. In Fernández, L. and González, L. M., editors, *Proceedings of the 2001 spanish relativity meeting*, volume 617 of *Lecture Notes in Physics*, pages 159–192. Springer, 2003.

[166] Isaacson, R. Gravitational radiation in the limit of high frequency. II. nonlinear terms and the effective stress tensor. *Phys. Rev.*, 166:1272–1280, 1968.

[167] Israel, W. and Stewart, J. M. Transient relativistic thermodynamics and kinetic theory. *Ann. Phys.*, 118:341, 1979.

[168] Jantzen, R. T. and York, James W., J. New minimal distortion shift gauge. *Phys. Rev. D*, 73:104008, 2006.

[169] Katz, J. I., Lynden-Bell, D., and Israel, W. Quasilocal energy in static gravitational fields. *Class. Quantum Grav.*, 5:971–987, 1988.

[170] Kerr, R. P. Gravitational field of a spinning mass as an example of algebraically special metrics. *Phys. Rev. Lett.*, 11:237–238, 1963.

[171] Kidder, L. E. and Finn, L. S. Spectral methods for numerical relativity. the initial data problem. *Phys. Rev. D*, 62:084026, 2000.

[172] Kidder, L. E., Scheel, M. A., and Teukolsky, S. A. Extending the lifetime of 3D black hole computations with a new hyperbolic system of evolution equations. *Phys. Rev. D*, 64:064017, 2001.

[173] Kidder, L. E., Scheel, M. A., Teukolsky, S. A., Carlson, E. D., and Cook, G. B. Black hole evolution by spectral methods. *Phys. Rev. D*, 62:084032, 2000.

[174] Kidder, L. E., Lindblom, L., Scheel, M. A., Buchman, L. T., and Pfeiffer, H. P. Boundary conditions for the Einstein evolution system. *Phys. Rev. D*, 71:064020, 2005.

[175] Kokkotas, K. D. and Schmidt, B. G. Quasi-normal modes of stars and black holes. *Living Rev. Relativity*, 2:2, 1999. http://www.livingreviews.org/lrr-1999-2.

[176] Komar, A. Covariant conservation laws in general relativity. *Phys. Rev.*, 113:934–936, 1959.

[177] Kreiss, H. O. and Lorenz, J. *Initial-boundary value problems and the Navier-Stokes equations.* Academic Press, New York, 1989.

[178] Kreiss, H. O. and Oliger, J. *Methods for the approximate solution of time dependent problems.* GARP publication series No. 10, Geneva, 1973.

[179] Kreiss, H. O. and Scherer, G. Finite element and finite difference methods for hyperbolic partial differential equations. In Boor, C. D., editor, *Mathematical Aspects of Finite Elements in Partial Differential Equations*, New York, 1974. Academica Press.

[180] Kreiss, H. O. and Scherer, G. Method of lines for hyperbolic equations. *SIAM J. Numer. Anal.*, 29:640–646, 1992.

[181] Krishnan, B. *Isolated Horizons in Numerical Relativity.* PhD thesis, Pennsylvania State University, 2002.

[182] Kruskal, M. D. Maximal extension of Schwarzschild metric. *Phys. Rev.*, pages 1743–1745, 1960.

[183] Landau, L. D. and Lifshitz, E. M. *The Classical Theory of Fields, Course of Theoretical Physics, Volume 2.* Elsevier Butterworth-Heinemann, Oxford, 2004.

[184] Landau, L. D. and Lifshitz, E. M. *Fluid Mechanics, Course of Theoretical Physics, Volume 6.* Elsevier Butterworth-Heinemann, Oxford, 2004.

[185] Lax, P. D. and Phillips, R. S. Local boundary conditions for dissipative symmetric linear differential operators. *Commun. Pure Appl. Math.*, 13:427–455, 1960.

[186] Lehner, L. Numerical relativity: A review. *Class. Quantum Grav.*, 18:R25–R86, 2001.

[187] Leveque, R. J. *Numerical Methods for Conservation Laws.* Birkhauser Verlag, Basel, 1992.

[188] Libson, J., Massó, J., Seidel, E., Suen, W.-M., and Walker, P. Event horizons in numerical relativity: Methods and tests. *Phys. Rev. D*, 53(8):4335–4350, 1996.

[189] Lichnerowicz, A. L'intégration des équations de la gravitation relativiste et la problème des n corps. *J. Math. Pures et Appl.*, 23:37, 1944.

[190] Lindblom, L. and Scheel, M. A. Dynamical gauge conditions for the Einstein evolution equations. *Phys. Rev. D*, 67:124005, 2003.

[191] Lindblom, L., Scheel, M. A., Kidder, L. E., Owen, R., and Rinne, O. A new generalized harmonic evolution system. *Class. Quantum Grav.*, 23:S447–S462, 2006.

[192] Lindquist, R. W. Initial-value problem on Einstein-Rosen manifolds. *Jour. Math. Phys.*, 4(7):938, 1963.

[193] Lousto, C. O. and Zlochower, Y. A practical formula for the radiated angular momentum. *Phys. Rev.*, D76:041502, 2007.

[194] Maartens, R. Dissipative cosmology. *Class. Quantum Grav.*, 12:1455, 1995.

[195] Maartens, R. Causal thermodynamics in relativity. Lectures given at the Hanno Rund Workshop on Relativity and Thermodynamics, South Africa, June 1996; astro-ph/9609119, 1996.

[196] Marsa, R. L. and Choptuik, M. W. Black hole–scalar field interactions in spherical symmetry. *Phys. Rev. D*, 54:4929–4943, 1996.

[197] Martel, K. and Poisson, E. Gravitational perturbations of the schwarzschild spacetime: A practical covariant and gauge-invariant formalism. *Phys. Rev.*, D71:104003, 2005.

[198] Martí, J. M., Ibáñez, J. M., and Miralles, J. M. Numerical relativistic hydrodynamics: Local characteristic approach. *Phys. Rev. D*, 43:3794, 1991.

[199] Martí, J. M. and Müller, E. The analytical solution of the riemann problem in relativistic hydrodynamics. *J. Fluid Mech.*, 258:317–333, 1994.

[200] Martí, J. M. and Müller, E. Numerical hydrodynamics in special relativity. *Living Rev. Relativity*, 6:7, 2003.

[201] Matzner, R. A., Huq, M. F., and Shoemaker, D. Initial data and coordinates for multiple black hole systems. *Phys. Rev. D*, 59:024015, 1999.

[202] Messiah, A. *Quatum Mechanics*. Dover Publications, New York, 1999.

[203] Metzger, J. Numerical computation of constant mean curvature surfaces using finite elements. *Class. Quantum Grav.*, 21(19):4625–4646, 2004.

[204] Misner, C. Wormhole initial conditions. *Phys. Rev.*, 118(4):1110–1111, 1960.

[205] Misner, C. W. The method of images in geometrostatics. *Ann. Phys.*, 24:102–117, 1963.

[206] Misner, C. W., Thorne, K. S., and Wheeler, J. A. *Gravitation*. W. H. Freeman, San Francisco, 1973.

[207] Mitchell, A. R. *The finite element method in partial differential equations*. J. Wiley and Sons, U.S.A., 1977.

[208] Mitchell, A. R. and Griffiths, D. F. *The Finite Difference Method in Partial Differential Equations*. Wiley, New York, 1980.

[209] Moncrief, V. Gravitational perturbations of spherically symmetric systems. I. the exterior problem. *Annals of Physics*, 88:323–342, 1974.

[210] Moreno, C., Nunez, D., and Sarbach, O. Kerr-Schild type initial data for two Kerr black holes. *Class. Quantum Grav.*, 19:6059–6073, 2002.

[211] Nagar, A. and Rezzolla, L. Gauge-invariant non-spherical metric perturbations of schwarzschild black-hole spacetimes. *Class. Quant. Grav.*, 22:R167, 2005.

[212] Nagy, G., Ortiz, O. E., and Reula, O. A. Strongly hyperbolic second order Einstein's evolution equations. *Phys. Rev. D*, 70:044012, 2004.

[213] Nakamura, T., Maeda, K., Miyama, S., and Sasaki, M. General Relativistic Collapse of an Axially Symmetric Star. I —The Formulation and the Initial Value Equations—. *Prog. Theor. Phys.*, 63:1229–1244, April 1980.

[214] Nakamura, T., Kojima, Y., and Oohara, K. A method of determining apparent horizons in three-dimensional numerical relativity. *Phys. Lett. A*, 106(5-6):235–238, 10 December 1984.

[215] Nakamura, T., Oohara, K., and Kojima, Y. General relativistic collapse to black holes and gravitational waves from black holes. *Prog. Theor. Phys. Suppl.*, 90:1–218, 1987.

[216] Nerozzi, A., Beetle, C., Bruni, M., Burko, L. M., and Pollney, D. Towards wave extraction in numerical relativity: The quasi-Kinnersley frame. *Phys. Rev. D*, 72:024014, 2005.

[217] Newman, E. T., Couch, E., Chinnapared, K., Exton, A., Prakash, A., and Torrence, R. Metric of a rotating charged mass. *J. Math. Phys.*, 6(6):918–919, 1965.

[218] Newman, E. T. and Penrose, R. An approach to gravitational radiation by a method of spin coefficients. *J. Math. Phys.*, 3(3):566–578, 1962. erratum in J. Math. Phys. 4, 998 (1963).

[219] Newman, E. T. and Penrose, R. Note on the Bondi-Metzner-Sachs group. *J. Math. Phys.*, 7(5):863–870, May 1966.

[220] Newmann, J. V. and Richtmyer, R. D. A method for the numerical calculation of hydrodynamical shocks. *J. Appl. Phys.*, 21:232, 1950.

[221] Nordström, G. On the energy of the gravitational field in Einstein's theory. *Proc. Kon. Ned. Akad. Wet.*, 20:1238–1245, 1918.

[222] O'Murchadha, N. and York, J. W. Gravitational energy. *Phys. Rev. D*, 10(8):2345–2357, 1974.

[223] Pais, A. *'Subtle is the Lord...' the science and the life of Albert Einsten.* Oxford University Press, Oxford and New York, 1982.

[224] Petrich, L. I., Shapiro, S. L., and Teukolsky, S. A. Oppenheimer-Snyder collapse with maximal time slicing and isotropic coordinates. *Phys. Rev. D*, 31(10):2459–2469, 15 May 1985.

[225] Petrov, A. Z. *Einstein Spaces.* Pergamon Press, Oxford, 1969.

[226] Pfeiffer, H. P. and York, J. W. Extrinsic curvature and the Einstein constraints. *Phys. Rev. D*, 67:044022, 2003.

[227] Poisson, E. *A Relativist's Toolkit: The Mathematics of Black-Hole Mechanics.* Cambridge University Press, 2004.

[228] Pollney, D. et al. Recoil velocities from equal-mass binary black-hole mergers: a systematic investigation of spin-orbit aligned configurations. *Phys. Rev.*, D76:124002, 2007.

[229] Pons, J. A., Martí, J. M., and Müller, E. The exact solution of the Riemann problem with non-zero tangential velocities in relativistic hydrodynamics. *J. Fluid Mech.*, 422:125–139, 2000.

[230] Press, W. H., Flannery, B. P., Teukolsky, S. A., and Vetterling, W. T. *Numerical Recipes.* Cambridge University Press, New York, 2nd edition, 1992.

[231] Pretorius, F. Evolution of binary black hole spacetimes. *Phys. Rev. Lett.*, 95:121101, 2005.

[232] Pretorius, F. Numerical relativity using a generalized harmonic decomposition. *Class. Quantum Grav.*, 22:425–452, 2005.

[233] Pretorius, F. Simulation of binary black hole spacetimes with a harmonic evolution scheme. *Class. Quantum Grav.*, 23:S529–S552, 2006.

[234] Racah, G. Theory of complex spectra II. *Phys. Rev.*, 62:438, 1942.

[235] Regge, T. and Wheeler, J. Stability of a Schwarzschild singularity. *Phys. Rev.*, 108(4):1063–1069, 1957.

[236] Reimann, B. Slice stretching at the event horizon when geodesically slicing the Schwarzschild spacetime with excision. *Class. Quantum Grav.*, 21:4297–4304, 2004.

[237] Reimann, B. How slice stretching arises when maximally slicing the Schwarzschild spacetime with vanishing shift. *Class. Quantum Grav.*, 22:4563–4587, 2005.

[238] Reimann, B. and Brügmann, B. Late time analysis for maximal slicing of reissner- nordström puncture evolutions. *Phys. Rev. D*, 69:124009, 2004.

[239] Reimann, B. and Brügmann, B. Maximal slicing for puncture evolutions of Schwarzschild and Reissner-Nordström black holes. *Phys. Rev. D*, 69:044006, 2004.

[240] Reissner, H. Über die eigengravitationn des elektrischen felds nach der Einsteinshen theorie. *Ann. Phys.*, 50:106–120, 1916.

[241] Reula, O. Hyperbolic methods for Einstein's equations. *Living Rev. Relativity*, 1:3, 1998.

[242] Richardson, L. F. The approximate arithmetic solution by finite differences of physical problems involving differential equations, with applications to the stresses in a masonry dam. *Phil. Trans. Roy. Soc.*, 210:307–357, 1910.

[243] Richtmyer, R. D. and Morton, K. *Difference Methods for Initial Value Problems.* Interscience Publishers, New York, 1967.

[244] Rinne, O. and Stewart, J. M. A strongly hyperbolic and regular reduction of Einstein's equations for axisymmetric spacetimes. *Class. Quantum Grav.*, 22:1143–11166, 2005.

[245] Ruiz, M., Alcubierre, M., and Nunez, D. Regularization of spherical and axisymmetric evolution codes in numerical relativity. *Gen. Rel. Grav.*, 40:159–182, 2008.

[246] Ruiz, M., Takahashi, R., Alcubierre, M., and Nunez, D. Multipole expansions for energy and momenta carried by gravitational waves. 2007. arXiv:0707.4654.

[247] Sachs, R. Gravitational waves in general relativity VIII. Waves in asymptotically flat space-time. *Proc. Roy. Soc. London*, A270:103–126, 1962.

[248] Salgado, M. General relativistic hydrodynamics: a new approach. *Rev. Mex. Fis.*, 44:1–8, 1998.

[249] Sarbach, O., Calabrese, G., Pullin, J., and Tiglio, M. Hyperbolicity of the BSSN system of Einstein evolution equations. *Phys. Rev. D*, 66:064002, 2002.

[250] Sarbach, O. and Tiglio, M. Exploiting gauge and constraint freedom in hyperbolic formulations of Einstein's equations. *Phys. Rev. D*, 66:064023, 2002.

[251] Sarbach, O. and Tiglio, M. Boundary conditions for Einstein's field equations: analytical and numerical analysis. *Journal of Hyperbolic Differential Equations*, 2:839–883, 2005.

[252] Sarbach, O. and Tiglio, M. Gauge invariant perturbations of Schwarzschild black holes in horizon-penetrating coordinates. *Phys. Rev. D*, 64:084016, 2001.

[253] Scheel, M., Baumgarte, T., Cook, G., Shapiro, S. L., and Teukolsky, S. Numerical evolution of black holes with a hyperbolic formulation of general relativity. *Phys. Rev. D*, 56:6320–6335, 1997.

[254] Scheel, M. A., Baumgarte, T. W., Cook, G. B., Shapiro, S. L., and Teukolsky, S. A. Treating instabilities in a hyperbolic formulation of Einstein's equations. *Phys. Rev. D*, 58:044020, 1998.

[255] Scheel, M. A., Pfeiffer, H. P., Lindblom, L., Kidder, L. E., Rinne, O., and Teukolsky, S. A. Solving Einstein's equations with dual coordinate frames. *Phys. Rev. D*, 74:104006, 2006.

[256] Scheel, M. A., Shapiro, S. L., and Teukolsky, S. A. Collapse to black holes in Brans-Dicke theory: I. horizon boundary conditions for dynamical spacetimes. *Phys. Rev. D*, 51(8):4208–4235, 1995.

[257] Schnetter, E., Diener, P., Dorband, N., and Tiglio, M. A multi-block infrastructure for three-dimensional time-dependent numerical relativity. *Class. Quantum Grav.*, 23:S553–S578, 2006.

[258] Schutz, B. F. *Geometrical methods of mathematical physics.* Cambridge University Press, 1980.

[259] Schutz, B. F. *A first course in general relativity.* Cambridge University Press, 1985.

[260] Schwarzschild, K. Über das Gravitationsfeld eines Massenpunktes nach der Einsteinchen Theorie. *Sitzungsber. Dtsch. Akad. Wiss. Berlin, Kl. Math. Phys. Tech.*, pages 189–196, 1916.

[261] Seidel, E. and Suen, W.-M. Towards a singularity-proof scheme in numerical relativity. *Phys. Rev. Lett.*, 69(13):1845–1848, 1992.

[262] Shapiro, S. L. and Teukolsky, S. A. Gravitational collapse of supermassive stars to black holes: Numerical solution of the Einstein equations. *Astrophysical J.*, 234:L177–L181, December 15 1979.

[263] Shapiro, S. L. and Teukolsky, S. A. Gravitational collapse to neutron stars and black holes: Computer generation of spherical spacetimes. *Astrophysical J.*, 235:199–215, 1980.

[264] Shapiro, S. L. and Teukolsky, S. A. *Black Holes, White Dwarfs, and Neutron Stars.* John Wiley & Sons, New York, 1983.

[265] Shapiro, S. L. and Teukolsky, S. A. Relativistic stellar dynamics on the computer. I. Motivation and numerical method. *Astrophys. J.*, 298:34–57, November 1 1985.

[266] Shapiro, S. L. and Teukolsky, S. A. Relativistic stellar dynamics on the computer. II. Physical applications. *Astrophys. J.*, 298:58–79, November 1 1985.

[267] Shibata, M. Time symmetric initial conditions of gravitational waves for 3D numerical relativity. *Phys. Rev. D*, 55:7529–7537, 1997.

[268] Shibata, M. and Nakamura, T. Evolution of three-dimensional gravitational waves: Harmonic slicing case. *Phys. Rev. D*, 52:5428, 1995.

[269] Shoemaker, D. M., Huq, M. F., and Matzner, R. A. Generic tracking of multiple apparent horizons with level flow. *Phys. Rev. D*, 62:124005, 2000.

[270] Smarr, L. *The Structure of General Relativity with a Numerical Illustration: The Collision of Two Black Holes.* PhD thesis, University of Texas, Austin, Austin, Texas, 1975.

[271] Smarr, L., Čadež, A., DeWitt, B., and Eppley, K. R. Collision of two black holes: Theoretical framework. *Phys. Rev. D*, 14(10):2443–2452, 1976.

[272] Smarr, L. and York, J. W. Kinematical conditions in the construction of spacetime. *Phys. Rev. D*, 17(10):2529–2552, 15 May 1978.

[273] Smarr, L. and York, J. W. Radiation gauge in general relativity. *Phys. Rev. D*, 17:1945, 1978.

[274] Sopuerta, C. F., Yunes, N., and Laguna, P. Gravitational recoil from binary black hole mergers: The close-limit approximation. *Phys. Rev.*, D74:124010, 2006.

[275] Stergioulas, N. Rotating stars in relativity. *Living Rev. Relativity*, 6:3, 2003.

[276] Szabados, L. B. Quasi-local energy-momentum and angular momentum in general relativity: A review article. *Living Reviews in Relativity*, 7(4), 2004.

[277] Szekeres, P. On the singularities of a Riemannian manifold. *Publ. Mat. Debrecen*, 7:285–301, 1960.

[278] Szilágyi, B., Gomez, R., Bishop, N. T., and Winicour, J. Cauchy boundaries in linearized gravitational theory. *Phys. Rev. D*, 62:104006, 2000.

[279] Szilágyi, B., Schmidt, B., and Winicour, J. Boundary conditions in linearized harmonic gravity. *Phys. Rev. D*, 65:064015, 2002.

[280] Szilágyi, B. and Winicour, J. Well-posed initial-boundary evolution in general relativity. *Phys. Rev. D*, 68:041501, 2003.

[281] Tadmor, E. Spectral methods for hyperbolic problems. In *Lecture notes delivered at Ecole des Ondes, "Méthodes numériques d'ordre élevé pour les ondes en régime transitoire", INRIA–Rocquencourt January 24-28.*, 1994.

[282] Taylor, J. H. and Weisberg, J. M. A new test of general relativity: Gravitational radiation and the binary pulsar PSR 1913+16. *Astrophys. J.*, 253:908–920, 1982.

[283] Teukolsky, S. On the stability of the iterated crank-nicholson method in numerical relativity. *Phys. Rev. D*, 61:087501, 2000.

[284] Thornburg, J. Coordinates and boundary conditions for the general relativistic initial data problem. *Class. Quantum Grav.*, 4(5):1119–1131, September 1987.

[285] Thornburg, J. *Numerical Relativity in Black Hole Spacetimes.* PhD thesis, University of British Columbia, Vancouver, British Columbia, 1993.

[286] Thornburg, J. A 3+1 computational scheme for dynamic spherically symmetric black hole spacetimes – II: Time evolution. gr-qc/9906022, 1999.

[287] Thornburg, J. Event and apparent horizon finders for 3+1 numerical relativity. *Living Rev. Relativity*, 2006. [Online article].

[288] Thorne, K. Multipole expansions of gravitational radiation. *Rev. Mod. Phys.*, 52(2):299, 1980.

[289] Tichy, W. and Brügmann, B. Quasi-equilibrium binary black hole sequences for puncture data derived from helical killing vector conditions. *Phys. Rev. D*, 69:024006, 2004.

[290] Tichy, W., Brügmann, B., and Laguna, P. Gauge conditions for binary black hole puncture data based on an approximate helical Killing vector. *Phys. Rev. D*, 68:064008, 2003.

[291] Tod, K. P. Looking for marginally trapped surfaces. *Class. Quantum Grav.*, 8:L115–L118, 1991.

[292] Toro, E. F. *Riemann Solvers and Numerical Methods for Fluid Dynamics.* Springer-Verlag, 1999.

[293] van Meter, J. R., Baker, J. G., Koppitz, M., and Choi, D.-I. How to move a black hole without excision: gauge conditions for the numerical evolution of a moving puncture. *Phys. Rev. D*, 73:124011, 2006.

[294] Varshalovich, D. A. and Kersonskii, V. K. *Quatum theory of angular momentum.* World Scientific, Singapore, 1989.

[295] Wald, R. M. *General relativity.* The University of Chicago Press, Chicago, 1984.

[296] Walker, P. *Horizons, Hyperbolic Systems, and Inner Boundary Conditions in Numerical Relativity.* PhD thesis, University of Illinois at Urbana-Champaign, Urbana, Illinois, 1998.

[297] Weinberg, S. *Gravitation and Cosmology: Principles and Applications of the General Theory of Relativity.* John Wiley and Sons, New York, 1972.

[298] Wheeler, J. A. *A journey into gravity and spacetime.* Scientific American Library, distributed by W. H. Freeman, New York, U.S.A., 1990.

[299] Wigner, E. P. *Group theory and its applications to the quantum mechanics of atomic spectra.* Academic Press, New York, 1959.

[300] Wilson, J. R. and Mathews, G. J. *Relativistic numerical hydrodynamics.* Cambridge University Press, 2003.

[301] Winicour, J. Characteristic evolution and matching. *Living Rev. Relativity*, 1:5, 1998. [Online article].

[302] Yo, H.-J., Baumgarte, T. W., and Shapiro, S. L. Improved numerical stability of stationary black hole evolution calculations. *Phys. Rev. D*, 66:084026, 2002.

[303] York, J. W. Gravitational degrees of freedom and the initial-value problem. *Phys. Rev. Lett.*, 26:1656–1658, 1971.

[304] York, J. W. Role of conformal three-geometry in the dynamics of gravitation. *Phys. Rev. Lett.*, 28:1082–1085, 1972.

[305] York, J. W. Kinematics and dynamics of general relativity. In Smarr, L. L., editor, *Sources of gravitational radiation*, pages 83–126. Cambridge University Press, Cambridge, UK, 1979.

[306] York, J. W. Energy and momentum of the gravitational field. In Tipler, F. J., editor, *Essays in General Relativity: A Festschrift for Abraham Taub*, pages 39–58. Academic Press, New York, 1980.

[307] York, J. W. Conformal 'thin-sandwich' data for the initial-value problem of general relativity. *Phys. Rev. Lett.*, 82:1350–1353, 1999.

[308] Zerilli, F. J. Effective potential for even-parity Regge-Wheeler gravitational perturbation equations. *Phys. Rev. Lett.*, 24(13):737–738, 1970.

[309] Zerilli, F. J. Gravitational field of a particle falling in a Schwarzschild geometry analyzed in tensor harmonics. *Phys. Rev. D.*, 2:2141, 1970.

INDEX